Skeletal Muscle

Form and Function

Alan J. McComas, MB
McMaster University

Human Kinetics

Library of Congress Cataloging-in-Publication Data

McComas, Alan J.
    Skeletal muscle form and function / Alan J. McComas.
      p. cm.
    Includes bibliographical references and index.
    ISBN 0-87322-780-8 (case)
    1. Striated muscle--Physiology. I. Title.
    [DNLM: 1. Muscle, Skeletal--physiology. 2. Neuromuscular
Junction--physiology. 3. Mechanoreceptors. WE 500 M129s 1996]
    QP321.M3376   1996
    612.7'4--dc20
    DNLM/DLC
    for Library of Congress   95-24525
                                         CIP

ISBN: 0-87322-780-8

Acquisitions Editor: Rik Washburn, PhD; **Developmental Editors:** Anne Heiles, Nanette Smith; **Assistant Editors:** Kirby Mittelmeier, Ann Greenseth, Henry Woolsey; **Editorial Assistants:** Andrew Starr, Coree Schutter; **Copyeditor:** Karen Bojda; **Proofreader:** Jim Burns; **Typesetting and Layout:** Julie Overholt; **Text Designer:** Judy Henderson; **Cover Designer:** Jack Davis

Printed in the United States of America   10 9 8 7 6 5 4 3

Human Kinetics
Web site: www.humankinetics.com

United States: Human Kinetics, P.O. Box 5076, Champaign, IL 61825-5076
800-747-4457
e-mail: humank@hkusa.com

Canada: Human Kinetics, 475 Devonshire Road, Unit 100, Windsor, ON N8Y 2L5
800-465-7301 (in Canada only)
e-mail: hkcan@mnsi.net

Europe: Human Kinetics, P.O. Box IW14, Leeds LS16 6TR, United Kingdom
+44 (0) 113 278 1708
e-mail: humank@hkeurope.com

Australia: Human Kinetics, 57A Price Avenue, Lower Mitcham, South Australia 5062
08 8277 1555
e-mail: liahka@senet.com.au

New Zealand: Human Kinetics, P.O. Box 105-231, Auckland Central
09-309-1890
e-mail: hkp@ihug.co.nz

To my teachers.

Contents

Preface

There was a time, perhaps only 30 years ago, when a neurophysiologist could claim to "know" the entire field of neuroscience, including muscle. Not any longer. So great has been the expansion of knowledge that previously coherent areas of scientific inquiry have become completely fragmented. In skeletal muscle, for example, the expert on ion channels may neither know nor care about the mechanism of contraction. Even a subject such as muscle fatigue has become so large that a specialist in Ca^{2+} kinetics may be totally unfamiliar with the changes in motoneuron firing rate or with the differences in susceptibility to fatigue among the various types of muscle fiber. And then, as in other tissues, there is the heavy imprint of molecular biology to be absorbed.

Yet, despite the fragmentation of knowledge, skeletal muscle continues to fascinate. Perhaps no other tissue in the body is so amenable to study, and it was muscle and its nerve supply that gave the first clues as to how synapses work and also revealed the twin phenomena of programmed neuronal death and transient hyperinnervation in the embryo. Again, it was muscle that provided the basis of our understanding of trophic dependencies among cells, and other examples of the primacy of discovery are not difficult to find.

The present book is an attempt to bring this huge diversity of information together, so that almost all aspects of muscle can be considered and, where appropriate, related to each other. While it is natural to reveal what is new, it is important to remember the classic experiments upon which much of our knowledge rests. So elegant and fundamental has been this work, that it is hardly surprising that several lifelong or erstwhile myologists have won Nobel Prizes—Charles Sherrington, Edgar Adrian, A.V. Hill, Andrew Huxley, Alan Hodgkin, Jack Eccles, Bernard Katz, Bert Sakmann, and Erwin Neher.

This book is intended for all those who need to know about skeletal muscle, from the undergraduate student beginning kinesiology to the physiotherapist, the physiatrist, the graduate student, and the electromyographer. To assist learning, the text has been interspersed with declarative subheadings and a large number of figures, many of them original. Points of special interest have been incorporated in "boxes," and an extensive glossary has been provided to explain various terms, symbols, and abbreviations. An important and novel feature of the book is the applied physiology sections at the ends of all chapters. The intention here has been to show how specific defects of muscle or nerve cells can result in certain clinical disorders that are known, at least by name, to most lay people. Although these sections are optional reading, their inclusion provides an introduction to the ''neuromuscular clinic'' for those students who are required to carry out special projects as part of their course work.

No book is perfect, but whatever features are praiseworthy in the present one are due to the efforts of many. Above all others, I wish to thank Jane Butler, my research and editorial assistant, for her dedicated efforts. Jane's deep knowledge of muscle and her scrupulous insistence on the highest standards are evident everywhere. Lynn McIntyre and Rebecca Newberry assisted with some of the illustrations, and Barbara McComas applied her formidable concentration to the final organization of the references and to the exhaustive copyright permission process. In this context I would like to thank all those authors and publishers, who graciously allowed us to reproduce illustrations and tables. At the same time, I recognize that there are many other scientists whose work could so easily have been included in this book, had not the demands of space and style proved insuperable. The typing, and repeated retyping, has been carried on the shoulders of three devoted secretaries—in chronological order, Patricia Holmes, Tracy Roketta, and Elizabeth LaRose.

My colleagues in the experimental work cited from our laboratory have been identified by name in the text; it may seem invidious to extract two names from a long list, yet I cannot help but acknowledge Roberto Sica and Adrian Upton for their ideas, friendship, and support over so many years. And then there are others, great men all, who heavily influenced me in my career: Fred Harper, who instilled a love of physiology; Sir John Gray, who tried hard to educate me in precision of thought and expression; Sir Andrew Huxley, who demonstrated that genius can be associated with humility; Jack Diamond, with his infectious enthusiasm for neurobiology; Moran Campbell, who invited me to work in his department at McMaster; and John Walton, now the Lord Walton of Detchant, who was a kind and enlightened patron in the early days. I would also like to thank the various granting agencies that have sustained my muscle research over a long period, and especially the Muscular Dystrophy Association of Canada, NSERC, and the Leman Brothers Muscular Dystrophy Foundation.

A word of thanks as well to Anne Heiles and Nanette Secor for their invaluable expertise in the editing process.

Last, a very big thanks to my wife, Kate, not only for her quiet encouragement, but also, during the long gestation period of the book, for her tolerance of the many tasks left untouched around our home.

Alan McComas

Credits

Figure 1.7 From *The Structure and Function of Muscle* (p. 309), by G.H. Bourne, 1972, New York: Academic Press. Copyright 1972 by Academic Press. Reprinted with permission.

Figure 2.4C From "Neurofilament Phosphorylation: A New Look at Regulation and Function," by R.A. Nixon and R.K. Sihag, 1991, *Trends in Neurosciences*, **14**, p. 502. Copyright 1991 by Elsevier Trends Journals. Adapted with permission.

Figure 2.5 From "Sensory Potentials of Normal and Diseased Nerves," by F. Buchthal, A. Rosenfalck, and F. Behse, 1975, *Peripheral Neuropathy*, **1**, p. 444. Copyright 1975 by W.B. Saunders Company. Adapted with permission.

Figure 2.6 From "The Relationship between Internodal Length and Growth in Human Nerves," by A.D. Vizoso, 1950, *Journal of Anatomy*, **84**, pp. 343 and 345. Copyright 1950 by Cambridge University Press. Reprinted with the permission of Cambridge University Press.

Figure 3.3 From "Scanning and Light Microscopic Study of Age Changes at a Neuromuscular Junction in the Mouse," by M.A. Fahim, J.A. Holley, and N. Rob-

bins, 1983, *Journal of Neurocytology*, **12**, p. 19. Copyright 1983 by Chapman and Hall. Reprinted with permission.

Figure 3.5 From "Synaptic Structure and Development: The Neuromuscular Junction," by Z.W. Hall and J.R. Sanes, 1993, *Cell*, **72**, p. 112. Copyright 1993 by Cell Press. Adapted with permission.

Table 3.1 From "Synaptic Structure and Development: The Neuromuscular Junction," by Z.W. Hall and J.R. Sanes, 1993, *Cell*, **72**, p. 103. Copyright 1993 by Cell Press. Adapted with permission.

Table 4.1 From *Mammalian Muscle Receptors and Their Central Actions* (p. 48), by P.B.C. Matthews, 1972, Kent: Edward Arnold Ltd. Copyright 1972 by P.B.C. Matthews. Reproduced by permission of Edward Arnold (Publishers) Limited.

Figure 4.3 From "The Effects of Stimulation of Static and Dynamic Fusimotor Fibres on the Response to Stretching of the Primary Endings of Muscle Spindles," by A. Crowe and P.B.C. Matthews, 1964, *Journal of Physiology*, **174**, p. 112. Copyright 1964 by The Physiological Society. Adapted with permission.

Figure 10.2 Reproduced from *The Journal of Cell Biology*, 1979, volume 81, p. 281 by copyright permission of the Rockefeller University Press.

Figure 10.3 From *Principles of Neural Science* (p.143), by E.R. Kandel and S.A. Siegelbaum, 1991, Norwalk: Appleton and Lange. Copyright 1991 by Appleton and Lange. Reprinted with permission.

Figure 10.8 From "An Electrophysiological Investigation of Neuromuscular Transmission in Myasthenia Gravis," by D. Elmqvist, W.W. Hofmann, J. Kugelberg, and D.M.J. Quastel, 1964, *Journal of Physiology*, **174**, pp. 420 and 422. Copyright 1964 by Cambridge University Press. Adapted with permission.

Figure 11.1(Top, Bottom) From *The Structure and Function of Muscle* (p. 316), by G.H. Bourne, 1972, New York: Academic Press. Copyright 1972 by Academic Press. Reprinted with permission.

Figure 11.3 (Top, Bottom) From *The Structure and Function of Muscle* (p. 312), by G.H. Bourne, 1972, New York: Academic Press. Copyright 1972 by Academic Press. Reprinted with permisison.

Figure 11.5 From "Tension Development in Highly Stretched Vertebrate Muscle Fibres," by A.M. Gordon, A.F. Huxley, and F.J. Julian, 1966, *Journal of Physiology*, **184**, p. 149. Copyright 1966 by The Physiological Society. Adapted with permission.

Figure 11.6 From "The Variation in Isometric Tension with Sarcomere Length in Vertebrate Muscle Fibres," by A.M. Gordon, A.F. Huxley, and F.J. Julian, 1966, *Journal of Physiology*, **184**, pp. 185-186. Copyright 1966 by The Physiological Society. Adapted with permission.

Figure 11.8 From "Structure of the Actin-Myosin Complex and Its Implications for Muscle Contraction," by I. Rayment et al., 1993, *Science*, **261**, p. 63. Copyright 1993 by The American Association for the Advancement of Science. Adapted with permission.

Figure 11.11 From "On the Relationships between Membrane Potential, Calcium Transient and Tension in Single Barnacle Muscle Fibres," by C.C. Ashley and E.B. Ridgway, 1970, *Journal of Physiology*, **209**, p. 111. Copyright 1970 by The Physiological Society. Adapted with permission.

Figure 11.12 From "Structure of Sarcoplasmic Reticulum," by C. Franzini-Armstrong, 1980, *Federation Proceedings* (1980) **39**, p. 2404. Copyright 1980 by FASEB. Reprinted with permission.

Figure 11.16 From "Influence of joint position on ankle plantarflexion in humans," by D.G. Sale, J. Quinlan, E. Marsh, A.J. McComas, and A.Y. Bélanger, 1982, *Journal of Applied Physiology*, **52**, p. 1637. Copyright 1982 by The American Physiological Society. Reprinted with permission.

Figure 11.17 From "Twitch Potentiation after Voluntary Contraction," by A.A. Vandervoort, J. Quinlan, and A.J. McComas, 1983, *Experimental Neurology*, **81**, p. 145. Copyright 1983 by Academic Press. Adapted with permission.

Figure 12.3 From "Electrophysiological Estimation of the Number of Motor Units within a Human Muscle," by A.J. McComas, P.R.W. Fawcett, M.J. Campbell, and R.E.P. Sica, 1971, *Journal of Neurology, Neurosurgery, and Psychiatry*, **34**, p. 128. Copyright 1971 by BMJ Publishing Group. Reprinted with permission.

Figure 12.4 From "Anatomy and Innervation Ratios in Motor Units of Cat Gastrocnemius," by R.E. Burke and P. Tsairis, 1973, *Journal of Physiology*, **234**, pp. 754-756. Copyright 1973 by The Physiological Society. Adapted with permission.

Figure 12.6 From "Fast and Slow Twitch Units in a Human Muscle," by R.E.P. Sica and A.J. McComas, 1971, *Journal of Neurology, Neurosurgery, and Psychiatry*, **34**, p. 118. Copyright 1971 by the British Medical Association. Adapted with permission.

Figure 12.7 From "Fast and Slow Twitch Muscles in Man," by A.J. McComas and H.C. Thomas, 1968, *Journal of the Neurological Sciences*, **7**, p. 304. Copyright 1968 by Elsevier Science. Reprinted with permission.

Table 12.7 From "Analysis of Myosin Light and Heavy Chain Types in Single Human Skeletal Muscle Fibres," by R. Billeter, C.W. Heizmann, H. Howald, and E. Jenny, 1981, *European Journal of Biochemistry*, **116**, p. 394. Copyright 1981 by the Federation of European Biochemical Societies. Adapted with permission.

Figure 12.8A,B From "Physiological Types and Histochemical Profiles in Motor Units of the Cat Gastrocnemius," by R.E. Burke, D.N. Levine, P. Tsairis, and F.E. Zajac III, 1973, *Journal of Physiology*, **234**, p. 735. Copyright 1973 by The Physiological Society. Reprinted with permission.

Figure 12.8C From "Motor Unit Organization of Human Medial Gastrocnemius," by R.A.F. Garnett, M.J. O'Donovan, J.A. Stephens, and A. Taylor, 1979, *Journal of Physiology*, **287**, p. 37. Copyright 1979 by The Physiological Society. Reprinted with permission.

Figure 13.2 From "Isometric Force Production by Motor Units of Extensor Digitorum Communis Muscle in Man," by A.W. Monster and H. Chan, 1977, *Journal of Neurophysiology*, **40**, p. 1434. Copyright 1977 by The American Physiological Society. Reprinted with permission.

Figure 13.3 From "Motor-Unit Discharge Rates in Maximal Voluntary Contractions of Three Human Muscles," by F. Bellemare, J.J. Woods, R. Johansson, and B. Bigland-Ritchie, 1983, *Journal of Neurophysiology*, **50**, p. 1386. Copyright 1983 by The American Physiological Society. Reprinted with permission.

Figure 13.5B From "The Orderly Recruitment of Human Motor Units during Voluntary Isometric Contractions," by H.S. Milner-Brown, R.B. Stein, and R. Yemm, 1973, *Journal of Physiology*, **230**, p. 365. Copyright 1973 by The Physiological Society. Reprinted with permission.

Figures 14.1, 14.3, 14.4, 14.9, 14.11 From *Molecular Biology of the Cell* (pp. 57, 70, 489, and 491), by Alberts et al., 1983, New York: Garland Publishing, Inc. Copyright 1983 by Alberts et al. Reprinted with permission.

Figure 14.5A,B From *Baillières Clinical Endocrinology and Metabolism* (pp. 500-501), by T. DeBarsy and H.G. Hers, 1990, Toronto: Baillière Tindall. Copyright 1990 by Baillière Tindall. Adapted with permission.

Figure 14.12 From "A Metabolic Myopathy Due to Absence of Muscle Phosphorylase," by C.M. Pearson, D.G. Rimer, and W.F.H.M. Mommaerts, 1961, *American Journal of Medicine*, **30**, p. 506. Copyright 1961 by Cahners Publishing Company. Adapted with permission from *American Journal of Medicine*.

Figure 15.4 From *The Effects of Use and Disuse on Neuromuscular Function* (p. 524), by E. Gutmann and P. Hník, 1963, Prague: Publishing House of the Czechoslovak Academy of Sciences. Copyright 1963 by the Nakladatelství Československé akademie vęd. Adapted with permission.

Figure 15.5 From "The Measurement of K⁺ Concentration Changes in Human Muscles during Volitional Contractions," by F. Vyskočil, P. Hník, H. Rehfeldt, R. Vejapada, and E. Ujec, 1983, *Pflügers Archive*, **399**, p. 236 (Figures 1, 2). Copyright 1983 by Springer-Verlag. Adapted with permission.

Figure 15.6 From "Increased Sodium Pump Activity Following Repetitive Stimulation of Rat Soleus Muscles," by A. Hicks and A.J. McComas, 1989, *Journal of Physiology*, **414**, p. 339. Copyright 1989 by The Physiological Society. Reprinted with permission.

Figure 15.7 From "Changes in Tetanic and Resting [Ca²⁺]ᵢ during Fatigue and Recovery of Single Muscle Fibres from Xenopus Laevis," by J.A. Lee, H. Westerblad, and D.G. Allen, 1991, *Journal of Physiology*, **433**, pp. 310 and 317. Copyright 1991 by The Physiological Society. Adapted with permission.

Figure 15.9 Reprinted with permission from *Nature*, "Muscular Fatigue Investigated by Phosphorus Nuclear Magnetic Resonance," by M.J. Dawson, D.G. Gadian, and D.R. Wilkie, 1978, *Nature*, **274**, p. 862. Copyright 1978 by Macmillan Magazines Limited.

Figure 15.10 From *Neuromuscular Fatigue* (p. 39), by A.J. McComas, V. Galea, R.W. Einhorn, A.L. Hicks, and S. Kuiack, 1993, North Holland: Royal Academy of Arts and Sciences. Copyright 1993 by the Royal Netherlands Academy of Arts and Sciences. Adapted with permission.

Figure 16.1 From "Axotomy-Induced Changes in Rabbit Hindlimb Nerves and the Effects of Chronic Electrical Stimulation," by T. Gordon, J. Gillespie, R. Orozco, and L. Davis, 1991, *Journal of Neuroscience*, **11**, pp. 2160-2161. Copyright 1991 by The Society for Neuroscience. Adapted with permission.

Figure 16.2 From *Peripheral Neuropathy* (p. 678), by P.J. Dyck, H. Nukada, A.C. Lais, and J.L. Karnes, 1984, Philadelphia: W.B. Saunders Company. Copyright 1984 by W.B. Saunders Company. Adapted with permission.

Table 16.1 From "Nerve Conduction during Wallerian Degeneration in the Baboon," by R.W. Gilliatt and R.J. Hjorth, 1972, *Journal of Neurology, Neurosurgery, and Psychiatry*, **35**, p. 339. Copyright 1972 by BMJ Publishing Group. Adapted with permission.

Figure 16.4 From "Normal and Denervated Muscle. A Morphometric Study of Fine Structure," by H.H. Stonnington and A.G. Engel, 1973, *Neurology*, **23**, p. 716. Copyright 1973 by New York Times Media Company. Adapted with permission.

Figure 16.6A,B,C From "Membrane Properties Underlying Spontaneous Activity of Denervated Muscle Fibres," by D. Purves and B. Sakmann, 1974, *Journal of Physiology*, **239**, p. 133. Copyright 1974 by The Physiological Society. Adapted with permission.

Figure 17.2 From "Changes in Motor Innervation and Cholinesterase Localization Induced by Botulinum Toxin in Skeletal Muscle of the Mouse: Differences between Fast and Slow Muscles," by L.W. Duchen, 1970, *Journal of Neurology, Neurosurgery, and Psychiatry*, **33**, p. 51. Copyright 1970 by BMJ Publishing Group. Reprinted with permission.

Figure 17.6 Reproduced from *The Journal of General Physiology*, 1973, volume 61, p. 11 by copyright permission of the Rockefeller University Press.

Figure 17.8 From "The Extent and Time Course of Motoneuron Involvement in Amyotrophic Lateral Sclerosis," by M. Dantes and A.J. McComas, 1991, *Muscle & Nerve*, **14**, p. 418. Copyright 1991 by John Wiley and Sons, Inc. Reprinted with permission.

Figure 17.9 From "Functional Compensation in Partially Denervated Muscles," by A.J. McComas, R.E.P. Sica, M.J. Campbell, and A.R.M. Upton, 1971, *Journal of Neurology, Neurosurgery, and Psychiatry*, **34**, p. 456.

Figure 17.10 From "Reorganization of Motor-Unit Properties in Reinnervated Muscles of the Cat," by T. Gordon and R.B. Stein, 1982, *Journal of Neurophysiology*, **48**, p. 1181. Copyright 1982 by The American Physiological Society. Adapted with permission.

Figure 18.2 From "The Influence of Activity on Some Contractile Characteristics of Mammalian Fast and Slow Muscles," by S. Salmons and G. Vrbová, 1969, *Journal of Physiology*, **201**, p. 542. Copyright 1969 by The Physiological Society. Adapted by permission.

Figure 18.3 From "Control of Contractile Properties Within Adaptive Ranges by Patterns of Impulse Activity in the Rat," by R.H. Westgaard and T. Lømo, 1988, *Journal of Neuroscience*, **8**, p. 4417. Copyright 1988 by The Society for Neuroscience. Adapted with permission.

Figure 18.4 Reprinted with permission from *Nature*, "Nerve Stump Length and Membrane Changes in Denervated Skeletal Muscle," by J.B. Harris and S. Thesleff, 1972, *Nature* (New Biology), **236**, p. 60. Copyright 1972 by Macmillan Magazines Limited.

Figure 18.8 From "Rescue of Motoneurons from Cell Death by a Purified Skeletal Muscle Polypeptide: Effects of the ChAT Development Factor, CDF," by J.L. McManaman, R.W. Oppenheim, D. Prevette, and D. Marchetti, 1990, *Neuron*, **4**, p. 893. Copyright 1990 by Cell Press. Adapted with permission.

Figure 18.9 From "Functional Changes in Motoneurones of Hemiparetic Muscles," by A.J. McComas, R.E.P. Sica, A.R.M. Upton, and N. Aguilera, 1973, *Journal of Neurology, Neurosurgery, and Psychiatry*, **36**, p. 185. Copyright 1973 by BMJ Publishing Group. Reprinted with permission.

Figure 19.1 From "Human Neuromuscular Adaptations That Accompany Changes in Activity," by A.J. McComas, 1994, *Medicine and Science in Sports and Exercise*, **26**, p. 1502. Copyright 1994 by Williams and Wilkins. Reprinted with permission.

Figure 19.2 From "Electrical and Mechanical Changes in Immobilized Human Muscle," by J. Duchateau and K. Hainaut, 1987, *Journal of Applied Physiology*, **62**, p. 2171. Copyright 1987 by The American Physiological Society. Adapted with permission.

Figure 19.4 From "Contrasting Effects of Suspension on Hindlimb Muscles in the Hamster," by K. Corley, N. Kowalchuk, and A.J. McComas, 1984, *Experimental Neurology*, **85**, p. 35. Copyright 1984 by Academic Press, Inc. Adapted with permission.

Figure 20.2 From "The Role of Learning and Coordination in Strength Training," by O.M. Rutherford and D.A. Jones, 1986, *European Journal of Applied Physiology*, **55**, p. 102 (Figure 2). Copyright 1986 by Springer-Verlag. Adapted with permission.

Table 20.1 From "Human Neuromuscular Adaptations That Accompany Changes in Activity," by A.J. McComas, 1994, *Medicine and Science in Sports and Exercise*, **26**, p. 1499. Copyright 1994 by Williams and Wilkins. Reprinted with permission.

Figure 20.4 From "Adaptations in Coactivation after Isometric Resistance Training," by B. Carolan and E. Carafelli, 1992, *Journal of Applied Physiology*, **73**, p. 914. Copyright 1992 by The American Physiological Society. Adapted with permission.

Tables 20.2, 20.3 From "Exercise Training Induces Transitions of Myosin Isoform Subunits Within Histochemically Typed Human Muscle Fibres," by H. Baumann, M. Jäggi, F. Soland, H. Howald, and M.C. Schaub, 1987, *Pflügers Archiv*, **409**, p. 353. Copyright 1987 by Springer-Verlag. Adapted with permission.

Figure 20.7 From "Neural Control of Gene Expression in Skeletal Muscle. Effects of Chronic Stimulation on Lactate Dehydrogenase Isoenzymes and Citrate Synthase," by U. Seedorf, E. Leberer, B.J. Kirschbaum, and D. Pette, 1986, *Biochemical Journal*, **239**, p. 117. Copyright 1986 by The Biochemical Society and Portland Press. Adapted with permission.

Figure 20.9 From "Perspectives on Molecular and Cellular Exercise Physiology," by F.W. Booth, 1988, *Journal of Applied Physiology*, **65**, p. 1462. Copyright 1988 by The American Physiological Society. Adapted with permission.

Figure 21.1 From "Injury to Skeletal Muscle Fibers of Mice Following Lengthening Contractions," by K.K. McCully and J.A. Faulkner, 1985, *Journal of Applied Physiology*, **59**, p. 125. Copyright 1985 by the American Physiological Society. Adapted with permission.

Figure 21.4 From "Electrophysiological Features of Muscle Regeneration," by A. Stuart, A.J. McComas, G. Goldspink, and G. Elder, 1981, *Experimental Neurology*, **74**, pp. 156-157. Copyright 1981 by Academic Press, Inc. Reprinted with permission.

Figure 21.5 From "Electrophysiological Features of Muscle Regeneration," by A. Stuart, A.J. McComas, G. Goldspink, and G. Elder, 1981, *Experimental Neurology*, **74**, p. 153. Copyright 1981 by Academic Press, Inc. Reprinted with permission.

Figure 22.1 Reprinted with permission from *Nature*, "A Study of Age Group Track and Field Records to Relate Age and Running Speed," by D.H. Moore II, 1975, *Nature*, **253**, p. 264. Copyright 1975 by Macmillan Magazines Limited.

Table 22.1 From ''Strength and Endurance of Skeletal Muscle in the Elderly,'' by A.A. Vandervoort, K.C. Hayes, and A.Y. Bélanger, 1986, *Physiotherapy Canada*, **38**, p. 168. Copyright 1986 by Physiotherapy Canada. Adapted with permission.

Figures 22.2, 22.3 From ''Contractile Changes in Opposing Muscles of the Human Ankle Joint with Aging,'' by A.A. Vandervoort and A.J. McComas, 1986, *Journal of Applied Physiology*, **61**, pp. 364-365. Copyright 1986 by The American Physiological Society. Reprinted with permission.

Figure 22.6 From ''Physiological Changes in Ageing Muscle,'' by M.J. Campbell, A.J. McComas, and F. Petito, 1973, *Journal of Neurology, Neurosurgery, and Psychiatry*, **36**, p. 177. Copyright 1973 by BMJ Publishing Group. Adapted with permission.

Figure 22.7 Adapted by permission of the publisher from (''Neuron Survival in the Aging Mouse'' by H.A. Johnson and S. Erner), *Experimental Gerontology*, **7**, p. 113. Copyright 1972 by Elsevier Science Inc.

Figure 22.8 From ''Growth and Degeneration of Motor End-Plates in Normal Cat Hind Limb Muscles,'' by A.R. Tuffery, 1971, *Journal of Anatomy*, **110**, p. 226. Copyright 1971 by Cambridge University Press. Reprinted with the permission of Cambridge University Press.

Figure 22.9 Reprinted from *International Journal of Developmental Neuroscience*, **8**, by J.L. Rosenheimer, ''Factors Affecting Denervation-Like Changes at the Neuromuscular Junction during Aging,'' p. 646, Copyright 1990, with kind permission from Elsevier Science Ltd, The Boulevard, Langford Lane, Kidington 0X5 1GB, UK.

Table 22.5 From ''Ultrastructural Studies of Young and Old Mouse Neuromuscular Junctions,'' by M.A. Fahim and N. Robbins, 1982, *Journal of Neurocytology*, **11**, p. 652. Copyright 1982 by Chapman Hall Limited. Reprinted with permission.

Figure 22.11 From ''Effects of Aging on Nerve Sprouting and Regeneration,'' by A. Pestronk, D.B. Drachman, and J.W. Griffin, 1980, *Experimental Neurology*, **70**, p. 75. Copyright 1980 by Academic Press. Adapted with permission.

Table 22.6 From ''Strength and Endurance of Skeletal Muscles in the Elderly,'' by A.A. Vandervoort, K.C. Hayes, and A.Y. Bélanger, 1986, *Physiotherapy Canada*, **38**, p. 170. Copyright 1986 by Physiotherapy Canada. Reprinted with permission.

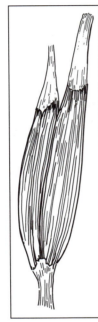

Part I

Structure and Development

The structure of a tissue reflects its function. In the case of striated muscle, that function is to generate force or to produce movement. To achieve these mechanical goals effectively, the muscle fiber is packed with contractile elements in the form of thick and thin filaments. In this first section, we will examine the structure of the filaments and that of the special molecules that hold the filaments together, thereby giving skeletal muscle its characteristic banded appearance.

But how is a signal brought to the filaments to make them interact with each other and produce movement or force? To understand the excitation of muscle, we must begin by examining the structure of the motoneurons (motor nerve cells) in the spinal cord and brainstem. We must then look at the structure of the nerve fibers (axons) which enable electrical impulses to be transmitted rapidly and efficiently from the motoneurons to the muscle. However, an impulse cannot simply leap across the narrow gap that separates an axon and a muscle fiber. Instead, the structure of the neuromuscular junction allows the impulse in the axon to release a chemical (acetylcholine) that then sets up a signal in the muscle

1

fiber. This impulse in the muscle fiber must somehow be conducted into the region of the contractile filaments as it travels along the fiber. Once again, there is a structure appropriate for the function in the form of the T-tubules. We will discover that the T-tubules make special contacts with sacs inside the fiber so that another chemical messenger, this time Ca^{2+}, can act as a link in the excitation pathway. Thus, at every step in the chain of events, the structure of the muscle or nerve fiber appears optimally suited to the task that must be performed. Nor, in the survey of muscle structure, must we forget the organelles and other cell parts that are common to all tissues—the nuclei, mitochondria, and cell membranes. From the familiar light microscope of the last century, the investigative tools have progressed to the interference and electron microscopes, to the technique of freeze-fracture, and to X-ray diffraction.

After the anatomy of the muscle fiber, motoneuron, and neuromuscular junction has been reviewed in the first three chapters, there comes a chapter devoted to muscle receptors. For convenience, the physiology of the muscle receptors is dealt with at the same time.

Although the muscle receptors intrude from time to time in our survey of skeletal muscle and its nerve supply, the real place for the receptors lies in the study of motor control and is therefore outside the scope of the present book. The last two chapters of this section (chapters 5 and 6) consider how muscle fibers and motoneurons develop in the embryo and how the growing nerve fibers can establish synaptic connections within the muscle fibers. This is an exciting and topical line of scientific inquiry, requiring the skills of the microscopist and those of the molecular biologist too. It is for the molecular biologist to identify the signals responsible for activating the genes of the muscle fiber in the correct sequence. Doubtless many signals and genes remain to be discovered, but enough are now known to give some idea of the way in which the genetic instructions are given during development.

At the end of each chapter, as in the two other parts of the book, there is a section on applied physiology. The intention here is to show how disease or dysfunction can result from a precise alteration in structure or from aberrant embryological development.

Chapter 1

The Muscle Fiber

In this chapter, the sizes and shapes of muscles are considered first. Next comes an account of the connective tissue that, among other functions, provides a scaffolding for the muscle fibers. These general aspects of the muscle can be referred to as its *architecture*. The structure of one of the many thousands of fibers in a muscle belly is then described, making use of observations from both the light and electron microscopes. The description begins with the thin membrane that envelops the fiber and then examines in turn the various features in the fiber interior. All of the parts of a fiber are concerned in some way with the production of movement or force, but it is the myofibrils that ultimately produce these effects, and it is the myofibrils that make up the bulk of the fiber.

However, the operation of the myofibrils depends on the electrical and chemical properties of the tubular systems in their neighborhood and on the availability of the energy-releasing compound, ATP; it is the mitochondria in the fiber interior that are the source of ATP. Last, but by no means least, are the nuclei of the muscle fiber that, through the expression of their genetic information, provide the instructions for the different proteins of the fiber to be synthesized in the myofibrils, tubular systems, and organelles. The structures of all

these components of the muscle fiber will be described in this chapter, but only after an account has been given of the general features of a muscle.

Muscle Architecture

For a human, as for other animals, to move is to survive. Apart from thinking, every human activity requires a movement, or at least a muscle contraction, whether it be for walking or running, catching or letting go, shouting or whispering, looking or listening, or even standing still. And, of course, no muscle contractions are more important than those responsible for the simple movements of breathing. This enormous diversity of muscle function is reflected in the sizes and shapes of the muscles.

Muscles, Like Movements, Are Remarkably Diverse

Let us consider the sizes of the muscles first. The tensor tympani, a small muscle responsible for adjusting the tension on the eardrum, contains only a few hundred muscle fibers, while the medial gastrocnemius, one of the principal calf muscles employed for walking and running, has approximately a million fibers. The approximate number of fibers in these two examples and in a selection of other human muscles is given in Table 1.1

The shapes of the muscles also vary. Some muscles are relatively thick—for example, the vastus muscles, which form most of the quadriceps, or the gluteal muscles, which abduct or extend the hip. Other muscles, however, form long, slender straps—such as the hamstrings, or the gracilis and sartorius. Yet other muscles, such as the flexors and extensors of the fingers and toes, are characterized by their very long tendons. This variation is suited to the different types of tasks that muscles must perform. The longer a muscle, the more it can shorten, and the higher its velocity of shortening; in contrast, the thicker a muscle, the more force (tension) it can develop. The long tendons of the finger flexors run beneath fibrous bands (*retinacula*), which serve as pulleys, and are inserted into the bases of the phalanges. This arrangement enables modest excursions of the tendons to be translated into large grasping movements of the fingers.

Muscle Fibers Have Striations

Regardless of their shapes and sizes, all muscles are made up of individual fibers, and these, in the large limb muscles of an adult, are roughly 50 µm in diameter. In infancy, the fibers are very much smaller than this, but strength training or physical labor in adults can cause the diameters to double. When viewed longitudinally under the microscope, each fiber appears as a series of alternating light and dark striations. The light

stripe is the *I-band*, and the dark stripe, with its higher refractive index, is the *A-band*. A dense line runs through the middle of the I-band; this is the *Z-line* or *Z-disc*. The central part of the A-band, which in a relaxed fiber appears rather lighter (the *H-zone*), is bisected by another dark structure, the *M-region* (Figure 1.1). The part of a fiber between two successive Z-lines is termed a *sarcomere* and in a relaxed fiber has a length of approximately 2.2 µm. The explanation for the striations will become apparent when the structure of the myofibril is considered later in the chapter. Table 1.2 gives the numbers of sarcomeres arranged end to end in single fibers of various human muscles; notice the very high values for the strap muscles of the thigh (sartorius, gracilis, semitendinosus) and the much lower ones for some of the other muscles (soleus, medial gastrocnemius, tibialis posterior).

Strap Muscles Are Divided Into Compartments

The fibers in the long strap muscles are often stated to run from one end of the muscle to the other, and indeed, the sarcomere numbers given in Table 1.2 are based on this premise. Under the microscope, however, the muscle bellies are seen to be divided into compartments by one or more transverse fibrous bands (*inscriptions*); the sartorius has four such compartments, the semitendinosus has three, and the biceps femoris and gracilis each have two. Because of these compartments, the longest human muscle fibers are approximately 12 cm, which

Table 1.1   Numbers of Muscle Fibers in Various Human Muscles

Muscle	Number of muscle fibers
First lumbrical	10,250[a]
External rectus	27,000
Platysma	27,000
First dorsal interosseous	40,500
Sartorius	128,150[a]
Brachioradialis	129,200[a]
Tibialis anterior	271,350
Medial gastrocnemius	1,033,000

Note. Results given to nearest 50.

[a]Average values. Value for sartorius from MacCallum (1898); all others from Feinstein et al. (1955).

Table 1.2   Numbers of Sarcomeres in Human Muscles

Muscle	Number of sarcomeres per fiber ($\times 10^4$)		
	I[a]	II[a]	III[a]
Tibialis posterior	1.1	1.5	0.8
Soleus	1.4	—	—
Medial gastrocnemius	1.6	1.5	1.5
Semitendinosus	5.8	6.6	—
Gracilis	8.1	9.3	8.4
Sartorius	15.3	17.4	13.5

[a]Refers to individual limbs analyzed. The values for the three thigh muscles at the bottom of the table do not take into account the fibrous inscriptions in the belly, and the true values will be rather lower than those stated.

Note. Data from "Muscle Architecture of the Human Lower Limb" by T.L. Wickiewicz, R.R. Roy, P.L. Powell, and V.R. Edgerton, 1983, *Clinical Orthopaedics and Related Research,* **179**, p. 277.

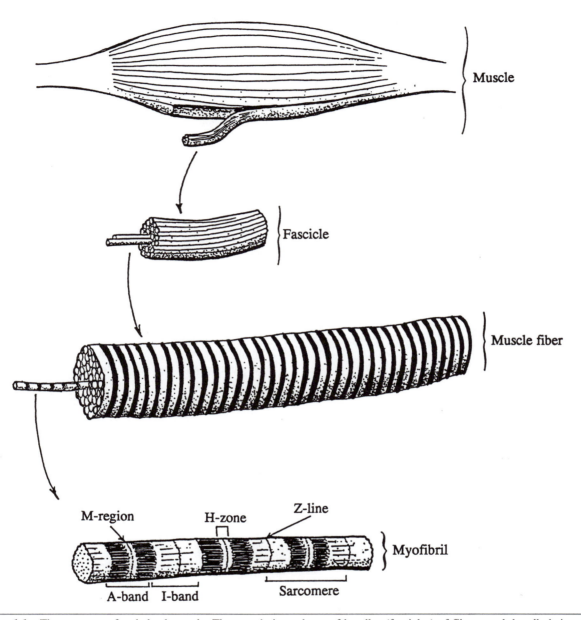

M-region   H-zone   Z-line

Myofibril

A-band   I-band   Sarcomere

Figure 1.1 The structure of a skeletal muscle. The muscle is made up of bundles (fascicles) of fibers, each bundle being lined by perimysial connective tissue (see Figure 1.3). The muscle fibers and the myofibrils appear striated, due to alternate dark (A) and light (I) bands; the bands are "in register" between all the myofibrils in the same fiber.

corresponds to 5.5×10^4 sarcomeres. Each compartment must necessarily have its own nerve supply, and individual nerve fibers often supply muscle fibers in adjacent compartments. By ensuring that contraction occurs fairly synchronously along the muscle belly, the compartmental innervation enables the muscle belly to shorten very rapidly. It may also be that compartments allow a more effective distribution of neurotrophic factors from the motoneurons to the muscle fibers (see chapter 18).

Pinnation Produces More Force

in most human muscles the fibers lie in the longitudinal axis of a fusiform muscle belly (Figure 1.2). In some

of the larger muscles, however, the fibers are obliquely inserted into the tendon and, because of the resemblance to a feather, this arrangement is termed *pinnation*. The muscle fibers in a pinnate muscle are obviously shorter than those in a fusiform belly and, because they pull on the tendon at an angle, not all their force reaches the tendon; it is the *cosine* of the angle of insertion into the tendon that determines the transmitted fraction of the force. The advantage of pinnation is that the effective cross-sectional area of the muscle is increased, since there are many more muscle fibers than in a fusiform belly of the same size. The force developed by the muscle, which depends on the cross-sectional area of all the fibers, is increased proportionately. A less obvious consequence of pinnation is that the distance moved by a central tendon during a contraction is actually *greater* than the amount of shortening of the muscle fibers. This effect can also be an advantage, for it enables the muscle fibers to function over the optimum part of their length-tension curves (see Gans & Gaunt, 1991; also Figure 11.6 in chapter 11).

Muscle Connective Tissue

The connective tissue of a muscle is almost as important as the muscle fibers, for without it there would be no structure to the muscle belly and no way for the movements and forces, produced in the fibers, to be transmitted to the tendon. It will be seen that the connective tissue in the muscle belly has several anatomical parts, associated with different sizes and orientations of collagen fibers.

Muscle Connective Tissue Is Organized at Three Levels

Like connective tissue elsewhere in the body, that of muscle consists of fibers embedded in an amorphous ground substance. Most of the fibers are *collagen*, of which there are at least five (I-V) immunologically distinct types; the remaining fibers are *elastin*. The connective tissue in muscle has three anatomical parts (Figure 1.3; see also Borg & Caufield, 1980; Rowe, 1981).

The *epimysium* is a particularly tough coat that covers the entire surface of the muscle belly and separates it from other muscles. It contains tightly woven bundles of collagen fibers, from 600 to 1800 nm in diameter, which have a wavy appearance and are connected to the perimysium.

The *perimysium* is also tough and relatively thick; it divides the muscle fibers into bundles, or *fascicles*, and it also provides the pathway for the major blood vessels and nerves to run through the muscle belly. Some of the large collagen bundles lie alongside the outer muscle fibers in the fascicles; others run around the fascicles so as to form a crisscross pattern. Underneath the coarse perimysial sheets of connective tissue is a looser and more delicate network in which collagen fibrils run in all directions, some being connected to the endomysium (discussed next). The arterioles and venules are found in these regions, often with intramuscular nerve branches (Figure 1.3, top). A cross-section through a fascicle shows that the muscle fibers have polygonal rather than circular outlines (Figure 1.3, bottom); the polygonal configuration enables the greatest number of fibers to be fitted within a fascicle. The interstitial spaces separating the fibers are usually no wider than 1 µm.

The *endomysium* invests each of the muscle fibers and is composed of a dense feltwork of collagen fibrils, from 60 to 120 nm in diameter, some of which are continuous with the fine fibrillar mesh of the perimysium. It is probable that the endomysium also makes connections to the basement membrane, a glycoprotein layer which lies on the outside of the muscle fiber membrane.

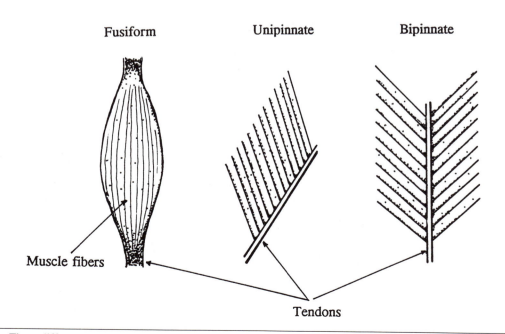

Figure 1.2   Three different arrangements of muscle fibers.

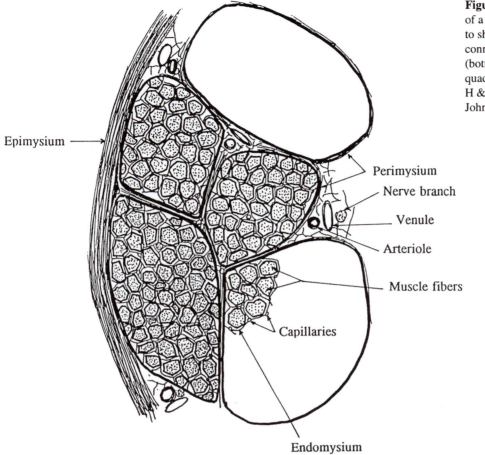

Epimysium

Perimysium

Nerve branch

Venule

Arteriole

Muscle fibers

Capillaries

Endomysium

Figure 1.3   (top) Cross section of a muscle drawn schematically to show the three types of connective tissue sheath. (bottom) Cross section of human quadriceps muscle, stained with H & E. (150x). (Courtesy of Dr. John Maguire.)

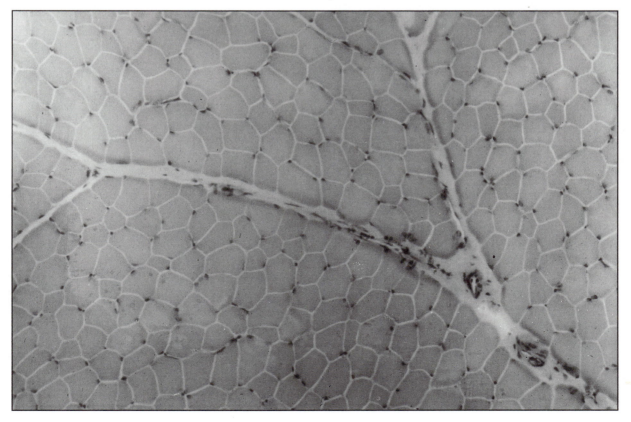

Muscle Connective Tissue Has Multiple Functions

1. In development, the connective tissue serves as a scaffolding upon which the muscle fibers can form. When muscle development is complete, the connective tissue continues to hold the muscle fibers together and largely determines the gross structure of the muscle belly.

2. The loose connective tissue of the perimysium provides a conduit for the blood vessels and nerves supplying the muscle fibers (Figure 1.3).

3. The connective tissue tends to resist passive stretching of the muscle and ensures that the forces are distributed in such a way as to minimize damage to the muscle fibers. Further, the elasticity associated with the elastin fibrils and the wavy collagen bundles enables the muscle belly to regain its shape when the passive forces are removed.

4. The endomysium, through lateral connections to the muscle fiber, conveys part of the contractile force to the tendon. For example, Street and Ramsey (1965) crushed single muscle fibers, causing the myofibrils to retract; when the intact portion of the fiber was stimulated, the full force was still transmitted through the empty "tube" to the far end of the fiber. Although some of the force could have been transferred through the cytoskeletal elements (spectrin, dystrophin, actin) under the plasmalemma (see p. 10), the endomysium and basement membrane would also have been involved, through structures (*costameres*) at the level of the Z-discs. The costameres connect myofibrils to the plasmalemma and basement membrane, and probably also to the endomysium (see Street, 1983).

The Ends of the Muscle Fibers Are Specialized for Transmitting Force to the Tendon

As the muscle fibers approach their tendon of attachment, they narrow considerably, decreasing their diameters by as much as 90% (Loeb, Pratt, Chanaud, & Richmond, 1987); the tapering of the fibers gives the muscle belly its typical fusiform shape. At the very ends of each fiber there is extensive folding of the plasmalemma, the folds interdigitating with connective tissue processes. This folding ensures that the contractile force will be distributed over a larger area, reducing the stress on the surface of the fiber. Also, because the force is transmitted at an angle, it causes a shearing stress, and adhesive structures, such as the different layers at the fiber ends, are more resistant to shearing stresses than they are to orthogonally applied forces (Tidball,

1983). With the electron microscope, it can be seen that the myofibrils do not extend all the way to the plasmalemma; instead, the thin filaments (*actin*) are inserted immediately beneath the plasmalemma into a dense layer of material which contains the attachment proteins *vinculin, talin, paxillin,* and *tensin.* Figure 1.4 shows a hypothetical model for the connection of the actin filaments to the plasmalemma, incorporating talin and vinculin. In this model, talin is joined to the fibronectin receptor; this receptor straddles the plasmalemma and, in the cell interior, is connected to bundles of fibronectin. The fibronectin receptor belongs to a class of proteins, the *integrins,* that span the surface membrane.

On the outer side of the surface membrane is a well developed *basement membrane,* which consists of several layers and is considered in more detail in the following section. Just as there are specialized connecting proteins on the internal face of the plasmalemma, so there are on the outer surface too. These include *fibronectin* and *tenascin,* and they link the integrin molecules in the plasmalemma to the *type I collagen fibers* in the tendon.

Basement Membrane

The basement membrane is a glycoprotein complex that surrounds the muscle fiber, lying between the endomysium and the plasmalemma. Although thin, the basement membrane consists of different layers, each with a distinct structure. Further, there are a number of specialized proteins associated with the basement membrane. Thus, in addition to providing, with the endomysium, a scaffolding for the muscle fiber, the basement membrane has enzymatic actions and also "trophic" functions during development and innervation.

The Basement Membrane Has Several Layers

With the electron microscope, it can be seen that the basement membrane consists of two parts, a basement lamina and a reticular lamina (Figure 1.5). The reticular lamina is composed of collagen and other fibrils within an amorphous ground substance; toward the ends of the muscle belly, it is connected to the collagen fibers in the muscle tendon. The basement lamina is itself made up of two layers, the thinner of which is 2.5 nm thick and, because of its electron translucence, is termed the *lamina lucida;* the other lamina, the *lamina densa,* is from 10 to 15 nm thick and is more easily seen with the electron microscope. The plasmalemma and the various layers of the basement membrane are collectively referred to as the *sarcolemma* (Figure 1.5). Some authors,

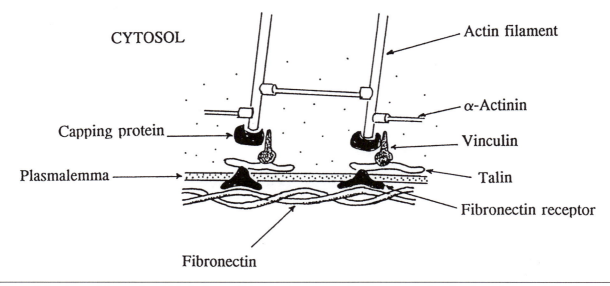

Figure 1.4 A possible way in which the intracellular attachment proteins are connected to the plasmalemma and to the extracellular supporting matrix. Not all the known attachment proteins are shown. Note how the α-actinin molecules brace the actin filaments. Adapted from Alberts et al. (1989, p. 636).

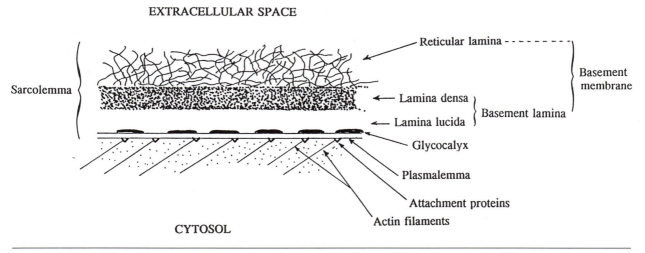

Figure 1.5 The different components of the basement membrane. The basement membrane and the plasmalemma constitute the sarcolemma, while the basement membrane is itself made up of the basement and reticular laminas.

however, prefer to use the term sarcolemma to describe the *plasmalemma* alone.

The Basement Membrane Contains Several Types of Protein and Carbohydrate

The composition of the basement membrane has been investigated with immunohistochemical techniques. Among the proteins identified are the following:

• *Acetylcholinesterase (AChE).* This is the only enzyme found so far in the basement membrane; it splits acetylcholine released from the motor nerve terminals (see chapter 10).

• *Collagen.* Several types of collagen are present, including the relatively thick type I collagen fibers. The collagen fibers and the thinner fibrils give strength to the basement membrane and also help to connect the muscle fiber to the endomysium.

• *Fibronectin* and *laminin.* Both these proteins link the integrin molecules of the plasmalemma to the collagen fibers of the basement membrane.

• *Agrin.* This protein is involved in the formation of the neuromuscular junction by instructing the motor nerve and muscle fiber membranes to undergo the necessary structural specializations.

Some of the carbohydrates in the lamina lucida consist of the *sugar residues* of various glycoproteins embedded in the plasmalemma. However, most of the carbohydrate is in the form of *glycosaminoglycans*. These molecules are long, unbranched disaccharide polymers and include hyaluronic acid, chondroitin sulfate, and heparin sulfate. These compounds coat the proteins in the basement membrane and also form the amorphous ground substance.

The Basement Membrane Has Several Important Functions

Far from being an inert covering for the muscle fiber, as was once thought, the basement membrane has a number of important and varied functions.

1. *Termination of synaptic transmission*, through hydrolysis of the transmitter, acetylcholine, by the enzyme acetylcholinesterase (see chapter 10).

2. *Attachment* of the muscle fiber to the endomysium, the motor nerve terminal, and, at the ends of the fiber, to the muscle tendons.

3. *Scaffolding* for muscle fiber regeneration, by ensuring that the satellite cells multiply within the confines of the damaged fiber (see chapter 21).

4. *Regulation of neuromuscular junction*. Not only does the basement membrane somehow guide a regenerating axon to the site of the former neuromuscular junction, but it also provides a signal for the growth cone to develop the specialized structures characteristic of a motor nerve terminal. Similarly, the basement membrane, even in the absence of a motor nerve terminal, is able to stimulate the muscle fiber to develop synaptic folds and to incorporate acetylcholine receptors into the plasmalemma (see chapter 17).

Plasmalemma

The contents of all living cells are bounded by a plasma membrane, the *plasmalemma*, some 7.5 nm thick; the contents themselves comprise an aqueous solution of inorganic ions, sugars, amino acids, peptides, and proteins that is termed the *cytosol*. The cytosol and the filaments and organelles within it are the *cytoplasm*; in the case of the skeletal muscle fiber, the cytoplasm is sometimes referred to as the *sarcoplasm*.

Apart from giving each cell its own anatomical identity, the main function of the plasmalemma is to enable the cytosol to have a chemical composition markedly different from that of the fluid surrounding the cell. The plasmalemma achieves this last function through populations of protein channels and pumps, which are embedded in the two lipid layers of the membrane. In nerve and muscle fibers, as distinct from most other cells, the plasmalemma has the additional property of excitability, enabling electrical impulses to be transmitted over the length of the cell (see chapter 9).

The Cell Membrane is a Lipid Bilayer and Has Fluid Properties

The plasmalemma is not a smooth structure over the entire muscle fiber. Rather, in the region of the innervation zone (motor end-plate), the plasmalemma is highly convoluted due to the presence of junctional folds (see chapter 3). Elsewhere along the surface of the fiber the membrane displays much shallower folds; these result from slackness of the membrane when the fiber is in its resting or contracted states, and they disappear if the fiber is passively stretched. There are also numerous small in-pocketings of membrane, the *caveolae*, which are connected to the surface membrane by narrow necks; their function is uncertain, though they can also act as reserve sources of membrane during stretching of the fiber (Dulhunty & Franzini-Armstrong, 1975). Biochemical and ultrastructural investigations indicate that the plasmalemma is largely composed of lipid molecules arranged perpendicularly to the surface of the fiber and forming two layers (Figure 1.6). The hydrophilic "heads" of the lipid molecules form the internal and external surfaces of the membrane, while the hydrophobic "tails" make up the interior of the membrane. The heads are composed of *choline*, *phosphate*, and *glycerol*; the tails consist of *fatty acid chains*, and there are two associated with each head. The plasmalemma also contains relatively large amounts of *cholesterol*, the molecules of which are interposed between the phospholipid and serve to stiffen the membrane. Since much of the membrane lipid has a melting point below body temperature, the plasmalemma would be expected to have fluid properties. By labeling a spot of membrane with a fluorescent dye and then observing its enlargement under the microscope, Fambrough, Hartzell, Rash, and Ritchie (1974) were able to show that this was indeed the case. The fluidity is due to the lipid molecules being able to exchange places with each other within their layer of the membrane.

Different Types of Protein Are Embedded in the Plasmalemma

The membrane also contains proteins, of which two types are generally recognized; these are referred to as extrinsic and intrinsic. Extrinsic ("peripheral") proteins are only

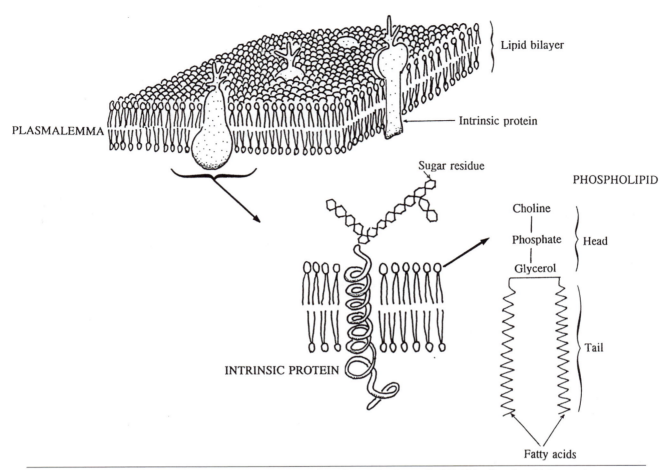

Figure 1.6 The plasmalemma of the muscle fiber, consisting of two closely applied lipid leaflets in which proteins are embedded. A glycosylated transmembrane protein, formed by a single α-helix of amino acids and a "sugar residue," is shown in the lower part of the figure. At bottom right is an enlargement of a single phospholipid molecule.

attached at the internal or external surface of the membrane and can be dislodged relatively easily by chemical means. In contrast, intrinsic ("transmembrane") proteins penetrate the full thickness of the membrane and are difficult to remove. Many of the proteins are glycosylated, that is, they have *sugar residues* which may extend as far as the lamina lucida of the basement membrane (Figure 1.6). Freely ending sugar residues may trap molecules in the extracellular fluid and direct them to the protein in the membrane. Among the special proteins known to be localized in the plasmalemma are the following:

• Transport systems for sugars, lipids, amino acids, and ions. The transport proteins fall into several classes. *Uniports* carry only one solute across the membrane. *Symports* function by moving two solutes in the same direction, for example, sugar molecules and H^+. *Antiports* act by transferring two solutes in opposite directions; the best known example is the N^+-K^+ pump, which extrudes Na^+ from the cell in exchange for K^+. Other transport proteins are the *ion channels*, which allow a specific ion to cross the membrane down its electrochemical gradient, once the channel has been opened by a change in voltage across the membrane. The *acetylcholine receptor* is a different type of ion channel (ligand-gated), in which the opening is brought about by combination with acetylcholine.

• *Adenylate cyclase*, responsible for synthesizing one of the second messengers, cyclic AMP. A GTP-binding protein (G protein) on the cytoplasmic surface of the plasmalemma is involved in the activation of adenylate cyclase.

• *Kinases* for phosphorylating various membrane proteins.

• *Hormone receptors*. The ability of the muscle fiber to respond to such varied hormones as thyroid hormone, insulin, and epinephrine depends on the presence in the plasmalemma of specific receptors for each hormone.

• *Integrins*. These proteins link the basement membrane and endomysium to the plasmalemma and to cytoskeletal structures. Two of the integrins are the fibronectin receptor and the dystrophin-associated glycoproteins (see Figures 1.4 and 1.17).

In comparison with the lipid molecules, the ability of proteins to move within the membrane is restricted, because they are anchored to intracellular or extracellular filaments through binding proteins such as *ankyrin* and *desmin*.

The Cytoskeleton Strengthens the Plasmalemma and Holds Intracellular Structures in Place

Like other cells in the body, the muscle fiber has a cytoskeleton that gives it strength and also holds the different intracellular structures in place. At the periphery of the fiber the cytoskeleton reinforces the plasmalemma, preventing it from tearing during contraction and relaxation. *Actin* and *spectrin* are especially important in this role, as is the recently discovered protein *dystrophin*. Other intermediate filaments, composed of *desmin, synemin,* and *vimentin*, are wrapped around the myofibrils at the Z-discs and bind the myofibrils together. The same types of filament serve to position some of the intracellular organelles, such as the nuclei and mitochondria. The intermediate filaments are considered again later.

Myofibrils

The most obvious structures within the muscle fiber are the myofibrils, which are the units responsible for contraction and relaxation of the fiber. Each myofibril is from 1 to 2 μm in diameter and is separated from its neighbors by mitochondria and the sarcoplasmic and transverse tubular systems; in a fiber of 50 μm diameter, there are up to 8,000 myofibrils. It will be seen that the myofibrils contain two types of protein filament, actin and myosin, and that it is the regular disposition of these filaments along the myofibril that gives the myofibril, and the muscle fiber, its banded appearance under the microscope. The two types of filament support each other through cross-bridges and are themselves supported by other specialized structures. Ultimately the filaments, through the supporting structures, are connected to the endomysial connective tissue and thence to the muscle tendon.

Actin and Myosin Filaments Correspond to Light and Dark Bands

It is the light and dark bands of the myofibrils that give the muscle fiber its striated appearance (Figure 1.7A). The dark, highly refractory, *A-bands* correspond to the presence of *myosin* (thick) filaments, while the light, less refractory, *I-bands* contain the *actin* (thin) filaments

(Figure 1.7B). The latter filaments are, of course, separate from the actin filaments, already discussed, which help to form the cytoskeleton of the muscle fibers. The molecular structures of the actin and myosin filaments are considered again in chapter 11, in relation to the contractile mechanism of the myofibrils.

In a resting muscle fiber, the actin filaments overlap the myosin filaments to some extent; the region without overlap constitutes the pale *H-zone* in the central part of the A-band. In the middle of the H-zone is a dark region, the *M-region*, formed by the presence of fine, filamentous structures which cross-connect the myosin filaments and give them a regular spacing from each other; the structure of the M-region is considered in more detail in the following section. There is another array of fine filaments, composed of *titin*, that stabilizes the myosin filaments in the longitudinal axis (see p. 13). Each of the myosin filaments is surrounded by a hexagonal lattice of actin filaments (Figure 1.7C). During a contraction, the myosin filaments contact the actin filaments by molecular cross-bridges and attempt to slide opposing actin filaments toward each other, thereby diminishing the width of the H-zone and the I-bands. The actin filaments contain, in addition to actin, two regulatory proteins, *troponin* and *tropomyosin*, and also a strengthening protein, *nebulin*. The interaction between the actin and myosin filaments through the cross-bridges is considered in chapter 11.

The Myosin Filaments Are Held Together in the M-Region

Even in the early studies with the light microscope, a dark line could be distinguished in the center of the muscle fiber A-band. With the electron microscope, however, it became apparent that there are actually several M-lines, and hence it is more appropriate to refer to the *M-region* rather than to the M-line. A further complication of the M-region was the finding of thin filaments (*M-filaments*), approximately 5 nm in diameter, that run parallel to the myosin filaments and appear to be connected to the latter as well as to each other. It is the cross-links between the two types of filament which are responsible for the appearance of multiple M-lines in each A-band. Antibody labeling and extraction studies have identified two of the proteins in the M-region, in addition to myosin. One of these proteins is the M-protein, or *myomesin*; it has a molecular weight (MW) of 165 kD and is thought to correspond to the M-filaments, as shown in Figure 1.8. The other protein is *creatine kinase*, and it is likely that these molecules form the struts holding the myosin filaments in a tight lattice and in register with each other. Other proteins, yet to be identified, evidently brace the M-filaments and produce additional M-lines.

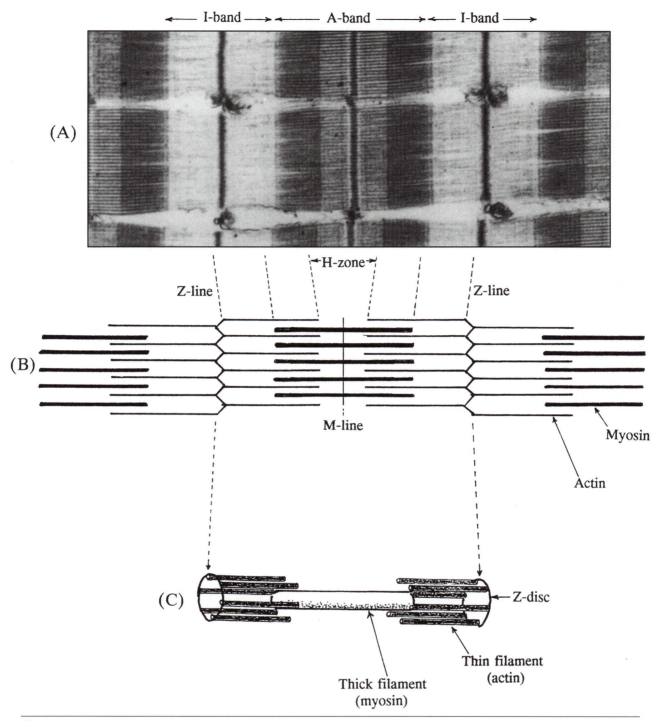

Figure 1.7 (A) Longitudinal electron micrograph of several myofibrils, showing the characteristic striations. (B and C) The overlapping arrangement of the thick (myosin) and thin (actin) filaments responsible for the striated appearance. (A) Reprinted from H.E. Huxley (1972).

Titin Also Helps to Stabilize the Myosin Filaments

If a solvent such as gelsolin is used to dissolve the actin filaments in a muscle fiber, it can be seen that the thick filaments of myosin are connected to the Z-discs by fine strands. These strands, which are approximately 5 nm in diameter and 1 μm in length, consist of an extremely large protein, *titin* (MW, 3,000 kD); it is probable that 6, or possibly 12, titin molecules make up each of the strands (Trinick, 1991). Part of the titin strand overlaps the myosin filament and is firmly attached to

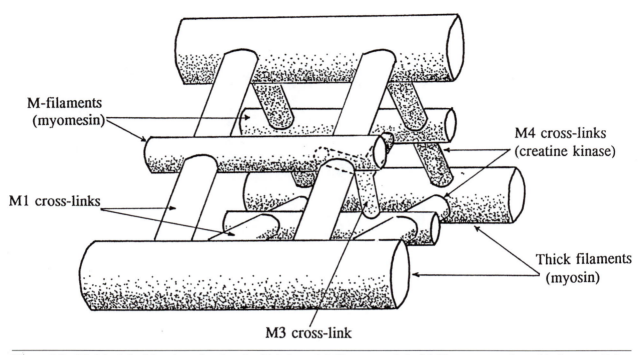

M-filaments
(myomesin)

M4 cross-links
(creatine kinase)

M1 cross-links

Thick filaments
(myosin)

M3 cross-link

Figure 1.8   Model of the M-region, showing the thick (myosin) filaments and the M-filaments, together with the cross-links holding them in position. Adapted from Luther and Squire (1978, p. 322) and Strehler, Carlsson, Eppenberger, and Thornell (1983, p. 154).

it, while the remainder extends through the I-band and is likely to have elastic properties (Figure 1.9). From its structure and position, each pair of titin strands is thought to function as a longitudinal stabilizer for the myosin filament, keeping it in the center of the sarcomere during contraction and relaxation. It is likely that the titin strands contribute much of the elasticity of the muscle fiber when the latter is stretched, either passively by forces applied to the ends of the muscle, or actively by unequal sarcomere shortening in the long axis of the fiber. Figure 1.9 also shows the strands of nebulin, which support the actin filaments.

The Actin Filaments Are Tethered at the Z-Disc

The filaments of actin, like those of myosin, require positional support; they achieve this partly from the cross-bridges and partly from their insertions into the *Z-disc* in the center of each I-band. The Z-disc is composed mainly of the proteins α-actinin, desmin, vimentin, and synemin. α-*Actinin*, a protein dimer of 180 kD, is thought to attach the ends of the actin filaments on one side of the Z-disc to those on the opposite side. The actin filaments, which do not themselves extend across the Z-disc, are converted from a hexagonal lattice around the myosin filaments to a square lattice at their insertions into the Z-disc. A further feature of the Z-disc is that the point at which an actin filament is attached on

one side is midway between two attachment points on the opposite side of the disc. *Desmin, vimentin,* and *synemin,* the other main protein components of the Z-disc, form intermediate filaments that are wrapped around the disc and also link adjacent discs together in the longitudinal and transverse axes (Figure 1.10). It is these intermediate filaments which, through their transverse connections from one Z-disc to another, keep all the myofibrils "in register" within a single muscle fiber. The Z-discs are also attached, through the intermediate filaments, to the cytoskeleton beneath the plasmalemma and to the plasmalemma itself. Through the attachment proteins of the plasmalemma, the Z-discs are ultimately connected to the basement membrane and the endomysium ensheathing the muscle fiber. Each transverse array of intermediate filaments constitutes a *costamere.*

Tubular Systems

Early in the present century, Veratti (1902) was able to stain a fine, interlacing network within the muscle fiber. It was only many years later, with the aid of the electron microscope, that the details of this structure could be resolved into a tubular system comprising two parts (see, for example, Franzini-Armstrong & Porter, 1964).

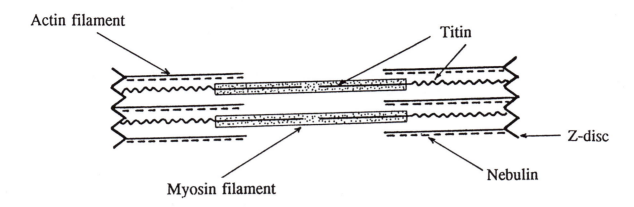

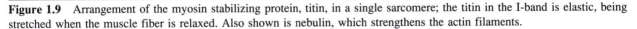

Figure 1.9 Arrangement of the myosin stabilizing protein, titin, in a single sarcomere; the titin in the I-band is elastic, being stretched when the muscle fiber is relaxed. Also shown is nebulin, which strengthens the actin filaments.

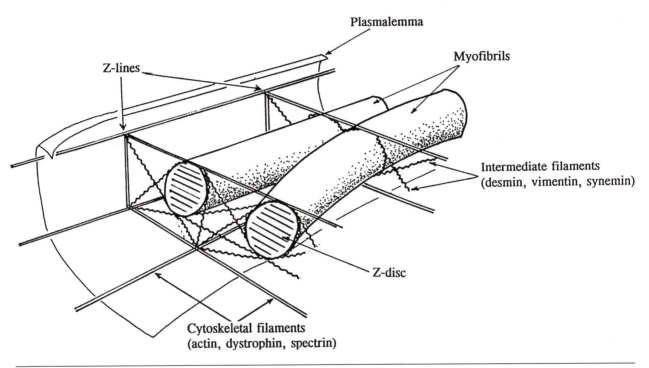

Figure 1.10 Scheme for the tethering of the Z-discs by intermediate filaments. The latter are also attached to the cytoskeletal elements under the plasmalemma. Adapted from Lazarides and Capetanaki (1986, p. 756).

The Sarcoplasmic Reticulum (SR) Surrounds the Myofibrils

An elaborate system of channels runs in the long axis of the muscle fiber and largely surrounds individual myofibrils (Figure 1.11). Through side branches, the longitudinal channels enveloping one myofibril are connected to each other and to those around other myofibrils. When the muscle fiber contracts, the longitudinal channels become shorter and wider. The function of the sarcoplasmic reticulum (SR) is to release Ca^{2+} into the cytosol around the myofibrils, where it combines with troponin C and allows contraction to take place. When

Ca^{2+} is pumped back into the SR, the contraction is terminated. In keeping with its role in "activating" the fiber, the SR has a membrane that contains both Ca^{2+} release channels (ryanodine receptors) and Ca^{2+} ATPase pumps.

The Transverse Tubular System Connects the Surface of the Fiber to the Interior

The transverse tubular system (T-system) lies perpendicular to the long axis of the muscle fiber in the form of narrow channels that encircle the myofibrils at regular intervals. In mammalian muscle fibers, including those

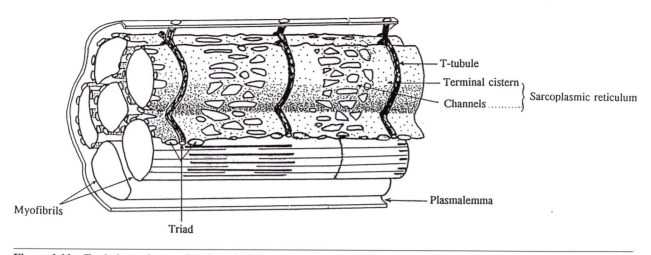

Figure 1.11 T-tubules and sarcoplasmic reticulum surrounding several myofibrils.

of humans, there are two zones of transverse tubules in each sarcomere; they lie at the junctions of the A- and I-bands (Figure 1.11). In cardiac muscle fibers and in the fibers of frogs, there is only one T-system for each sarcomere, and this is situated at the Z-line. As the T-tubules encircle the myofibrils, they interrupt the longitudinal channels of the sarcoplasmic reticulum. At these points of contact, the sarcoplasmic reticulum is dilated to form *terminal cisterns* (or *lateral sacs*); neighboring sacs are connected together. Figure 1.11 shows that each of the relatively narrow T-tubules is embraced on either side by a cistern; the three elements, which surround the myofibril, are referred to as a *triad*. Electron microscopy reveals that the membranes of the T-tubules and sarcoplasmic reticulum, although closely apposed at the triads, remain intact. According to Peachey (1965), approximately 80% of the transverse tubular system in a frog muscle fiber is surrounded by the sarcoplasmic reticulum. At the surface of the muscle fiber, the T-tubules form small openings at the junctions of the A- and I-bands, and their membranes become confluent with the plasmalemma. By examining muscle fibers bathed in solutions containing electron-dense material or fluorescent dyes, it has been shown that the T-tubules contain extracellular fluid. The principal function of the tubules is to conduct impulses from the surface to the interior of the muscle fiber, thereby stimulating the release of Ca^{2+} from the sarcoplasmic reticulum.

Nuclei and Mitochondria

The muscle fiber contains numerous nuclei (myonuclei) which in health are dispersed along the inner surface of the plasmalemma, particularly in the region of the motor end-plate; each nucleus is bounded by two membranes (Figure 1.12). The function of the nucleus is to prepare and send instructions for protein synthesis out into the cytoplasm. The instructions come from the genes contained in the chromosomes.

Other than the nuclei, myofilaments, and tubular systems, the main structures in the cytoplasm are the mitochondria. During evolution, these organelles have become specialized to form ATP, the major energy-producing compound of living cells. Although most of the cell enzymes are housed in the mitochondria, in relation to ATP synthesis, others lie freely in the cytosol. In addition, the cytoplasm contains two of the chemical fuels that the mitochondria can degrade to produce ATP; these are the glycogen granules and the lipid droplets. The nuclei and the mitochondria will now be considered.

Nuclear DNA is Transcribed Into RNA

In humans there are 23 pairs of chromosomes, and it is thought that altogether they may hold 100,000 or so genes. When the DNA of a gene is to be expressed, the message is first *transcribed* into RNA. While still in the nucleus, the RNA is then processed, with noncoding sequences of nucleic acids being cut out. Within a special part of the nucleus termed the *nucleolus*, the messenger RNA is packaged with proteins to form relatively *large ribosomes*; the latter are made smaller and then exported through pores in the nuclear membrane into the cytosol of the muscle fiber. The ribosomes, aided by transfer RNA, then start to assemble the sequence of amino acids that is characteristic of a particular protein; for each of the amino acids there is a genetic code in the form of a triplet of nucleotides (*codon*). Particularly in developing or regenerating muscle fibers, the ribosomes are attached in the cytoplasm to sheets of membrane enclosing a flattened cavity; this is the *endoplasmic reticulum*.

With the light microscope, the myonuclei are indistinguishable from the nuclei of the *satellite cells*. The latter

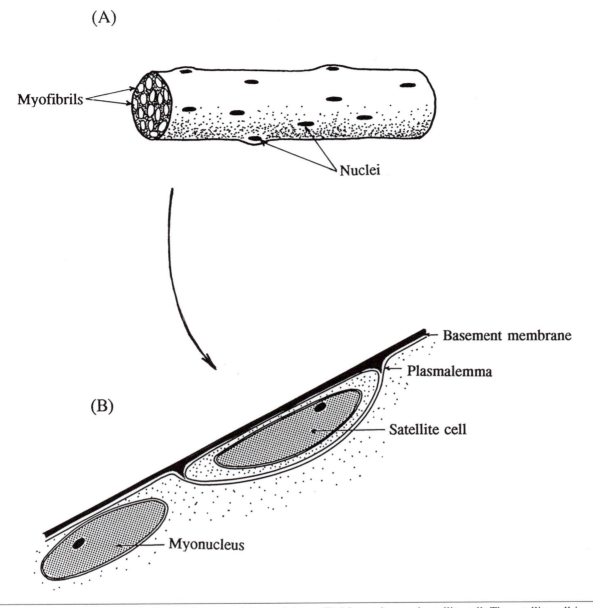

(A)

Myofibrils

Nuclei

Basement membrane

Plasmalemma

(B)

Satellite cell

Myonucleus

Figure 1.12   (A) Single muscle fiber with myonuclei at periphery. (B) Myonucleus and satellite cell. The satellite cell is separated from the fiber by its own plasmalemma and that of the fiber.

cells probably account for less than 1% of the muscle fiber nuclei in the adult. They can only be differentiated from the myonuclei by the electron microscope, which reveals the presence of twin membranes separating the cytoplasm of the statellite cell from that of the muscle fiber (Figure 1.12B). The satellite cells are of particular importance for the regeneration of muscle following disease or injury.

Mitochondria Supply the Muscle Fiber With ATP

The mitochrondria (sarcosomes) are ovoid structures, measuring from 1 to 2 μm in their longest diameters.

Each mitochondrion has a double membrane, of which the inner one is repeatedly folded to form *cristae*; these folds bulge into the central compartment of the mitochondrion (Figure 1.13). The matrix between the cristae houses the enzyme systems required for the Krebs *tricarboxylic acid cycle*, which breaks down pyruvate to carbon dioxide and water. A large part of the energy released by the successive reactions is captured by a chain of *cytochromes*, which enable ATP to be formed. There is now good evidence that the cytochrome chain is located on the inner membrane of the cristae, together with the ATP-forming enzyme (*ATP synthetase*). The matrix also contains several copies of the mitochondrion's own DNA. The outer membrane of the cristae has

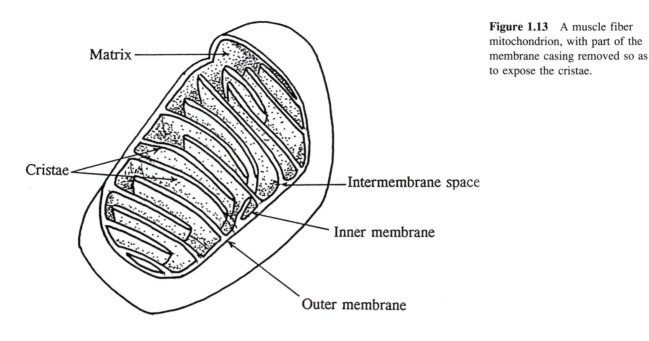

Figure 1.13 A muscle fiber mitochondrion, with part of the membrane casing removed so as to expose the cristae.

enzymes necessary for lipid synthesis. Mitochondria are found especially between the myofibrils in the region of the Z-line and also in relation to the nuclei and the motor end-plate. Although the mitochondria appear oval in longitudinal sections of muscle fibers, some are not ovoid but are much more complex; extremely bizarre shapes may occur in disease.

Before finishing the description of the cytoplasmic components, mention should be made of the glycogen granules. The granules, some 25 to 40 nm in diameter, are scattered throughout the cytosol and are the major sources of energy for the muscle fiber. Lipid droplets are often to be found close to mitochondria and contain fatty acids or triglycerides; they form important supplementary sources of energy.

APPLIED PHYSIOLOGY

In this section, which concludes the chapter, we will first consider how the absence of a single type of protein at the surface of the muscle fiber results in a fatal disease.

The Absence of Dystrophin Causes Muscular Dystrophy

In the middle of the last century there were two reports of a disease, apparently confined to male children, in which death resulted from progressive weakness of skeletal muscles (Duchenne, 1861, 1872; Meryon, 1852). This disease, subsequently given the eponym of Duchenne muscular dystrophy (DMD), affects one in every 3,000 male births; on average, one of three cases is the result

of a new mutation in the ovum of the mother or grandmother. In many instances the diagnosis of DMD is made in the 4th or 5th years of life because the boy is clumsy and late in walking, while attempts to run are frustrated by frequent falls. It is interesting that the calf muscles may be considerably enlarged, as Duchenne clearly recognized (Figure 1.14). Walking, when acquired, is rather ponderous; due to weakness of the paraspinal muscles, the pelvis tends to sag when the foot is raised, and this is compensated for by tilting the body to the opposite side. By the age of 12 years, walking is usually no longer possible, and a wheelchair existence begins. Weakness, having started in the larger, more proximal muscles, spreads to involve the smaller, more distal ones. Eventually there may only be movements of the fingers, face, and tongue and respiration left. Death occurs in the late teens or early 20s from bronchopneumonia or else from heart failure, since cardiac and smooth muscle are affected in addition to skeletal muscles.

For many years, there were keenly held hypotheses as to the pathogenesis of DMD, with some authorities postulating primary involvement of the intramuscular blood vessels or, since a third of the cases are mentally retarded, of the nervous system. The first real clue as to the correct mechanism was the finding, with the electron microscope, of small discontinuities of the plasmalemma at a time when the remainder of the muscle fiber appeared normal (Mokri & Engel, 1975). It is now thought that these tears in the membrane allow the entry of Ca^{2+} from the interstitial fluid, with subsequent disruption of enzymatic process in the fiber and local necrosis (see Figure 1.15). The necrotic area, usually wedge-shaped, is invaded by macrophages that ingest and remove the necrotic debris. The damaged sector may then be sealed

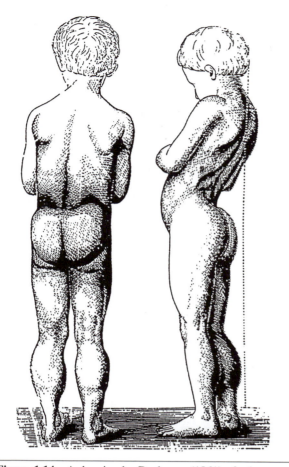

Figure 1.14 A drawing by Duchenne (1861) of a boy with muscular dystrophy. The drawing emphasizes the curious combination of muscle wasting in the arms and shoulders, with muscle enlargement in the calves and buttocks. There is increased anteroposterior curvature of the lumbar spine (lordosis). Not only was Duchenne a fine artist, but he was probably the first medical photographer and electromyographer; he also invented the muscle biopsy needle and several ingenious orthotic devices (Reicke & Nelson, 1990).

off from the remainder of the fiber by newly formed plasmalemma, as in Figure 1.15. Satellite cells may also become active and aid in the repair process. Alternatively, the necrosis may extend longitudinally, causing death of most or all of the muscle fiber. Some of the microscopic changes in DMD are shown in Figure 1.16. In an advanced case of DMD, all the fibers within a large region of the muscle belly may be destroyed.

How are the membrane tears to be explained? The answer came independently from two molecular biology laboratories. Since only males are affected by DMD, the gene for the condition must be recessive residing on the X-chromosome. In the puzzling condition of a Belgian woman with what appeared to be DMD, it was shown by Worton and colleagues that

part of the X-chromosome had switched places (translocated) with part of chromosome 21 (Ray et al., 1985). In the other laboratory, Kunkel's group found that, in approximately 5% of DMD cases, there were major deletions in the X-chromosome; in one patient, in whom DMD was associated with two other inherited diseases, the deletion was large enough to be seen with the light microscope (Francke et al., 1985). It was argued that the deletions, like the translocation in the female case of DMD, must have involved the site of the DMD gene and by the technique of chromosome walking, the gene was quite rapidly identified and cloned. The DMD gene proved to be the largest yet discovered, with 2×10^6 base pairs, and it coded for a previously unrecognized protein, *dystrophin*, having a MW of 400 kD (Figure 1.17A).

At first it was not clear where dystrophin was located in the muscle fiber, but it is now agreed that the molecule not only lies close to the plasmalemma but is attached to the latter through glycoproteins; outside the fiber, the glycoproteins are connected to laminin in the basement membrane (Ibraghimov-Beskrovnaya et al., 1992; Matsumura & Campbell, 1994; Figure 1.17B). Most of the subplasmalemmal portion of the molecule consists of a long rod, 120 nm long, formed by a triple helix of 2,700 amino acids (Figure 1.17A); it is possible that the rod molecules form an interlocking lattice (Hoffman & Kunkel, 1989). In view of its location under the membrane and also on account of its resemblance to the cytoskeletal protein, spectrin, it is likely that the dystrophin reinforces the plasmalemma and prevents it from tearing during contraction and relaxation. It is, of course, quite possible that such a large molecule serves more than one function, particularly since it is found in many neurons of the central nervous system (see Matsumura & Campbell, 1994) and there are increased amounts at the neuromuscular junction.

In Becker Dystrophy, the Abnormal Dystrophin Is Partly Effective

Becker muscular dystrophy (BMD) resembles DMD in many respects, but it declares itself later and runs a much slower course. In this condition also, the gene for dystrophin is the culprit; however, whereas in DMD hardly any dystrophin-like proteins can be found, in BMD there are significant amounts of abnormal, smaller molecules. At the level of the gene, the DMD deletions probably cause "frame shifts" that result in premature termination of mRNA translation and the synthesis of very small, unstable fragments. In BMD, however, the reading frame is preserved, allowing the production of rather larger, and apparently useful, portions of dystrophin (Monaco, Bertelson, Liechti-Gallati, Moser, & Kunkel, 1988).

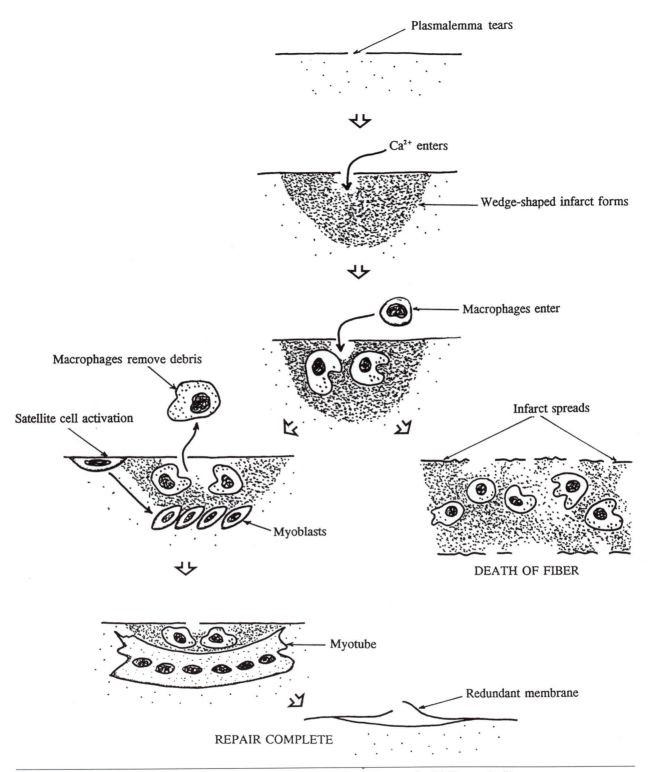

Figure 1.15   Possible sequence of events following a tear in the plasmalemma of a DMD muscle fiber.

Some Simple Biomechanics

The concluding part of the applied physiology section opens the door into the world of muscle mechanics, beginning with the effects of muscle pinnation and closing with an overview of muscles, bones, and joints as lever systems. This part is only an introduction, however, and more detailed treatments of muscle biomechanics can be found in other texts (e.g., Enoka, 1994).

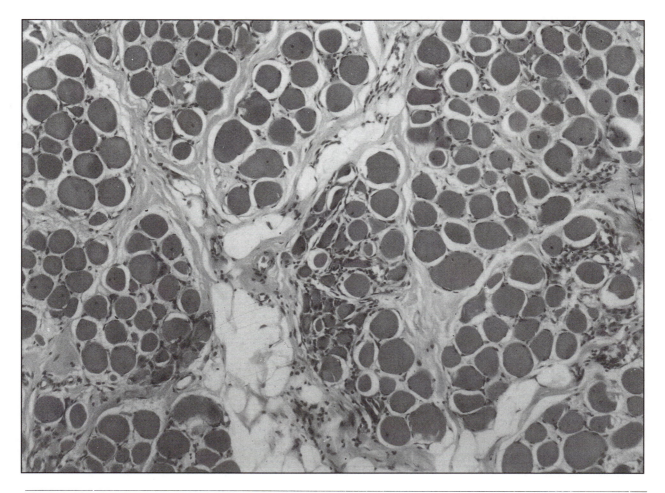

Figure 1.16   Microscopic changes in DMD muscle. In this cross section, the muscle fibers are rounded and of very different sizes; some fibers have central nuclei. There are several infiltrations of inflammatory cells where fibers are undergoing necrosis or, in some cases, starting to regenerate. The perimysial and endomysial connective tissues are greatly increased and some of the muscle has been replaced by fat (white areas). H & E stain. (150x).

In Pinnate Muscles, Not All the Force Developed by the Fibers Is Transmitted in the Axis of the Tendon

In Figure 1.18A, consider the muscle fiber emphasized by stippling. When the contractile filaments engage, a force, P, will be developed in the axis of the fiber (represented by z in Figure 1.18B); this force has both a vertical and a horizontal vector (y and x in Figure 1.18B). The vertical vector is the force that would be needed to support a weight attached to the tendon (W in Figure 1.18A). With reference to the triangle of forces in Figure 1.18C, the size of the vertical vector will be

$$P \times \frac{y}{z}$$

or P multiplied by the *cosine* of the angle of pinnation, θ. If the angle of pinnation is 30°, cosine θ is 0.87. In other words, 87% of the force developed by the fibers

is transmitted in the axis of the tendon. Were the angle of pinnation larger, at 60°, only half the force would be transmitted, since cosine 60° is 0.50.

In Bipinnate Muscles, the Movement of the Tendon is Greater Than the Shortening of the Muscle

At first sight, it is surprising that a tendon could move more than the muscle fibers shorten, but this is true of bipinnate muscles, as Figure 1.19 shows. In the relaxed condition, suppose that the angle of pinnation is, once again, 30°. Imagine now that the fibers on each side of the tendon contract and are allowed to shorten, so that the angle of pinnation becomes 60°. A fiber that was 10 cm long will now be 4.2 cm shorter, while the tendon will have moved through 5.8 cm—provided that the tendon is kept in the same vertical axis through contraction of the contralateral fibers.

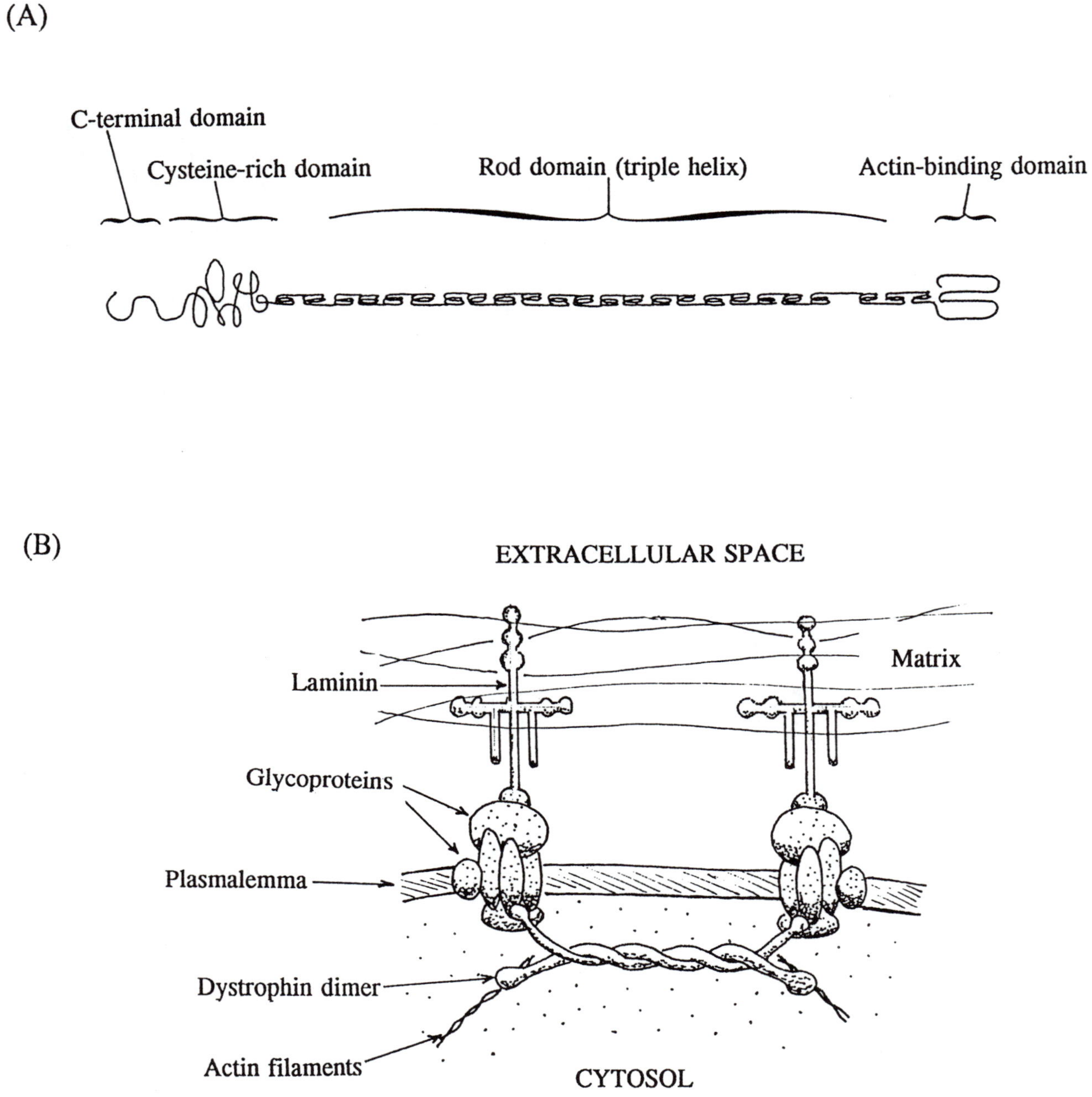

(A)

(B)

Figure 1.17   (A) The molecular structure of dystrophin, as determined by Koenig, Monaco, and Kunkel (1988). (B) Attachment of dystrophin to glycoproteins in the plasmalemma. The dystrophin molecules are almost certainly connected to each other, and possibly to other elements of the fiber skeleton (actin, spectrin). Adapted from Matsumura and Campbell (1994, p. 4).

Bones and Joints Act as Lever Systems

When a muscle tendon is inserted onto a bone, the bone acts as a lever pivoting around the joint. Thus, the far end of the bone will move more than the point of tendon attachment. However, there is a disadvantage to the lever system, in that a weight fastened to the end of the bone will have to be supported by a proportionately greater muscle force; this is because the product of force and distance is constant throughout the length of the lever. Thus, in Figure 1.20,

$$P_1 \times d_1 = P_2 \times d_2$$

where P and d represent force and distance respectively. Suppose that a weight of 1 kg is placed in the hand, that the distance from the palm to the center of the elbow joint is 32 cm, and that the biceps tendon is inserted 4.0

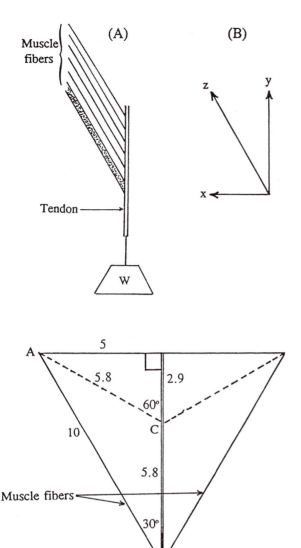

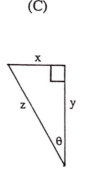

Figure 1.19 In a bipinnate muscle, the movement of the tendon exceeds the amount of muscle fiber shortening. Thus, when the fiber, *AB*, contracts to a new length, *AC*, the tendon moves from *B* to *C*. Numbers indicate relative lengths.

cm from the joint. If the biceps was the only elbow flexor supporting the weight in the hand (it isn't; it is assisted by the brachialis and the brachioradialis), it would have to develop $\frac{32}{4} \times 1 = 8$ kgf (8 kilograms-force), or $8 \times 9.8 = 78.4$ N (newtons, the SI unit of force).

The product of force and distance is *torque*. In the example given above, the biceps would be exerting a torque of 4 cm × 78.4 N = 313.6 N · cm = 3.14 newton · meters. As noted earlier, the torque will be the same,

regardless of the point along the forearm at which it is measured.

Because of the lever action, if the biceps shortens by 2 cm, the hand will move by $\frac{32}{4} \times 2 = 16$ cm (Figure 1.20B).

The same principles of lever action apply if the muscle is inserted on the side of the joint opposite to the load, as in the case of the triceps at the elbow or the gastrocnemius-soleus complex at the ankle. Once again, $P_1 \times d_1 = P_2 \times d_2$ (Figure 1.21A), and torque will be the same wherever it is measured in the longer lever arm. Note, however, that the torque on one side of the joint will be in the opposite direction of that on the other side. Consider the calf muscles and the force that they must develop in order for a subject to stand on the ball of the foot. Suppose that the distance between the insertion of the muscle (via the Achilles tendon) and the center of the ankle joint is 5 cm, the distance between the ankle joint and the ball of the foot is 12.5 cm, and the weight of the body is 70 kg (Figure 1.21A). Then, if the subject puts all his or her weight on the heel of one foot and *momentarily* lifts that heel from the ground by pressing down on the ball of the foot, the calf muscles must develop

$$70 \times \frac{12.5}{5.0} = 175 \text{ kgf (kilograms-force)}$$

or

$$175 \times 9.8 = 1715 \text{ N (newtons)}$$

as depicted in Figure 1.21B.

Now that the structure of a muscle and one of its constituent fibers have been described, we can move on to consider a very different type of cell. This cell, which interacts so closely with the muscle fiber, is the motoneuron.

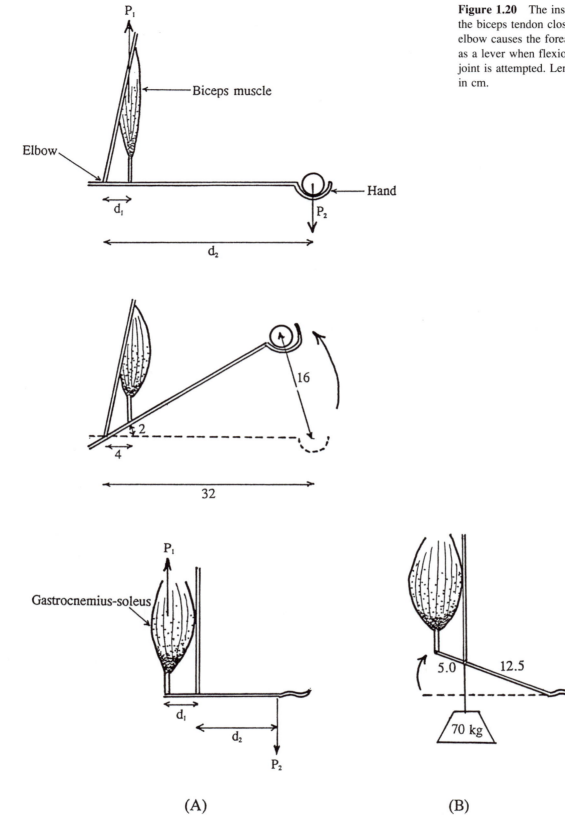

Figure 1.20   The insertion of the biceps tendon close to the elbow causes the forearm to act as a lever when flexion of the joint is attempted. Lengths are in cm.

(A)   (B)

Figure 1.21   Lever action of a joint extensor muscle. In (A), the muscle is the gastrocnemius-soleus complex, inserted onto the calcaneum at the back of the heel. In (B), a contraction of the muscle transiently raises the body, while the center of gravity remains close to the heel. Lengths are in cm.

Chapter 2

The Motoneuron

When a muscle is required to contract, it is sent the necessary instructions in the form of nerve impulses (action potentials) by large cells lying in the ventral gray matter of the spinal cord (or in a corresponding region of the brainstem). These cells are the *motoneurons*, of which there are usually a hundred or more for each muscle. A motoneuron consists of a cell body, or soma, and special processes termed the dendrites and axon (Figure 2.1).

As we shall see later (chapter 8), the motoneurons also send chemical signals to the muscle fibers, via a relatively slow, energy-dependent mechanism called axoplasmic transport. For the moment, however, it is necessary to consider the general features of the motoneuron, including its processes. After this comes a more detailed examination of the structure of the axon, including the fatty coat that surrounds it.

General Features of Motoneurons

The soma of the motoneuron contains the nucleus of the cell together with various structures which will be considered later. The function of the *dendrites* is to receive signals from other neurons, while the major role of the *axon* is to transmit a resulting message to the muscle fibers in the form of electrical impulses. The axons also transport chemical messages and organelles between the motoneurons and the muscle fibers.

Some Human Axons Are Extremely Long

The lengths of the axons vary considerably. In the case of a man 180 cm tall, an axon running from the lumbosacral region of the spinal cord to one of the plantar muscles would be about 125 cm long. In contrast, the fibers passing in the cranial nerves to the external ocular muscles, or to the muscles of the face or tongue, would only measure about one tenth of this length. As the axon nears the muscle, it splits, with each branch dividing further, so that eventually a single parent axon may contact several hundred muscle fibers. The contact region between an axonal twig and a muscle fiber is termed a *synapse*, or *neuromuscular junction*. The synapse possesses special structural and functional features that enable the impulse in the nerve fiber to be translated into an impulse in the muscle fiber through the action of a chemical link, the transmitter *acetylcholine* (ACh).

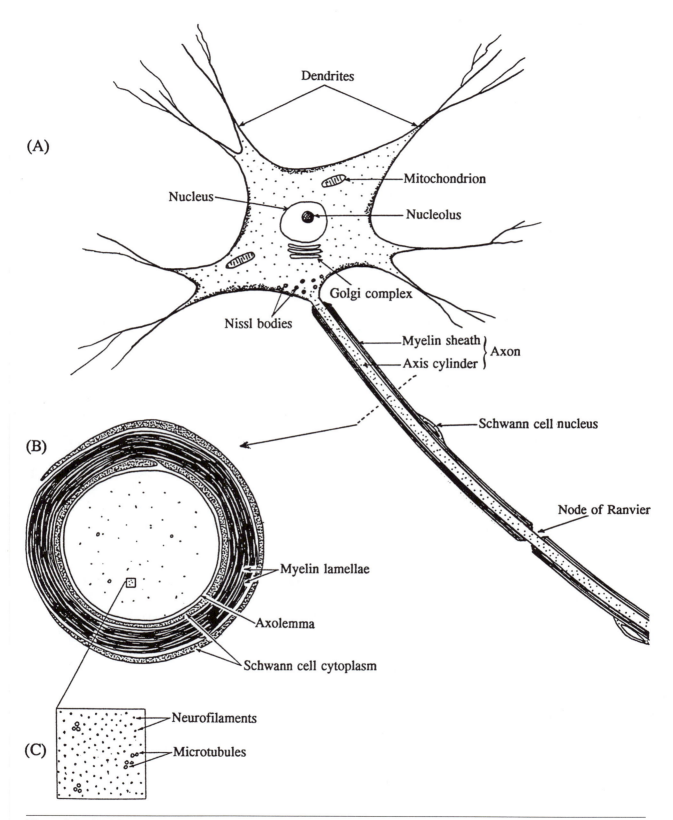

Figure 2.1   (A) Motoneuron cell body (soma) and proximal axon. (B) Transverse section through the axon to show the arrangement of the myelin lamellae in relation to the axis cylinder and the cytoplasm of the Schwann cell. (C) Part of the axis cylinder, enlarged to show the microtubules and neurofilaments.

In the normal adult mammal, most muscle fibers have only one neuromuscular junction and are innervated by a single motoneuron. Sherrington (1929) pointed out that a single motor nerve fiber and the colony of muscle fibers which it supplies could be considered as a functional entity, because each time the nerve fiber discharges an impulse, the muscle fibers of the colony are excited together; Sherrington described this entity as a *motor unit* (see Chapter 12).

Several Methods Have Been Used to Study the Structure of the Motoneuron

Haggar and Barr (1950) examined serial sections of cat spinal cord and were able to make three-dimensional models of the soma and dendrites. Another technique, devised by Chu (1954), has been to dissect pieces of ventral horn from recent autopsy specimens of human spinal cord and to make them into a crude suspension with physiological saline. Within this suspension, some motoneurons are completely dissociated from other tissue and yet retain considerable lengths of axons and dendrites suitable for examination. Somewhat later a technique for injecting the dye *procion yellow* through a micropipette into the soma of a neuron was applied to motoneurons (Kellerth, 1973); the dye diffuses into the dendrites and proximal axon, enabling the full extent of the cell to be visualized. Horseradish peroxidase (HRP) can be used for the same purpose; even the fine dendritic branches are permeated by this material, which can then be stained (Figure 2.2). For information concerning the fine structure of the motoneuron, it has been necessary to turn to the electron microscope, and several comprehensive reports are available (see, for example, Bodian, 1964). The various parts of the motoneuron will now be considered in more detail.

A Muscle Is Innervated by a Column of Motoneurons in the Spinal Cord

Within the ventral gray matter of the spinal cord, the cell bodies of the motoneurons are arranged in columns parallel to the long axis of the cord. The cells in each column are usually connected to the same muscle, and, as a consequence, the muscle receives its nerve supply from two or more adjacent segments of the spinal cord (Figure 2.3). The cells supplying the extrafusal muscle fibers are termed α-motoneurons in order to distinguish them from the γ-motoneurons (fusimotoneurons) innervating the small fibers within the muscle spindles.

Motoneuron Soma

The cell bodies of the α-motoneurons are much larger than those of the γ type and may have diameters of 100

µm; an average value is about 70 µm. Large though these cell bodies are, their volumes are very much less than those of their axons. Thus, the axon of a lumbosacral motoneuron contains approximately 100 times as much cytoplasm as the soma. Indeed, if the motoneuron soma was enlarged to the size of a tennis ball, the axon would stretch the length of seven football fields!

Genetic Instructions Are Sent From the Nucleus Via the Nucleolus

The most prominent feature of the soma is the relatively large *nucleus*, which is bounded by a wavy double membrane and contains nucleoplasm and a nucleolus. Within the nucleoplasm are the 23 pairs of chromosomes, which together contain the 100,000 or so genes characteristic of the human genome. Only a fraction of the genes are expressed in the motoneuron, however, and many of these will be different from those expressed in the muscle fiber. It is the genes that are ultimately responsible for directing the metabolic activities of the neuron. They do so by sending instructions, in the form of messenger RNA, out into the cytoplasm of the cell, possibly through the pores in the nuclear membrane, which can be seen with the electron microscope. Since human motoneurons do not multiply after birth, the chromosomes cannot be distinguished in the nucleoplasm of the motoneuron; all that can be seen in the nucleoplasm are irregularly distributed particles, from 1 to 2 µm in diameter. Within the nucleus, the *nucleolus* appears as a large structure, measuring about 4 µm across and consisting mostly of ribosomal RNA. This RNA is made on the DNA templates of the genes (*transcription*) and is eventually passed out from the nucleolus into the cytoplasm to take part in the synthesis of proteins.

Several Specialized Structures are Found in the Motoneuron Cytoplasm

The cytoplasm of the motoneuron soma contains several different types of organelles, the most prominent of which are the *Nissl bodies*. These bodies, which are especially numerous near the base of the axon, range from 0.5 to 3 µm in diameter, and electron micrographs show them to be composed of tightly packed arrays of endoplasmic reticulum. The membranes of the reticulum are studded with ribosomes, giving them a rough appearance. Acting on instructions received through messenger RNA, the ribosomes engage in the synthesis of proteins (*translation*). Their behavior during the reactions of the cell following injury to the axon is of great interest (see chapter 18).

In addition to the Nissl bodies, the cytoplasm of the motoneuron contains many *mitochondria*. There are

Dendrites

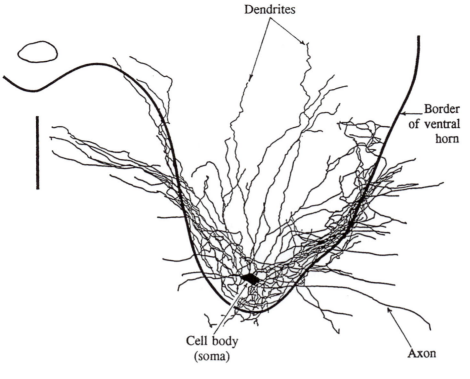

Border
of ventral
horn

Cell body
(soma)

Axon

Figure 2.2   Cell body and processes of a motoneuron. The motoneuron is from the cervical region of the cat spinal cord and has been injected with horseradish peroxidase. Note the extensive territory covered by the dendritic branches, including some of the white matter surrounding the ventral horn. (Courtesy of Dr. Ken Rose.)

250 μm

also *microtubules* and *neurofilaments*, most of which enter the dendrites and the axon. The *Golgi apparatus* is the name given to a system of flattened tubular channels to be found near the nucleus; one of its functions is thought to be the packaging of proteins prior to their delivery into the axon. Finally, with advancing age, the motoneuron cytoplasm contains increasingly large masses of pigment, the *lipofuscin granules*; the significance of this material is unknown.

Cytoskeletal Proteins in the Motoneuron

The motoneuron, like other cells in the body, contains cytoskeletal proteins. In the motoneuron, three of these are especially plentiful and serve important functions; they are *actin, tubulin,* and the *neurofilament proteins*.

Three of the Cytoskeletal Proteins Have Multiple Functions

The actions of the three cytoskeletal proteins are the following:

• By forming the cytoskeleton of the motoneuron, they not only strengthen the cell and give it a characteristic shape, but they also anchor the nucleus and some of the major structures within the interior of the cell.

• By a process of elongation, they help to create new segments of axis cylinder, and so play a critical role in nerve regeneration.

• By mechanisms described in chapter 8, two of the proteins provide a pathway for the conveyance of cellular material from the soma to the axon, dendrites, Schwann cells, and muscle fiber; this process is termed *axoplasmic transport*.

Actin is a pear-shaped globular protein with a MW of 42 kD and a diameter of 4 nm; each molecule is stabilized by being tightly bound to one Ca^{2+} ion. In the presence of ATP, the globular (G) actin polymerizes to form filamentous (F) actin. Each actin filament appears as if two strands of globular molecules are twisted round each other to form a helix, though, in fact, actin strands cannot exist on their own. The spacing between successive turns of the helix is approximately 37 nm (Figure 2.4A). Actin is also considered in chapter 11, in which a description is given of its role in the sliding filament mechanism of the muscle contraction.

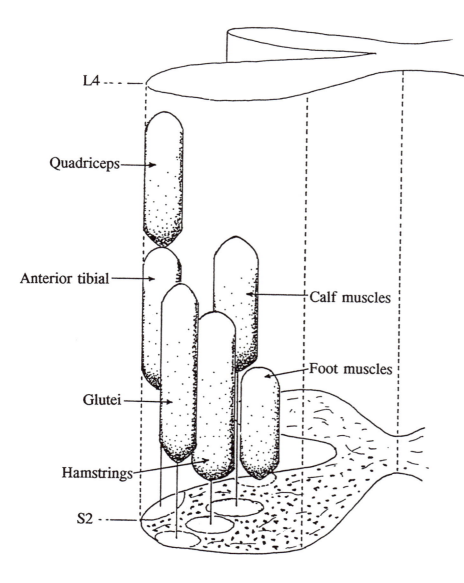

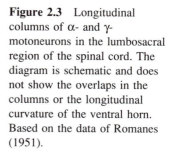

L4

Quadriceps

Anterior tibial

Calf muscles

Glutei

Foot muscles

Hamstrings

S2

Figure 2.3 Longitudinal columns of α- and γ-motoneurons in the lumbosacral region of the spinal cord. The diagram is schematic and does not show the overlaps in the columns or the longitudinal curvature of the ventral horn. Based on the data of Romanes (1951).

Tubulin is another globular protein and has a MW of 50 kD; slightly different forms of the protein, α-tubulin and β-tubulin, link together to form a dimer. The tubulin dimers can remain in solution in the cytoplasm or can polymerize under the influence of ATP to form protofilaments, with the α-tubulin of one dimer attached to the β-tubulin of another. Characteristically, 13 of the protofilaments then link up to form each helical loop of a hollow structure, the microtubule (Figure 2.4B).

Actin Filaments and Microtubules Can Lengthen or Shorten

Both actin filaments and microtubules are polar structures, since their respective subunits have specific orientation within the polymers. The ends of the actin filaments and microtubules can grow by polymerization or shorten by depolymerization. One end, the + *(plus) end*, is capable of lengthening much more rapidly than

the other, the − *(minus) end*. *Capping proteins* can bind to one end or the other of the polymer, stabilizing that end and perhaps also attaching it to another structure at the surface or in the interior of the cell. Several drugs are known that bind to actin and tubulin and thereby interfere with their structure and function. *Colchicine* causes depolymerization of the microtubules by inhibiting the addition of further tubulin molecules, while the *cytochalasins* have a similar effect on actin. *Vinblastine* and *vincristine*, two drugs derived from the periwinkle and used in cancer treatment, induce the formation of tubulin crystals. In contrast, *taxol* stabilizes the microtubules, and *phalloidin* does the same to the actin filaments.

Neurofilaments Consist of Three Types of Proteins

There are three neurofilament (NF) proteins, NF-L, NF-M, and NF-H, with MWs of 70, 160, and 200 kD,

(A)

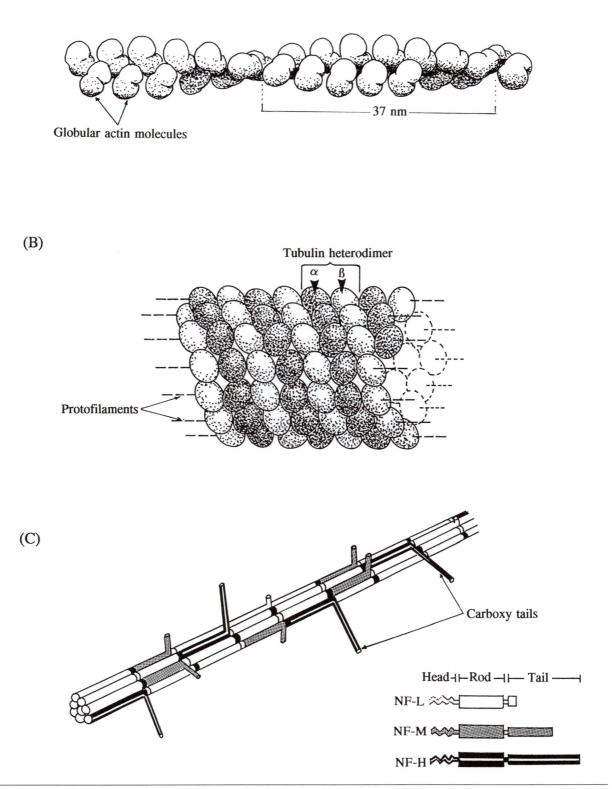

Figure 2.4   (A) Part of an actin filament, formed by globular actin molecules packed into a tight helix, giving the appearance of two strands twisted round each other. (B) A short length of microtubule; the microtubule is formed by 13 parallel protofilaments, each of which consists of a series of αβ-tubulin dimers. Adapted from Amos, Linck, and Klug (1976, p. 853). (C) A single neurofilament, consisting of three types of interlocking subunits. The carboxy (COOH) tails of the subunits are protruding from the neurofilament. Adapted from Nixon and Sihag (1991).

respectively. Each filament protein molecule has an amino *head* and a carboxy-terminal *tail*; in between is the *rod* domain, in the form of an α-helix. The three types of protein line up with each other to produce neurofilaments (Figure 2.4C), 7 nm in diameter. The carboxy-terminals of the NF-M and NF-H molecules have large numbers of sites that can be phosphorylated; phosphorylation causes the tails to stick out, where they may connect to other neurofilaments and form a lattice (Figure 2.4C). At the same time, the projecting tails will delay the movement of the neurofilaments down the axon, hence the slow component of axoplasmic transport (see chapter 8). The effect of phosphorylating the amino head-terminal is quite different, for it promotes the disassembly of the filament network by depolymerization (see Nixon & Sihag, 1991).

Axon, Dendrites, and Glia

We shall now consider in more depth the special processes that radiate from the cell body of the motoneuron and also the glial cells that surround the cell body in the spinal cord. Of the special processes, it is the axon that will receive the most attention, in part because it has been studied more extensively than the dendrites. Of particular interest are the structure and physiological properties of the fatty layers (myelin sheath) that form the insulating coat of the axon.

The Dendrites of Motoneurons Branch Extensively

The motoneurons possess several dendrites that radiate from the cell body in dorsal, superior, and inferior directions; the dendrites become progressively narrower and also divide into several branches. The remarkable complexity of the dendritic arborization is evident in Figure 2.2, which shows a single motoneuron injected with dye. It can be seen that the dendrites extend over considerable distances in the gray matter of the spinal cord, enabling them to receive information from the axons of a variety of other neurons. The connections between the axon and dendrites are made at *synapses*, which, in their fine structures, have certain features in common with the neuromuscular junction (chapter 3); the terminations of the axon twigs are expanded to form *boutons*, inside which many synaptic vesicles can be distinguished. The packing density of the boutons on the dendritic membrane increases as the dendrites narrow; more proximally, there are gaps into which processes from the glial cells project. Some synapses are also found on the membrane of the soma.

Glial Cells Provide Structural and Metabolic Support of the Motoneurons

The glial cells of the central nervous system do not conduct impulses and therefore have no direct role in signaling. Their importance lies in various supportive functions; they provide a structural matrix for the neurons and also control the passage of substances from the capillaries into the neuronal milieu. In addition, it appears that the metabolism of the glial cells is at least partly linked to that of the neurons. In the case of the motoneuron, several oligodendroglial cells cluster around its periphery and occupy any spaces available between the synaptic knobs (boutons) of the incoming fibers. One of the metabolic activities of the glial cells becomes particularly important when the motoneuron is discharging impulses; it is the removal of K^+ from the narrow interstitial spaces by means of the Na^+-K^+ pump (see chapter 7). It has been suggested that, within the central nervous system, the oligodendroglia provide the myelinated axons of neurons with Na^+ channels at the nodes of Ranvier (Bevan, Chiu, Gray, & Ritchie, 1985).

Motoneuron Axons Have Myelin Sheaths Interrupted by Nodes

The axon arises from a conical protrusion of the motoneuron soma known as the *axon hillock* and extends to the muscle as a long *axis cylinder*. The cylinder is bounded by a membrane, the *axolemma*, that has structural and functional features similar to those of the muscle fiber plasmalemma. Within the cytoplasm of the axis cylinder are arrays of small *microtubules* and *neurofilaments* that run in the long axis of the fiber between the soma and the many neuromuscular junctions. The microtubules are about 20 nm wide, while the neurofilaments are rather smaller, being approximately 7 nm thick (see p. 29). An interlacing system of rather wider tubules forms the *endoplasmic reticulum*.

All the motor axons have lipid coverings surrounding the axis cylinders; these are the *myelin sheaths*, and they are derived from a type of satellite cell known as the *Schwann cell*. The motor and sensory axons in humans are not quite as thick as those in the cat and monkey; if the myelin sheaths are included in the measurements, their diameters range from approximately 2 to 14 μm (Figure 2.5). In the ventral roots, which contain motor fibers only, the axon diameters fall into a bimodal distribution with a separation at about 7 μm. The population of fibers with large (7-14 μm) diameters corresponds to the α-motor axons supplying the extrafusal muscle fibers; the smaller axons are the γ-axons innervating the intrafusal muscle fibers inside the muscle spindles.

Myelin Sheaths Are Formed by Schwann Cells

The myelin sheath has been studied with the electron microscope and has been shown to consist of spiral wrappings of tightly packed membranes laid down by the Schwann cell (Geren, 1954). The myelin sheath is not a continuous structure but is interrupted at regular intervals by the nodes of Ranvier. It will be seen in chapter 9 that the nodes of Ranvier are the sites at which the action potential is generated as it travels down the axon. The distance between two successive nodes is an *internode* and corresponds to the territory of a single Schwann cell. At birth, the internodes measure about 230 μm in human motor nerves; as the limb grows, the lengths of the internodes increase, because the number of Schwann cells remains the same. Vizoso (1950) found that the longest internodes in the ulnar and anterior tibial nerves of an 18-year-old subject were 1,100 and 980 μm respectively, indicating that there had been a fourfold increase in their lengths. If, during the course of a disease process, the myelin sheath is destroyed and then remade, or if the axon is damaged and then regenerates, many of the newly formed internodes are much shorter than in the normal adult (Figure 2.6). The explanation for this discrepancy is that during the recovery process the Schwann cells divide and thereby increase their number; each new cell occupies a smaller length of axon.

Schwann Cell Fingers and Gap Substance Surround the Axis Cylinder at the Nodes

The electron microscope has been invaluable in revealing the ultrastructure of the nodes of Ranvier. Figure

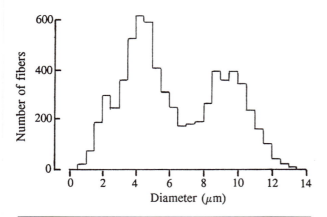

Figure 2.5   Diameters of myelinated nerve fibers in a specimen of sural (cutaneous) nerve from a 14-year-old boy. The range of diameters and the bimodal distributions of values are typical of both motor and sensory nerves in humans. Adapted from Buchthal, Rosenfalck, and Behse (1975).

2.7 summarizes, in diagram form, the findings of Williams and Landon (1967). On the left side of the figure, the axon has been bisected by a vertical incision and a segment removed; to the right of the node, some of the basement membrane has been peeled away. In the bisected segment, it is apparent that each myelin lamella is attached to the axolemma by a loop. On the other side of the node (*paranodal* region) the myelin sheath is seen to be indented by columns of Schwann cell cytoplasm. As the columns approach the node, they first become confluent and then send finger-like processes toward the nodal portion of the axolemma.

The processes are embedded in an amorphous extracellular material known simply as *gap substance*. Studies by Landon and Langley (1971), among others, have demonstrated that the gap substance is composed of mucopolysaccharides, and that the anionic groups of these molecules exert a powerful electrostatic attraction for cations. Landon and Langley suggest that the gap substance may serve to maintain a high concentration of Na^+ ions available for flow across the axolemma during the action potential. Similarly, the gap substance might limit the diffusion of K^+ from the vicinity of the node following an impulse and hold it in readiness for active transportation back into the fiber.

Cytoplasmic Material Is Exchanged Between the Axis Cylinder and Schwann Cells

It is tempting to assign a functional role to the Schwann cell processes that project to the nodal axolemma. It is possible that these convey ATP to the nerve membrane from the mitochondria in the columns of Schwann cell cytoplasm; the ATP might then be used to supply the energy required for Na^+ and K^+ pumping (see Landon & Langley, 1971). Another possible role for the Schwann fingers, analogous to that suggested by Ritchie (see Bevan et al., 1985) for glial cells in the central nervous system, is to provide the axolemma with Na^+ channels.

An interesting feature of the myelin sheath is the *Schmidt-Lanterman incisures (clefts)*. At each of these regions a narrow cytoplasmic process is sent from the Schwann cell to the axis cylinder. The projection does not actually penetrate the myelin wrappings but instead winds its way around the axon, following the plane of separation between the myelin lamellae. Using time-lapse cinematography, Gitlin and Singer (1974) have observed that the Schmidt-Lanterman clefts are not static structures but can be open or closed at different times. It is possible that the open phase enables materials, including trophic molecules (see chapter 18), to be passed between the axis cylinder, on the one hand, and the Schwann cell and the endoneurial space of the nerve

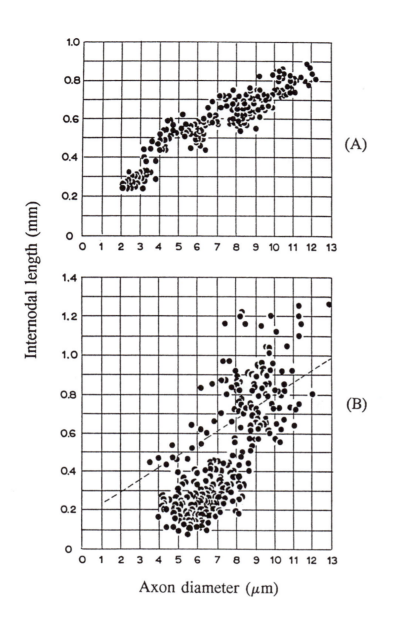

(A)

(B)

Internodal length (mm)

Axon diameter (μm)

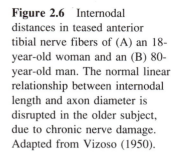

Figure 2.6 Internodal distances in teased anterior tibial nerve fibers of (A) an 18-year-old woman and an (B) 80-year-old man. The normal linear relationship between internodal length and axon diameter is disrupted in the older subject, due to chronic nerve damage. Adapted from Vizoso (1950).

fiber, on the other. Even the remainder of the myelin sheath is not stationary but can be seen to develop small indentations that then regress; possibly this type of movement is related to some component of axoplasmic flow (see chapter 8 and Weiss, 1969).

Axons Are Held Together by Connective Tissue

An account of the axon would not be complete without consideration of the way the axons (nerve fibers) are supported by the connective tissue within a peripheral nerve. Under the microscope, each nerve fiber is seen to be enclosed in a sheath of connective tissue, or *endoneurium*. Within the nerve trunk, the fibers are collected into a number of bundles or *fasciculi*, each being bounded by a condensation of connective tissue termed

the *perineurium*. Last, there is a relatively thick coat of connective tissue that invests the whole nerve trunk; this is the *epineurium* (Figure 2.8). Surrounding the nerve trunk are the vessels conveying blood to and from the fibers; these are the *vasa nervorum*. Small arterioles and venules pierce the epineurium at intervals, giving rise to an intraneural system of capillaries.

APPLIED PHYSIOLOGY

There are literally hundreds of causes of peripheral nerve degeneration, but the disorder we shall consider now is hereditary. To make matters more interesting, the abnormal gene has been located, even though the function of the protein product is not yet known.

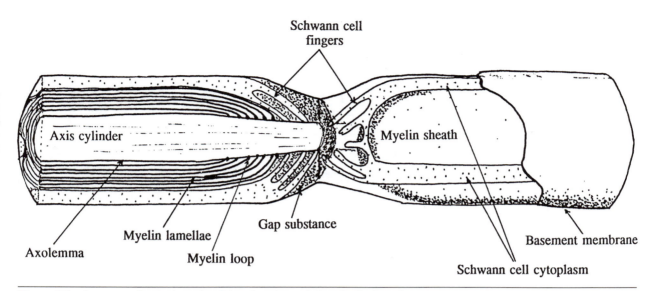

Figure 2.7   Structure of a single node of Ranvier, shown in longitudinal section.

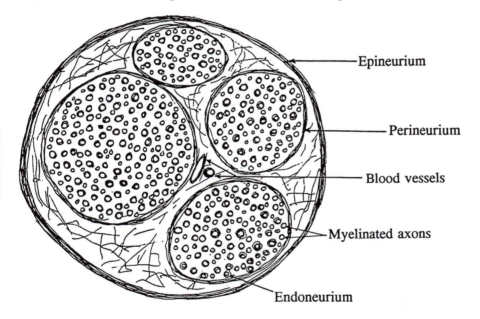

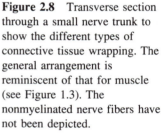

Figure 2.8   Transverse section through a small nerve trunk to show the different types of connective tissue wrapping. The general arrangement is reminiscent of that for muscle (see Figure 1.3). The nonmyelinated nerve fibers have not been depicted.

Charcot-Marie-Tooth Disease Is Described

In 1886 two neurologists in France and one in England described patients with what appeared to be the same condition—severe weakness and wasting of muscles below the knees and in the hands, together with loss of sensation in the hands and feet. This condition, known as Charcot-Marie-Tooth (CMT) disease or peroneal muscular atrophy, is hereditary and produces its effects through degeneration of the peripheral nerves.

CMT Nerves Show Demyelination and Schwann Cell Hypertrophy

Usually clinical evidence of CMT disease appears in the late teens, and the first signs are in the feet, which

lose muscle tissue and toe movement and are highly curved (pes cavus). In most cases, the legs have been likened to inverted champagne bottles or to stork legs because of the extreme thinning below the knee, together with some wasting of the lower thigh (Figure 2.9A). Several forms of the condition are now recognized, the distinctions depending on the mode of inheritance, the presence or absence of sensory nerve fiber involvement, and, especially, whether or not there is loss of myelin (demyelination) in the axons. In the latter cases, the nerves are often thickened and can be readily felt by an examiner; nerve conduction studies in the electromyography (EMG) laboratory make the diagnosis easy, for the impulse conduction velocities are greatly reduced (see chapter 9).

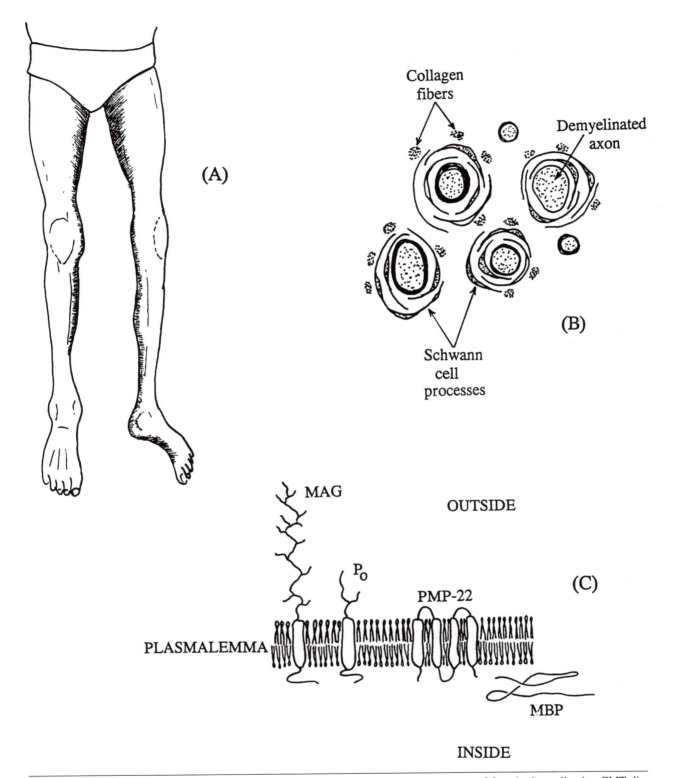

Figure 2.9 (A) Typical thinning of muscles in lower part of leg, and prominent curvature of feet, in demyelinating CMT disease. (B) Cross section through peripheral nerve in demyelinating CMT disease. There is loss of some axons and thin or absent myelin sheaths in others; some axons have an "onion bulb" appearance (see text). (C) Protein components of normal myelin sheath. *MAG* = myelin-associated glycoprotein; *MBP* = myelin basic proteins; *PMP-22* = peripheral myelin protein-22; P_0 = protein zero. See also Suter, Welcher, and Snipes (1993).

Under the microscope, many nerve fibers can be seen to have abnormal myelin sheaths, and some are completely demyelinated. A striking observation is hypertrophy of the Schwann cells, so that individual axons are surrounded by concentric layers of Schwann cell cytoplasm, giving the impression of an onion bulb (Figure 2.9B). The bulb is thickened further by connective tissue in the form of collagen fibrils that run in the same direction as the nerve fiber. As the disease advances, more and more axons disappear, until some muscles become totally denervated.

Is the Primary Defect in the Axis Cylinder or Schwann Cell?

In the demyelinating form of CMT disease, the question arises: Is the genetic abnormality one that affects the motor and sensory neurons primarily, with secondary involvement of the Schwann cells and their myelin sheaths, or is it the other way around? One experimental approach has been to study a strain of mouse (Trembler) that also has a demyelinating peripheral neuropathy. It was possible to transplant a segment of nerve from a Trembler mouse into the sciatic nerve of a normal littermate. When the nerve fibers of the host animal regenerated through the graft, they became ensheathed by Schwann cells from the donor animal and developed abnormal myelin sheaths (Aguayo, Attiwell, Trecarten, Perkins, & Bray, 1977). Clearly, in the Trembler animals it is the Schwann cell rather than the axis cylinder that is at fault. Subsequently, grafts of nerve from CMT patients into immunodeficient, but otherwise normal, mice gave similar results (Aguayo, Perkins, Bray, & Duncan, 1978; see, however, Dyck, Lais, & Low, 1978).

More recently, molecular genetics studies have identified a gene on chromosome 17 as responsible for demyelinating CMT in most of the families. Further, it is now known that this gene codes for one of the proteins associated with the myelin sheath, peripheral myelin protein (PMP-22; see Figure 2.9C). Since the involved region of chromosome 17 is duplicated in CMT, it would appear likely that there is too much PMP-22, but the function of the protein in relation to the myelin sheath is not understood at present (Suter, Welcher, & Snipes, 1993).

Chapter 3

The Neuromuscular Junction

Now that the motoneuron has been described, we can explore the area of contact that its axon makes with the muscle fiber. This small but highly developed region is the *neuromuscular junction*, or *synapse*, and it is here that excitation spreads from the axon to the muscle fiber. Since insufficient current flows from the axon to stimulate the muscle fiber directly, the nerve ending releases a chemical, *acetylcholine* (ACh), to bring about the necessary amplification of the axonal signal.

General Features of the Neuromuscular Junction

The structure of the nerve ending reflects its ability to release ACh, and the underlying muscle fiber is constructed so as to capture as many of the ACh molecules as possible. In keeping with the organization of the first part of this book, this chapter deals with the *structure* of the neuromuscular junction; the opportunity to visualize the *activity* of the junction will come in chapter 10.

A Narrow Space Separates the Membranes of the Muscle Fiber and Nerve Ending

As the motor axon approaches the muscle fiber, it loses its myelin sheath and divides into several small twigs, the outermost ones in mammals often forming an incomplete ring (Figure 3.1A). The twigs lie in grooves, or *gutters*, on the surface of the muscle fiber and run short distances before terminating; the twigs have irregular expansions (*boutons*) corresponding to the sites of transmitter release. The region of the muscle fiber under these twigs is termed the *motor end-plate*; it includes the plasmalemma and also a mound of sarcoplasm, the *sole-plate*. Within the sole-plate are collected a number of muscle fiber nuclei as well as many mitochondria, ribosomes, and pinocytic vesicles. Lying over the whole end-plate are membrane-covered cytoplasmic processes derived from the Schwann cells. With the electron microscope, it can be seen that the membrane of the axon terminal is separated from the muscle fiber plasmalemma by a distinct gap measuring about 70 nm; this is the *primary synaptic cleft* (Figures 3.1 and 3.2). The primary cleft is interrupted by repeated invaginations of the plasmalemma into the sole-plate; each junctional fold forms one *secondary synaptic cleft*. The secondary clefts are approximately 0.5 to 1 µm deep and are rather wider at their terminations than at their necks. In amphibians, the junctional folds are consistently perpendicular to the long axis of a motor nerve twig, but in mammals, the arrangement is less regular because of the frequent bending of the twigs (Figure 3.3).

The Synaptic Clefts Increase the Number of Receptors Available for Combination With Acetylcholine

The muscle membrane lining the primary cleft and the upper parts of the secondary clefts is thickened due to the presence of closely packed acetylcholine receptors (AChRs). The AChRs are held in position by 43-kD protein molecules which link them, via spectrin, to a network of actin filaments in the superficial cytoplasm of the muscle fiber. The crests of the junctional folds contain other proteins that also serve to tether the plasmalemma to the fiber cytoskeleton and to the overlying basement membrane, thereby maintaining the structure of the junctional folds; these additional proteins include *ankyrin*, β-*integrin*, *talin*, α-*actinin*, and *vinculin* (see Hall & Sanes, 1993), as well as *dystrophin* (Huard, Fortier, Dansereau, Labrecque, & Tremblay, 1992). The functional benefit in having a folded plasmalemma is that the arrangement increases the surface area of the muscle fiber that can combine with acetylcholine (ACh) released from the nerve terminals. Although the surface area is expanded 8 to 10 times, the paucity of AChRs in the deeper parts of the secondary clefts makes for a threefold increase in the AChR-bearing portion. The folding of the plasmalemma also increases the amount of acetylcholinesterase (AChE) that can be accommodated at the synapse, since this enzyme is located throughout the primary and secondary clefts, including the farthest reaches of the latter.

Muscle Fiber Acetylcholine Receptors (AChRs)

The acetylcholine receptor (AChR) was not only the first transmembrane ion channel to be visualized and purified, but also an ion channel that could be studied in detail by a variety of electrophysiological techniques. One of the factors facilitating the early investigations was the availability of naturally occurring toxins that bind to the receptor and can be labeled.

AChRs Can Be Seen With the Electron Microscope and Can Be Labeled Using α-Bungarotoxin

The most frequently used AChR-binding toxin is α-*bungarotoxin*, derived from the highly poisonous venom of an Asian snake, the banded krait. This toxin, labeled with either [^{3}H] or horseradish peroxidase, has been used to map the distribution of the AChRs at the neuromuscular junction and to determine their prevalence in the muscle plasmalemma. In the receptor-rich areas of the muscle plasmalemma, electron microscopy has shown that the latter contains particles some 6 to 12 nm in diameter occurring at intervals of 10 to 15 nm (Rosenbleuth, 1974). These particles are the AChRs themselves, and there are approximately 10,000 per µm^2 of membrane; since each receptor has two sites for binding ACh, the density of the binding sites is about 20,000 per µm^2.

High-power electron micrographs show the receptor to have a central pore with several rounded structures around it; at extreme magnification, assisted by computer processing, the receptor can be seen to span the muscle plasmalemma and to project into the synaptic cleft (Figure 3.4A).

The AChR Has Five Subunits, Two of Which Are Identical

The purification of the AChR protein was first accomplished using the electric ray, *Torpedo*. This fish has an organ on each side of its head that is capable of delivering shocks of a thousand volts or so, sufficiently strong to stun prey or aggressors. The electric organ is itself composed of modified muscle fibers with an abundance of nerve-muscle contacts and AChRs. The purified receptor protein was found to have a MW of 268 kD and to consist of five glycopeptide subunits, α, α, β, γ, δ (Raftery, Hunkapiller, Strader, & Hood, 1980); the two α subunits are identical, and each contains a single ACh binding site (Figure 3.4B). The mRNAs corresponding to the four different types of subunit are, like the AChRs, abundant in the electrical organ of *Torpedo*, and provided one approach to cloning the four receptor genes. Another technique was to construct and label a nucleotide coding for part of the AChR and to allow it to hybridize with fragments in the cDNA library (Noda et al., 1984). Once the genes had been cloned, the complete amino acid sequences of the subunits could be easily determined, and inspection of these suggested that each subunit contained four regions that spanned the muscle fiber membrane (Figure 3.4C). One of these regions, M2, is thought to form the lining of the central pore. Three rings of negatively charged amino acid residues, aspartate or glutamate, flank the M2 region and make the pore selective for cations.

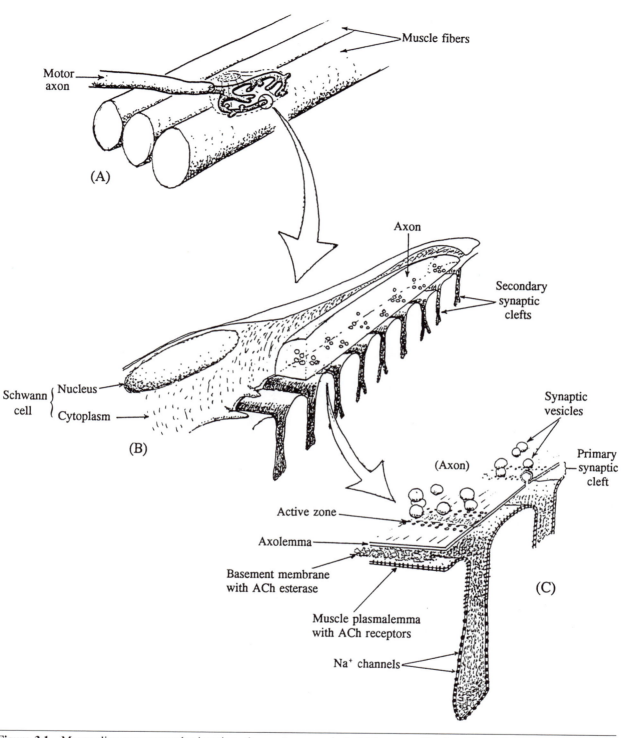

Figure 3.1 Mammalian neuromuscular junction, shown at progressively higher magnification. The different structures in the junction have not been drawn to scale. Note the synaptic vesicle emptying in (C). See also Figure 9.9.

ACh Receptors Are Replenished in the Plasmalemma

It was something of a surprise to discover that AChRs are not fixed in the muscle fiber membrane but are being continually renewed and replaced. The receptors are first prepared on internal membranes and are then transported to the fiber plasmalemma where, in cultured chick myotubes, they remain for an average period of 22 hr (Devreotes & Fambrough, 1975); in adult neuromuscular junctions, however, they stay in place for considerably longer, the average half-life being 8 to 11 days (Fambrough, 1979). They are then taken from the membrane and carried back into the interior

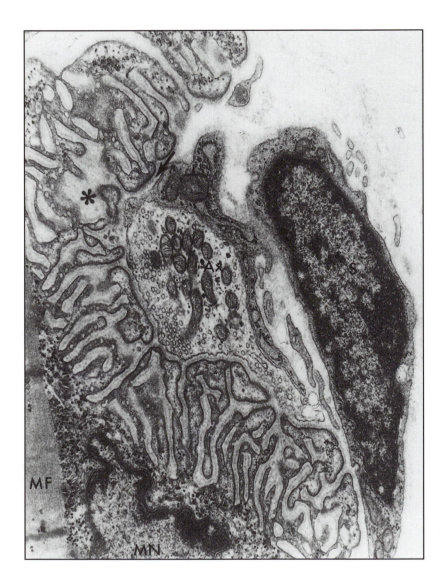

Figure 3.2 Electron micrograph of a human neuromuscular junction showing an axon terminal knob (*Ax*) lying beneath a Schwann cell (*S*, Schwann cell nucleus) and covering an array of secondary synaptic clefts (*); the latter are quite deep and often branch. The muscle fiber also displays a myonucleus (*MN*) and a superficial myofibril (*MF*). The arrow indicates the primary synaptic cleft. The various structures in the junction can be identified by referring to Figure 3.1. Note the synaptic vesicles in the axon terminal. 30,600×. (Courtesy of Dr. A.G. Engel.)

of the fiber within secondary lysosomes (degradation vacuoles).

The synthesis of AChRs is regulated by those myonuclei lying in the sole-plate of the muscle fiber, and the nuclei, in turn, appear to respond to neurotrophic factors released from the motor nerve terminal; two such factors are *calcitonin gene-related peptide* (CGRP; New & Mudge, 1986) and *ACh receptor-inducing activity* (ARIA; Usdin & Fischbach, 1986). In contrast to the synaptic myonuclei, those nuclei outside the neuromuscular junction have their AChR genes repressed by muscle activity, probably through a chain of molecular signals that include Ca^{2+} and protein kinase C (PKC; Figure 3.5).

Basement Membrane

The basement of the muscle fiber was considered in chapter 1. In the neuromuscular junction, the membrane

transverses both the primary and the secondary clefts; at the edges of the junction it becomes continuous with the basement membrane over the rest of the muscle fiber and with that covering the Schwann cell. In electron micrographs, the basement membrane appears as a fuzzy structure from 10 to 15 nm thick; fine processes connect this to the plasmalemmae of the axon terminal and muscle fiber.

The Basement Membrane Contains Proteins With Different Functions

In chapter 1, attention was drawn to the different types of protein that the basement membrane contains in addition to heparin sulfate proteoglycan. In the neuromuscular junction, these proteins include *agrin*, collagens III and IV, laminin, fibroblast growth factor (FGF), and a protease (Table 3.1). Of these, agrin has been shown to promote the clustering of AChRs in the muscle fiber

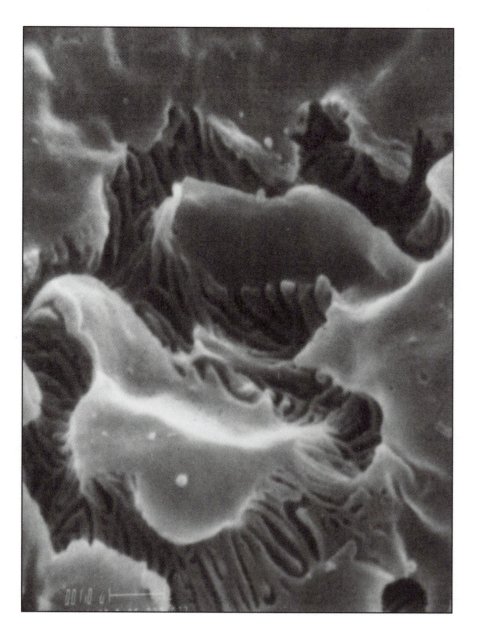

Figure 3.3 Junctional folds and secondary synaptic clefts in a mouse neuromuscular junction. The axon terminal has been removed and the exposed muscle fiber surface examined with a scanning microscope. 10,000×. Reprinted from Fahim, Holley, and Robbins (1983).

plasmalemmae of newly formed neuromuscular junctions during embryogenesis and presumably during reinnervation. Indeed, agrin fulfills the description of a neurotrophin (see chapter 18) in that it is synthesized in the motoneuron soma and transported to the nerve endings before being inserted into the basement membrane and exerting its influence on AChRs (Magill-Solc & McMahan, 1990b). FGF, also found in the basement membrane, has been shown to have a similar action to agrin in controlling the positions of AChRs at the neuromuscular junction (Baker, Chen, & Peng, 1992). The actions of agrin and FGF help to explain how the basement membrane exerts such a powerful influence on the morphological differentiation of the nerve terminal and the muscle fiber at the site of a prospective neuromuscular junction (Sanes, Marshall, & McMahan,

1978; see chapter 17). Of the other proteins associated with the basement membrane, the collagens provide strength, while the laminins appear to have a trophic role which includes axon guidance during embryogenesis. *Dystrophin* is one of the proteins that anchors the basement membrane to the muscle fiber plasmalemma, and it does so through attachments to laminin (Ibraghimov-Beskrovnaya et al., 1992).

Acetylcholinesterase in the Basement Membrane Hydrolyzes ACh

Situated within the basement membrane that runs through the primary and secondary synaptic clefts is an enzyme that hydrolyzes ACh and thereby terminates synaptic

(A)

(B)

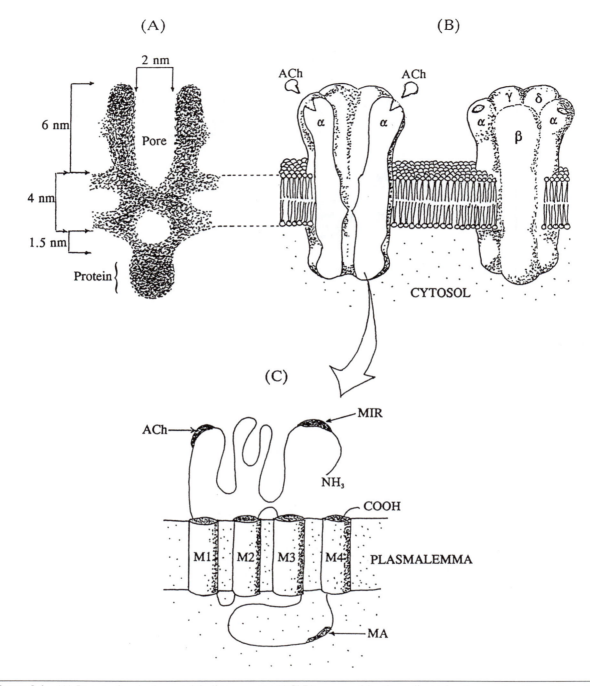

(C)

Figure 3.4   (A) Image of a single acetylcholine receptor (AChR), prepared by three-dimensional reconstruction. Adapted from Toyoshima and Unwin (1988, p. 250). (B) Model AChR showing five subunits. The receptor on the left has been cut through to show the internal pore that admits Na^+ and K^+ ions, following binding of ACh to the α subunits. (C) Molecular structure of one of the α subunits, showing four membrane-spanning regions. *MA* = amphipathic helix; *MIR* = main immunogenic region. Adapted from Beeson and Barnard (1990, p. 164).

transmission. This enzyme, *acetylcholinesterase* (AChE), is specific for ACh, in contrast to *cholinesterase* (ChE), which splits additional choline esters and is also found in the basement membrane. There is good evidence that the AChE in the axon terminal has been transported there from the motoneuron soma by axoplasmic flow. Although the muscle fiber is also able to synthesize this enzyme,

it is apparent from denervation experiments that much of this synthesis is under the trophic control of the motoneuron. Using labeled substrate, Barnard and colleagues (1975) have found that there are about 3,000 AChE molecules per μm^2 in the postsynaptic membranes at mouse end-plates. Unlike the AChRs, the AChE molecules are distributed evenly throughout the primary and secondary

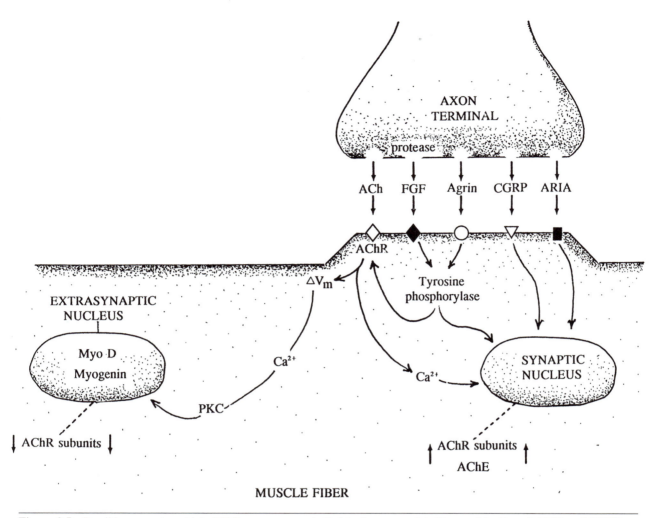

Figure 3.5 Possible means by which AChR genes are controlled in the myonuclei. The combination of ARIA and CGRP with their receptors and the influx of Ca^{2+} cause the genes to be expressed in the synaptic nuclei. Elsewhere in the muscle fiber, the AChR gene is repressed by protein kinase C (*PKC*) and muscle transcription factors (*Myo-D, myogenin*). Agrin and fibroblast growth factor (*FGF*) make the newly formed AChRs cluster in the synaptic clefts. Adapted from Hall and Sanes (1993).

synaptic clefts. The AChE molecules also differ in that they can be easily removed from the synaptic clefts following digestion with proteolytic enzymes. It is therefore probable that the AChE molecules lie within the cleft substance and that their protein tails have only weak attachments to the muscle plasmalemma. Like the AChRs, the AChE molecules are replaced with newly synthesized ones, their half-life being approximately 3 weeks (Kasprzak & Salpeter, 1985).

Recently, it has been suggested that AChE molecules, in addition to hydrolyzing transmitter, may be the precursors of protein-splitting enzymes (*proteases*) and have a trophic role (Small, 1990).

Axon Terminal

Now that descriptions have been given of the muscle fiber membrane and basement membrane at the neuro-muscular junction, it remains to examine the structure of the axon (motor nerve) terminal. As stated in the introduction to this chapter, the terminal is designed for the release of ACh and, of course, for preparing and storing the transmitter as well.

The Motor Nerve Endings Contain Four Types of Vesicles

The axon resembles the muscle fiber in being highly specialized at the neuromuscular junction. The axon terminations contain large numbers of small spheres with membranous linings; the majority measure some 50 to 60 nm across and are typically clustered opposite the secondary synaptic clefts. These spheres are the *synaptic vesicles* and contain the transmitter substance, ACh, as well as ATP and a proteoglycan (Figures 3.1 and 3.2). It has been suggested that the vesicles are derived from

Table 3.1   Protein Molecules Concentrated at the Neuromuscular Junction

Location within the synapse	
Postsynaptic membrane	*Nerve terminal membrane*
AChR $\alpha_2\beta\gamma\delta$ and	Latrotoxin-R
$\alpha_2\beta\epsilon\delta$	Ca^{2+} channel
N-CAM	Syntaxin
N-acetylgalactosaminyl	*Large or small synaptic vesicles*
transferase	Choline acetyltransferase[a]
Integrin $\beta1$ subunit	CGRP
Voltage-sensitive Na^+	Synaptophysin
channel	Synapsin
GM2-like lipid	p65B
Cytoskeleton	SV2
43K protein	*Basement membrane*
Vinculin	Agrin
Talin	Collagens $\alpha3$(IV), $\alpha4$(IV)
Paxillin	Laminin A
Filamin	S-laminin
α-Actinin	AChE
Tropomyosin 2	Heparan sulfate proteoglycan
58K protein	N-acetylgalactosaminyl-
87K protein	terminated carbohydrate
Dystrophin	Protease nexin 1
Dystrophin-related	Fibroblast growth factor (FGF)
protein	
Acetylated tubulin	
Ankyrin	
Laminin B	
Desmin	
Actin	
β-Spectrin	
3G2 antigen	

[a]May be present in nerve terminal cytoplasm and/or associated with small vesicles.
Adapted from Hall and Sanes (1993).

microtubules or endoplasmic reticulum within the axon terminal (Blumcke & Niedorf, 1965). However, it is now clear that most synaptic vesicles are formed by a recycling process in which the plasmalemma (axolemma) of the nerve terminal becomes invaginated and then pinched off (Heuser & Reese, 1973).

At first the new vesicles are encased by small polygonal particles of the protein *clathrin* (Ungewickell, 1984) and become the *coated vesicles* that can be seen in electron micrographs. The clathrin heavy and light chains are then dissociated by an ATPase in the axoplasm, giving the smooth appearance characteristic of most of the vesicles in the nerve ending (Schmid, Braell, Schossman, & Rothman, 1984). Occasionally, much larger smooth-walled vesicles can be seen (*giant vesicles*), and these are thought to contain proportionately greater amounts of ACh. A fourth type of vesicle is in a minority but is of great theoretical interest; this is the *dense-cored vesicle*, which

is from 70 to 110 nm in diameter. Similar vesicles are a prominent feature of adrenergic nerve endings in the central and autonomic nervous systems, and it is possible that, at the neuromuscular junction also, they contain monoamine transmitter molecules. Other possible contents are the neurotrophic factors secreted by the motoneuron; one such factor could be CGRP, which is known to be released from motor nerve terminals. The different types of vesicle contain special *docking proteins* in their walls, which enable the vesicles to attach themselves to the axolemma and then discharge their contents into the synaptic cleft (see chapter 10).

Spontaneous Evacuation of Synaptic Vesicles Causes MEPPs

The possibility that the vesicles contain ACh was first considered when electrophysiological experiments indicated that the transmitter was released from the axon terminal in small packets, or *quanta*. Even at rest, there is a continuous random release of ACh, each quantum producing a small depolarization of the muscle fiber membrane (Figure 3.6); the depolarizations were referred to as *miniature end-plate potentials* (MEPPs) by Fatt and Katz (1952), who described them first. By using raised concentrations of Mg^{2+} in the bathing fluid so as to reduce the release of ACh from the axon terminal, it can be shown that the end-plate potential following a nerve impulse is normally the summation of many MEPPs. It has also been shown that the spontaneous release of ACh probably reflects the resting level of Ca^{2+} ions within the nerve terminal and that this is regulated by Ca^{2+} uptake and release from the mitochondria (Alnaes & Rahamimoff, 1975).

The Vesicle Hypothesis of ACh Release Is Also Supported by Two Other Types of Evidence

The validity of the vesicle hypothesis is strengthened by the results of combined biochemical and electron microscopical studies on tissue homogenates; thus, both in the brain and in the electric organ of the electric eel, it has been possible to prepare cell fractions that are rich in vesicles and also contain large amounts of ACh (Whittaker, Michaelson, & Kirkland, 1964). Unfortunately, a similar experiment cannot be performed on motor nerve terminals because of the difficulty in obtaining satisfactory preparations of vesicles without contamination from subcellular components of muscle.

A second important test of the vesicle hypothesis has been to demonstrate a reduction in their number following massive release of ACh by the nerve terminal. In the experiments of Jones and Kwanbunbumpen

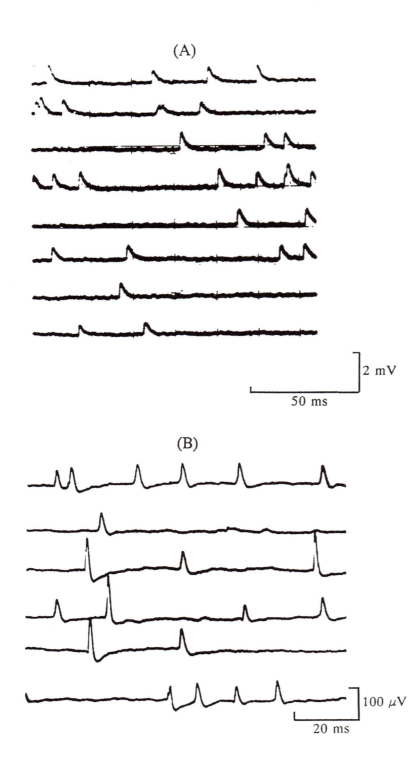

(A)

2 mV

50 ms

(B)

100 μV

20 ms

Figure 3.6 Miniature end-plate potentials (MEPPs) recorded *in situ* (A) with an intracellular microelectrode from a mouse gracilis muscle fiber and (B) with a coaxial needle electrode from the vastus medialis muscle of a normal subject. Because in the latter instance the recording electrode is effectively extracellular, the potentials appear much smaller (note amplification).

(1970), the rat phrenic nerve was stimulated repetitively and resynthesis of ACh was prevented by hemicholinium. The authors found that a significant depletion of vesicles had occurred in the zone of axoplasm bordering the synaptic cleft and that the residual vesicles were reduced in size.

Another, and rather interesting, technique for depleting the nerve terminals of ACh is to expose them to the venom of the black widow spider; with this method also, a reduction in the number of vesicles within the nerve terminal can be demonstrated (Clark, Mauro, Longenecker, & Hurlbut, 1970). A still more demanding technique is to freeze the muscle very rapidly, at a time when unusually large amounts of ACh are being released from the nerve endings, as for example, following electrical stimulation of the nerve twigs in the presence of 4-aminopyridine. It is then possible, in electron micrographs, to see vesicles that have been captured while opening up after fusing with the axolemma (Heuser & Reese, 1981).

Each Synaptic Vesicle Contains Approximately 10,000 Molecules of ACh

The number of ACh molecules inside a single vesicle has been determined in various ways. For example, Katz and Miledi (1972) analyzed membrane potential "noise" during the delivery of ACh from a micropipette; they estimated that a single molecule of ACh produced a depolarization of about 0.3 μV and that at least 1,000 molecules would be necessary to account for a MEPP of 0.3 mV in a frog muscle fiber. However, the number of molecules in a vesicle must be rather larger than this, for additional molecules will be lost by diffusion out of the synaptic cleft and by hydrolysis with AChE. Potter (1970), using radioactively labeled ACh, calculated that a single impulse in a phrenic motor nerve fiber released about 4 $\times 10^6$ molecules of transmitter from the axon terminal; he considered that a typical vesicle might contain from 4,000 to 9,000 molecules of ACh.

An especially accurate (and elegant) estimate of quantal size was made by Kuffler and Yoshikami (1975), who devised a highly sensitive bioassay system to determine the amount of ACh delivered from a micropipette onto the exposed subsynaptic membrane in the frog and snake. These authors concluded that a quantum of transmitter contained fewer than 10,000 molecules of ACh. This value agrees well with that of Miledi, Molenaar, and Polak (1983), who collected the ACh released spontaneously in frog sartorius muscles in the presence of an anticholinesterase drug, and compared the amount with the numbers of MEPP currents in the same period of time; they estimated that each vesicle contained, on average, approximately 12,000 molecules of ACh.

The functional advantage of containing the transmitter in a vesicle, rather than in the axoplasm, is probably twofold. First, the ACh is protected from the small amount of ChE that is normally to be found within the nerve terminal. Second, the influx of Ca^{2+} during the impulse is only able to activate a certain number of sites on the inner surface of the axolemma, and it is therefore advantageous to deliver an optimal quantity of ACh to each site (see the following section).

The Axolemma Contains Active Zones for the Release of ACh

The axolemma facing the synaptic cleft contains regular structures that can be seen by electron microscopy in regular ultrathin sections, but are best visualized after the membrane has been sheared into two leaflets by the freeze-fracture technique. These structures, termed *active zones*, consist of parallel rows of small particles arranged in two pairs. In the amphibian neuromuscular junction, in which the motor nerve terminals are oriented in the long axis of the muscle fiber, the active zones run transversely across the terminals and lie opposite the secondary synaptic clefts (Figure 3.1); each pair of rows is separated from the other pair by a ridge. In mammalian neuromuscular junctions, the active zones have a more variable orientation to each other and to the axis of the nerve terminal; in addition, the rows contain fewer particles and there are no intervening ridges between the pairs of rows. The particles in the active zones are thought to be the Ca^{2+} channels. Their presence, together with that of the docking proteins, enables the synaptic vesicles to fuse with the axolemma (see chapter 10).

The Nerve Ending Also Contains Endoplasmic Reticulum, Microtubules, Neurofilaments, and Mitochondria

In addition to the synaptic vesicles, the nerve ending contains large numbers of mitochondria, the main function of which is to provide ATP, through oxidative metabolism, for various synthetic and pumping activities. Thus, ATP is required for the conjugation of acetate and choline to form ACh and for fueling the Na^+-K^+ pump in the axolemma. In addition, ATP is consumed when clathrin is removed from the coated vesicles, as part of the membrane recycling process (see p. 44). Massive stimulation of the motor nerve endings can produce degenerative changes in the mitochondria (Jones & Kwanbunbumpen, 1970).

Electron microscopy of the nerve ending shows that it also contains smooth endoplasmic reticulum, microtubules, and neurofilaments. The microtubules and neurofilaments act as channels for the axoplasmic transport of material between the motoneuron soma and nerve terminals (see chapter 8). It is possible that some of the neurofilaments direct the synaptic vesicles to the active zones in the axolemma.

APPLIED PHYSIOLOGY

There are many fascinating clinical disorders that result from alterations in function at the neuromuscular junction. However, these disorders are better understood after the physiology of the junction has been studied and have therefore been deferred until chapter 10.

The account of muscle structure is almost complete. In the next chapter, we examine the final element—the sensory nerve endings in the muscle and tendon. It is these endings that signal the state of the muscle to the brain and spinal cord.

Chapter 4

Muscle Receptors

For muscles to work effectively, the motoneuron must be able not only to commence discharging at the right instant, but also to adjust its firing rate from moment to moment, depending on the nature of the task and the changing loads imposed on the moving part. These adjustments in firing rate, in turn, depend on the continuous flow of information to the central nervous system from the muscles, joints, and skin. It is the nature of the receptors (sense organs) in the muscles that must now be considered.

Muscles Contain Morphologically Distinct Receptors as Well as Free Nerve Endings

It is convenient to classify muscle receptors into four types:

- Muscle spindles
- Golgi tendon organs
- Paciniform corpuscles
- Free nerve endings

Of these receptors, the muscle spindles and tendon organs are by far the most important in signaling information about mechanical events. The muscle spindles are exquisitely designed to sense the *lengths* of the muscle fibers during contraction, while the tendon organ is concerned more with the *forces* developed by the fiber. The paciniform corpuscle is another morphologically distinct receptor; it has a nerve ending that is ensheathed in thin wrappings of connective tissue, rather like the layers of an onion. The corpuscle is exquisitely sensitive to rapid small deformations, and is widely distributed in such tissues as skin, periosteum, and abdominal viscera; in muscle it probably plays a minor role in providing information about contraction and relaxation. The free nerve endings are important, in that one of their functions is to signal the metabolic status of the interstitial fluid and hence of the muscle fibers; these endings respond to changes in oxygen tension, $[H^+]$, $[K^+]$, and probably other signals as well.

The Muscle Spindle

The first of the muscle receptors that we will consider in detail is the muscle spindle. The spindle is remarkable for several reasons, one being the complexity of its structure and another being the way its responsiveness can be adjusted by the central nervous system.

Muscle Spindles Are Long Thin Structures Surrounded by Extrafusal Fibers

A muscle spindle is many times longer than it is wide. Thus, although a spindle is only 80 to 250 µm in diameter at its thickest part, the capsule, it is usually several mm long and may even extend for 10 mm. The spindles are surrounded by the (extrafusal) muscle fibers that make up most of the muscle belly; most spindles occur singly, but others are found in an end-to-end "tandem" arrangement. The number of spindles in a muscle belly varies greatly from one muscle to another, even when allowances are made for the size of the muscle (Table 4.1). The muscles at the back of the neck and the small muscles of the hand have the richest supply of spindles, and the large muscles of the arm and leg are least well endowed. This difference in density is probably related to the ability to carry out small movements of the head and fingers rapidly and accurately.

The Muscle Spindle Contains Two Types of Muscle Fiber

The characteristic fusiform shape of the spindle is due to the presence, in its middle part, of a connective tissue

Table 4.1   Values of Spindle Densities in Certain Human Muscles

Muscle	No. of spindles	Weight (g)	Density (spindles/g)
Obliquus capitis superior	141	3.3	42.7
Rectus capitis posterior major	122	4.0	30.5
Abductor pollis brevis	80	2.7	29.3
First lumbrical, foot	36	1.7	21.0
Second lumbrical, hand	36	1.8	19.7
Opponens pollicis	44	2.5	17.3
First lumbrical, hand	51	3.1	16.5
Masseter, deep portion	42	3.8	11.2
Biceps brachii	320	164	2.0
Pectoralis major	450	296	1.5
Triceps brachii	520	364	1.4
Latissimus dorsi	368	246	1.4
Teres major	44	123	0.4

Note. From *Mammalian Muscle Receptors and Their Central Actions*, (p. 48) by P.B.C. Matthews, 1972, London: Edward Arnold. Copyright 1972 P.B.C. Matthews. Reprinted by permission. Values rounded off. Data from Voss (1937, 1956, 1958); Schulze (1955); Freimann (1954); Körner (1960).

capsule containing fluid. The composition and function of the fluid are still unknown, though it has long been thought that the fluid is lymph. In the longitudinal axis of the spindle, the central and outer regions are often referred to as the equatorial and polar regions respectively.

A remarkable feature of the muscle spindle is that its sensitivity as a mechanoreceptor can be adjusted; this is possible because the sensory nerve endings are attached to *intrafusal* (inside the spindle) muscle fibers that can be made to contract and relax. Two types of intrafusal fiber can be recognized on the basis of the numbers and distributions of their nuclei (Figure 4.1). The *nuclear bag* fibers have an extraordinary number of nuclei in their middle part, within the capsule; indeed, in this region the muscle fiber appear to consist of little else but nuclei. The *nuclear chain* fibers, in contrast, have their nuclei distributed more evenly along their lengths, and the nuclei are found in the centers of the fibers. The two types of fiber differ in other respects, for the bag fibers are usually thicker and rather longer than the chain fibers. In addition, the bag fibers are less numerous, for there are usually only 2 (between 1 and 4) in each spindle, whereas there may be as many as 12 chain fibers (Table 4.2). Although the two nuclear bag fibers appear very similar, they can be distinguished by histochemical staining for myosin ATPase and presumably have other differences in contractile proteins; the two fibers are referred to as bag_1 and bag_2 respectively.

The Intrafusal Muscle Fibers Have Their Own Motor Nerve Supply

Two types of motor nerve ending are found on the intrafusal muscle fibers, in the polar regions (Figure 4.1). One type, the *plate ending*, resembles that found at the neuromuscular junction on the extrafusal fiber; this type of ending is associated mainly with the nuclear bag fibers. The other type of motor ending is more extensive, running along the greater part of the fiber outside the central region; this type is the *trail ending*

Table 4.2   Characteristics of Nuclear Bag and Nuclear Chain Intrafusal Fibers

	Nuclear bag	Nuclear chain
No. per spindle	2	2-12
Diameter (µm)	20-25	10-12
Motor nerve endings	Plate	Trail
Sensory response	Dynamic	Static
Sensory axon	Ia, II (bag_2)	Ia, II

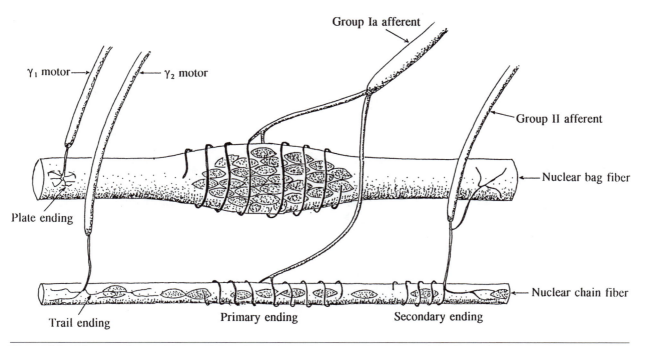

Figure 4.1   Diagram of central (equatorial) regions of a nuclear bag and a nuclear chain fiber, showing the sensory and motor axons that are characteristic of the two types of fiber. Adapted from Matthews (1964, p. 232).

and is found mostly on the nuclear chain fibers. Both types of ending are attached to γ-*motor* (fusimotor) axons, which can be identified in the ventral roots of the spinal cord by their small diameters. For most spindles there are two γ-motor axons, one for the nuclear bag fibers and one for the nuclear chain fibers (Porayko & Smith, 1968). A recently discovered refinement of the situation is that, while the bag₁ fiber has its own γ-motor axon, the bag₂ fiber shares the other γ-axon with the chain fibers (Kucera, 1985). In a minority of spindles, however, there may be only a single γ-axon, and this is then responsible for innervating all the intrafusal muscle fibers. In addition to the γ-motor axons, a spindle may receive a branch from an α-motor axon. Such branches, termed β-*axons*, appear to supply only plate endings on the nuclear bag fibers.

The Contractile and Electrical Properties of the Intrafusal Fibers Differ

On the basis of the differences in structure and innervation, it might be expected that the intrafusal muscle fibers would respond differently to excitation through their axons, and such is the case. By stimulating intrafusal fibers directly (Smith, 1966) or through their motor axons (Boyd, 1966; Bessou & Pagès, 1972), it is possible to show that the contractions of the bag fibers are localized and appear slower and weaker than those of the chain fibers; this distinction is especially true of the bag₁ intrafusal fibers. The electrical properties of the

two types of fiber also differ. Bessou and Pagès (1972) collaborated with Barker (1974) in showing that the bag fibers usually developed graded, nonpropagated electrical potentials, while the nuclear chain fibers responded to nerve stimulation with propagated action potentials, like extrafusal muscle fibers.

Since the intrafusal fibers receive their motor innervation on either side of their central regions, and since, in the case of the bag fibers, the central region consists mainly of nuclei rather than myofibrils, it is the central region that is stretched by the contracting ends of the fibers. This stretching increases the sensitivity of the primary sensory nerve endings (see p. 50).

The Primary and Secondary Sensory Endings Are Arranged on and Around the Intrafusal Muscle Fibers

Each spindle has one large afferent nerve fiber, the *Group Ia axon*, which terminates in spirals around each of the intrafusal fibers within the equatorial region; this type of ending is known as the *primary ending* (Figures 4.1 and 4.2). In human spindles the primary endings take the form, not of spirals, but of multiple C-shaped bands that partly surround the intrafusal fibers (Kennedy, 1970).

There is also a rather thinner sensory nerve fiber for each spindle, the *Group II axon*, and this terminates in *flower spray endings* on the nuclear chain and bag₂ fibers; these constitute the *secondary endings*.

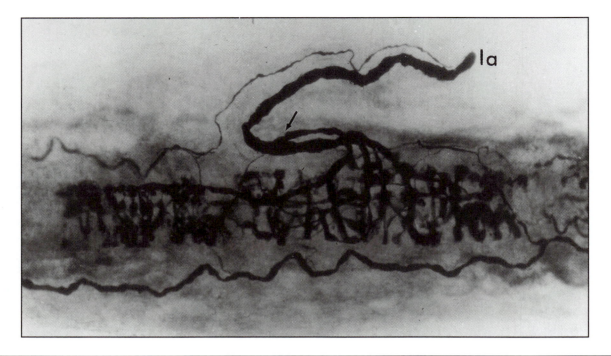

Figure 4.2 The central (equatorial) region of a human muscle spindle, showing the branching of the Ia axon (*arrow*) to supply annulospiral-like endings on the (unstained) nuclear bag and chain fibers. (Courtesy of Dr. Dennis Harriman.)

Activation of the Muscle Spindle

It is convenient to deal with the functional properties of the muscle receptors at this stage so that, apart from a small section in chapter 15, the examination of these sense organs can be completed.

Passive Stretch of a Spindle Produces Both a Rapid Firing and a Sustained Discharge in the Sensory Nerve Axons

In the laboratory, the responses of the muscle spindles to passive stretch can be studied by recording the impulse discharges in the Ia and II sensory axons; in most experiments, stretch is applied in the form of a ramp. The impulse discharge in the two types of axon differs. The Ia axon fires a high-frequency burst of impulses while the stretching takes place (*dynamic response*) and a steady discharge of impulses when the muscle spindle is held at its new length (*static response*) (Figure 4.3); the II axon has only a static response. A further distinction is that the primary endings of the Ia axons have much lower thresholds to stretching than the secondary endings of the II axons. Matthews (1972) suggested that the complex response of the primary endings was due to the Ia axon being connected to both a nuclear bag and a nuclear chain fiber (Figure 4.1); the bag fiber could contribute the dynamic discharge and the chain fiber the static one. The two types of response are reflected in the *generator potential*, that is, the depolarization of the sensory nerve endings caused by the opening of the stretch-activated ion channels (Hunt, Wilkinson, & Fukami, 1978). There is still uncertainty as to how these channels are made to open, but presumably it has something to do with the extension of the nerve spirals around the intrafusal muscle fibers.

There is also uncertainty about the mechanism of the dynamic response and why the impulse frequency should be proportional to the *rate of stretching*. One factor seems to be the unusual viscous properties of the nuclear bag fiber, a factor that was first appreciated when single muscle spindles were dissected out and inspected under the light microscope while being stretched. Thus, if a bag fiber is pulled to a new length, the equatorial region (containing annulospiral endings) is first extended and then slides back to its original length over several hundred milliseconds (Smith, 1966); this behavior could be due to the equatorial region being predominantly elastic and the polar regions being viscous (Matthews, 1964; Figure 4.4).

Two Types of γ-Motor Axon Increase the Sensitivity of the Spindle Endings to Stretch

Just as there are two types of sensory discharge for the spindle, dynamic and static, so there are two types of fusimotor axon. Stimulation of the *dynamic* fusimotor axon enhances the impulse frequency in the primary

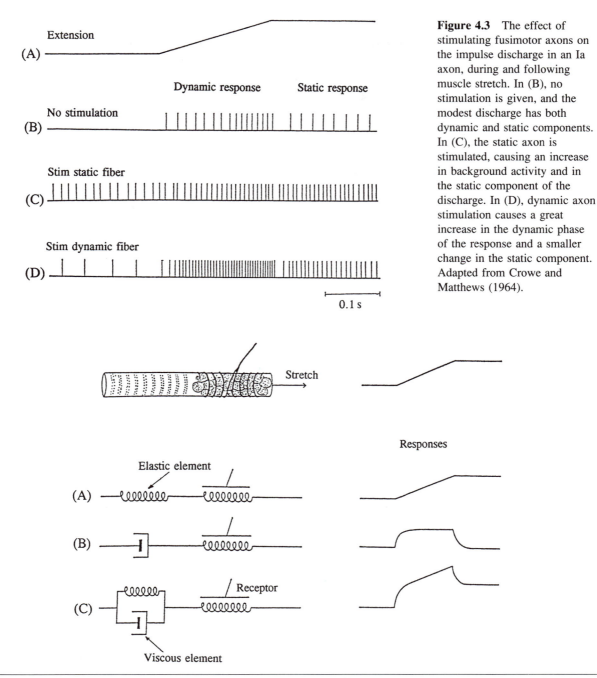

Figure 4.3 The effect of stimulating fusimotor axons on the impulse discharge in an Ia axon, during and following muscle stretch. In (B), no stimulation is given, and the modest discharge has both dynamic and static components. In (C), the static axon is stimulated, causing an increase in background activity and in the static component of the discharge. In (D), dynamic axon stimulation causes a great increase in the dynamic phase of the response and a smaller change in the static component. Adapted from Crowe and Matthews (1964).

Figure 4.4 Models to illustrate how the Ia impulse discharge would be affected by the elastic and viscous properties of the intrafusal muscle fibers. The annulospiral (primary) ending (*receptor*) is shown as being in a purely elastic region of the spindle, and its response will depend on the stretch transmitted to this region. In (A), stretching the spindle evokes a discharge that is entirely proportional to the instantaneous length of the receptor region; that is, there is only a static component. (B) Replacement of the elastic element by a viscous one introduces a dynamic component in the discharge but abolishes the static one. (C) The introduction of elastic and viscous elements in parallel would be expected to give both dynamic and static phases of the response. Because these are the responses actually observed, this model is the best one. Adapted from Matthews (1964).

endings during the time that the spindle is being lengthened (Matthews, 1962), while stimulation of the *static* fusimotor fibers increases the impulse frequency in both primary and secondary endings once the spindle is at its new length. The simplest explanation for the two types of motor response is that the static fusimotor axons supply the nuclear chain fibers within the spindle, and the dynamic fusimotor axons innervate the nuclear bag fiber (Matthews, 1962). In both cases, however, the fusimotor fibers produce their effects in the same way.

By causing the intrafusal fibers to contract, especially in their polar regions, they simultaneously extend the equatorial regions in which the annulospiral endings are stretched and enhance the sensitivity of the latter to passive stretch. The spindle can therefore be regarded as an exquisitely engineered length sensor; not only can it signal both the rate and the extent of stretching, but its sensitivity can be enhanced by the central nervous system, as the need arises, by contraction of the intrafusal muscle fibers.

Muscle Spindle Axons Project to the Spinal Cord and Brain

Part of the flow of impulses from the muscle receptors is directed to the brain, where it may or may not be appreciated at a conscious level. In addition, there are powerful synaptic connections to motoneurons, especially those innervating the same muscle (homonymous motoneurons). Indeed, there is good evidence that the strongest connections of all are made with motoneurons supplying motor units in the vicinity of the spindles (compartmentalization of muscle).

Group Ia Axons Excite Homonymous Motoneurons

The group Ia axons that project to motoneurons supplying the same muscle end in knobs, or boutons, on the dendrites; sometimes an additional neuron is involved in the pathway (Figure 4.5). Regardless of the length of the neural pathway, the effect of the Ia axons is to tend to excite the motoneurons. These connections provide a means for rapidly executed reflex adjustments in muscle length following any stretching or change in muscle load. One can think of many situations when such adjustments might come into play—for example, restraining a fractious dog on a lead, holding a tray while objects are added or removed, standing in an accelerating or slowing bus, and so on. The maintenance of posture is also largely under the direction of the muscle spindles. Since the neural pathways are in the spinal cord and are relatively short, the change in motoneuron firing can be made more rapidly than if intervention by the brain were needed. Although the Ia axons are excitatory to motoneurons serving the same muscles or its synergists, they are inhibitory, through an interneuron, to motoneurons of the antagonist muscles (Figure 4.5). Group II axons resemble the Ia axons in being excitatory to homonymous motoneurons, though the connections are usually through one or more interneurons.

Muscle Spindles Are Involved in Muscle Tone and in Tendon Jerks

It is the muscle stretch receptors that are responsible for the impression of "tone" that can be elicited by passive manipulation of a limb; in this case, extension of a joint will activate stretch endings in the flexor muscle and will evoke a reflex contraction of the muscle. If the endings are compromised by disease or injury, so as to be relatively inexcitable, then the reflexes will be diminished, and the tone perceived by an examiner will be decreased. Conversely, an increase in spindle activity, as in patients with certain types of brain or spinal cord lesion, will give an impression of greater resistance when the limb is moved (*spasticity*).

Muscle spindles are also responsible for the so-called *tendon reflexes*. A sharp tap on a tendon will stretch the muscle belly and the spindles contained within it, evoking an impulse volley in the Ia axons; these axons, in turn, evoke a brisk reflex contraction of the muscle (Figure 4.6B), causing the limb to move. The *H-reflex*, named after Paul Hoffmann (1918), is the electrical

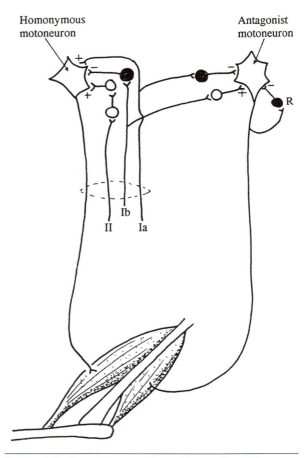

Figure 4.5 Synaptic connections of group Ia and Ib axons in the spinal cord. The inhibitory interneurons are shown by filled circles.

equivalent of the tendon reflex; in this case, the impulse volley in the Ia axons is set up by an electrical stimulus to the motor nerve, rather than by a mechanical perturbation (Figure 4.6C).

The Golgi Tendon Organ

Named after their discoverer, Camillo Golgi (1903), the Golgi tendon organs are fewer in number than the muscle

spindles; a further difference is that the tendon organs are found at the musculotendinous junction, rather than in the muscle belly. Also, a tendon organ has a much simpler structure than a muscle spindle, and its physiological properties are easier to comprehend.

Golgi Tendon Organs Are Attached to Extrafusal Muscle Fibers

The tendon organs are encapsulated structures, typically from 0.5 to 1.0 mm long and from 0.1 to 0.2 mm

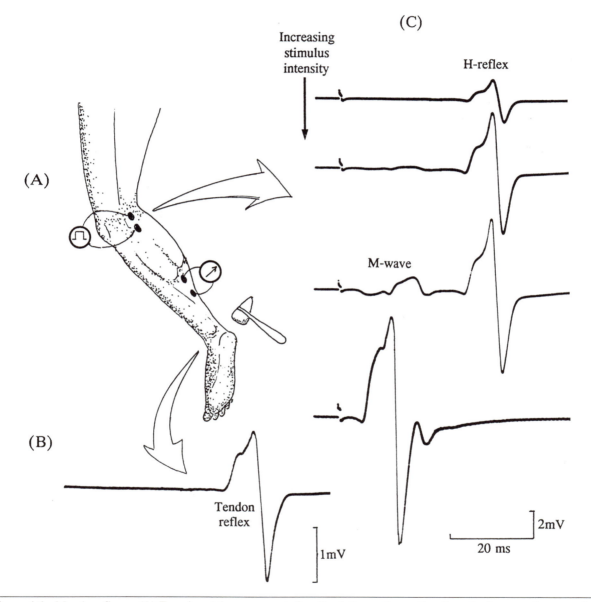

Figure 4.6 Muscle reflexes. (A) Experimental arrangements for stimulating and recording from the human soleus muscle. (B) The tendon reflex. An EMG response has been evoked in a human soleus muscle by a tap to the Achilles tendon. (C) H-reflexes elicited in the same muscle by progressively stronger electrical stimuli to the tibial nerve (from above downward). Beyond a certain value, an increase in stimulus strength is associated with a decline in the H-reflex, due to collision between antidromic and reflexly elicited impulses in the motor axons. Eventually the H-reflex disappears altogether, and only the indirectly evoked muscle compound action potential (M-wave) remains (lowest trace).

in diameter. A single nerve axon, of relatively large diameter, conducts the sensory impulses to the spinal cord and is termed the Ib axon; it appears to arise from nonmyelinated sprays in contact with the tendon fascicles (Figure 4.7A). High-power microscopy reveals that the fine nerve twigs weave their way between the bundles of collagen fibers (Figure 4.7B). When the muscle fibers contract, the collagen bundles straighten and compress the nerve twigs, setting up an impulse discharge in the latter (Swett & Schoultz, 1975). Through the collagen bundles, a tendon organ is connected to 4 to 25 muscle fibers.

Tendon Organs Are Excited by Motor Unit Contractions

In contrast to muscle spindles, tendon organs are not very sensitive to passive stretch of the muscle and its tendon, though there is some variation among muscles. When muscles contract, however, the tendon organs discharge, and their sensitivity is such that they can respond to a single motor unit through the one or two

muscle fibers that attach the unit to the tendon organ. Each tendon organ is influenced by 4 to 15 motor units (Houk & Henneman, 1967). The Ib axons, arising from the tendon organs, differ from the spindle axons in being inhibitory to motoneurons of the same muscle and excitatory to those of antagonists (Figure 4.5). Through their reflex connections, they enable the force developed by a muscle to be held constant, if this is desirable.

Free Nerve Endings

One last type of muscle receptor remains to be discussed, a type that is both the simplest in its structure and the most abundant within the muscle. This is the free nerve ending.

Free Nerve Endings May Respond to Specific Stimuli or to Tissue Damage

The free nerve endings relay their information to the spinal cord through the smallest myelinated nerve fibers

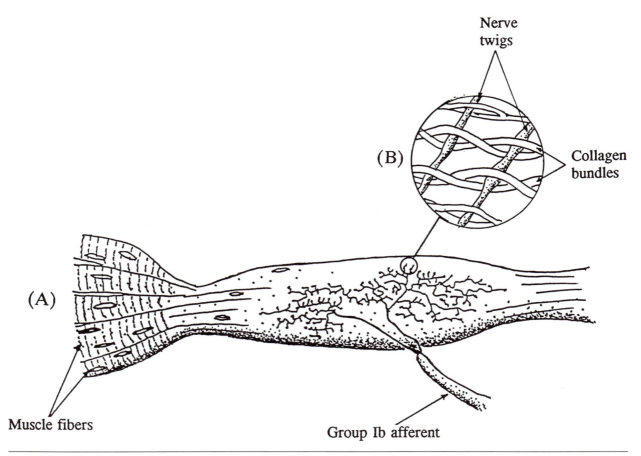

Figure 4.7 (A) Structure of a Golgi tendon organ, modified from Cajal (1909, Figure 199); the capsule has been removed. (B) Interrelationship of nerve fiber twigs and collagen bundles in part of a tendon organ, seen in higher magnification than in (A).

(group III afferents) and the nonmyelinated fibers (group IV afferents). The free nerve endings innervate almost all structures within the muscle belly: the extrafusal fibers, the different types of connective tissue—including the muscle sheath (epimysium), the larger blood vessels, and even the muscle spindles and tendon organs (Stacey, 1969). Electrophysiological recordings from group III and IV afferents show that many of the free nerve endings are sensitive to mechanical stimuli, such as those associated with muscle contraction or with pressure or stretching (Mense & Meyer, 1985). Other free nerve endings respond specifically to changes in temperature or in their chemical environment; for example, elevations in K^+ or lactic acid concentrations may provoke discharges (Thimm & Baum, 1987). Yet other free nerve endings have high thresholds and only discharge impulses in response to stimuli that can cause tissue damage (Mense & Stahnke, 1983). These are the nociceptive endings, and it is possible that they are sensitized by the release of neuropeptides (e.g., bradykinin) or other stimulants (e.g., arachidonic acid) at the site of injury.

Some nonmyelinated endings are quiescent when a muscle is resting but become increasingly active during the course of ischemic contractions. These fibers, and those responsive to rises in K^+ or lactic acid concentrations, could be responsible not only for muscle discomfort during exercise, but also for cardiovascular and respiratory reflexes (Coote, Hilton, & Perez-Gonzalez, 1971). The endings could also provide the afferent arm of the reflex that has been postulated by Bigland-Ritchie, Dawson, Johansson, and Lippold (1986) to depress motoneuron excitability during sustained effort (see chapter 15).

APPLIED PHYSIOLOGY

Instead of attending the neuromuscular clinic, we shall visit the physiology laboratory and try to answer a question that is as puzzling as it is fascinating.

Do Signals From Muscle Receptors Reach Consciousness?

There is no doubt that impulse messages from the nociceptive free nerve endings can be perceived; an example is the momentary pain experienced by a patient when an electromyographer penetrates the epimysial sheath of a muscle with a needle electrode. Also, during and after severe exercise, there is a discomfort, poorly localized in the muscle, which is accentuated by pressure over the belly. But what about the messages from the muscle spindles and Golgi tendon organs—can these

be perceived? Over the years, scientific opinion has tilted first in one direction and then in the other, and even today, despite many ingenious experiments, the answer to the question is still uncertain. The earliest view was that information from receptors in the muscles and joints was indeed required for awareness of the position of the limb (Sherrington, 1900). This view then succumbed to the results of experiments in which passive movements of a toe or finger could no longer be appreciated after impulses in joint nerve fibers had been blocked by local anesthesia of the digit (Browne, Lee, & Ring, 1954; Provins, 1958), despite the same changes in muscle length taking place as before. An even more direct experiment was that of Gelfan and Carter (1967), who exposed tendons in the ventral forearms of conscious subjects. These authors found that there was no sensation of movement when a tendon was pulled so as to stretch the muscle; when the tendon was pulled in the opposite direction so as to flex the finger, without disturbing the muscle belly, the new positions of the fingers were instantly detected.

Against these observations is the clear demonstration that impulse volleys in large-diameter muscle afferents evoke responses in the mammalian cerebral cortex, including that of the monkey (Albe-Fessard, Liebeskind, & Lamarre, 1965). Further, when the tendons of the human biceps brachii or triceps brachii are vibrated, so as to excite the muscle spindles, the position of the elbow joint is misjudged; this discrepancy suggests that the Ia axon discharge has been misinterpreted by the brain as muscle stretch (Goodwin, McCloskey, & Matthews, 1972).

More recently still, the pendulum appears to be swinging back toward the lack of conscious awareness. For example, when single muscle spindle afferents are stimulated electrically through tungsten microelectrodes inserted into the peripheral nerve, there is no detectable sensation; in contrast, electrical stimulation of most joint afferents does evoke sensation (Macefield, Gandevia, & Burke, 1990).

The situation cannot be regarded as closed, however (Gandevia, McCloskey, & Burke, 1992). For one thing, the excitation of a single spindle afferent by itself does not occur under natural circumstances. Also, as Gandevia and McCloskey (1976) showed, the sensations of movement and of joint position are normally required by the body during active contractions. Correspondingly, they have observed that after a finger has been anesthetized, there is a marked improvement in the detection of passive movements when a subject is allowed to tense the long finger flexor muscle.

The examination of the structure of skeletal muscle and its nerve supply is finished. In the next two chapters we shall see how the muscles and motoneurons develop in the embryo.

Chapter 5

Muscle Formation

The muscles of the human body are formed long before birth. Indeed, fetal imaging has shown that the muscles of the trunk start to contract as early as the 8th week of gestation. By 20 weeks, a remarkably rich repertoire of apparently purposeful movements involving the hands, feet, and head has become established (Table 5.1).

At least five major interrelated processes are involved in muscle development. The gross anatomical features of muscle development have been known for many years, both in humans and in animal species. Special insights have come from tissue transplants and other experiments in animal embryos, as well as from dissections, microscopic examination of embryonic tissue, and studies of muscle cells in culture. Recently, an understanding of muscle development has reached a new level, following the identification of some of the genes and peptide growth factors that are involved in muscle cell determination and differentiation as well as in the laying-down of the body plan. It will be shown that the formation of muscles depends on five interrelated processes:

- Induction of mesoderm from presumptive ectoderm

- Commitment of a portion of the mesoderm to develop into skeletal muscle
- Proliferation and differentiation of myogenic (muscle-forming) cells
- Laying-down of the body plan, including the outgrowth of the limb buds
- Assembly of muscles

Step 1: Mesoderm Is Induced From Ectoderm

As the oocyte develops in the ovary, it not only becomes larger and accumulates yolk, but it also forms animal and vegetal "poles" (Figure 5.1), which can be distinguished by the uneven distribution of their cytoplasmic inclusions (e.g., lipid droplets and yolk granules). Within the cytoplasm of the oocyte are large amounts of mRNAs; although some of the mRNAs, such as those for histone proteins, remain freely distributed, other mRNAs start to move along the cytoskeleton and to accumulate at one or other of the two poles. Some of the mRNAs collecting at the vegetal pole will subsequently

Table 5.1   Appearance of Movements in Human Fetus[a]

Gestational age (weeks)	Movement
8	Quick flexion and extension movements of the trunk.
10	Extension of arms and legs accompany trunk movements.
12	Head rotation. Hands brought up to face. Sudden jerks elicited by mechanical stimuli ("startle" response).
13-14	Creeping and climbing movements. Mouth opening and tongue protrusion. Swallowing and breathing movements appear.
15	Thumb put in mouth and sucked.
16	Limb movements now well coordinated. Hands "explore" uterine and placental surfaces.
18	Exploration extended to parts of body (head, trunk, and legs).
20	More discrete movements, involving individual fingers, eyelids, feet.

[a]Based on Ianniruberto & Tajani, 1981. Another particularly detailed study is that by de Vries and her collaborators (de Vries, 1987; de Vries, Visser, & Prechtl, 1982).

produce inductive signals; as the oocyte has not yet been fertilized, these mRNAs are entirely maternal in origin.

Following invasion of the oocyte by a sperm, a dorsoventral polarity is imposed on the egg, in addition to the animal-vegetal one already present. The fertilized egg now divides repeatedly to form the multicellular *blastula*; the cells in the animal pole are destined to become *ectoderm*, and those in the vegetal pole *endoderm* (Figure 5.1).

In the equatorial (marginal) zone of the blastula, the vegetal and animal cell masses are in contact, and inductive signals now pass from the vegetal mass, instructing the adjacent "animal" cells to develop into *mesoderm* (Nieuwkoop, 1973). The internal cavity in the blastula, the *blastocoel*, prevents the inductive signals from reaching the main part of the animal cap, which, as already stated, will become ectoderm (Figure 5.1).

Peptide Growth Factors Provide Inductive Signals

The inductive signals are carried by peptide growth factors (PGFs); one of these, *activin A* (MW, 24 kD), was first identified in the medium bathing a *Xenopus*

cell line (Smith, Price, Van Nimmen, & Huylebroeck, 1990). Activin A appears to be the most important of the inducing peptide growth factors in amphibians, but its effects in mammalian development are less striking (Matzuk et al., 1995). Other inducers, possibly acting later and in a cascade, include *fibroblast growth factor* (FGF) and *transforming growth factor-β* (TGF-β).

The next major event in the development of the embryo is *gastrulation*, during which the blastula is invaginated by the expansion of the dorsal and ventral mesoderm so as to form a *gastrula* (Figure 5.1; Ruiz i Altaba & Melton, 1990). As the dorsal mesoderm comes to lie beneath the dorsal ectoderm, it induces the ectoderm to form the central nervous system (Figure 5.2; Gurdon, Mohun, Sharpe, & Taylor, 1989). The anteroposterior axis of the embryo is related to the extent to which the dorsal mesoderm has migrated under the ectoderm. The cells that have moved farthest will induce the brain and eyes, while the posterior structures will be induced by the mesodermal cells that have traveled least. It now appears that the peptide growth factors, *activins A* and *B*, in addition to inducing mesoderm as previously described, are responsible for establishing the anteroposterior axis in the embryo. For example, if activin B mRNA is injected into a *Xenopus* embryo, it causes a second body axis to form (Thomsen et al., 1990). The inducers appear to travel among the tissues of the embryo by passive diffusion and, within a few hours, can exert effects on many cells distant from their sites of origin (Gurdon, Harger, Mitchell, & Lemaire, 1994).

Step 2: A Portion of the Mesoderm Forms Somites and Then Develops Into Skeletal Muscle

The cells in the equatorial zone of the blastula, which have been instructed to become mesoderm, are influenced in their further development by their positions in the dorsoventral axis. Following gastrulation, the dorsal cells ultimately give rise to notochord and skeletal muscle, while the ventral cells form blood and mesenchyme (Figure 5.3). That part of the dorsal mesoderm which is responsible for muscle formation divides, under the influence of activin B and the homeobox genes (see p. 66), into segmental blocks of tissue called *somites* (Figure 5.4). It is now known that *all* the skeletal muscle in the body, whether in the trunk or in the limbs, is derived from the somites (for review, see Kenny-Mobbs, 1985).

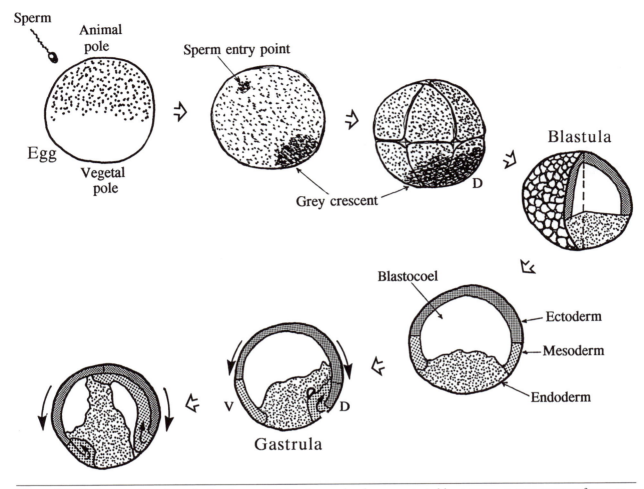

Figure 5.1 Early steps in embryonic development. After the egg has been penetrated by a sperm, a grey crescent forms on the opposite side and becomes the dorsal side of the embryo. In the blastula, mesoderm is formed by inductive signals passing from vegetal to animal cells, and then the mesoderm invaginates the embryo (gastrulation). The ectodermal cells spread from the animal pole to the vegetal pole, as indicated by the arrows in the last two drawings of the series. The formation of the nervous system is initiated by genetic instructions from the migrated mesoderm to the overlying ectoderm. For convenience, the drawing does not indicate the gradual increase in the size of the embryo during development. Adapted from Ruiz i Altaba and Melton (1990, p. 57).

The somitic mesoderm is responsible for certain other tissues beside muscle, however; these are the vertebral column and ribs, and the deeper part of the skin (dermis). Nonsomitic mesoderm gives rise to the notochord and to the heart and blood vessels (Figure 5.4). The decision of somitic mesoderm to form muscle, rather than one of the other tissues, is made by the expression of one or more *myogenic master regulatory genes* in the mesodermal cells (for review, see Olson, 1990). These genes code for the following regulatory factors: *Myo D*, *myogenin*, *myf-5*, and *myf-6* (MRF 4). The similarities in the amino acid sequences among these polypeptides are shown in Figure 5.5. Each of these factors, once expressed, combines with myogenic genes in the nuclei of the mesodermal cells. Of the four genes, that for myogenin appears to be the most important in that very little skeletal muscle forms in experimental embryos in which this gene has been disrupted (Hasty et al., 1993; Nabeshima et al., 1993). However, the fact that the other myogenic genes appear to be less indispensable (Braun, Rudnicki, Arnold, & Jaenisch, 1992; Rudnicki, Braun, Hinuma, & Jaenisch, 1992) does not exclude the possibility that they have important roles to play in muscle formation. Indeed, it is now thought that the genes act in a cascade, with *Myo D* and *myf-5* regulating the expression of *myogenin* (Emerson, 1993), and *myf-6* (MRF 4) being involved at a later stage. So powerful is this family of regulatory factors that they have the ability to cause differentiated cells of other types (fibroblasts, adipocytes, liver and melanoma cells) to switch partially or completely to muscle development (Weintraub et al., 1989).

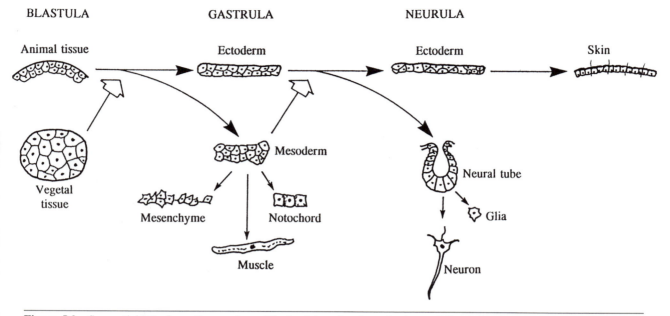

Figure 5.2   Sequential transformations and possible destinies of embryonic tissues following inductive signals (open arrows). The ectoderm is also responsible for forming the lens of the eye, under the influence of part of the neural tube. Adapted from Gurdon, Mohun, Sharpe, and Taylor (1989).

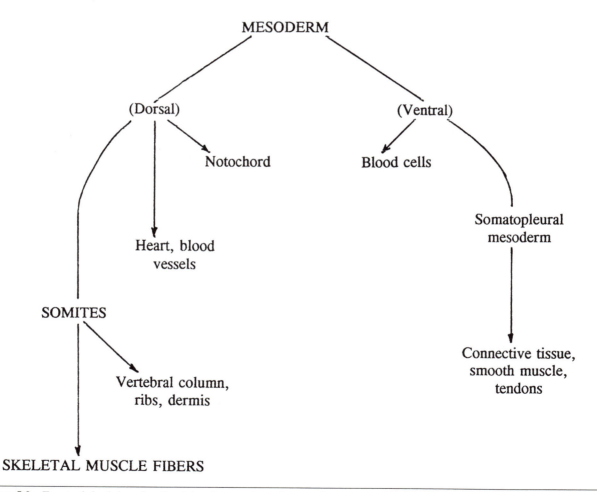

Figure 5.3   Eventual destinies of cells arising from embryonic mesoderm.

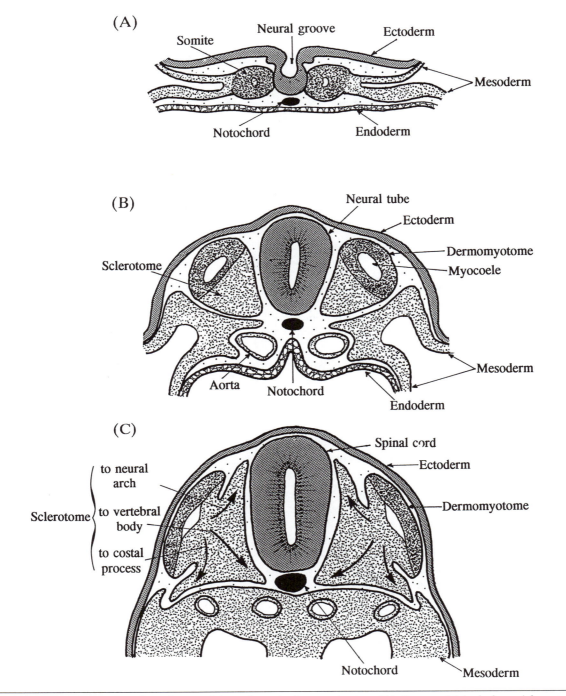

Figure 5.4 Transverse sections through part of the embryo; in the neurula stage (A), the neural tube is formed from ectoderm. Subsequently (B, C), the vertebral column develops from the sclerotomal parts of the somites, and skeletal muscle is seen to originate in the dermomyotomal parts.

Step 3: The Myogenic (Muscle-Forming) Cells Proliferate and Then Differentiate

The formation of a single human muscle requires that millions of muscle cells must be created before joining together into muscle fibers. This enormous cellular proliferation starts in the somites, where the mesodermal cells multiply under the influence of activins A and/or B and the other peptide growth factors (Figure 5.6). The proliferation continues among the *myoblasts*, which are the earliest muscle-forming cells to appear, following the expression of the master regulatory muscle genes in a proportion of the mesodermal cells. Whether an

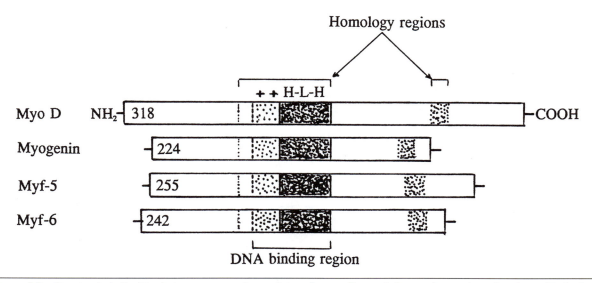

Figure 5.5 Structural similarities between myogenic regulatory factors. For each factor, the number of amino acids in the polypeptide is given at the left. The two regions of homology are shown at the top of the figure; the larger region contains a basic sequence (++) and a helix-loop-helix (HLH) arrangement. Only part of the larger homology region is necessary for the myogenic action of each polypeptide; this is the DNA-binding segment shown at the bottom. Adapted from Olson (1990).

individual myoblast will continue replicating or will start to differentiate further depends, on the one hand, on the local concentrations of the peptide growth factors expressed by the inducing genes (activins, FGF, TGF-β; see p. 58) and by the "immediate-early" gene family (*fos*, *myc*, and *jun*). On the other hand, a myoblast will come out of the proliferative cycle and will continue its differentiation into a muscle cell if there is a relatively high concentration of the proteins expressed by Myo D and the other myogenic master regulatory genes. It is likely that the ability of the growth factors to suppress differentiation is partly mediated by a protein called *Id* (inhibitor of DNA binding), which is found in many cell types and is known to be down-regulated during muscle differentiation (Benezra et al., 1990).

Myoblasts Resemble Mesodermal Cells and Have Simple Structures

When first formed, the myoblast is a spindle-shaped cell with a central nucleus and is indistinguishable in appearance from its mesodermal ancestors. Within the ovoid nucleus is a prominent nucleolus, indicating that the cell is actively synthesizing RNA. The cytoplasm of the cell is also packed with RNA in the form of ribosomes. Scattered mitochondria are present, and a small Golgi apparatus is evident; the endoplasmic reticulum is poorly developed. Electron microscopy also reveals thin (60 nm) filaments lying in the periphery of the cell under the plasma membrane. Some of these filaments are actin and are probably involved in movements of the myoblast and its appendages. Thick filaments are not seen in mammalian myoblasts except just

prior to cell fusion, and there is therefore no evidence of myofibrils in most of the cells.

From microelectrode studies of muscle developing in tissue culture, it appears that the resting membrane potential of the myoblast is low (around −20 mV; Fambrough, Hartzell, Rash, & Ritchie, 1974). Since the internal K^+ concentration is normal, the low potential presumably results from leakiness of the plasma membrane for Na^+ (see chapter 9); the fact that a basement membrane has not yet developed may also be relevant in the control of permeability. Functional acetylcholine receptors (AChRs) do not appear to be present in the plasma membranes of most myoblasts (Fambrough et al., 1974).

Structures Resembling Gap Junctions Form Between Fusing Myoblasts

The myoblasts divide repeatedly by mitosis and, both in tissue culture and in regenerating muscle, are seen to be mobile. After these *proliferative mitoses*, each cell undergoes a final division, the *quantal mitosis*, and then prepares for fusion into myotubes (Figure 5.7). In culture the cells now extend cytoplasmic processes that explore the surfaces of other myoblasts and of myotubes (see p. 65) lying nearby. Once the cells have recognized each other as myoblasts (or as myoblast and myotube), they line up together with their long axes parallel. It is probable that *fibronectin*, one of the extracellular glycoproteins involved in cell guidance, serves as a substrate upon which the myoblasts can move during these preparatory events. After alignment the myoblasts

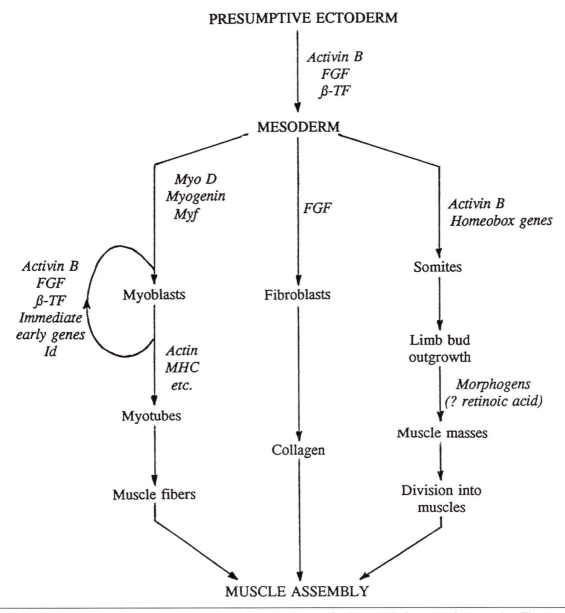

Figure 5.6 Transformation of presumptive ectoderm into mesoderm and subsequently into complete muscles. The genes and regulatory factors involved in the various steps are indicated by italics.

adhere to each other and eventually develop electron-dense structures under their respective plasmalemmae, which resemble the gap junctions found in other types of cell (Rash & Fambrough, 1973). At this stage, *connexin 43*, a gene that codes for one of the gap junction proteins in various tissues of the body, is expressed in the developing skeletal muscle.

Ca²⁺ Entry, Following Depolarization, Promotes Cell Fusion

The formation of new proteins is also suggested by studies in which antibodies have been raised against membrane preparations made from cells prior to fusion.

It is known that the entry of Ca^{2+} into the myoblasts is essential if fusion is to proceed, because the process is halted if these ions are removed from the culture medium (Shainberg, Yagil, & Yaffe, 1969), or if Ca^{2+} entry is blocked pharmacologically; conversely, precocious fusion occurs if a Ca^{2+} ionophore is included in the medium (David, See, & Higginbotham, 1981). In turn, the entry of Ca^{2+} is dependent on transient depolarization of the myoblast, which increases the permeability of the plasmalemma to Ca^{2+}.

Several naturally occurring agents can induce depolarization of myoblasts in culture and Entwistle and co-workers (Entwistle, Zalin, Bevan, & Warner, 1988; Entwistle, Zalin, Warner, & Bevan, 1988) suggest that,

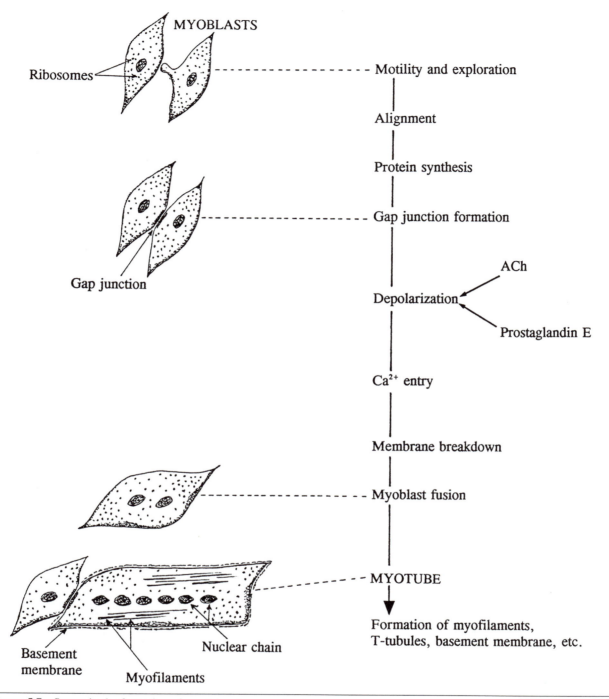

Figure 5.7   Stages in the formation of myotubes from myoblasts.

in vivo, the developing muscle may have several independent mechanisms available, so as to provide the fusion process with a large safety factor. One of these mechanisms is likely to be the binding of prostaglandin E with prostanoid receptors. In those myoblasts that have already formed AChRs, another mechanism may be the combination of the receptors with ACh released from the motor nerve terminals (Figure 5.7).

Apposed Cell Membranes Break Down During Myoblast Fusion

While the recognition and adhesion phases may last several hours, in culture at least, the fusion process itself is relatively brief, occupying 8 to 10 minutes (Bischoff & Holtzer, 1969) or even less (Rash & Fambrough, 1973). Experiments using cycloheximide show that, although new proteins may be synthesized during the adhesion phase, this is not a requirement for the fusion stage (Bischoff & Lowe, 1974). It is interesting that, in tissue culture, fusion may occur between myoblasts derived from different species and between cardiac and skeletal myoblasts, but not between myoblasts and other types of cell (e.g., from liver, kidney, or cartilage).

As part of the fusion process there is, at the surfaces of the cells, breakdown of the contiguous membranes and, possibly, the incorporation of membrane from closely applied cytoplasmic vesicles (Kalderon & Gilula, 1979). The nuclei come into apposition in the center of the newly formed myotube, and adjustments are made to the cytoskeleton so as to strengthen the structure of the syncytium. Additional details of myoblast fusion may be found in the reviews of Inestrosa (1982) and Wakelam (1985).

New Proteins, Including Those for Myofilaments, Are Synthesized After Fusion

Once a myotube has been formed, further myoblasts will add themselves to it, as described on p. 62. The nuclei of the new cells continue to line up with preexisting nuclei in the center of the myotube so as to form a chain; at the same time, they become somewhat larger and rounded. These nuclei will not partake in any further mitoses. Meanwhile, the cytoplasm of the myotube increases in amount and contains increasing numbers of thick and thin *filaments* (myosin and actin) that have been synthesized by polysomes. The myofilaments are grouped in bundles; these become thicker and form *myofibrils*. Soon the A- and I-bands of the myofibrils can be distinguished, indicating that the component myosin and actin filaments are now in register; the Z-lines appear subsequently.

The sequence in which the special proteins of muscle appear during embryogenesis has been determined by immunohistochemistry; the first protein is *desmin*, and then follows *titin*, muscle-specific *actin*, *myosin* heavy chains and, last, *nebulin* (Fürst, Osborn, & Weber, 1989). Some of these proteins can already be detected at the myoblast stage, though in small amounts. In the myotube, the myofibrils are formed in two parts; one of these is the Z-disc, with filaments attached at either side, and the other structural unit is the thick filaments of the A-band. Both structural units appear to form on bundles of microfilaments, which act as templates; once formed, the thick filaments and Z-disc units are linked together by the long titin filaments (see Figure 1.9; Fürst et al., 1989).

Aside from the appearance of myofibrils, other changes become evident in the myotube. *Mitochondria* become more numerous and are elongated with densely packed cristae. A rudimentary *transverse tubular system* (T-system) forms from invaginations of the plasma membrane; similarly, *sarcoplasmic reticulum* is derived from the outer nuclear envelope. A fuzziness on the surface of the myotube denotes the development of a *basement membrane*.

The Myotube Plasmalemma Becomes More Permeable to K⁺ and Develops ACh Sensitivity

Intracelluar microelectrode studies of cultured cells show that the properties of the *plasmalemma* are changing. The resting membrane potential rises as the myotube becomes larger and reaches maximum values of about -60 mV (Fischbach, Nameroff, & Nelson, 1971); this change is probably caused by a rise in the relative permeability of the membrane for K^+ as opposed to Na^+ (see chapter 9). The membrane is now capable of firing action potentials both spontaneously or following direct stimulation. In tissue culture, the myotubes can be seen to twitch spontaneously and synchronously. It is possible that the cells are electrically coupled to each other; the presence of *tight junctions* between adjacent myotubes, visible upon electron microscopy, would provide a possible mechanism.

By iontophoretic application of ACh or the use of ^{125}I-labeled α-bungarotoxin, it can be shown that the membrane now contains AChRs, some of which have been formed in the cytoplasm and then transported to the surface of the myotube. Although the whole surface of the cell is sensitive to ACh, there are certain "hot spots" corresponding to particularly high receptor densities. Fambrough et al. (1974) have estimated that in 1 hr roughly 100 AChRs may be formed per μm^2 of membrane. The myotubes also contain *cholinesterase*, and, like the ACh receptors, much of this appears to be manufactured in the vicinity of the nuclei and carried to the surface, where it is incorporated in the basement membrane.

Myotubes Enlarge and Continue Their Differentiation Into Muscle Fibers

As the myotube enlarges, it forms more myofilaments, and both the sarcoplasmic and T-tubular systems become better differentiated. At the same time, for most species, the nuclei move along the cytoskeleton from the center of the cell to take up new positions under the plasmalemma. The myotube has now completed its transition into a muscle fiber.

Step 4: The Body Plan Is Laid Down

While the somitic cells are pursuing their myogenic destinies, the shape of the embryo becomes increasingly well defined. The genes that control the shape of the embryo, and the appearance of the major body features, are called *homeotic*.

Homeotic Genes Regulate Early Embryonic Development

The homeotic genes are the master regulators of development and code for proteins which, in turn, bind to chromosomal DNA and thereby control the activities of other genes. The latter are those regulating the expression of yet other genes, including those involved in such cellular processes as signaling between cells, programmed cell movements, and the production of surface proteins necessary for cell-to-cell recognition. Among the homeotic genes are some that contain the same region of DNA; this region is the *homeobox*, and it codes for a 60-amino acid sequence that binds the remainder of the homeotic protein to the DNA strand.

It appears that the first expression of the homeotic genes is brought about by the same growth factors responsible for inducing mesoderm from the animal cap (i.e., activins A and B, FGF, TGF-β). Some of the homeotic genes are involved in the inducing action of mesoderm on ectoderm that results in the development of the head, body, and tail in the anteroposterior axis (see p. 58). Other homeotic genes control the segmentation of the embryo and the creation of the limb buds. If the embryo is to form properly, it is obvious that the homeotic genes must be switched on in the correct sequence; such an arrangement implies that each gene affects the transcription of the next one in the developmental program (Ruiz i Altaba & Melton, 1990).

Different Concentrations of Morphogens Provide Positional Cues

Other cues for gene expression depend on the *positional values* of the cells. Thus, throughout the embryo, each cell can be defined in terms of its relationship to three orthogonal axes: dorsoventral, anteroposterior, and proximal-distal. Several workers, including Wolpert (1969) and Crick (1970), have proposed that, for each axis, there is a chemical *morphogen*; the morphogen would diffuse from the site of secretion so as to establish a concentration gradient in that axis. A cell would recognize its position in the embryo by sensing the strength of the different morphogens and would then express its genome appropriately. In the limb bud, the morphogen that determines the *anteroposterior axis* (thumb to little finger in the human hand) is derived from a small region on the posterior margin, the *zone of polarizing activity*. Tissue grafting experiments indicate that the *proximaldistal axis* of the limb may be determined by the time spent by each cell in the *progress zone*; this region, situated at the base of the limb bud below the apical epidermal ridge, would send signals of different durations to the cells in the limb bud, depending on when they moved out into the elongating limb bud. One remarkable consequence of the positional cuing is that, if somites are transplanted from the thoracic region of a chick embryo to the brachial area, the muscles that form are those of the shoulder and upper arm rather than those of the chest wall (Butler, Cosmos, & Cauwenbergs, 1988; Chevallier, Kieny, & Mauger, 1977).

Although once regarded as fanciful, the concept of diffusible morphogens has gained ground. In the *Drosophila* (fruit fly) embryo, the proteins expressed by some of the homeotic genes can be detected by immunofluorescence (for review, see St. Johnston & Nüsslein-Volhard, 1992), and quite striking gradients exist for some of the proteins at particular steps of development (Figure 5.8). In the mammalian embryo, similar evidence has not been available; recently, however, a marked gradient of gene expression in the rostrocaudal axis has been demonstrated, presumably in response to the corresponding gradient of a morphogen (Donoghue, Morris-Valero, Johnson, Merlie, & Sanes, 1992; Grieshammer, Sassoon, & Rosenthal, 1992; Figure 5.8). In vertebrates, one possible morphogen is *retinoic acid*, a vitamin A derivative. This substance, normally found in different concentrations in the limb bud, has been shown experimentally to be associated with the establishment of all three body axes in the amphibian limb and of the anteroposterior axis in the chick limb (Summerbell & Maden, 1990). Another morphogen is *fibroblast growth factor* (FGF), which may act initially in conjunction with retinoic acid and then independently (Niswander, Jeffrey, Martin, & Tickle, 1994).

Step 5: Muscles Are Assembled

As far as the development of the limb musculature is concerned, the first step is the appearance of dorsal and ventral muscle masses in the limb bud. These blocks of tissue then split.

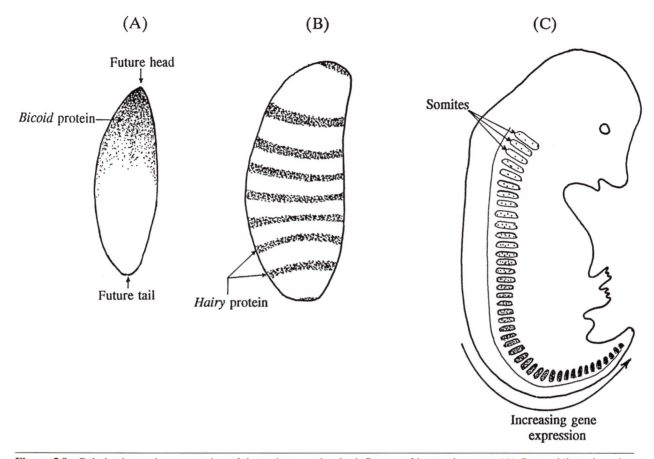

Figure 5.8 Polarization and segmentation of the embryo, under the influence of homeotic genes. (A) *Drosophila* embryo in which the protein product of the *bicoid* gene has been shown by immunofluorescence, shortly after fertilization. The high concentration of bicoid protein at one end of the embryo is critical for the later development of the head in this region. (B) Three hours after fertilization, another gene, *hairy*, is expressed in seven stripes across the embryo; these stripes serve as boundaries that eventually divide the embryo into 14 segments. Both *bicoid* and *hairy* are examples of homeotic genes. (C) A mouse embryo in which an artificially constructed gene (*myosin light chain 1-chloramphenicol acetyltransferase*) has been expressed in skeletal muscle within body segments. The expression is strongest at the tail end of the embryo and progressively weaker toward the head end. This difference in gene expression has been attributed to the presence of a morphogen having a concentration gradient in the rostrocaudal axis. The drawings in (A), (B), and (C) are based on the work of Nüsslein-Volhard (1991), Langeland and Carroll (1993), and Grieshammer et al. (1992) respectively and are not to the same scale.

Muscle Masses Split Repeatedly

After the first split, each of the newly formed muscle masses is cleaved again, and further subdivisions occur until individual muscles are established (Figure 5.9). Each muscle is then assembled within a connective tissue scaffolding that defines the tendons at each end of the muscle, as well as the fibrous sheets that envelop the belly and also group the muscle fibers into bundles (*fascicles*). Like the muscle fibers, the connective tissue is derived from mesoderm, though from a different region, the *somatopleure*. The somatopleural mesodermal cells destined to develop into connective tissues multiply under the influence of the peptide growth factors (including FGF) and the proteins transcribed from the immediate-early genes. As

yet, a master regulatory gene remains to be identified, but clearly some signal must commit the mesodermal cells into becoming fibroblasts.

Successive Populations of Muscle Fibers Form, Containing Different Myosins

Once formed, the fibroblasts replicate and migrate into appropriate positions within and around the developing muscle, presumably in response to positional cues. In the human fetus, the first (primary) myotubes are formed at 7 weeks of gestation; their proliferation and development have been investigated in the quadriceps muscle by Draeger, Weeds, and Fitzsimons (1987), who employed antibodies to detect the appearance of different types

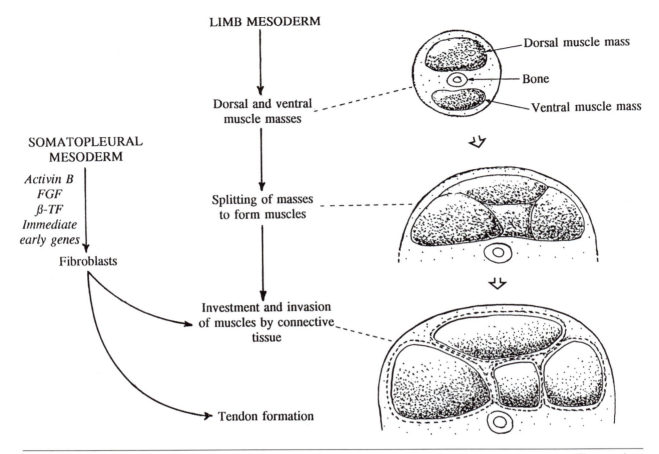

Figure 5.9   Splitting of dorsal and ventral muscle masses, formed from limb mesoderm, into separate muscles. The muscles are then surrounded and penetrated by connective tissue and attached to tendons, all derived from somatopleural mesoderm.

of myosin. According to Draeger et al., at 8 to 10 weeks gestation, the primary myotubes contain "slow" myosin (Figure 5.10). Because the motor axons have already penetrated the muscle masses (Toop, 1975), the expression of the slow myosin gene may be nerve-dependent. In other species, the initial fiber typing can occur even in the absence of motor innervation (Butler, Cosmos, & Brierley, 1982; Condon, Silberstein, Blau, & Thompson, 1990). By 14 to 15 weeks gestation, the myotubes become surrounded by smaller, secondary ones containing a "fast" myosin. Both the primary and secondary myotubes lie beneath the same basement membrane so as to form a cluster, separated from other clusters by connective tissue (Figure 5.10). By 16 to 17 weeks gestation, some of the secondary myotubes have moved out from under the basement membrane and have started to attract smaller, tertiary myotubes around them. The tertiary myotubes can be recognized not only by their smaller size but also by the presence of a different type of fast myosin. Because of their increased numbers, the myotubes are now more closely apposed and no longer give the impression of clusters. With further development, the secondary and tertiary myotubes enlarge, and by 31 to 34 weeks gestation, approximately half of the

secondary myotubes acquire slow myosin. Thus, the slow-twitch fibers appear to be derived from the primary and half of the secondary myotubes, while the fast-twitch fibers come from the remainder of the secondary myotubes and from the tertiary population (Figure 5.10).

Muscle development in the human fetus has also been carefully studied by Stickland (1981), who counted the numbers of myotubes and muscle fibers in sartorius muscles. From his data, it appears that there are large increases in the numbers of myotubes and muscle fibers until about 25 weeks gestational age, after which cell proliferation slows down and may even have ceased by the time of birth. Beyond 25 weeks gestational age, most of the increase in muscle cross-sectional area is due to enlargement of existing fibers. The interstitial connective tissue, initially accounting for 60% of the muscle volume, decreases to about 20% by 36 weeks gestation.

Postnatal Development of Muscle

At birth the muscle fibers are still relatively small, most no larger than 20 μm in diameter (Figure 5.11). In

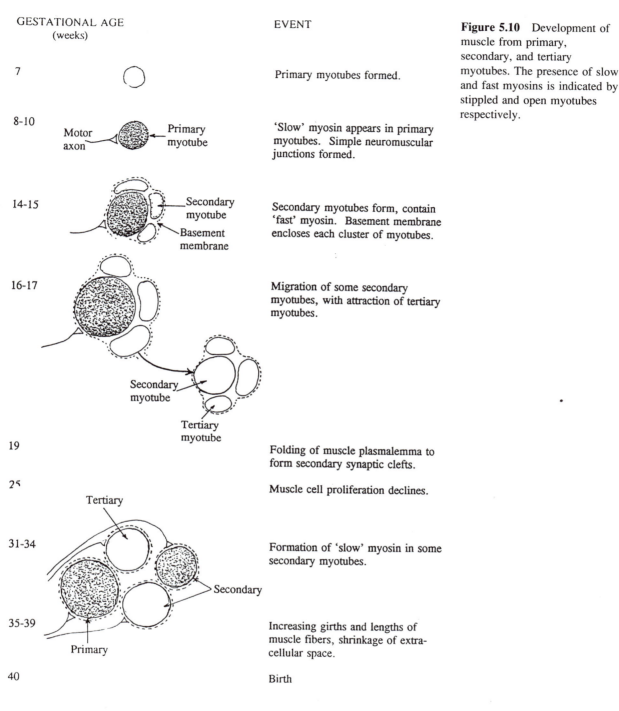

GESTATIONAL AGE
(weeks)

EVENT

7 — Primary myotubes formed.

8-10 — Motor axon / Primary myotube — 'Slow' myosin appears in primary myotubes. Simple neuromuscular junctions formed.

14-15 — Secondary myotube / Basement membrane — Secondary myotubes form, contain 'fast' myosin. Basement membrane encloses each cluster of myotubes.

16-17 — Secondary myotube / Tertiary myotube — Migration of some secondary myotubes, with attraction of tertiary myotubes.

19 — Folding of muscle plasmalemma to form secondary synaptic clefts.

25 — Muscle cell proliferation declines.

31-34 — Tertiary / Secondary — Formation of 'slow' myosin in some secondary myotubes.

35-39 — Primary — Increasing girths and lengths of muscle fibers, shrinkage of extra-cellular space.

40 — Birth

Figure 5.10 Development of muscle from primary, secondary, and tertiary myotubes. The presence of slow and fast myosins is indicated by stippled and open myotubes respectively.

human muscle, however, a 20-fold enlargement of the muscle belly must take place during childhood and puberty, assuming that the proportional contribution of muscle to the total body weight is roughly the same. In this section, we consider the various stimuli for muscle growth and the mechanisms involved.

New Sarcomeres Are Added to the Ends of the Growing Muscle Fibers

Obviously, much of the increase in muscle size will take place in an axial direction through the formation of new sarcomeres, and this process has been demonstrated at the ends of growing fibers by Williams and Goldspink (1971). The nature of the stimulus for this longitudinal expansion is not known for sure, though *growth hormone* is certainly involved (Figure 5.12). Thus, when a deficiency of growth hormone occurs, as in panhypopituitarism of childhood, there is a failure of all tissues to develop, including the somatic musculature. In experimental animals in which the pituitary glands have been ablated, it can be shown that muscle growth is restored by the injection of growth hormone

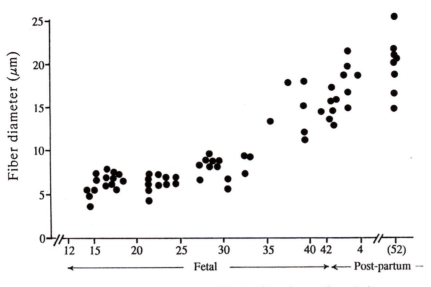

Figure 5.11 Increasing sizes of muscle fibers in human fetus and infant. The values are for fibers that have not yet differentiated into type I or type II. Adapted from Colling-Saltin (1978).

(Goldberg & Goodman, 1969). *Insulin* is another hormone necessary for muscle growth; at the level of the muscle fiber, it has been shown to stimulate amino acid transport, enhance protein synthesis, and inhibit protein degradation (Goldberg, 1968; Fulks, Li, & Goldberg, 1975).

Local mechanical factors are also of major importance in determining muscle length and in ensuring that the number of sarcomeres remains appropriate for the distance separating the attachments of the muscle—irrespective of whether the subject is a dwarf or a giant. Tabary, Tabary, Tardieu, Tardieu, and Goldspink (1972) have shown that if a cat soleus muscle is maintained in a shortened position, by fixing the leg in a plaster cast, there is a 40% loss of sarcomeres within 3 weeks. Conversely, prolonged extension results in the formation of new sarcomeres. The functional advantage of an adjustable fiber length is clear, for the muscle will always be able to work at, or below, the optimal region of the length-tension curve (see chapter 11); overstretching of the muscle and damage to the cross-bridges are less likely to occur.

In Males, Muscle Strength is Increased at Puberty, Presumably by Testosterone

At the same time that the fibers are growing longitudinally, their diameters are increasing so that, in man, mean values of about 50 μm eventually will be achieved (Brooke & Engel, 1969). It is this change in muscle growth that is responsible for the increased muscle strength during development.

McComas, Sica, and Petito (1973) measured isometric twitch tensions in the extensor hallucis brevis muscle in boys of different ages and found that muscle strength increased in two phases (Figure 5.13). The first phase lasted until puberty, and during this time there was a gradual increase in strength, such that an approximately linear relationship could be demonstrated between maximum twitch tension and age. The second phase occurred at puberty and was more dramatic in that a doubling of contractile force took place within a comparatively short time, certainly no longer than 2 years. It is probable that this spurt in muscle strength is due to the direct action of *testosterone* on muscle fibers, for the serum levels of this hormone are raised at puberty, and it is known that testosterone has a direct anabolic effect on muscle fibers. This anabolic effect has been demonstrated by experiments on the levator ani muscle of the rat, which is particularly hormone dependent. If an animal is castrated, marked atrophy of this muscle occurs; the atrophy can then be reversed by the administration of testosterone (Wainman & Shipounoff, 1941). In man, synthetic steroid substances closely related to testosterone, the *anabolic steroids*, have been found to increase muscle bulk and strength and have been used therapeutically. As recent Olympic Games have shown, anabolic steroids have also been taken by athletes anxious to improve their performance not only in such "heavy" field sports as shot-putting and discus-throwing, but also in sprinting; indeed the fastest time ever recorded for the 100 m, 9.78 s at the Seoul Games, was disallowed for just this reason.

As Muscles Enlarge, Myofibrils Thicken and Split

According to MacCallum (1898) and Stickland (1981), the number of muscle fibers does not increase in a human

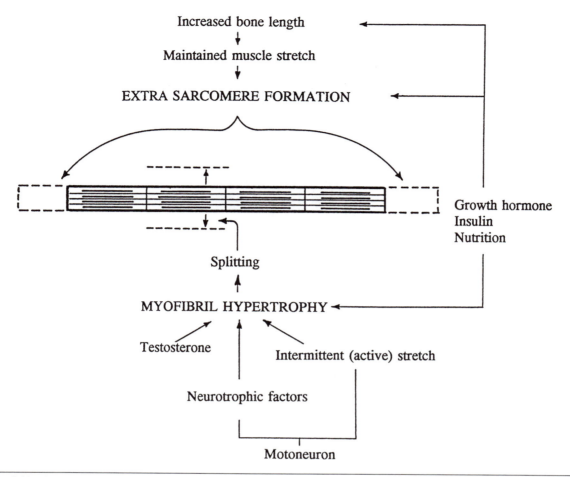

Figure 5.12   Summary of factors stimulating muscle fiber enlargement.

muscle (sartorius) after birth. If this is true of all human muscles, then the increased strength at puberty can only have resulted from the synthesis of new myofibrils in existing muscle fibers. This last process has been studied in mice during the vigorous postnatal growth phase by Goldspink (1970). He found that in some fibers the numbers of myofibrils increase by as much as 15 times. On the basis of electron microscopical observations on individual myofibrils, Goldspink suggested that the myofibrils first enlarge and then split in two, with the rupture commencing in the Z-discs. Myofibril enlargement and splitting also occurs during work hypertrophy and in recovery from the atrophy induced by starvation; the cycle may be completed in the remarkably short time of 1.5 to 2 days (Goldspink, 1965).

Stretching of Sarcomeres Promotes Muscle Fiber Growth

From observations on body-building courses and weight-lifting exercises in man (MacDougall, 1986), it is clear that muscle contraction is a far more potent stimulus for muscle fiber hypertrophy if it is eccentric or isometric rather than concentric (see chapter 11).

Since the action potential is common to both types of contraction, the difference must be related to the degree of muscle stretch. In concentric contractions, all the sarcomeres are allowed to shorten; in an isometric contraction, however, some sarcomeres will shorten, but others must be stretched, because the two ends of the muscle remain almost the same distance apart. In an eccentric contraction, the stretching of the sarcomeres is even greater, since the ends of the muscle are separated further. Hence stretch has two anabolic effects on muscle. Maintained stretch, as in the plaster cast experiments by Tabary and colleagues (1972), induces the formation of new sarcomeres, including fresh myofilaments, at the extremities of the fiber. Severe intermittent stretch, as in forceful isometric or eccentric contractions, results in the synthesis of new myofilaments around existing myofibrils, but there is no increase in sarcomeres.

Since protein is being formed during myofibrillar enlargement, one might expect some change in the number or disposition of the ribosomes to become evident. Galavazi and Szirmai (1971) were able to demonstrate an increase in intermyofibrillar polyribosomes in electron micrographs of levator ani muscles from castrated rats treated with testosterone. Surprisingly, the increase

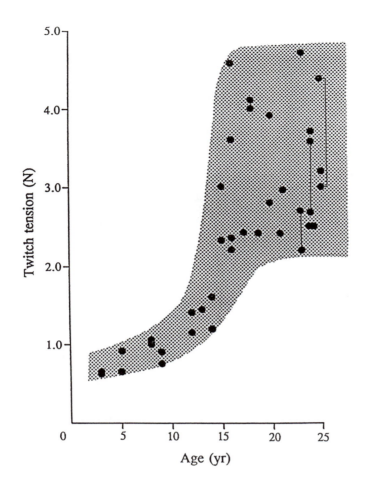

Figure 5.13   Maximal isometric twitch tensions of extensor hallucis brevis muscles in male subjects of different ages. Bilateral observations have been linked together. Note the marked increase in tension during puberty (usually 14-16 yr). Data from McComas, Sica, and Petito (1973).

in myofilaments occurs rather earlier than that of the polyribosomes. That the ribosomes in such animals are being driven by myonuclear DNA is suggested by the rise in DNA, increased incorporation of [³H]-thymidine, and proliferation of nuclei within the fibers.

Splitting Improves the Nutrition of the Myofibrils

It is of interest to consider what the advantage might be of myofibrillar splitting as opposed to having thicker myofibrils, because the contractile force should be no different in the two situations. One possibility is that the biochemical support systems for the myofibrils, operating through the mitochondria and sarcoplasmic reticulum, may be relatively ineffective in nourishing large cylinders of contractile material and that the myofibrils have an optimal surface:volume ratio. Figure 5.12 summarizes the factors that are known to influence muscle growth; they are both hormonal (growth hormone, insulin, and testosterone) and mechanical (muscle stretch). The ability to produce muscle atrophy by either denervation or starvation at any age indicates that an intact nerve supply and adequate nutrition must be considered as additional factors necessary for normal growth.

Twitch Speeds of Human Muscle Fibers Alter During Infancy

Isometric twitch studies enable the speed of muscle contraction and relaxation to be measured. In newborn kittens, all muscles have *slow* twitches, but in those destined to become *fast*, a speeding-up occurs during the next few weeks under the controlling influence of the motoneuron (Buller, Eccles, & Eccles 1960a). In the human triceps surae however, the contraction time is rather short during the neonatal period, but by 3 years it has reached its adult value, having increased by 50% (Gatev, Stamatova, & Angelova, 1977). In the human extensor hallucis brevis muscle also, the contraction time has entered the adult range by 3 years, though the relaxation phase remains slow for a rather longer time (McComas, Sica, & Petito, 1973). Recent studies by Elder and Kakulas (1993) show that 6 to 12 months is a critical period in the development of leg muscles, as reflected in the twitch durations (Figure 5.14); this finding suggests that the support of body weight may be an important controlling factor.

Finally, the electrical properties of the muscle fiber membranes also change during growth. In the 1-week-old mouse, the resting potentials of the fibers are still well

(A)

(B)

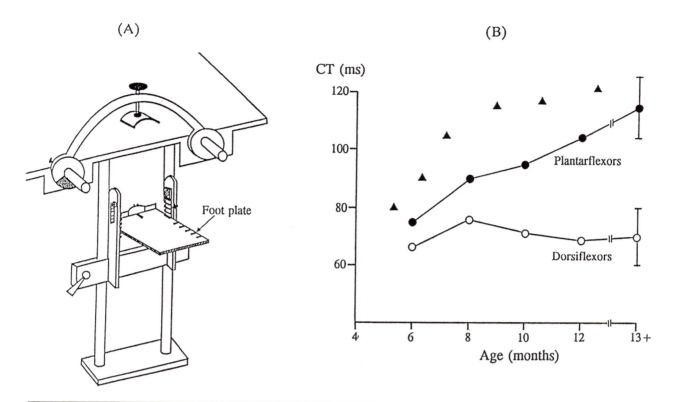

Figure 5.14 (A) Apparatus for recording isometric twitches in ankle dorsiflexor and plantarflexor muscles of infants. With the child lying supine, the knee is flexed over the end of the examining table, and the foot is strapped to the horizontal plate below. (B) Mean contraction times (CT) of dorsiflexor and plantarflexor muscles in 35 infants; vertical bars around the points for the eldest infants indicate standard deviations. The triangles show the successive values obtained for the plantarflexor muscles of a single infant. From "Histochemical and contractile property changes during human muscle development," by G.C.B. Elder and B.A. Kakulas, 1993, *Muscle and Nerve*, **16**, p. 1259. Copyright 1993 by John Wiley & Sons, Inc. Adapted with permission.

below those of the adult. Subsequently, the potentials rise in both "fast" and "slow" muscles, with the former achieving rather higher values (Harris & Luff, 1970).

APPLIED PHYSIOLOGY

In this section, we will discover that muscles may sometimes fail to develop normally in the embryo and may occasionally be missing altogether. Although the mechanisms involved are not fully understood, the eventual explanations are likely to involve some of the factors that have just been considered.

Muscles May Be Missing at Birth

Very occasionally, otherwise normal infants are born lacking a muscle or whole muscle group; the pectoralis major, with or without the other pectoral muscles, is probably the most frequent example, and in such cases the deficiency is nearly always unilateral. It is likely that

the abnormality results from the failure of the pectoral muscle mass to form as the cervical somites differentiate; in some cases there appears to be a genetic basis.

In Thalidomide Babies and Wingless Chickens, the Limb Buds Fail to Form Properly

In the 1950s, *thalidomide* was a commonly prescribed sleeping drug and one that apparently had few side-effects. It came as a surprise, therefore, when McBride (1961) raised the possibility that thalidomide, taken during early pregnancy, was responsible for serious skeletal abnormalities in the fetus. The most striking of these abnormalities was the failure of the arms and/or legs to develop fully (*amelia*), so that an infant could be born with no more than stumps to which the fingers or toes were attached. X-rays of the arms showed that the radius was aplastic; this observation, and the fact that the thumb was poorly developed or absent, suggested that segmental defects had occurred in the embryo. In turn, the segmental lesions were attributed to lesions of the neural crest or sensory neurons; either type of lesion could have deprived the limb bud of neurotropic signals necessary for its growth and differentiation (McCredie, 1975; McBride, 1978). In key experiments, however, Strecker and Stephens (1983) inserted tantalum foil between the neural tube and somites of chick embryos, so as to block nerve outgrowth, and found that the wings still formed normally. At present, then, the neural explanation for thalidomide appears unconvincing; the possiblity that thalidomide interferes with DNA transcription (Koch, 1990) does not remove the need to know how the embryonic tissues are subsequently affected.

In contrast to thalidomide embryopathy, the pathogenesis of the limb defects in genetically amelic chickens is better understood. *Limbless* is a mutant autosomal recessive gene that causes complete absence of the wings and legs (Prahlad, Skala, Jones, & Briles, 1979). In this condition, the exchange of tissue grafts between normal and mutant chick embryos has established that the primary defect is a failure of the *apical epidermal ridge* to form in the ectoderm; hence the signals necessary for the normal growth of the limb buds are missing (Carrington & Fallon, 1988).

More discrete lesions of the apical epidermal ridge can be produced by feeding *retinoic acid* to pregnant mice. In the embryo, the ridge cells undergo excessive cell death and cause various types of digital abnormality to appear—missing toes, split toes, or even extra toes (Sulik & Dehart, 1988). These experimental findings have considerable clinical significance, because retinoic acid has been widely used in the treatment of recalcitrant acne.

In Centronuclear Myopathy, the Myotubes Do Not Mature

The congenital myopathies are a group of disorders which can cause variable weakness in infancy and are due to primary abnormalities of the muscle fibers. Among these disorders is *centronuclear myopathy*, in which the muscles are composed of small, centrally nucleated fibers (Figure 5.15). All the evidence suggests that the latter are myotubes that have failed to continue in their development, possibly because of a defect in their innervation (Elder, Dean, McComas, Paes, & De Sa, 1983). Curiously, the muscle fibers inside the muscle spindles are unaffected and consequently remain larger than the extrafusal fibers.

The study of muscle development has been completed, and our attention must now turn to the formation of the motoneurons and of their subsequent connections to the muscle fibers.

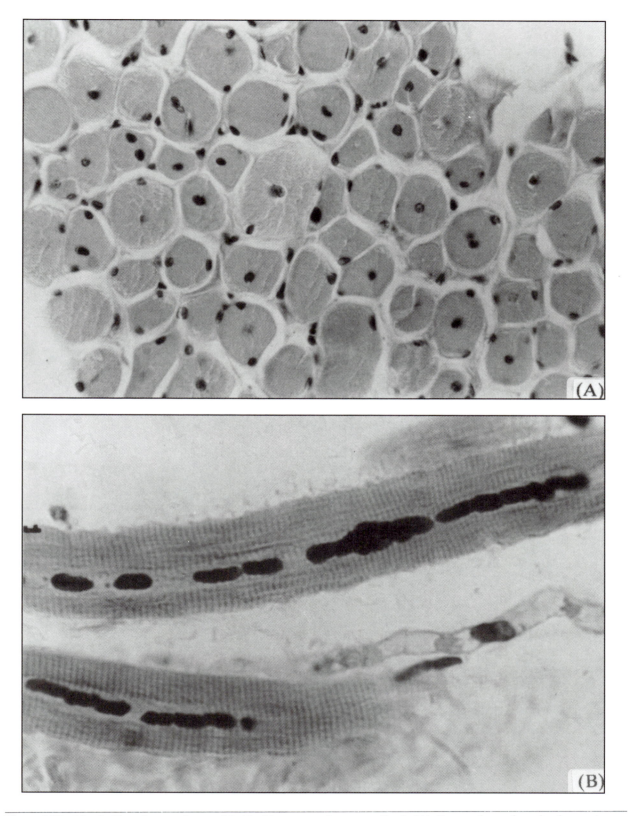

Figure 5.15 Muscle fibers from an 8-year-old girl with centronuclear myopathy. In (A), the muscle specimen has been cut transversely; almost every fiber has a central nucleus, and the muscle fibers are much smaller than normal. (420x). In (B), two muscle fibers have been sectioned longitudinally to show rows of nuclei (nuclear chains) in the centers of the fibers. (583x). From *The Congenital Myopathies* (p. 1548), by B.Q. Banker, 1986, New York: McGraw-Hill, Inc. Copyright 1986 by McGraw-Hill, Inc. Reprinted with permission.

Chapter 6

Development of Muscle Innervation

As with muscle formation, the development of the motor innervation in the embryo can be visualized as occurring in stages:

1. Passage of inductive signals from mesoderm to ectoderm
2. Formation of the neural tube
3. Neuronal proliferation and migration
4. Axonal outgrowth
5. Formation of the neuromuscular junction
6. Synapse elimination and motoneuron death

Step 1: Inductive Signals Pass From Mesoderm to Ectoderm

The motoneurons, like all mammalian cells, are ultimately derived from embryonic ectoderm. It will be recalled that ectoderm is itself descended from the cells making up the animal cap of the blastula (see Figure 5.1, chapter 5).

The Nervous System Arises from Ectoderm, Under the Influence of Mesoderm

The transition of part of the ectoderm into neural tissue is the result of a quite remarkable inductive mechanism,

first demonstrated by Spemann and Mangold (1924). In a classic experiment performed on a salamander, they removed the dorsal lip of the cleft in the blastula (*blastopore*) through which invagination of the mesoderm normally commences (Figure 6.1). This tissue, when transplanted to a different site in a second embryo, resulted in the formation of a second primitive nervous system. Subsequent experiments, using pigmented and nonpigmented embryos as donor and host embryos respectively, established that the nervous system was the consequence of a powerful inductive action of the mesoderm in the dorsal lip of the blastopore upon the ectoderm.

Spemann and Mangold referred to the dorsal lip as the "organizer," and in the light of recent advances in molecular biology, it is likely that this tissue releases one or more peptide growth factors that cause the epidermal genes to be turned off, and the neural genes turned on, in the ectodermal cells (Figure 6.2). Thus, during gastrulation, the gene for the *neural-cell adhesion molecule* (N-CAM) is expressed and is restricted to the neural plate (Jacobsen & Rutishauser, 1986); other neural genes activated are those that code for β-tubulin, vesicle-associated membrane protein (VAMP), and a nervous system-specific β-unit of Na$^+$, K$^+$-ATPase (for review,

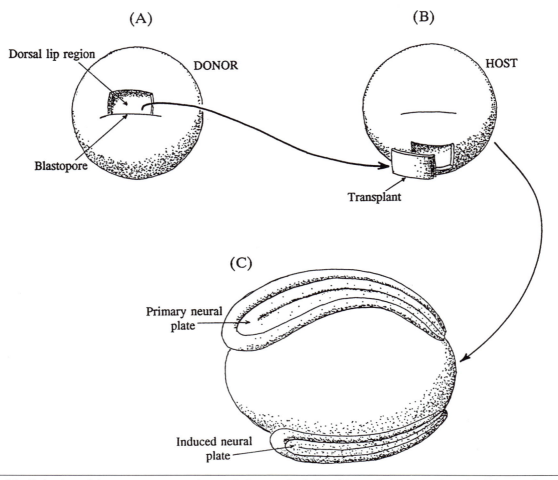

Figure 6.1   Induction of the nervous system. A neural plate can be induced in a salamander embryo by (A) removing a piece of tissue from the upper lip of the blastopore and (B) grafting it onto the ventral aspect of a second embryo. (C) A second induced plate then forms at the site of the graft. This famous experiment by Spemann and Mangold (1924) provided evidence for some kind of "organizer" in the mesoderm. Based on Hamburger (1947).

see Good, Richter, & David, 1990). By analogy with muscle formation, it is likely that the transcription of those "constitutive" neural genes follows the switching-on of a master neural regulatory gene.

Step 2: The Neural Tube Forms From Thickening and Invagination of the Dorsal Ectoderm

The first visible change as a result of the inductive process is the formation of an anteroposterior *neural groove* in the ectoderm. As this groove becomes more pronounced, the ectoderm lying on either side of it thickens so as to form two *neural plates* (Figure 6.3A). Because of these prominent changes in the dorsal surface, the embryo is now referred to as a *neurula*. On each side, the neural plate develops a fold of tissue

along its lateral margin; starting anteriorly, the two neural folds meet in the midline so as to form a cylinder of cells, the *neural tube* (Figure 6.3C). It is the neural tube that will ultimately form the brain and spinal cord. The peripheral nervous system arises from a part of the neural plate which is not included in the neural tube but comes to lie dorsally and laterally to it; this is the *neural crest*. In addition to forming the spinal and autonomic ganglia, the neural crest is the origin of the Schwann cells in the peripheral nerves; it also gives rise to a variety of nonneural tissues.

Step 3: Nerve Cells Proliferate and Then Migrate

In this next phase, the developing nervous system rapidly increases the number of future neurons, and shortly after moves them to appropriate locations in the brainstem and spinal cord.

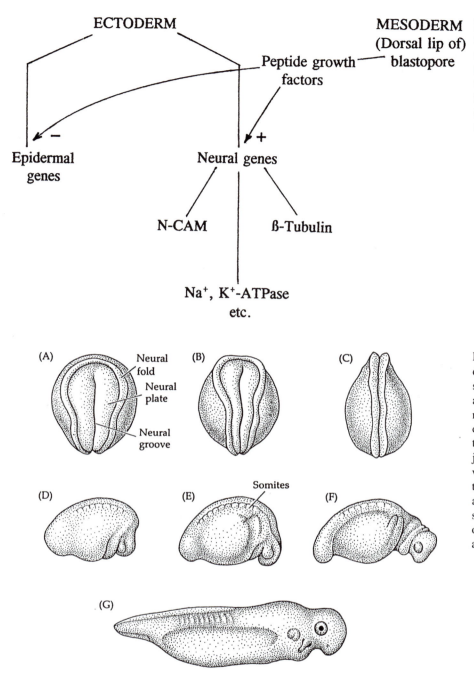

Figure 6.2 Genetic basis for induction of the nervous system. Mesodermal tissue releases peptide growth factors that influence ectodermal tissue; the epidermal genes are turned off, and the neural genes are switched on.

Figure 6.3 Early stages in the development of the nervous system (neurulation) in an amphibian (*Ambystoma*). The neural folds, seen in (A), enlarge and come closer together (B, C), subsequently joining in the midline. When viewed from the side (D to G), the embryo is seen to lengthen, and the head, somite, and tail structures become more obvious. Reprinted from Purves and Lichtman (1985).

Neuroblasts Multiply Around the Ventricular Cavity

The epithelium lining the cavity, or *ventricle*, inside the neural tube appears to have several layers. In fact, there is only a single thickness of cells, and the "pseudostratification" is due to the nuclei occupying different positions in relation to the ventricular cavity (Figure 6.4D). The proliferation of the primitive nerve cells, or *neuroblasts*, is undertaken by cells that have their nuclei close to the ventricle. The nuclei then move away from the cavity along cytoplasmic processes that initially extend into the neural tube. As the nuclei approach their final destinations, the cytoplasmic processes lose their terminal connections.

The First Motor Neuroblasts Are Pushed Outward by Later Ones

Initially, the entire neuroepithelium of the ventricular zone is actively proliferating, as deduced by labeling with [³H]-thymidine (Nornes & Das, 1974). The first cells to migrate, however, are the most ventral ones

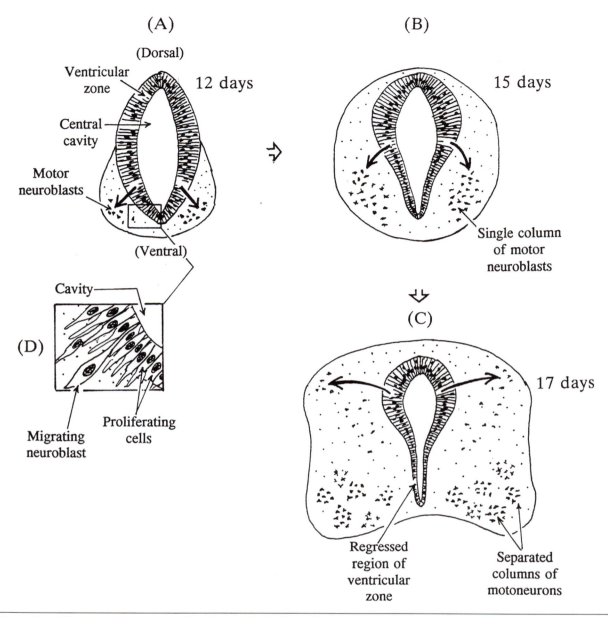

Figure 6.4   Migration of motor neuroblasts in the developing spinal cord of the fetal rat. (A) The cord is shown at 12 days of gestation, and the migrating neuroblasts (thick arrows) are seen leaving the ventral region of the ventricular zone. (B) At 15 days gestation, successively more dorsal populations of neuroblasts migrate outwards, pushing the earlier cells in front of them. The ventral region of the ventricular zone is now becoming thinner as it regresses. (C) At 17 days gestation, the last cells to leave the ventricular zone are those that will form the dorsal horn; by this time the developing cord is much wider, and the motor column has broken up into six smaller ones. Based on Nornes and Das (1974).

(thick arrows in Figure 6.4A) and are motor neuroblasts. These first cells are then pushed outward by later migrating cells, which include some that arise from a more dorsal region of the ventricular zone (thick arrows in Figure 6.4B). The motor neuroblasts come to form one continuous column in the rostrocaudal axis of the developing spinal cord. As the neuroblasts differentiate into motoneurons, and as their axons reach the muscle masses in the limb buds, the motor column begins to divide.

The migration of motor neuroblasts diminishes as the ventral region of the ventricular zone begins to shrink. The process of neuroblast migration, followed by regression of the corresponding region of the ventricular zone, proceeds in a dorsal direction. These later neuroblasts will form the intermediate grey region of the spinal cord and the different laminae of the dorsal horn. The last cells to leave the ventricular zone (thick arrows in Figure 6.4C) thread their way past the earlier arrivals and form the cap (substantia gelatinosa) on the dorsal horn.

In addition to the ventrodorsal gradient of progression, described previously, there is also a rostrocaudal one, such that neuroblast proliferation and migration occur rather earlier in the cervical region of the neural tube than in the lumbosacral one.

At the end of the migratory process, there are six main clusters of motoneurons for each of the limb buds. These clusters correspond to the dorsal and ventral locations of the muscle groups and also to the position of muscle groups in the proximo-distal axis of the limb bud. In general, the motoneurons that will supply dorsal muscles in the limb are situated laterally to those innervating the ventral muscles (Romanes, 1941, 1951).

Step 4: Axons Grow Out From the Spinal Cord Along the Extracellular Matrix

An axon starts to grow out from each motoneuron even while the cell is still in the process of migrating to its final position. The trajectories of the growing axons are probably determined by chemical guidance factors that either attract or repel them; such factors (*netrins*) have been identified for other types of spinal cord neuron (Serafini et al., 1994; Kennedy, Serafini, de la Torre, & Tessier-Lavigne, 1994). Outside the cord, the axon enters the limb bud at a time when the individual muscles have yet to form and when only the dorsal and ventral muscle masses can be distinguished. As they leave the embryonic spinal cord, the axons are grouped in bundles which, in the case of those destined for the limb buds, merge to form plexuses and are then redistributed into peripheral nerve trunks. If a segment of neural tube is transplanted from the thoracic to the brachial region, the outgrowth of axons forms a brachial plexus, indicating that the axons are affected, directly or indirectly, by their positions in the anteroposterior axis of the embryo (Butler, Cauwenbergs, & Cosmos, 1986). In the trunk and limb buds, the axons appear to follow paths that are marked out in the extracellular matrix of the trunk and limb buds by guidance factors, rather different from those in the spinal cord, such as *laminin* and *fibronectin*.

Axons Find Their Correct Muscle Destinations Reliably

In the limb buds, the peripheral nerve trunks give off main branches which enter the dorsal and ventral muscle masses and then break into smaller branches, corresponding to each of the main subdivisions of the muscle masses. The routing process that ensures that the axons from a particular motoneuron pool reach the correct muscle belly is remarkably accurate; it is also resilient in that it can compensate for different types of experimental perturbation by redirecting the axons (Lance-Jones & Landmesser, 1980a, 1980b). Naturally occurring examples of successful rerouting are found in those human subjects with anomalous branching of their peripheral nerves. Thus, axons for the short abductor and opponens muscles of the thumb normally arrive in the median nerve. In some people, however, the axons cross over to join the ulnar nerve in the forearm (Martin-Gruber anastomosis) but still find their proper targets after the ulnar nerve has entered the palm. Figure 6.5 shows axonal outgrowths in the zebrafish, in which there are only three motoneurons to innervate the muscles in each ipsilateral segment of the trunk. Each axon follows a path diverging from those of the other two and supplies a particular muscle mass.

Growth Cones Explore the Surroundings of the Growing Axons

When the motor axons first arrive in the belly of the developing muscle, they are still nonmyelinated since the Schwann cells, which are derived from the neural crest, follow later. At the tip of the axon is a specialized expansion, the *growth cone*, which consists of a leaflet of axoplasm from which fingers (*filopodia*) project. The interior of the growth cone is packed with actin filaments and various types of organelle, including cisternae, vesicles, and lysosomes (Figure 6.5E). Microscopic studies of axonal growth, either in tissue culture or in transparent regions of embryos, reveal that the growth cone has extraordinary mobility, extending and withdrawing, bending first in one direction and then in another, and all the while appearing to explore the surface of the tissue.

Fluorescence and differential interference microscopy, following the introduction of Ca^{2+}-sensitive dyes into the axoplasm, have shown that movements of the growth cone are associated with transient increases in Ca^{2+} concentration, which are presumably translated into sliding of the actin filaments. The first contact between a growth cone and a developing muscle fiber appears to be random, although it is always close to the entry of the nerve fibers into the muscle mass. Because the muscle fibers grow by additions to the two ends, the neuromuscular junctions remain in the approximate centers of the fibers as the latter lengthen.

These first axons are exploratory ones; they form a neuromuscular junction on each of the myotubes that has developed by fusion of the myoblasts. Subsequently, other axons arrive and establish neuromuscular junctions in the same regions of the developing fibers (Bennett & Pettigrew, 1974). Although, as already seen, the initial contacts between motor axons and developing

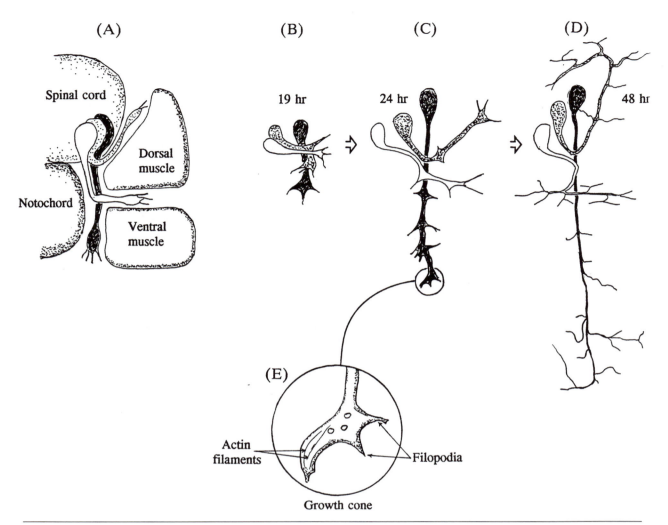

Figure 6.5   Axon outgrowth in the zebrafish. (A) is a cross section made in a 19-hour-old embryo and shows the three motoneurons responsible for innervating all the mucles in the trunk segment on that side. The axons have started to invade the spaces between the muscle masses. (B, C, and D) are side views of the same three motoneurons and illustrate the divergent paths, elongation, and branching of the respective axons. (E) is an enlargement of a growth cone from one of the axon endings. Adapted from Westerfield and Eisen (1988, pp. 20-21).

muscle fibers are randomly located, the same sites will be preferred in later life if axons regenerate or sprout after a period of muscle denervation (see chapter 17).

Step 5: The Axons Establish Connections With the Muscle Fibers

Once the nerve and muscle fibers have made their first contacts, synaptic transmission becomes established within 1 to 2 hr (Chow & Cohen, 1983). The reason for this rapid onset is that the exploratory growth cone already contains ACh and can release it spontaneously. The release of ACh increases after the first nerve-muscle

contact, because of signals from the developing muscle fiber to the nerve terminal (Xie & Poo, 1986).

Differentiation of the Growth Cone Results From Signals Passing From the Developing Muscle Fiber

One of the signals from the muscle fiber appears to be due to the adhesion between the fiber and the terminal; possible candidate molecules are *neural-cell adhesion molecule* (N-CAM) and *cadherin*. The formation of transient gap junctions (Allen & Warner, 1991; Fischbach, 1972) could provide a second mechanism. Last, the myotubes could secrete molecules such as arachidonic acid and growth factors (Hall & Sanes, 1993). Regardless of its nature, the effect of the molecular signal is to cause a rise in the Ca^{2+} concentration in the growth cone (Dan & Poo, 1992), an event that promotes the differentiation of the growth cone into a presynaptic terminal. The vesicle-associated proteins may play an accessory role.

The Motor Nerve Terminal Controls Differentiation of the Synaptic Region of the Muscle Fiber

Just as the muscle fiber supervises the conversion of the nerve ending from a growth cone into a presynaptic terminal, so the presynaptic terminal influences the differentiation of the underlying muscle fiber. One of the best understood changes is the clustering of AChRs in the muscle plasmalemma. The clustering is made possible by the release of *agrin*, which is synthesized by the motoneuron soma, transported to the axon terminals, and inserted into the basement membrane within the synaptic cleft (Magill-Solc & McMahan, 1990a). The agrin molecules combine with receptors in the muscle fiber plasmalemma and are then able to phosphorylate tyrosine in the β-subunits of the AChRs (Figure 3.5, chapter 3); the phosphorylation is evidently needed to tether the AChRs to the plasmalemma (Wallace, Qu, & Huganir, 1991). A molecule with a similar effect to agrin is *FGF*, which is also found in the synaptic basement membrane and is released by proteolysis.

Not only is the spatial distribution of the AChRs controlled by the nerve terminal, but the number of receptors is increased following enhanced transcription of AChR genes in the synaptic myonuclei. One of the mediators in this effect is *calcitonin gene-related peptide* (CGRP), which is released from the dense-core vesicles in the nerve terminal (Uchida et al., 1990). Another possible mediator is *AChR-inducing activity* (ARIA), which is present in the spinal cord though not yet shown to be released from the presynaptic endings.

At the same time that AChRs are accumulating at the new neuromuscular junctions, their numbers are decreasing elsewhere along the developing fibers, due to down-regulation of gene transcription in the nonsynaptic myonuclei. This reduction is due to electrical activity, in the form of action potentials, propagating along the fibers. ·This effect of electrical activity has been revealed by the ability of electrical stimulation to suppress the appearance of extrajunctional AChRs in denervated muscle fibers (Lømo & Rosenthal, 1972); conversely, different types of neuromuscular blockade in innervated fibers can induce extrajunctional AChRs.

Neuromuscular Junctions Appear in Human Fetal Muscle at 8 to 9 Weeks

In the human embryo, the neuromuscular junctions are first seen in the intercostal muscles at the age of 8.6

weeks (corresponding to a crown-rump length of 3.2 cm); their appearance in the tibialis anterior occurs rather later, at 10 weeks (4.3 cm; Juntunen & Teravainen, 1972). Initially, newly formed junctions have a primitive appearance, consisting of a simple (primary) *cleft* between the apposed nerve and locally thickened myotube membranes (Kelly & Zacks, 1969). At about 19 weeks of human fetal development, folding of the muscle membrane starts to take place, with the formation of the secondary clefts. The density of AChRs increases in the postsynaptic membrane, and this process is followed by the insertion of AChE into the basement membrane, usually one or more days after synaptic transmission has been established.

Synaptic Vesicles Increase in the Nerve Terminal

As the neuromuscular junction continues to develop, the myonuclei migrate into its vicinity and begin to heap up on each other, increasing the complexity of the *sole-plate*. Meanwhile, in the axon terminal, increasing numbers of synaptic vesicles are to be seen; initially very few of these (e.g., one to four) are available for release following a single impulse (Robbins & Yonezawa, 1971), and the resulting depolarization is too small to initiate an action potential. Within a relatively short time, however, the quantal release increases and successful transmission becomes established. Some of the vesicles in the miniature axon terminal are relatively large and have dense cores; these vesicles contain CGRP and other molecules that may have special trophic functions. As these changes are taking place at the neuromuscular junction, the motor axon becomes thicker and acquires a myelin sheath from the Schwann cells which have followed the axon outward from the neural crest.

Step 6: Redundant Synapses and Motoneurons Are Eliminated

Unlike the situation in the adult, there is a phase in embryonic development when several axons come to innervate the same muscle fiber. In the mouse and rat, the surplus innervation is present at birth and is relinquished during the next 2 weeks (Box 6.1). In the human embryo, however, the results of cholinesterase staining suggest that the withdrawal of the motor axon terminals takes place before birth, probably between 25 weeks of gestation and term (Toop, 1975).

These important events are not the only regressive changes to take place in the motor innervation of the

BOX 6.1   POLYNEURONAL INNERVATION

For many years, it had been accepted that each skeletal twitch muscle fiber was innervated by a single axon, and it was assumed that this arrangement was established as soon as embryonic muscles made their first contacts with exploratory axons. It was therefore something of a surprise when Paul Redfern, working in Thesleff's laboratory in Lund, Sweden, showed that the muscle fibers of newborn rat pups were multiply innervated (Redfern, 1970).

Redfern's crucial finding was that the sizes of end-plate potentials evoked in muscle fibers of the diaphragm depended on the intensity of the stimulus to the phrenic nerve, indicating that more than one axon must have supplied each neuromuscular junction (Figure 6.6). It later transpired that anatomists, using the light microscope to examine specimens of muscle in which the nerve fibers had been stained with silver, had observed multiple axons ending on muscle fibers of newborn mice many years earlier (cf. Purves & Lichtman, 1985). Redfern's important results were quickly confirmed, not only by recordings of end-plate potentials, but also by measurements of twitch and tetanic tensions. Other aspects of the multiple innervation were also described. For example, Brown, Jansen, and Van Essen (1976) reported that, at birth, all rat soleus muscle fibers were multiply innervated, and Dennis, Ziskind-Conhaim, and Harris (1981) demonstrated that an average of three axons were present at each neuromuscular junction. Dennis et

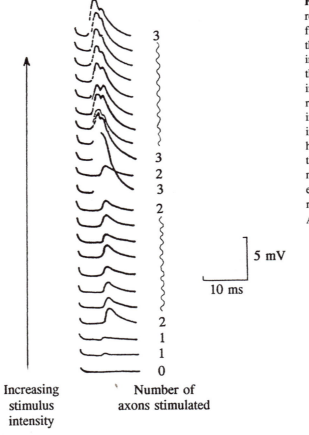

Figure 6.6   Microelectrode recordings from a single muscle fiber in a newborn rat, revealing the presence of multiple innervation. As the stimulus to the motor nerve is gradually increased, the end-plate responses can be seen to enlarge in three steps; this finding indicates that three axons must have been present, with different thresholds for excitation. The number of axons contributing to each response is given at the right.

Adapted from Redfern (1970).

5 mV

10 ms

Increasing
stimulus
intensity

Number of
axons stimulated

al. also measured the time taken for the surplus axons to be removed and found, in rat intercostal muscle fibers, only single motor axons remained by 13 days postpartum.

Further insights into synapse elimination have come from repeated inspection of the same, identified, neuromuscular junctions, using the vital stain 4-Di-2-ASP to visualize motor axons, and fluorescently labeled α-bungarotoxin to show AChRs. In sternomastoid muscles of newborn mice, AChRs begin to disappear before regressive changes are observed in the motor nerve terminals (Lichtman & Balice-Gordon, 1990; Balice-Gordon & Lichtman, 1994). At this time, there is a reduction in the sizes of the miniature end-plate potentials, which would be expected from the loss of receptors.

What is the mechanism for the withdrawal of the surplus axons? One clue is that the process is inhibited by paralyzing the muscle fibers (Benoit & Changeux, 1978) and is accelerated by chronic stimulation (O'Brien, Ostberg, & Vrbová, 1978); further, the most active nerve-muscle connections promote the disappearance of the least active ones (Balice-Gordon & Lichtman, 1994). Although the answer to the question is not available at present, several explanations have been advanced (Figure 6.7; see also Balice-Gordon & Lichtman, 1994). One of these is that the muscle fibers produce a limited supply of a trophic factor that can only be taken up by an active motor nerve terminal and is essential for the maintenance of that terminal (Figure 6.7A; Purves & Lichtman, 1980). Support for this view comes from experiments in which muscle fibers were crushed on both sides of their neuromuscular junctions. Although the junctions were not damaged, there was a disappearance of axon terminals as the muscle fibers degenerated, followed by a reappearance when the muscle fibers began to regenerate (Rich & Lichtman, 1989).

An alternative suggestion, by Changeux and Danchin (1976), is that impulses in muscle fibers convert AChRs from a labile to a stable state and hence would tend to preserve the synaptic connections to the most active motor axons (Figure 6.7B). Dan and Poo (1992) have provided evidence that, following transmission at one neuromuscular junction, there is a Ca^{2+}-mediated release of a toxin from the presynaptic ending (Figure 6.7C). A Ca^{2+} mechanism has also been proposed by Vrbova and Lowrie (1989). These authors postulate that K^+, released into the interstitial spaces by muscle fiber impulses, opens Ca^{2+} channels in the motor nerve terminals, allowing the entry of Ca^{2+}. The Ca^{2+} then activates proteases that, in turn, digest the protein cytoskeleton of the axon terminal, causing the terminal to withdraw (Figure 6.7D). This last suggestion is consistent with the observation of Vrbova and Lowrie that loss of terminals is prevented by the application of protease inhibitors in vitro, or by lowering extracellular $[Ca^{2+}]$, while the process is hastened by raising $[K^+]$.

There is one further aspect of synapse elimination which must also be accounted for in a satisfactory hypothesis, and this concerns the architecture of the motor units. When a single motor axon is stimulated repetitively, the muscle fibers belonging to that motor unit are depleted of glycogen and are then seen to be widely scattered in the muscle belly, overlapping with the territories of 20 to 30 other motor units (see chapter 12). It would appear that, during synapse elimination, there is a mechanism that discourages motor axons from innervating neighboring muscle fibers (Willison, 1978). Is extracellular K^+ the key? The rise in interstitial $[K^+]$ would clearly be high in a region of the muscle where there were many muscle fibers supplied by the same motor axon. In keeping with the hypothesis of Vrbova and Lowrie (previously discussed), this rise could promote

continued

POLYNEURONAL INNERVATION   *(continued)*

destruction of the majority of axon terminals until a stage was reached in which excitation of the same axon would produce changes in [K$^+$] tolerated by the remaining terminals.

The process of synapse elimination, first demonstrated in muscle fibers, has since proved to be a widespread, though not universal, phenomenon in the central and autonomic nervous systems, and it is possible that the mechanisms involved are similar to those at the neuromuscular junction (cf. Purves & Lichtman, 1985).

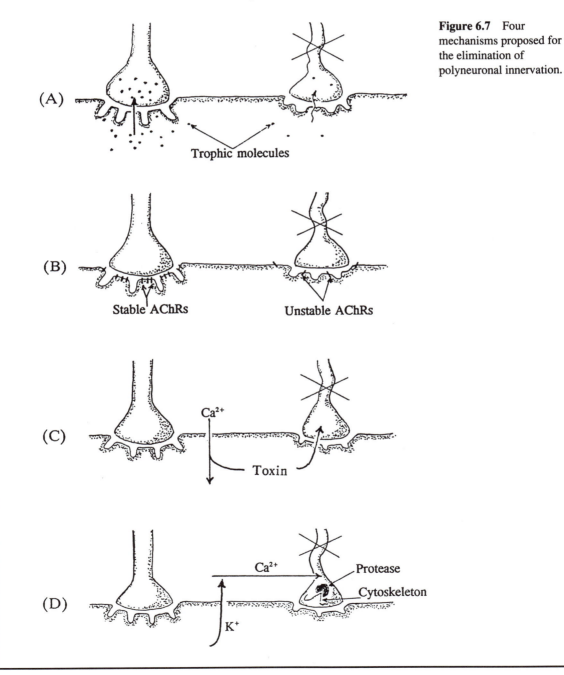

Figure 6.7 Four mechanisms proposed for the elimination of polyneuronal innervation.

(A)

Trophic molecules

(B)

Stable AChRs   Unstable AChRs

(C)

Ca^{2+}

Toxin

(D)

Ca^{2+} Protease

Cytoskeleton

K$^+$

embryo. Thus, well before the period of synapse elimination is over, there is a marked degeneration and disappearance of motoneurons in the spinal cord; this is an example of the neuronal cell death that is found widely throughout the central nervous system.

Surplus Motoneurons Die

The number of motoneurons in the embryonic spinal cord is influenced by the amount of muscle that has to be innervated. For example, if a limb bud is removed from a chick embryo, the number of motoneurons is reduced (Shorey, 1909). Conversely, the motoneuron population is increased if an additional limb bud is grafted onto the trunk of an amphibian embryo (Detwiler, 1920). In early studies, it was thought that the size of the "target," that is, the amount of muscle tissue, was determining the extent of neuronal *proliferation* in the spinal cord. However, Hamburger (1958) counted spinal motoneurons in chick embryos at different ages and found that approximately half of the cells died as part of the pattern of *normal* development (Figure 6.8). Later studies showed that this cell death was complete before the withdrawal of polyneuronal innervation from the muscle fibers commenced (Box 6.1).

The probable significance of motoneuron death is that an abundance of motor axons is necessary to ensure that a sufficient number reach the muscle. Once all the muscle fibers have become innervated, initially by several axons on each fiber (Box 6.1), many of the motoneurons become redundant. Indeed, studies of adult human muscle show that, despite cell death in the embryo, there is still an appreciable safety margin in muscle innervation. For example, muscle strength is not compromised in muscles that have been partially denervated, as a result of chronic disease or old nerve injury, if half or even three-quarters of the motoneurons are lost (cf. McComas, Sica, Campbell, & Upton, 1971).

The nature of the signal responsible for motoneuron death is presently unknown. One possibility is that the motoneurons that die are those that reach the "wrong" muscles, but this is not borne out by experiments in which axons are deliberately misrouted (Lance-Jones & Landmesser, 1980b). An alternative proposal, suggested by limb-bud grafting studies, is that motoneurons compete for trophic factors produced by the muscle fibers (see chapter 18). However, there are experimental situations in which motoneuron death is not affected by the amount of muscle available for innervation (discussed next). Moreover, motoneuron death, like synapse elimination, is strongly influenced by muscle activity. If activity is blocked by the application of curare or α-bungarotoxin, motoneuron death is abolished (Laing & Prestige, 1978); on the other hand, the rate and extent

of cell death are enhanced if chick embryo muscle is stimulated electrically (Oppenheim & Nunez, 1982).

A more subtle proposal, which reconciles these discordant results, is that muscle-derived trophic factors are indeed required by the embryonic motoneurons during activity; however, the limiting factor is not production of trophic factors by the muscle fibers, but rather the *access* to the factors by the motoneurons through their synaptic connections (Oppenheim, 1989). Thus, a motoneuron that had more synaptic connections than another would be at a competitive advantage for survival. Such an explanation also accounts for certain observations on rats that have been subjected to crushing of their sciatic nerves (Albani, Lowrie, & Vrbová, 1988; Krishnan, Lowrie, & Vrbová, 1985). Although parts of the muscles in these animals do not become reinnervated, there are no losses of motoneurons; there is, however, increased branching of the axons and polyneuronal innervation of the muscle fibers. The question remains as to whether such a scheme could account for the remarkable range in motor unit sizes of normal muscles, which may vary at least 100-fold (see chapter 12).

APPLIED PHYSIOLOGY

In this section, we return to the neuromuscular clinic to examine two conditions. In the first condition, the gross structure of the spinal cord is abnormal, while in the other, there is a selective deficiency of motoneurons. Although the second disorder might be expected to be the less serious of the two, the effects in an infant can be fatal within days.

In Spina Bifida, the Neural Tube Is Improperly Formed

Occasionally a baby is born with a soft bulge over the lower part of the spine; the walls of the swelling are formed by the spinal meninges, and the contents consist of cerebrospinal fluid. Since the laminae and spines are missing from the lumbar vertebrae, the fluid-filled sac or *meningocele* is able to distend the skin and may reach a considerable size. Exploration of the sac may reveal that the spinal cord is open, that is, the nervous tissue which would normally form the roof of the central canal and become the dorsal part of the spinal cord lies to each side (*myelocele*). In some cases the lumbosacral nerve roots enter the sac, which is then termed a *myelomeningocele*. More commonly the sac is absent, even though the spinal cord is malformed, and in these infants the neurological consequences are less severe. In patients with myelomeningoceles, however, there is usually total paralysis of muscles below the knee, loss of

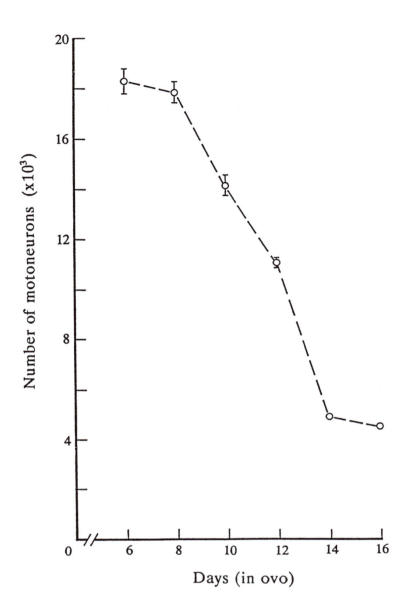

Figure 6.8 Natural motoneuron death in the brachial spinal cord of the chick embryo. The segments counted included spinal nerves 12 to 18, inclusive. (Courtesy of Dr. E. Cosmos and Dr. P. Cauwenbergs.)

sensation over the feet, and lack of bowel and bladder control.

Regardless of their extents, these congenital disorders are loosely referred to as *spina bifida* and are usually attributed to a failure of the two neural folds to meet in the midline during embryogenesis. In fact this explanation may be incorrect, since in the human fetus, as in the chicken and the rat, the caudal part of the spinal cord forms by a different process. Thus, in the early stages, there is an amorphous mass of mesenchymal and neuroectodermal tissue that subsequently forms one or more internal canals. As the cells proliferate and segregate, as part of the differentiation of the spinal cord, the multiple canals normally coalesce to form a single central canal; the latter then fuses with the canal formed rostrally by closure of the neural tube (Lemire, 1969). It is not clear why these processes may go awry in the caudal neural tube. From epidemiological studies in humans and by extension of observations made in certain mouse mutants (Chong-Hiyo, Pruitt, & Bennett, 1989), it is evident that a genetic factor may be responsible for an unknown proportion of cases. In addition, it is possible that some instances may be the result of environmental teratogens. *Retinoic acid* is one potential agent, as it may be used inadvertently to treat severe acne in a pregnant woman; valproic acid and folic acid antagonists have also been incriminated (Campbell, Dayton, & Sohal, 1986).

Too Few Motoneurons Are Present in Spinal Muscular Atrophy

In the hereditary disorder spinal muscular atrophy, there is a reduced number of spinal motoneurons in the spinal cord and brainstem. The severity of the depletion in any one motoneuron pool, and the number of pools affected, varies from one patient to another. At one extreme is the infantile form, in which the resulting weakness affects not only the movements of the arms and legs, but swallowing and breathing as well. At the other extreme is the middle-aged adult who, for the first time, notices some weakness of the limb girdle muscles, for example, in climbing the stairs or in raising the arms above the head. Regardless of the age of onset of symptoms, serial motor unit estimates (see chapter 12) indicate that the reduction in motor neurons is present at birth. The condition is evidently due to excessive neuronal death (apoptosis), for a spinal muscular-atrophy gene localized to chromosome 5q (Melki et al., 1990) has now been shown to code for an *apoptosis-inhibiting protein* (Roy et al., 1995).

Specimens of muscle taken from affected infants typically show a few large fibers among expanses of much smaller, noninnervated ones. In an adult, one of the most striking findings is the presence of groups of histochemically similar fibers, instead of the normal mosaic (see Figure 17.7, chapter 17).

We have now finished studying the structure of skeletal muscle and its nerve supply, and also the way that these very different types of cell develop in the embryo and come into contact with each other.

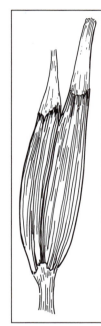

Part II

Putting Muscles to Work

Now that the structure of the muscle fiber and its innervation have been described, it is time to look at function. In chapter 7, we start at the level of the cell membrane by examining the molecules that act as ion channels and pumps, but we will also look at molecules that can temporarily capture an ion for some particular purpose. Although the concept of muscle and nerve activity is usually associated with electrical impulses and muscle contraction, it must not be forgotten that there is a slower, chemical signaling that takes place between muscle and nerve in the form of axoplasmic transport; this is the subject of chapter 8. In the next chapter, we consider the ionic basis of the resting membrane potentials of nerve and muscle fibers, and the brief perturbations that form the impulses. Some of this story was learned from the squid, because of the presence of a large axon suitable for study, but it will be seen that recordings of impulse activity can be readily made in human subjects.

Chapter 10 is largely about acetylcholine, the chemical which acts as an intermediary between the impulse in the axon and that in the muscle fiber. The extraordinary structure of the neuromuscular junction enables the release of acetylcholine from the nerve terminal, and its capture by receptors in the muscle fiber, to be performed with remarkable speed and efficiency. Next comes an account of the activation of the myofilaments, with their subsequent contraction, and of the key role played by Ca^{2+} in this coupling process.

In chapter 12, the focus of interest moves away from molecular events to the properties of whole colonies of muscle fibers, each colony being supplied by a single motoneuron so as to form a motor unit. It will be seen that the properties of the motor units differ greatly, and that each unit is specialized for the type of task in which it is called upon to participate. The study of the motor units is carried into the next chapter, which deals with their thresholds for excitation and their impulse firing patterns.

The final chapter of this part describes the biochemical changes that take place in the muscle fiber, as energy in the form of ATP is first synthesized and then consumed as the muscle begins to contract.

Chapter 7

Ion Channels, Pumps, and Binding Proteins

Ions such as Na^+, K^+, and Ca^{2+} play vital roles in nerve and muscle fibers through their ability to move across cell membranes. The simplest movement, that of diffusion, results from the opening up of channels in the membrane, either in response to a change in the transmembrane voltage or to the combination of the channel with a neurotransmitter (as in the acetylcholine receptor). To ensure that the correct distributions of ions are maintained across the membrane, prior to channel opening, the ions are pumped or otherwise transported against their respective concentration gradients by special molecules in the membrane. For Ca^{2+}, there is an additional type of molecule that binds the ions in the cytosol or sarcoplasmic reticulum and functions either as a store or as the initiator of a chemical reaction.

General Properties of Channels and Pumps

During the last decade particularly, there have been great advances in understanding the structure and molecular mechanisms of the ion channels, pumps, and binding proteins. Much of this new information has come from molecular biology experiments in which purification of the various proteins has led to cloning of the respective genes. Once the order of nucleotide bases in a gene is known, it is but a short step to establish the corresponding sequence of amino acids in the protein. This knowledge, in turn, makes it possible to deduce which parts of a molecule are in the membrane or in the cytosol or extracellular space.

Ions Diffuse Rapidly Through Channels in the Membrane; Pumping is a Slower Process

It is now clear that the ion channels have aqueous pores through which the appropriate ions can flow, and it is probable that most pumps also have pores. However, there is a striking difference between the two types of molecule in their abilities to transfer ions. If it were to stay open for one second, a single channel could allow several million ions through the membrane, whereas in the same period of time, a pump could only manage to move a hundred or so ions through. This great disparity in ion transfer capacity requires that pump molecules be very plentiful in membranes if they are to cope with rising concentrations of ion within the cytosol. On the other hand, because they can transmit ions so readily, some of the ion channels need only remain open for a fraction of a second in order to achieve their effects.

There is another important distinction between channels and pumps that is a consequence of the fact that, whereas ions flow down their respective electrochemical gradient in a channel, in a pump the ions must move against their concentration gradient. To make the latter possible, metabolic work must be performed by the pump, and this, in turn, requires that energy must be expended. The energy for the pumps is provided by the hydrolysis of ATP; each pump molecule has a site for combining with ATP as well as a channel for admitting the ion.

Channels Select Ions on the Basis of the Charges on the Ions and the Sizes of Their Water Shells

How is it that channels can discriminate between ions, so that they can pass one species but not another? The first point to make is that each of the cations is hydrated in solution; that is, the metallic atom is surrounded by a shell of water molecules. Since the hydrated molecules of different species of ion differ in size, the pores of one type of ion channel may be too narrow to pass ions of another species.

In addition, the ions are required to give up part of their water shell in order to pass through the pore, and the various species differ in their readiness to do so. Last, the ions must momentarily bind to the amino acid residues lining the pore, and one species of ion may be attracted more strongly than another (cf. Jan & Jan, 1989). It is now known that the pore inside an ion channel is long enough to accommodate several ions at any instant.

Na+ Channels Increase Membrane Excitability, While K+ and Cl- Channels Reduce It

In reviewing this fascinating field, the channels and pumps for a particular ion species will be considered together, before passing on to those for a different species. It will be seen that the Na+ channels and pumps are required to give membranes, such as those of nerve and muscle fibers, the property of excitability. Thus, the opening of the Na+ channels at successive points along the membrane enables an electrical signal, the action potential, to be transmitted from one region of the fiber to another. K+ channels have the opposite effect

in that they reduce membrane excitability; they terminate the action potential, but also ensure that the membrane does not become electrically unstable and generate action potentials spontaneously. Although they have been studied less extensively, Cl- channels are also present in the membranes of excitable cells and have the same effects as K+. Ca2+ channels have evolved to serve a quite different purpose; by allowing Ca2+ ions through the membrane, they initiate many of the chemical reactions in the interior of the cell. These reactions are remarkably diverse and are necessary not only for much of the "housekeeping" metabolism to be found in any cell, but also for such special functions as muscle fiber contraction and the release of acetylcholine from the motor nerve terminal.

The K+ Channel Evolved Before the Ca2+ and Na+ Channels

How did these various functions of the ion channels come about? The answer is that the cells must have found uses for the channels as the different types evolved. It is probable that the K+ channels were the first to appear in the earliest living cells and that the Ca2+ channels came next, following mutations of the K+ channel gene. The last channel to appear was that for Na+ ions, presumably by mutation of the Ca2+ channel gene (cf. Hille, 1992). One of the remarkable aspects of these events is that, once the gene evolved for a new type of channel, the nucleotide base sequence became extremely well conserved. The Na+ channels in the electroplax of an electric eel are very similar to those in a human nerve fiber, for example. Similarly, the mammalian gene for the K+ delayed rectifier channel was only found after the same gene had been cloned in the fruit fly. Indeed, yeasts, worms, and even a unicellular organism such as *Paramecium*, all have K+ and Ca2+ channels similar to those found in humans.

Sodium Channels

In mammalian skeletal muscle, the Na+ channel consists of two subunits, designated α and β (Figure 7.1; Barchi, 1988; Catterall, 1988). The larger, α, subunit has a molecular weight (MW) of 260 kD and contains substantial amounts of carbohydrate, mostly sialic acid, and lipid. The β subunit has a MW of 38 kD and also contains appreciable carbohydrate. The function of the β subunit is not clear, since the α subunit by itself can act as a Na+ channel; possibly the β subunit facilitates the insertion of the α subunit into membranes or else protects it from proteolysis (cf. Auld et al., 1988). In the Na+ channels of the brain, each molecule contains

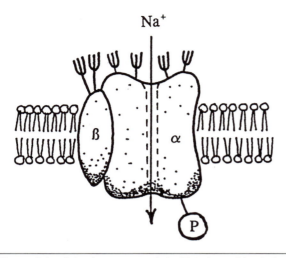

Na⁺

Figure 7.1 Subunit structure of the Na⁺ channel in skeletal muscle. Sugar residues shown as Ψ, and phosphorylation by P. Adapted from Catterall (1988, p. 52).

a second β subunit, while in the electroplax organ of the eel, the β subunit is absent.

The α Subunit Has Four Domains

Using a cDNA cloning technique, the amino acid sequence of a Na⁺ channel was first determined for the electroplax organ of the eel (Noda et al., 1984) and subsequently for the α subunits of rat brain and skeletal muscle (Trimmer et al., 1989). Regardless of its source, the α subunit was found to consist of approximately 2,000 amino acids; most of the latter are contained in four domains (Figure 7.2, A & C), which can be recognized by the similarity of their amino acid sequences. Within each domain, there are six regions (S1 to S6) that span the cell membrane and are linked together (Figure 7.2B). There is an additional membrane region, between S5 and S6, that forms the walls of the channel pore (see p. 98).

A Small "Gating" Current Is Produced Whenever a Channel Opens

In their classic study of the squid giant axon, Hodgkin and Huxley (1952b) were the first to suggest how ion channels *might* work: They did so by measuring the ionic currents that flowed when a potential difference was applied suddenly across the axon membrane (see chapter 9). Hodgkin and Huxley recognized that the action potential was due to a large inward current carried by Na⁺ ions; this current was able to flow because the membrane became highly permeable to Na⁺ ions. Following the transient Na⁺ current was an outward current, carried by K⁺ ions. Both currents were activated by

membrane depolarization, as shown in the current-voltage curves in Figure 7.3.

Having derived equations to describe the observed time-courses of the sodium and potassium currents (see Figure 9.4, chapter 9), Hodgkin and Huxley (1952b) suggested that some of the mathematical terms could have physical counterparts in the structure of the respective ion channels. Thus, each K⁺ channel might have four independent "gating" particles, all of which would have to move into a new position, under the influence of an electrical field, before the channel could open. The situation for the Na⁺ channel was rather more complex because, in addition to the gating particles that allowed the channel to open, there was thought to be another particle responsible for "inactivating" (closing) the channel; the two types of gate were referred to as *m* and *h* respectively.

One of the predictions from this classic work was that small charges, or gating currents, would momentarily flow across the membrane whenever a channel opened. Such currents were detected in Na⁺ channels some 20 years later by Armstrong and Bezanilla (1973) and were subsequently found for other voltage-gated channels.

A simple conceptual model of a voltage-gated channel is shown in Figure 7.4. In the model, there is a critically narrowed region of the aqueous pore, the selectivity filter, that only permits hydrated ions of a certain size to pass through. In the case of the Na⁺ channel, the filter can allow Na⁺ ions, with a diameter 0.1 nm, to pass through but bars entry to the slightly larger K⁺ ion (diameter 0.13 nm). The *gate*, which allows ions to leave the aqueous pore, is connected to a part of the channel, the *sensor*, that gauges the electrical field across the membrane. For the Na⁺ channel, the model would also include an inactivating gate (the h gate) at its cytoplasmic end. Na⁺ ions would only flow through the channel when this gate, as well as the m gate, was open (see chapter 9).

The S4 Region of Each Domain Contains the Voltage Sensor

Which parts of the α subunit correspond to the gating mechanism of the Na⁺ channel? There is general agreement that each of the four domains contributes one of the gating particles postulated by Hodgkin and Huxley (1952b) and that the channel becomes fully open only when all four have been activated. The sensor would be expected to have a number of charged groups that would move as the electrical potential changed across the membrane. Examination of the transmembrane regions in each of the four domains (Figure 7.2B) suggests that the S4 region is the most probable sensor,

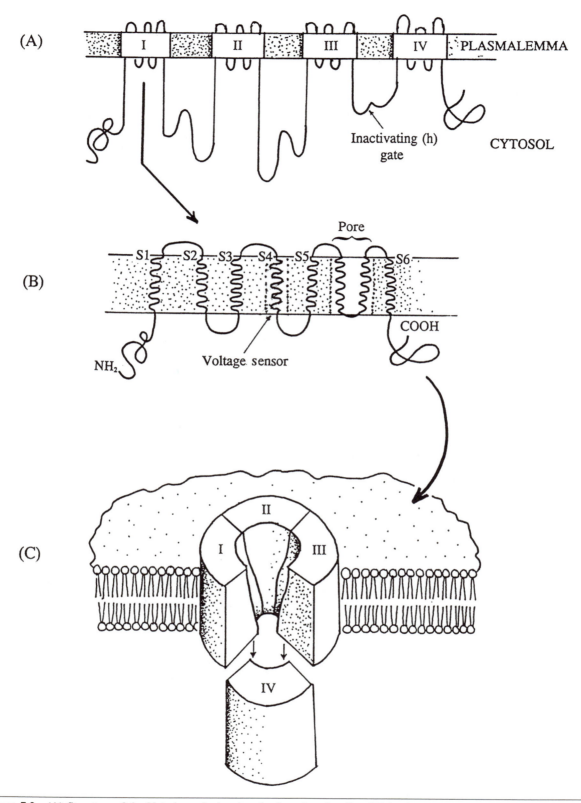

Figure 7.2   (A) Structure of the Na⁺ channel, showing the four domains (I to IV) in the α subunits. (B) One of the domains has been enlarged to reveal the transmembrane segments. S4 contains the voltage sensor controlling the gating mechanism, and the pore region lies between S5 and S6. (C) Three of the four domains have been assembled into a channel. Note that the basic structure of this channel is the same as those of the K⁺ and Ca²⁺ channels.

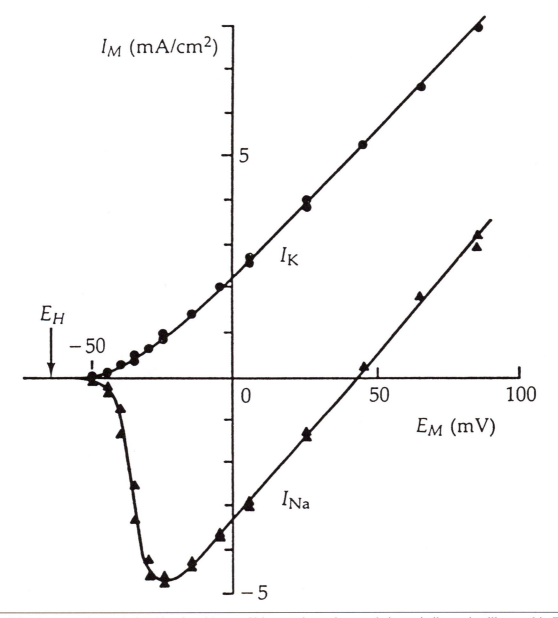

Figure 7.3 Current-voltage relationship of squid axon. Using a voltage clamp technique, similar to that illustrated in Figure 9.4, the membrane potential is instantaneously changed from the "holding" potential (E_H) to a series of less negative, and positive, values. Depolarizing the membrane in this way from a holding potential of −50 mV to −20 mV induces a large inward current following the opening of the Na$^+$ channels (I_{Na}); there is also a smaller outward current due to the K$^+$ channels opening (I_K). The net current flowing through the membrane at any instant will be the sum of the Na$^+$ and K$^+$ currents. Adapted from Cole and Moore (1960).

because it consists of repeated motifs in which a positively charged amino acid (usually arginine) is followed by two hydrophobic residues. If, as seems likely, the S4 region is arranged in an α-helix, then there would be a 60° rotation and a small (0.5 nm) outward movement of the helix in the membrane as the latter depolarized. This conformational change could allow the Na$^+$ ions to enter the pore.

An alternative proposal is that of Widdas and Baker (1992), who suggest that the gate may not be a moving part of the channel protein, but rather a bistable electrical barrier guarding the water-filled pore. The barrier would be composed of either 8 or 10 arginine residues (for a K$^+$ or a Na$^+$ channel respectively) joined together by guanido groups; the structure would contain an inner and an outer ring of N atoms (Figure 7.5). If positive charges were present on the inner ring, cations would be repelled and the channel would be closed. If, however, the positive charges moved to the outer ring, under the influence of an electrical field, the

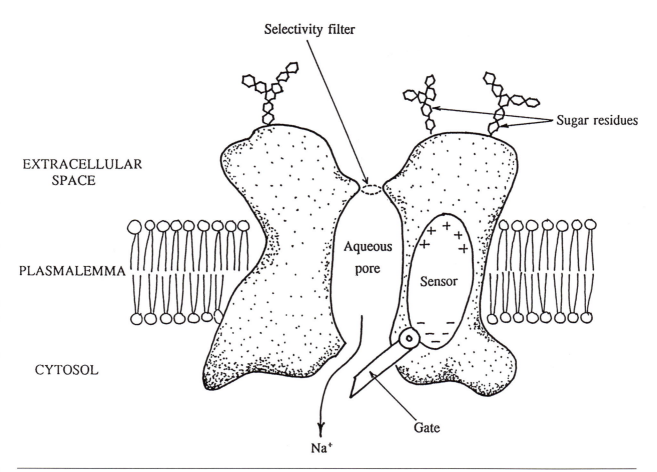

Figure 7.4 Model of an ion channel. The model has all the working parts that need to be incorporated into a functional channel, but the appearance of the channel is purely speculative. Adapted from Hille (1992, p. 66).

increased separation of the charges would allow cations to enter the pore.

Regardless of the detailed molecular structure of the gating mechanism, there is good experimental evidence that S4 is the transmembrane segment involved, since substitution of a single amino acid (phenylalanine for leucine) in this region produces a 25 mV shift in the membrane potential at which the Na$^+$ channel opens (Auld et al., 1990).

The S5 and S6 Regions Line the Channel Pore

It is not so easy, on theoretical grounds, to decide which part of the domain forms the pore, because the only clue is that the wall should be lined by hydrophilic amino acids. One experimental approach has been to investigate the function of the Na$^+$ channel after the amino acid sequence has been changed in different parts of the domain. This can be done using molecular biology techniques that alter the nucleotide base sequence in the cDNA clone (site-directed mutagenesis). Recent

studies, employing this technique, strongly favor that part of the domain linking the S5 and S6 regions as the pore (Figure 7.2B; Stevens, 1991).

Finally, there is the site of the inactivating mechanism (h gate) to consider. Armstrong, Bezanilla, and Rojas (1973) suggested that this gate was at the cytoplasmic end of the Na$^+$ channel, because inactivation could be abolished by the intracellular application of peptide-digesting enzymes. More precise localization has come from the use of antibodies directed against fragments of the channel; these experiments indicate that the h gate is probably formed by the polypeptide chain connecting domains III and IV (Figure 7.2B; Vassilev, Scheuer, & Catterall, 1988).

The Ion Current Flowing Through a Single Channel Can Be Measured With a Patch-Clamp Electrode

With the advent of patch clamping, it became possible to measure the flow of ionic current through a single channel. This important technique was described by

GATE CLOSED   GATE OPEN

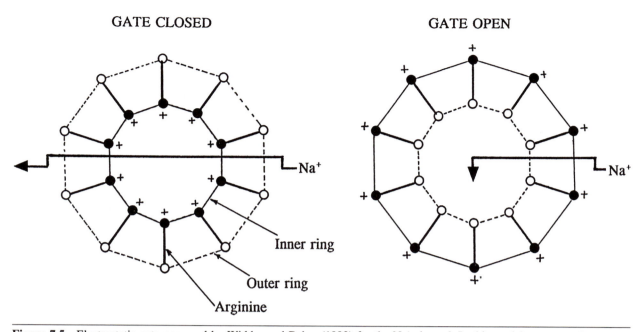

Inner ring

Outer ring

Arginine

Figure 7.5 Electrostatic gate proposed by Widdas and Baker (1992) for the Na⁺ channel. In this model, access to the aqueous pore is guarded by 10 arginine residues. The residues are radially oriented, and protons can travel between the inner and outer N atoms. At the resting membrane potential, the protons are associated with the inner ring of N atoms, and they repel Na⁺ ions, keeping the gate closed. When the membrane depolarizes, the protons move to the outer ring of N atoms, and there is now sufficient distance between opposing charges for Na⁺ ions to enter the pore. In this figure, each arginine residue is shown as a thick line, with filled and open circles representing N atoms with and without a proton, respectively.

Neher and Sakmann (1976b), who devised methods for sealing the tips of glass microelectrodes against the membranes of living cells (Figure 7.6A). Since the seal was made extremely tight, the background thermal noise in the circuit became sufficiently low so that the small ionic currents could be recognized. In Figure 7.6B, it can be seen that the ionic currents, in this case flowing through a Na⁺ channel, start and stop abruptly; in some cases the current repeatedly flickers on and off. In the membrane of a rat muscle fiber, Sigworth and Neher (1980) found that the Na⁺ channels have a mean conductance of 18 pS (18×10^{12} Ω^{-1}). In keeping with the Hodgkin-Huxley current-voltage curves, the time that a channel spends in the open state is determined by the potential across the membrane, so that nearly all channels are open after a depolarization to -50 mV (Figure 7.6C; Hartshorne, Keller, Talvenheimo, Catterall, & Montal, 1985).

Radioactive Neurotoxins Have Been Used to Count Na⁺ Channels in Membranes

The density of Na⁺ channels in different types of excitable membrane can be determined by labeling the channels with radioactively labeled neurotoxins and counting the radioactivity; most studies have been made with [³H]-saxitoxin or [³H]-tetrodotoxin (see the Applied Physiology section at the end of this chapter). In the case of mammalian skeletal muscle, the density values have been estimated at 209 to 557 Na⁺ channels/μm² of surface membrane (Hansen Bay & Strichartz, 1980; Ritchie & Rogart, 1977). The true density will be rather lower, however, for the estimates will have included Na⁺ channels in the transverse tubules; nor is the distribution at the surface uniform, because the density in the secondary synaptic clefts at the neuromuscular junction is 10 times higher than in the remainder of the fiber (Beam, Caldwell, & Campbell, 1985).

The Na⁺ channels in the membranes of the motor axons are concentrated at the nodes of Ranvier, where the densities may be several times higher than in muscle fiber membranes. The high densities at the nodes are essential both for the rapid depolarization of the axolemma during excitation and for giving a sufficiently high safety factor to the action potential mechanism.

The Sodium Pump (Na⁺-K⁺ Pump)

The existence of a pump capable of extruding Na⁺ was first postulated for cells in the renal tubule by Dean (1941) and was later shown to be a feature of most, if not all, living cells. In the simplest organisms, the pump, by expelling cations, prevents cells from becoming dangerously swollen through osmosis.

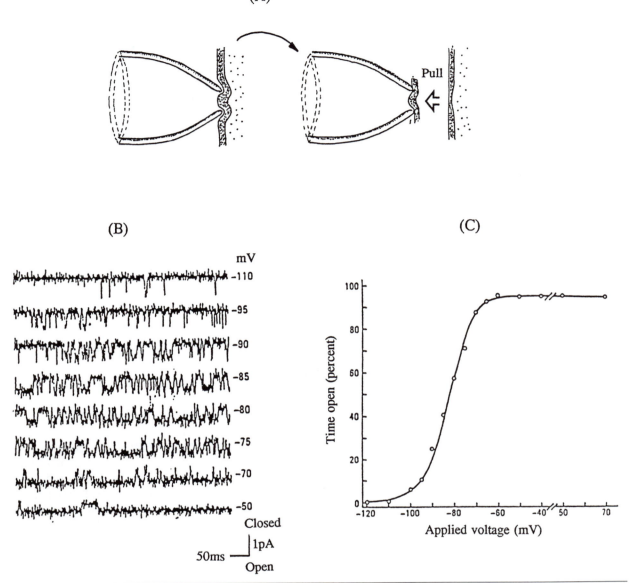

Figure 7.6 Patch-clamping technique applied to Na⁺ channels. (A) The tip of a glass micropipette is pressed against the surface of a cell and is then withdrawn, removing a small piece of membrane. The membrane of the cell reseals spontaneously. (B) Openings of a single Na⁺ channel inserted into an artificial membrane. As the negative potential across the membrane is reduced (depolarization), the channel opens more frequently until, at −50 mV, it is open nearly all the time. (C) Open time plotted as a function of membrane potential. (B) and (C) reprinted from Hartshorne, Keller, Talvenheimo, Catterall, and Montal (1985).

Ion Gradients, Established by the Na⁺-K⁺ Pump, Are Used for Many Purposes

During evolution the presence of the pump gave important new properties to cells. By extruding Na⁺ from the cell in exchange for K⁺, it created an internal milieu in which the concentrations of the two cations were very different from those outside the cell. Whereas Na⁺ is the most prevalent cation in the extracellular fluid, K⁺ is much more concentrated in the cytosol (see chapter 9). However, the pump also creates an electrical potential across the cell membrane, because in each cycle of activity only two K⁺ ions are traded for three Na⁺ ions. Thus, with each cycle, there is one surplus negative charge in the cytosol bordering the plasmalemma. This electrical potential serves as a source of potential energy and is employed by cells for the transport of different types of molecules across their membranes. For example, the uptake of many sugars and amino acids by cells is driven by the electrical potential and the concentration gradient of Na⁺ ions across the membrane; this force also runs the various cation antiport systems. In addition, in special cells such as nerve and muscle fibers, the electrical potential confers the ability to propagate electrical signals from one region to another and, indirectly, to release transmitter from the motor nerve terminals and to initiate contraction of the myofibrils.

Figure 7.7 shows a cross section through part of a mammalian muscle that has been exposed to a labeled antibody against the Na⁺-K⁺ pump. The pump molecules are seen to be associated with the surface membranes of the muscle fibers, as expected, and some fibers (type II) contain more than others. The section also includes bundles of nerve fibers, and rather surprisingly, there is a high density of pump molecules in the myelin sheaths. Last, both the muscle fibers and the nerve fibers contain appreciable numbers of pumps in the cytosol; presumably these are molecules either undergoing degradation or awaiting incorporation into the surface membranes.

The α Subunit Contains Binding Sites for Na⁺, K⁺, and ATP; There Is Also a β Subunit

Each Na⁺-K⁺-pump molecule is formed by the combination of two subunits (Figure 7.8). The larger of these, the α subunit, has a MW of 112 kD, is mostly intracellular, and is so folded on itself as to span the cell membranes from 6 to 8 times. The Na⁺ and K⁺ binding sites are part of this subunit, as is the site for ATP. It is now known that there are several isoforms of the α subunit, and these are expressed to different extents in the various tissues of the body. The β subunit appears to exist in only one form and is a smaller molecule (35 kD) which is mainly external to the cell membrane and is glycolyzated at three sites. The blocking agent, *ouabain*, combines with both the α subunit, as shown in Figure 7.8, and possibly with the β subunit as well.

A mechanism for the enzymatic reaction that redistributes Na⁺ and K⁺ across the membrane and splits ATP has been suggested by Albers (1967) and by Post, Kume, Tobin, Orcutt, and Sen (1969). In this scheme the Na,K-ATPase has two molecular conformations, E_1 and E_2 (Figure 7.9). When the E_2 conformation is phosphorylated by cytosolic ATP, it is converted to the E_1 form and releases two K⁺ ions into the interior of the cell in exchange for three Na⁺ ions. A positive charge is translocated across the membrane as the phosphorylated enzyme changes back to the E_2 conformation. Three

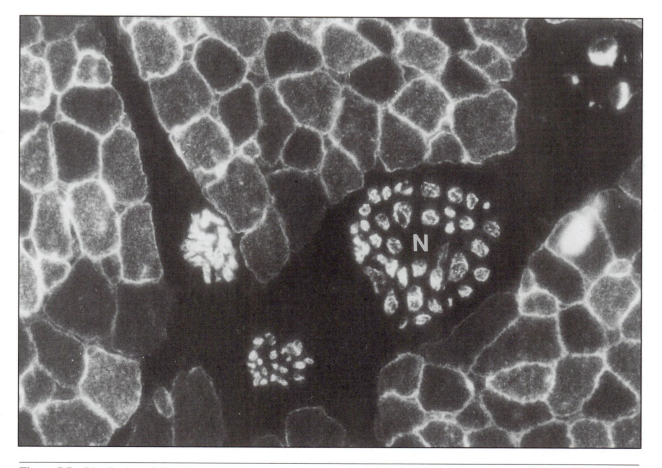

Figure 7.7 Distribution of Na⁺-K⁺-pump molecules in a cross section of skeletal muscle. The specimen includes three intramuscular nerve branches, the largest of which is labeled *N*. (Courtesy of Dr. Douglas Fambrough.)

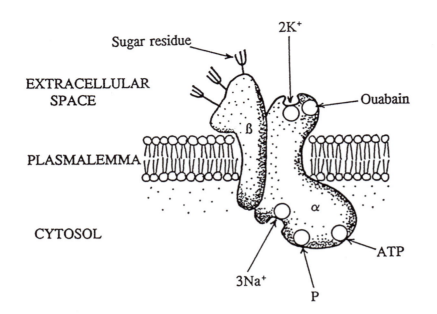

Figure 7.8 Model for Na⁺-K⁺ pump, showing α and β subunits, together with different binding sites. Adapted from Horisberger, Lemas, Kraehenbühl, and Rossier (1991, p. 567).

Na⁺ ions are then released into the extracellular fluid in exchange for two K⁺ ions. The initial step in the cycle, the electroneutral transition of the enzyme from the E_2 to E_1 form through phosphorylation, is then repeated (Figure 7.9). It has been estimated that a single cycle of the pump occupies approximately 10 ms (Fambrough, Wolitzky, Tamkun, & Takeyasu, 1987), but the minimum time may be only half as long, because the maximum transport capacity for a single pump molecule may be as high as 16,000 K⁺ ions/min. Under resting conditions, however, skeletal muscles only use 2% to 6% of their maximal pumping capacity (Clausen, 1986).

Pumping Increases After Impulse Activity

A rise in Na⁺-K⁺-pump activity occurs within a few seconds of stimulated or voluntary muscle contractions, and because of the electrogenic nature of the pump, the muscle fiber resting potentials may increase; this increase, in turn, causes the evoked muscle action potentials to enlarge (Hicks & McComas, 1989; Galea & McComas, 1991). Other evidence of enhanced pump activity comes from measurements of the uptake of radioactive Rb⁺, an ion that is treated like K⁺. In the rat, the enhancement is greater in the soleus than in the extensor digitorum longus muscle and is apparently due to the rise in intracellular Na⁺ concentration and the β-adrenergic effects of epinephrine and norepinephrine (Everts, Retterstøl, & Clausen, 1988). The elevation in pump activity affects not only the contracting fibers but quiescent ones in their vicinity (Kuiack & McComas, 1992). The cellular mechanisms responsible for the β-adrenergic changes in pump activity involve the cyclic AMP second messenger system.

Pump Density is Regulated by Usage

The density of pump sites in the surface membranes can be adjusted over a period of several hours; under physiological conditions, such modulation would presumably occur in response to altered activity levels of the muscle fibers. In cultured myoblasts, an increase in pumps at the fiber surfaces can be induced by exposure to veratridine, a drug that maintains the voltage-gated Na⁺ channels in an open state. Fambrough et al. (1987) have postulated that the accumulation of Na⁺ inside the fibers provides the necessary stimulus both in this model and in response to increased contractile activity; the same workers have shown that the higher density at the fiber surface is achieved by three mechanisms—increased synthesis and reduced turnover of pump molecules, and the transfer of pumps from the fiber interior. It now appears that there may normally be an excess of the α subunits in the endoplasmic reticulum and that, when additional Na⁺-K⁺ pumps are required, there is increased transcription of the gene coding for the β subunits. The newly formed β subunits combine with the α subunits in the interior of the fiber and are then conveyed to the surface of the fiber, where they are inserted into the plasmalemma (Taormino & Fambrough, 1990). The consequent changes in Na⁺-K⁺-pumping capacity will persist if an increase in physical activity is maintained, as has been shown in dogs (Knochel, Blachley, Johnson, & Carter, 1985) and in human subjects (Tibes, Hemmer, Schweigart, Bóning, & Fotescu, 1974).

Longer term changes in pump regulation, occurring over a period of several weeks, can be induced by alterations in thyroid activity. For example, pump activity may be as much as 10 times greater in muscles from rats

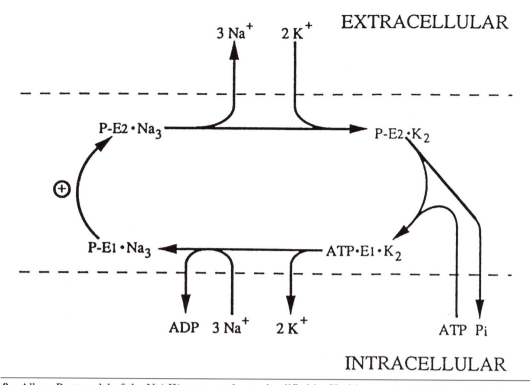

Figure 7.9 Albers-Post model of the Na^+-K^+-pump cycle, as simplified by Horisberger et al. (1991, p. 567). For explanation of steps, see text. Reprinted from Horisberger, Lemas, Kraehenbühl, and Rossier (1991).

made hyperthyroid, as opposed to those of hypothyroid animals (Kjeldsen, Everts, & Clausen, 1986). Up-regulation of the pump is also a feature of developing muscles, at least in the first few weeks of age in rats and mice. In contrast, both starvation and diabetes produce down-regulation of pump activity in experimental animals (Clausen & Everts, 1989).

Potassium Channels

Potassium channels are difficult to isolate and purify. Their molecular structure could only be deduced after the technique of chromosome walking had been used to clone the gene responsible for the *Shaker* mutation in the fruit fly, *Drosophila* (Papazian, Schwarz, Tempel, Jan, & Jan, 1987; Kamb, Iverson, & Tanouye, 1987; Box 7.1). Previous work had suggested that the leg shaking characteristically exhibited by this mutant under ether anesthesia was due to a failure of the voltage-gated K^+ channel to cut short the action potential in the motor nerve terminals (Jan, Jan, & Dennis, 1977).

The K^+-Channel Gene Codes for a Single Domain

After cloning, it was found that the protein encoded by the K^+-channel gene contained only 616 amino acids

and was therefore much smaller than the main subunits of the Na^+ and Ca^{2+} channels; rather, it corresponded to only one of the four domains of two other channels. A complete K^+ channel is evidently formed by the fusion of four of the K^+-channel proteins. Similar to the Na^+ and Ca^{2+} channels, each protein (domain) forms part of the central pore and each contains a voltage-sensitive region (S4). The four proteins may be identical, or they may differ from each other due to alternative splicing of the mRNA transcribed from the same K^+-channel gene. It is the alternative splicing that accounts for the remarkable diversity of K^+ channels, even in the same cell. The first K^+ channel to be described was the one associated with the action potential in the squid giant axon by Hodgkin and Huxley (1952b), and it remains the best understood of the K^+ channels.

More than 30 Different Types of K^+ Channel Exist

Over 30 different K^+ channels have been identified so far, on the basis of their different kinetic and pharmacological properties. The main K^+ channel in excitable cells, such as α-motoneurons and striated muscle fibers, is that responsible for terminating the action potential; its kinetic properties were described by Hodgkin and Huxley (1952b). In view of the slight delay before the

BOX 7.1   CLONING A POTASSIUM-CHANNEL GENE

The first voltage-gated ion channel for which the DNA could be cloned was the Na^+ channel. This was a very considerable achievement, and it took advantage of the fact that the concentration of Na^+ channels is especially high in the electroplax organ of fish, allowing the channels to be isolated and purified. Once part of the amino acid structure of the channel had been established, it was possible to construct the corresponding mRNA nucleotide sequence and to use the sequence to recognize the complementary DNA in the genomic library (Noda et al., 1984). In comparison, the cloning of a K^+-channel gene was a much greater challenge. For one thing, K^+ channels are not found in high concentrations in nervous tissue, and for another, K^+ channels are remarkably diverse in their pharmacological properties and have few ligands that bind with them.

The solution to the cloning difficulty was to make use of one of the neurological mutations exhibited by the fruit fly, *Drosophila melanogaster*. This mutant, *Shaker*, was originally noticed because its legs trembled when the fly was exposed to ether. Micro-electrode recordings from pupal and larval muscle fibers showed that the membranes appeared to lack the transient outward K^+ current responsible for repolarizing the membrane and terminating the action potential (Figure 7.10). It was subsequently shown that the same channel was probably deficient in nerve fibers and that the broadened action potential caused excessive amounts of transmitter to be released at neuromuscular junctions, provoking repetitive firing of the muscle fibers (Jan et al., 1977; Tanouye & Ferris, 1985). The next step was to look for breakpoints in the X-chromosome of flies in which the *Shaker* mutation had been induced by X-irradiation, and to undertake a chromosome walk until the affected gene could be identified by DNA probes. This elegant work was undertaken independently by Mark Tanouye and by Lily and Yuh Jan and their respective colleagues. Simultaneously, both teams reported the cloning of the *Shaker* DNA and thereby the first isolation of a K^+-channel gene (Papazian et al., 1987; Kamb et al., 1987). The definitive proof that the gene did indeed code for a K^+ channel came from experiments in which synthetic RNA, prepared by transcribing the cloned DNA, was injected into the *Xenopus* oocyte. Within a few days, voltage-gated K^+ currents could be detected in the oocyte membrane (Timpe et al., 1988).

channel opens in response to depolarization, and because of the shape of the current-voltage relationship (Figure 7.3), this channel became known as the delayed rectifier. However, since other K^+ channels also have delayed openings and have conductances that increase more slowly, the delayed rectifier channel is often referred to as the fast K^+ channel. The channel can admit Rb^+ and NH_4^+ ions, in addition to K^+, but is blocked by larger cations such as Cs^{2+} and Ba^{2+}; another very effective blocking agent is tetraethylammonium (TEA).

Another voltage-gated K^+ channel found in striated muscle fibers has the properties of an inward rectifier, since it opens when the membrane potential is larger than the K^+ equilibrium potential and allows net entry of K^+ ions into the fibers along their electrical gradients.

Most of the other K^+ channels are like the fast channels in being voltage gated, but there are some that are activated by intracellular messengers. Two of the latter are channels that open following rises in the intracellular concentration of Ca^{2+} ions and are therefore given the symbol: $I_{K(Ca)}$. One has a low conductance (*SK [small] channel*) and is blocked by apamin, a peptide found in bee venom; the other (*BK [big] channel*) has a very large conductance and is blocked by TEA. Both channels, when open, tend to stabilize the membrane and prevent further influx of Ca^{2+} into the muscle fiber. Another K^+ channel in striated muscle fibers is normally inhibited by intracellular ATP and opens when the ATP concentration declines. Yet another K^+ channel is activated by a rising concentration of intracellular Na^+ ions.

K^+ Channels Stabilize the Membrane Potential

In all cells, the basic function of the K^+ channels is to stabilize the membrane potential at its resting level and

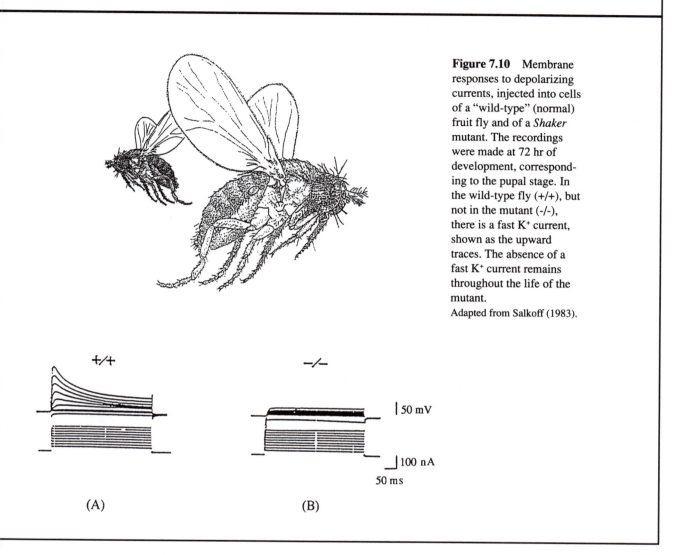

Figure 7.10 Membrane responses to depolarizing currents, injected into cells of a "wild-type" (normal) fruit fly and of a *Shaker* mutant. The recordings were made at 72 hr of development, corresponding to the pupal stage. In the wild-type fly (+/+), but not in the mutant (-/-), there is a fast K⁺ current, shown as the upward traces. The absence of a fast K⁺ current remains throughout the life of the mutant.
Adapted from Salkoff (1983).

thereby to counteract the excitatory effects of the Na⁺ (and Ca²⁺) channels. In skeletal muscle fibers, most of the K⁺ channels are not in the surface membrane but in the transverse tubules, and there is another type of channel localized to the sarcoplasmic reticulum. In motor nerve axons, the greatest density of fast K⁺ channels is found in the paranodal region, that is, in the axolemma lying beneath the attachments of the myelin loops on either side of the node of Ranvier (see Figure 2.7, chapter 2). Presumably, the K⁺ channels cannot be accommodated at the node itself, in view of the necessity of having a high concentration of Na⁺ channels instead. The densities of fast K⁺ channels in the membranes of the node and the remainder of the internode are approximately only one-sixth those of the paranodal regions (Black, Kocsis, & Waxman, 1990; Figure 7.11).

The properties of the different types of K⁺ channel in striated muscle and motor nerve fibers are summarized in Table 7.1. It is worth reiterating that they all serve to maintain or restore the membrane to its resting condition. Some of the K⁺ channels would be expected to become especially important in muscle fatigue; thus, the BK and SK channels would open as the intracellular concentration of Ca²⁺ increased, as would the K_{ATP} channel when the cytosolic store of ATP in the vicinity of the membrane declined. Rising concentrations of Na⁺ would open another type of K⁺ channel. If the membrane potential of the muscle fibers was higher than the K⁺ equilibrium potential, under the influence of the electrogenic Na⁺-K⁺ pump, K⁺ would be attracted from the interstitial spaces back into the muscle fibers. The inward rectifier K⁺ channel would have such an effect also.

Calcium Channels and Pumps

We have seen that Na⁺ channels confer excitability on cell membranes and that K⁺ channels have an opposite

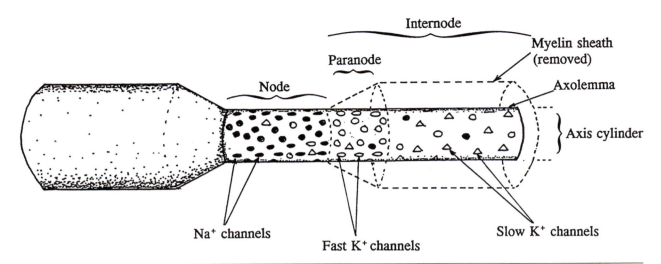

Figure 7.11   Myelinated nerve fiber, showing the distributions of Na⁺ channels (●), fast K⁺ channels (○), and slow K⁺ channels (△) at the node, paranode, and internode. Based on the data of Black, Kocsis, and Waxman (1990).

Table 7.1   Potassium Channels in Striated Muscle[a]

Name(s)	Gating mechanism	Conductance	Blocking agents	Actions
Fast (delayed outward rectifier)	Depolarization	17-64 pS	TEA, Ba^{2+}, Cs^+	Terminates action potential; contributes to refractory period
Inward (anomalous rectifier)	Membrane potentials positive to E_K	5-28 pS	TEA, Ba^{2+}, Cs^+	Stabilizes resting membrane; allows entry of K⁺ into fibers during fatigue
Sarcoplasmic reticulum channel	Depolarization of SR	150 pS	TEA	?
Calcium-activated (small [SK], intermediate [IK], big [BK])	Increased intracellular $[Ca^{2+}]$	6-250 pS	Apamin, charybdotoxin	Stabilizes membrane and diminishes Ca^{2+} entry; increases K⁺ entry into fibers during fatigue (?)
ATP-sensitive [K_{ATP}]	Decreased intracellular [ATP] and pH	20-90 pS	Tolbutamide, 4-aminopyridine	Stabilizes membrane and increases K⁺ entry into fibers during fatigue (?)
Na⁺-activated	Rise in intracellular $[Na^+] > 20$ mM	220 pS	TEA, 4-aminopyridine	Stabilizes membrane and increases K⁺ entry into fibers during fatigue (?)

[a]See Castle, Haylett, and Jenkinson (1989); Rudy (1988).

effect, by stabilizing the membrane potential. In contrast to both, Ca^{2+} channels, which are found in all living cells, are required to link the electrical signal at the surface membrane with molecular events within the cell. This linkage is made possible by the entry of Ca^{2+} ions into the interior of the cell; the Ca^{2+} ions are then seized by special binding proteins, and this combination, in turn, activates protein kinases that phosphorylate certain key molecules.

Ca^{2+} Ions Act as Second Messengers

If the depolarization of the surface membrane constitutes the initial instruction to a cell, then Ca^{2+} ions can be considered as second messengers. Among other actions, they are responsible for initiating the contractile process in the muscle fibers and for releasing acetylcholine from the motor nerve terminals. The signaling action of Ca^{2+} is made possible by the ability of the resting cell to reduce the internal concentration of this ion to a very low level, of the order of 10^{-7} M. Consequently, even a relatively small influx of Ca^{2+} will change the internal concentration appreciably; for example, during a cell activity such as muscle contraction, the level rises to 10^{-5} M, and similar perturbations are found in other types of specialized cell.

Calcium is important to the muscle fiber for another reason. If the plasmalemma is torn as a result of injury or disease, Ca^{2+} ions will flow into the fiber down their electrochemical gradient and disrupt both the structure and the function of the cytoplasmic contents. The disruption is due to the arrest of ATP production by the mitochondria and to activation of proteolytic enzymes; these enzymes, by degrading the intermediate filament proteins, weaken the scaffolding of the fiber.

Several Different Mechanisms Reduce Ca^{2+} in the Cytosol

How are the low resting levels of Ca^{2+} in the cytoplasm achieved? The answer is that some Ca^{2+} ions are removed from solution by binding to special proteins, and other Ca^{2+} ions are pumped or attracted across the various membranes of the cell. These membranes include the surface membrane, the membrane enclosing the endoplasmic reticulum (and sarcoplasmic reticulum in the case of the muscle fiber), and the inner membrane lining the mitochondria (Figure 7.12).

At the cell surface (Figure 7.13), the Ca^{2+} ions are pumped out in return for H^+, a process that, as in the case of other pumps, is fueled by ATP. The surface membrane also contains another mechanism which exchanges Na^+ ions for Ca^{2+} ions; the energy for this exchange comes from the Na^+ concentration gradient, established across the cell membrane by the Na^+-K^+ pump. Since three Na^+ ions are admitted for each (divalent) Ca^{2+} ion, there is a net gain of one positive charge by the negative cell interior.

In the inner membrane of the mitochondrion, there is another type of Na^+-Ca^{2+} exchanger, which expels one Ca^{2+} ion for two Na^+ ions and is therefore electrically neutral (Figure 7.13). There is also a "uniporter" molecule that allows Ca^{2+} to move into the mitochondrion down the large electrical gradient (-180 mV) across the inner membrane. This last Ca^{2+} removal mechanism is potentially the most powerful in most cells, because the foldings of the mitochondrial inner membrane provide a large surface for Ca^{2+} removal. Finally, if the cytosolic Ca^{2+} level rises to very high levels, for example, after injury to the cell surface, the mitochondria can admit phosphate ions and precipitate the Ca^{2+} as crystals of calcium phosphate (hydroxyapatite).

The third type of membrane that removes Ca^{2+} from the cytosol is that of the endoplasmic reticulum, much of which has become specialized in the muscle fiber to form the sarcoplasmic reticulum (SR). Approximately 90% of the SR membrane is occupied by a Ca^{2+}-ATPase; this enzyme acts swiftly and sensitively in response to rises in intracellular Ca^{2+} concentration (Figure 7.13).

The relative activities of the different Ca^{2+} transporting mechanisms within the muscle fiber depend on

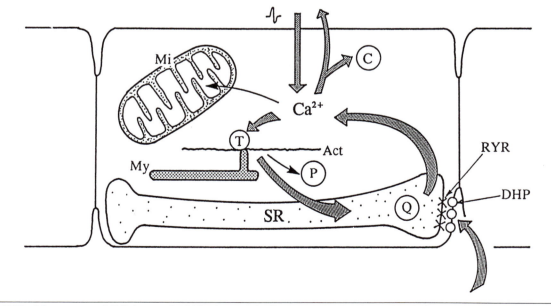

Figure 7.12 Movements of Ca^{2+} into and within the muscle fiber. *Act* = actin, *C* = calmodulin, *DHP* = dihydropyridine receptor, *Mi* = mitochondrion, *My* = myosin, *P* = parvalbumin, *Q* = calsequestrin, *RYR* = ryanodine receptor (channel), *SR* = sarcoplasmic reticulum, *T* = troponin-C.

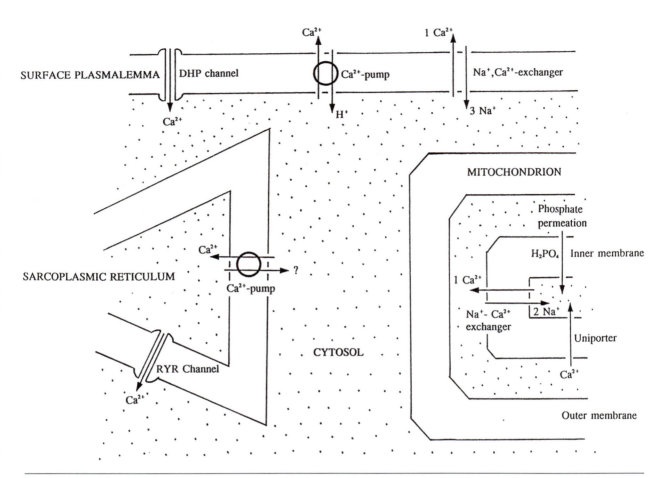

Figure 7.13 Calcium transport systems in three types of muscle fiber membrane: the cell surface, the sarcoplasmic reticulum (SR), and the inner membrane of the mitochondrion.

the concentration of Ca^{2+} in the cytosol. In *resting muscle*, approximately 70% of the Ca^{2+} is removed from the cytosol by the Ca^{2+} pump in the SR, and the remainder is extruded across the cell surface, mainly by the Na^+-Ca^{2+} exchanger. During *muscle contraction*, the Ca^{2+} pump in the SR becomes even more important and now expels up to 90% of the ions. If the intracellular Ca^{2+} concentration rises still higher (10^{-5} M), as when the fiber is damaged by *injury*, *exercise*, or *disease*, the uniporter in the mitochondrion becomes increasingly active and removes most of the ions (see Carafoli & Penniston, 1985, for review).

There Are Different Types of Ca^{2+} Channel

Calcium channels are classified as the *T* (tiny or transient), *L* (large or long-lasting), and *N* (neither T nor L, or neuronal). The best understood of these is a type of L-channel found in high density in the transverse tubular membranes of skeletal muscle fibers; this is

the *DHP channel*, so named because it is bound, and blocked, by *dihydropyridine*. A fourth Ca^{2+} channel, discovered more recently in skeletal muscle fibers, is the *ryanodine* channel; this last channel is localized to the sarcoplasmic reticulum and is probably connected to the DHP channel in the transverse tubules (see chapter 11).

The DHP Channel Has Five Subunits

The purified *DHP channel* is composed of five polypeptide chains (Figure 7.14; Catterall, 1988; Hofmann, Flockerzi, Nastainczyk, Ruth, & Schneider, 1990). The α_1 subunit is the central component of the complex and has a MW of 165 kD. It strongly resembles the Na^+ channel in having four domains, each of which has at least six regions that span the membrane. By analogy with the Na^+ channel, it seems probable that the positively charged, hydrophilic S4 region acts as the voltage sensor and that the segment linking the S5 and S6 regions forms part of the pore for Ca^{2+} ions. The Ca^{2+} channel also differs in containing more subunits; in addition to the α_1 subunit, there are α_2, β, and γ subunits (MW, 135 kD, 55 kD, & 28 kD respectively; Hofmann et al., 1990). Unlike the α_1 subunit, the α_2 and γ subunits are heavily glycosylated. It is probable that the β subunit serves the same function as the β subunit of the Na^+ channel; Catterall (1988) has suggested that it is attached to the α_1 subunit (Figure 7.14). The Ca^{2+} channel can be phosphorylated on both the α_1 and β subunits, and this increases its functional activity significantly.

The Ryanodine Receptor Releases Ca^{2+} Ions From the Sarcoplasmic Reticulum

The *ryanodine receptor* (RYR) is found in the terminal cisternae of the sarcoplasmic reticulum of skeletal muscle fibers and acts as a very effective Ca^{2+} channel. Its function is to release Ca^{2+} ions rapidly into the cytosol, where they will combine with troponin-C and trigger the contractile mechanism of the muscle fiber. The channel is composed of four large subunits, each of which has a MW of 564 kD. It is probably coupled mechanically to the DHP channel in the transverse tubules (see Chapter 11), and it can be locked in the open state by ryanodine. Under normal circumstances, the channel is inactivated by phosphorylation induced by a protein kinase (Wang & Best, 1992).

Ca^{2+} Channels Can Be Opened by Membrane Depolarization and by Intracellular Mechanisms

Like the Na^+ and K^+ channels, the Ca^{2+} channels are activated by membrane depolarization, and it is possible,

too, that a part of the Ca^{2+}-channel complex helps to inactivate the channel during depolarization, as in the case of the Na^+ channels. However, a more important source of inactivation is the rise in intracellular Ca^{2+} concentration that follows channel opening. In the case of RYR, however, the increase in Ca^{2+} concentration has the opposite effect and facilitates channel opening (cf. Fabiato & Fabiato, 1975).

The Ca^{2+} channel can also be opened by signals transmitted from the cell interior. For example, epinephrine binds to a surface receptor, activates adenyl cyclase, and produces cyclic AMP; the latter activates a protein kinase which then phosphorylates the DHP channel and causes it to open.

In skeletal muscle, the DHP channel has a peak conductance of approximately 20 pS but is much slower to activate and inactivate than the same type of channels in smooth and cardiac muscle. It also differs in that, unlike the other channels, it does not appear to be inhibited by a rise in intracellular Ca^{2+} concentration.

DHP Channels Can Be Blocked by Divalent Cations and by Drugs

DHP channels are blocked by divalent metallic ions such as Ni^{2+}, Cd^{2+}, Co^{2+}, and Mn^{2+}. In contrast, the channel is fully permeant to Sr^{2+} and Ba^{2+} ions because of their smaller sizes. The DHP channels are also blocked by a number of drugs that are used in medicine to control heart arrhythmias and to relax smooth muscle. These drugs fall into three distinct classes, each of which acts at a separate site on the Ca^{2+} channel. These drugs are the dihydropyridines (e.g., nifedipine and nimodipine), the phenylalkylamines (e.g., verapamil and D-600), and the benzothiazepines (e.g., diltiazem).

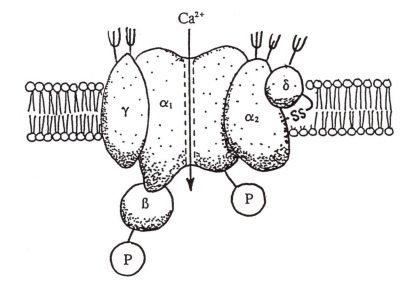

Figure 7.14 Subunit structure of the Ca^{2+} channel. Sugar residues and phosphorylation sites shown as Ψ and P respectively. Adapted from Catterall (1988, p. 52).

Calcium-Binding Proteins

The Ca^{2+}-binding proteins are necessary for the signaling actions of Ca^{2+} ions in the cell interior. Thus, the combination of Ca^{2+} with a binding protein initiates a conformational change in the latter and, in turn, triggers a chemical reaction (Hiraoki & Vogel, 1987). There are several known binding proteins, and four of these are especially important in skeletal muscle fibers.

The Combination of Ca^{2+} With Calmodulin Activates Enzymes

Calmodulin is a binding protein with a MW of 16.8 kD and contains four Ca^{2+}-binding domains. Each domain contains 12 amino acid residues that form a "pocket" for a Ca^{2+} ion; the four pockets are separated from each other by amino acid α-helices (Figure 7.15). Once Ca^{2+} is bound by calmodulin, the latter molecules are able to attach themselves to a wide variety of enzymes in the cytosol and thereby activate them. As yet it is not clear how the Ca:calmodulin complex is able to select the appropriate target enzyme among the many other types present. One of the enzymes activated is the Ca^{2+}-ATPase in the surface membrane; by increasing the extrusion of Ca^{2+} from the cell, the enzyme completes a negative feedback loop across the membrane. Other calmodulin-activated enzymes in muscle fibers include adenylate cyclase, cyclic nucleotide phosphodiesterase, a multifunctional calmodulin-dependent protein kinase, phosphorylase kinase, myosin light chain kinase, and calmodulin-dependent phosphoprotein phosphatase (England, 1986). Certain other enzymes, such as the mitochondrial dehydrogenases and protein kinase C, can be directly activated by Ca^{2+} ions without the intervention of calmodulin (Figure 7.16).

Parvalbumin Reduces [Ca^{2+}] in Fast-Twitch Fibers

A second Ca^{2+}-binding protein found in skeletal muscle is *parvalbumin*, which has a MW of 12 kD and contains two Ca^{2+}-binding pockets in each molecule. Parvalbumin is found in appreciable amounts only in fast-twitch skeletal muscle fibers. It is thought to remove Ca^{2+} ions when their concentration is significantly elevated, as in a tetanic contraction.

Ca^{2+}-Troponin Interactions Allow Cross-Bridge Engagement

Unlike parvalbumin, a third Ca^{2+}-binding protein in skeletal muscle *is* essential, being necessary for the contractile process to take place. This is *troponin*, a globular protein (MW, 80 kD) composed of three nonidentical subunits, *T*, *I*, and *C. Troponin-C* has a MW of 18 kD and acts as the calcium receptor; like calmodulin, it has four pockets for attaching Ca^{2+} ions, and these are separated by α-helices. *Troponin-T* interacts with *tropomyosin*; the latter protein, to which troponin is attached, is a long, rod-like molecule running alongside the actin filaments (see Figure 11.9). In the absence of Ca^{2+} but in the presence of troponin-T and troponin-I, tropomyosin prevents the myosin cross-bridges from attaching themselves to the actin filaments. However, the conformational change in the troponin molecule, following the combination of Ca^{2+} with troponin-C, lifts the tropomyosin rod away from the actin filaments and allows the myosin cross-bridges to engage.

Calsequestrin Binds Ca^{2+} in the SR

The fourth binding protein, *calsequestrin*, is also extremely important in the skeletal muscle fiber and is responsible for collecting Ca^{2+} ions in the sarcoplasmic reticulum when the fiber is not contracting. It is found in highest concentration in the terminal cisternae and is attached to that part of the membrane facing the transverse tubules. Calsequestrin has a MW of approximately 40 to 50 kD. The Ca^{2+} binding of calsequestrin is impressive, since MacLennan and Wong (1971) have estimated that a single molecule can capture 970 nmol of Ca^{2+}. Calsequestrin is therefore considerably more powerful than the other Ca^{2+}-binding proteins in the skeletal muscle fiber. When the muscle fiber becomes activated, the protein gives up its Ca^{2+} ions for release through the ryanodine channels of the sarcoplasmic reticulum.

Table 7.2 summarizes the main features of the major Ca^{2+}-binding proteins in the skeletal muscle fiber.

Calcium Pumps

The surface membrane of the skeletal muscle fiber, like that of other vertebrate and invertebrate cells, has a pump that expels Ca^{2+} from the interior of the fiber (Schatzmann, 1989). Because of the essential role of Ca^{2+} ions in the contractile process, the muscle fibers have a second type of pump in the membrane of the sarcoplasmic reticulum (Figure 7.13). This pump removes Ca^{2+} ions from the cytosol, thereby terminating the activation of the contractile filaments, and transfers them into the lumen of the sarcoplasmic reticulum.

Ca^{2+} Pumps in the Surface and SR Membranes of the Muscle Fiber Are Similar

The two types of Ca^{2+} pump have extensive similarities in their molecular structures, and each consists of two

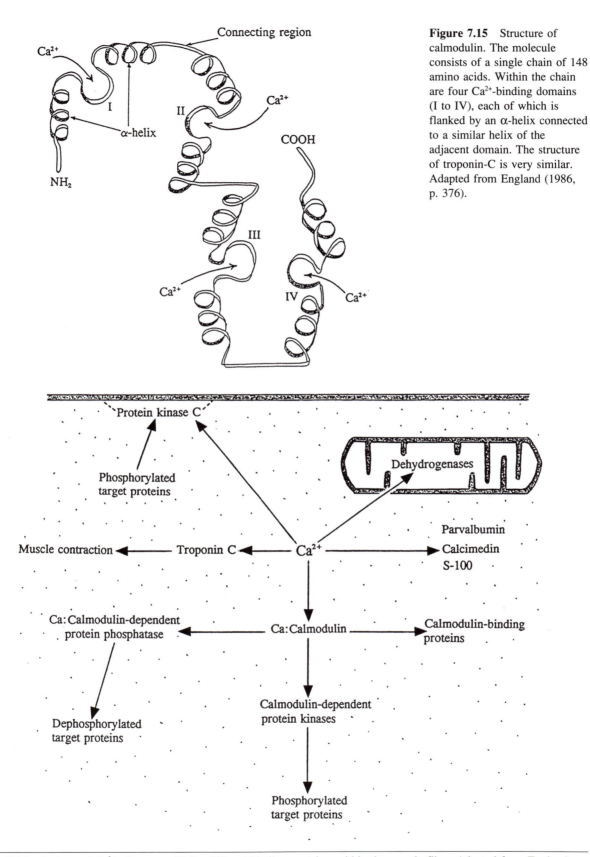

Figure 7.15 Structure of calmodulin. The molecule consists of a single chain of 148 amino acids. Within the chain are four Ca^{2+}-binding domains (I to IV), each of which is flanked by an α-helix connected to a similar helix of the adjacent domain. The structure of troponin-C is very similar. Adapted from England (1986, p. 376).

Figure 7.16 Actions of Ca^{2+}, alone or with its different binding proteins, within the muscle fiber. Adapted from England (1986, p. 375).

Table 7.2   Major Ca²⁺-Binding Proteins of Skeletal Muscle

Protein	MW (kD)	Location	Function
Calmodulin	16.8	Cytoplasm	Ubiquitous Ca^{2+}-dependent modulator of protein kinase and other enzymes
Parvalbumin	12	Cytoplasm (mainly fast-twitch muscle fibers)	Unknown (possible protective role in prolonged contraction by decreasing $[Ca^{2+}]_i$)
Troponin-C	18	Actin filaments	Trigger muscle contraction
Calsequestrin	40-50	Interior of SR	Sequester Ca^{2+} within SR

identical polypeptide chains; the *surface membrane pump* (SMP) has a MW of 140 kD and is rather larger than the *sarcoplasmic reticulum pump* (SRP) (MW, 110 kD). Part of the reason for the greater size of the SMP is that the molecule requires a Ca:calmodulin binding site in order for the pump to be activated. In contrast, the SRP responds to Ca^{2+} ions directly. Both pumps move Ca^{2+} ions aginst large concentration gradients (approximately 1:10,000), and both are stimulated by the presence of Mg^{2+} ions.

As expected of such ionic pumps, there is a higher affinity for Ca^{2+} ions in the cytosol than in the interstitial fluid or in the lumen of the sarcoplasmic reticulum. As with other pumps, the energy comes from the hydrolysis of ATP; for each molecule of ATP, two Ca^{2+} ions are transported across the membrane. It now appears that the pump moves H^+ ions in the opposite direction. The exchange is not electrically neutral, however, since only one H^+ ion is returned for each divalent Ca^{2+} ion; like the Na^+-K^+ pump, the Ca^{2+} pump increases the negativity of the cytosol bordering the membrane. It is probable that the phosphorylation of the pump, using ATP, takes place in a large region extending into the cytosol. The resulting perturbation is transmitted to parts of the molecule within the membrane, and Ca^{2+} ions are then displaced (Inesi & Kirtley, 1990). As in the case of the Na^+-K^+ pump, the membrane portion of the Ca^{2+} pump is arrayed in several helices that traverse the membrane and are linked by amino acid chains; indeed, there may be as many as 10 helices (Clarke, Loo, Inesi, & MacLennan, 1989). Among four of the transmembrane segments are six regions that contain glutamate or other acidic amino acid residues and that have been identified as the sites of calcium binding; it is likely that the four helices form a pore, rather like that inside an ion channel. Compared with an ion channel, however, the maximum rate that ions can be transferred by the pump is quite low. Thus, whereas 10×10^8 ions can flow through a single cation channel in one second, a Ca^{2+} pump can only deliver 20 ions in the same time. The sarcoplasmic reticulum compensates for this low pumping capacity by having the Ca^{2+}-pump molecules packed tightly together in the membrane of the terminal cisternae.

Anion Channels

Although the more dramatic events of cell life are mediated by cations, the surface membrane also contains channels for anions. In the case of skeletal muscle fibers, the most abundant anion in the extracellular fluid, Cl^-, evidently has more channels available to it than does K^+. Thus, the permeability of the resting membrane is several times higher to Cl^- than to K^+ (Hutter & Noble, 1960).

Plentiful Cl⁻ Channels Stabilize Membranes

The Cl^- channels resemble the K^+ ones in that they tend to stabilize the membrane potential. In other tissues, Cl^- channels are involved in cell volume regulation, signal transduction, and transport across epithelial surfaces. The channels are of several types, some being ligand gated and others voltage gated, although the cystic fibrosis channel appears to have a different mechanism. In skeletal muscle, it is the voltage-gated channel that is especially important; abnormal functioning of this channel, as the result of a genetic mutation, leads to hyperexcitability of the muscle fiber plasmalemma with spontaneous contraction (*myotonia*; see Box 7.2).

The structures of the Cl^- channels have been determined quite recently, using molecular biology techniques. The first channel to be cloned was that of the electric organ of the electric ray. *Torpedo marmorata* (Jentsch, Steinmeyer, & Schwarz, 1990), and this, in turn, led to cloning of the skeletal muscle channel in the same laboratory (Steinmeyer, Ortland, & Jentsch, 1991). The skeletal muscle channel (ClC-1) probably contains four subunits (Steinmeyer, Lorenz, Pusch, Koch, & Jentsch, 1994), each with a MW of 100 kD. The subunits are unlike those of the cation channels in having a greater number of transmembrane domains (12), with a further domain in the cytoplasm.

It is likely that the Cl^--channel density in the muscle fiber membrane is relatively high, since the current admitted by each channel is small (1 pS); for the same reason it has not been possible to carry out patch clamping. Nevertheless, it is known that channel opening is greatest at membrane potentials slightly higher than the normal resting potential and that it decreases when the membrane depolarizes. The Cl$^-$ channel can be blocked by Zn^{2+} and by a number of aromatic monocarboxylic acids.

APPLIED PHYSIOLOGY

In this section, there is rather a lot of ground to cover, because, as we shall see, the Na$^+$ channels can be affected by a number of agents, and these and other channels can be at fault in various hereditary disorders.

Na^{2+} Channels Are Blocked by Local Anesthetics and Biological Toxins

It has long been known that cocaine, when applied to peripheral nerves or mucous membranes, can cause anesthesia and paralysis. Derivatives of cocaine, such as procaine and xylocaine, have been employed as *local anesthetics* for many years and produce their effects by blocking the Na$^+$ channels. Powerful though these agents are, they are much less effective than some of the naturally occurring toxins, all of which may cause death. *Tetrodotoxin* (TTX) is secreted by the Japanese puffer fish. Its poisonous effects were well-known to the Chinese, who referred to it in the first herbal pharmacopoeia, *Pen-T'so Chin* (written during the life of the Emperor Shun Nung, 2838-2698 B.C.). One of the victims of tetrodotoxin poisoning was the circumnavigator, Captain James Cook, who fortunately survived and was able to provide a fascinating account of the paralysis and loss of sensation that had overtaken him (Cook, 1777). *Saxitoxin* (STX), another Na$^+$-channel blocker, is produced by the marine alga, *Gonyalaux catanella* (Kao, 1966). When plentiful in the sea, the algae form a ''red tide'' and infest shellfish; the latter, when eaten, induce paralysis and can prove fatal. *Batrachatoxin* also causes severe paralysis and is secreted in the skin of a small South American frog. Other toxins interfere with Na$^+$ channels by keeping them open and can be equally deadly. Thus, a sting from the tail of a scorpion causes local swelling and a sharp, burning sensation, followed by sweating and salivation. Numbness and muscle twitching develop and are associated with restlessness and increasing respiratory distress. There is pain in the abdomen and chest, together with vomiting. Finally, in severe cases, convulsions and loss of consciousness lead to death.

Aconitine is another Na$^+$-channel-opening poison. The alkaloid is contained in the monkshood, a blue-flowered, wild plant found in Europe and North America. Ingestion causes tingling and numbness in the mouth, throat, and skin. There is also restlessness, incoordination, vomiting, diarrhea, and convulsions. Rather similar symptoms result from swallowing *veratridine*, one of the alkaloids produced by the false hellebore plant.

TTX-Resistant Na$^+$ Channels Appear After Denervation

Following denervation, a new type of Na$^+$ channel is synthesized by skeletal muscle fibers and is incorporated in the surface membrane. This channel can be distinguished from the normal Na$^+$ channel by its resistance to tetrodotoxin and can be detected as early as 36 hr after denervation (Harris & Thesleff, 1972).

Episodes of Paralysis Can Occur in a Rare Familial Disorder

The condition *familial hyperkalemic periodic paralysis* (also known as *adynamia episodica hereditaria* or *Gamstorp's syndrome*) is one in which those affected are prone to attacks of muscle weakness. The disorder is inherited by an autosomal dominant gene, and therefore affects both sexes. The attacks of paralysis tend to occur during daytime and can be provoked by resting after heavy exercise, by fasting, or diagnostically, by ingesting K$^+$. The weakness may be preceded by muscle stiffness, due to spontaneous firing of action potentials by the muscle fibers (*myotonia*). Associated with the weakness is a rise in the K$^+$ concentration in the plasma, such that values of 6 to 7 mM are reached; the attacks characteristically last from 1 to 3 hr and do not progress to complete paralysis of the affected muscles. In one patient with this disorder, Brooks (1969) was able to record resting membrane potentials during an attack and found that the muscle fibers were severely depolarized (Figure 7.17); further, they could not respond to stimuli applied through the intracellular microelectrode.

Subsequently, in a specimen of excised intercostal muscle from another patient, Lehmann-Horn et al. (1987) found that the defect in the muscle fibers was due to the presence of an abnormal Na$^+$ channel. This channel resembled the normal Na$^+$ channel by opening in response to muscle depolarization or to small elevations in K$^+$ concentration, but it differed in failing to inactivate; that is, the channel stayed open. More recently, very tight linkage has been found in a large pedigree between the susceptibility to the episodic

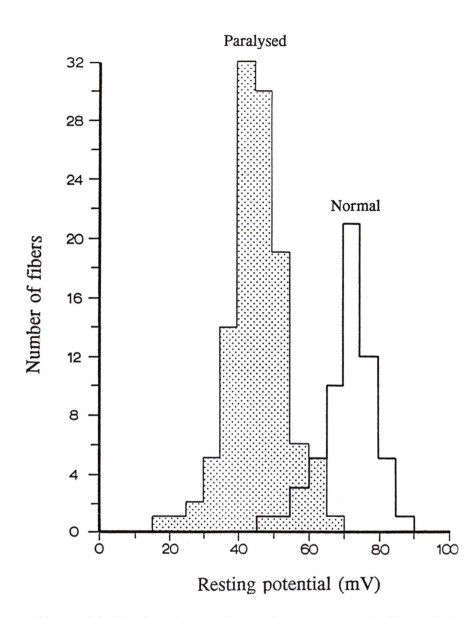

Figure 7.17 Measurements of resting membrane potentials in a patient with the hyperkalemic type of familial periodic paralysis. In between attacks of paralysis, the values were normal (open histogram), but during an episode of weakness, depolarization occurred (shaded histogram), and the fibers became inexcitable on direct stimulation. Adapted from Brooks (1969).

weakness and the Na^+-channel gene on human chromosome 17 (Fontaine et al., 1990). This linkage indicates that the abnormal Na^+ channel has the molecular defect in this condition; recent work has shown that there is an amino acid substitution in that part of the protein joining the S5 and S6 segments (Figure 7.2). This and other types of Na^+ channel disorder have recently been reviewed by Rüdel, Ricker, and Lehmann-Horn (1993).

Myotonic Muscular Dystrophy: A Na^+-K^+-Pump Defect?

So far there are no clinical disorders that are definitely known to arise from defective Na^+-K^+ pumping. One possibility, however, is the myotonic form of *muscular dystrophy* (dystrophia myotonica; Steinert's disease). This is a hereditary disorder, transmitted by an autosomal dominant gene, in which there is degeneration of muscle fibers and also a characteristic stiffness of the muscles. After making a fist, for example, patients may require several seconds before they can straighten their fingers again. As in familial hyperkalemic periodic paralysis (see the preceding section), the stiffness, or *myotonia*, is due to the spontaneous firing of action potentials by the muscle fibers. What makes this disease especially fascinating is the great variety of associated clinical defects—frontal baldness, hyperostosis of the skull, cataracts, ovarian and testicular atrophy, and insulin-resistant diabetes. Even the central nervous system is affected, for patients may undergo progressive mental deterioration and eventually become demented; excessive sleepiness is also a problem for many. The gene for this condition is found on chromosome 19 and codes for a serine-threonine protein kinase (Fu et al., 1992). Several observations suggest that one of the consequences of an abnormality in this enzyme is impaired

Na$^+$-K$^+$ pumping. Among the various clues are the several-fold reduction in Na$^+$-K$^+$-pumping activity in skeletal muscle biopsies, the increased intracellular Na$^+$ concentration in resting muscle fibers, and the inability of the muscle fibers to potentiate their action potentials promptly during contractions (a Na$^+$-K$^+$-pump mechanism; see Hicks & McComas, 1989; Fenton, Garner, & McComas, 1991).

Abnormal Cl$^-$ Channels Also Cause Muscle Stiffness

As already noted, myotonia is the clinical term used to describe an involuntary stiffness of muscles, caused by hyperexcitability of muscle fibers. The membrane defect can be due to prolonged opening of Na$^+$ channels (in familial hyperkalemic periodic paralysis), or, possibly, to defective Na$^+$-K$^+$-pump activation, allowing K$^+$ to accumulate in the interstitial spaces and to depolarize the muscle fibers (myotonic muscular dystrophy). In *myotonia congenita*, however, there is a third cause for the muscle stiffness—the muscle fiber membranes have a reduced Cl$^-$ permeability. An abnormality of Cl$^-$ channels was first detected in a strain of goats with hereditary myotonia. These curious animals, all descended from a small flock in Tennessee, are prone to develop rigidity of their legs and to topple over when startled (see Box 7.2). Bryant (1969) showed that the membrane resistance in myotonic goat muscle fibers at rest was considerably increased; this elevation was later shown to be due to the absence of the chloride component of the membrane conductance (Bryant & Morales-Aguilera, 1971).

How could a Cl$^-$ defect in the membrane cause myotonia? It will be recalled that in a resting mammalian muscle fiber, there are probably more open Cl$^-$ channels than K$^+$ ones, because the membrane conductance is several times greater for Cl$^-$ than for K$^+$. Further, both types of channel are responsible for stabilizing the membrane potential by counteracting *small* increases in Na$^+$ permeability, which might otherwise become regenerative. This stabilizing effect of K$^+$ and Cl$^-$ channels is evident from the inclusion of terms for permeability and concentration ratio for both these ions in the Goldman-Hodgkin-Katz equation for the resting membrane potential (see chapter 9). In a later study of goat myotonia, Adrian and Bryant (1974) pointed out that if Cl$^-$ permeability is reduced in the surface membrane of the muscle fiber, then the accumulation of K$^+$ ions in the transverse tubules during contractile activity will exert a powerful depolarizing effect on the fiber. This depolarization, in turn, will activate the Na$^+$ channels and cause impulses to be initiated.

Following the recognition of a Cl$^-$-channel defect in the myotonic goat, Lipicky and Bryant (1973) were able to demonstrate the same phenomenon in four of six patients with myotonia congenita. The final proof that a defective chloride channel is responsible for myotonia congenita has come from molecular biology studies, in which an abnormal gene on chromosome 7 has been found. In two families with the autosomal recessive form of the condition (Becker's disease), this gene coded for a Cl$^-$ channel with a phenylalanine-to-cysteine substitution in one of the transmembrane domains (Koch et al., 1992). A different mutation of the same gene accounts for those patients with autosomal dominant inheritance (Thomsen's disease; Steinmeyer, et al. 1994). In neither type of myotonia congenita is there any degeneration of the muscle fibers, so that the disorder cannot be classified as a muscular dystrophy. If anything, the muscles are usually well developed, presumably because the prolonged contractions cause muscle fiber hypertrophy. An interesting aspect of the disorder is that the stiffness is reduced if repeated voluntary contractions are made; this is the *warm-up phenomenon*.

A DHP-Channel Abnormality: Muscular Dysgenesis

There is a very interesting disorder of mice, caused by a mutant gene, in which the animals are almost totally paralyzed. This condition, named *muscular dysgenesis*, is considered in chapter 11.

Although a large amount of material has been covered in this chapter, it is essential for a proper understanding of the ionic basis of the resting and action potentials (chapter 9) and for biochemical events underlying excitation-contraction coupling (chapter 11). Before embarking on these topics, however, there is one more aspect of the motoneuron to examine.

BOX 7.2   THE MYOTONIC GOAT

The first identification of a disorder caused by an abnormal ion channel was made in a strain of goats. These animals, introduced into the southern United States during the 1880s under mysterious circumstances (McComas, 1977), attracted attention by their remarkable behavior when startled. Instead of running away, their legs became so stiff that the goats were unable to move and often toppled over. The stiffness could last for up to 1 min, after which the muscles gradually relaxed and the animals were able to move about once more. Some engine drivers, aware of this phenomenon, delighted in observing the effects of the train whistle as they approached a field containing these animals. An amusing anecdote was given by Mayberry (see McComas, 1977):

> I've heard a story about a new hired man who, a few days before a barbecue, was given a scatter gun and told to go into the back pasture and kill one of the goats. He was advised that the goats were very shy and that he should be very careful not to apprise them of his presence until ready to shoot. After crawling up cautiously, he picked out a nice fat kid, took careful aim, and fired. Goats dropped in every direction, some thirty animals collapsing simultaneously. Aghast, and without waiting for the resurrection, he ran back to the house. "I don't know how it happened," he panted. "I only fired once but I killed every damn one of them goats."

Why did the muscles of these goats become stiff?

The first investigation of the myotonic goat was made by Brown and Harvey in 1939. They recorded from affected muscles with needle electrodes and found that the stiffness was associated with rapid discharges of action potentials. Further, the electrical activity could still be elicited after nerve section and after curarization. This last observation excluded the possibility that the discharges could have arisen at the end-plate following spontaneous release of acetylcholine. Rather, it appeared that there was abnormal excitability of the muscle fibers themselves. The story of these fascinating animals was taken up some 30 years later by Bryant and his colleagues in Cincinnati. By injecting current through an intracellular microelectrode, Bryant found that the resistances of the muscle fiber membranes were much higher in myotonic animals than in normal ones (Bryant, 1969). The increased resistance was later shown to be due to a reduced Cl$^-$ permeability of the membrane; further, normal fibers became hyperexcitable if placed in a chloride-free bathing medium.

Bryant later took some of the myotonic goats to Cambridge, England, and carried out further studies with Richard Adrian. They showed that the repetitive discharges in the muscle fibers arose from unusually large negative after-potentials that followed the spike potentials (Figure 7.18). The after-potentials and the repetitive activity were abolished by disrupting the necks of the transverse tubules with glycerol. They reasoned that the negative after-potentials were due to depolarization of the muscle fibers by the accumulation of K$^+$ ions in the transverse tubules following impulse activity. In a normal fiber, the after-potentials were much smaller because the high Cl$^-$ permeability of the surface membrane acted as a stabilizing influence; that is, the membrane potential of the fiber would be intermediate between the equilibrium potentials for Cl⁻ (at the surface membrane) and for K⁺ (at the tubular membrane). This point can be appreciated by referring to the Goldman-Hodgkin-Katz equation (see chapter 9).

It was through the study of the myotonic goat that Lipicky and Bryant (1973) were able to identify a similar Cl⁻ impermeability in the muscles of patients with the genetic disorder, *myotonia congenita*, considered earlier.

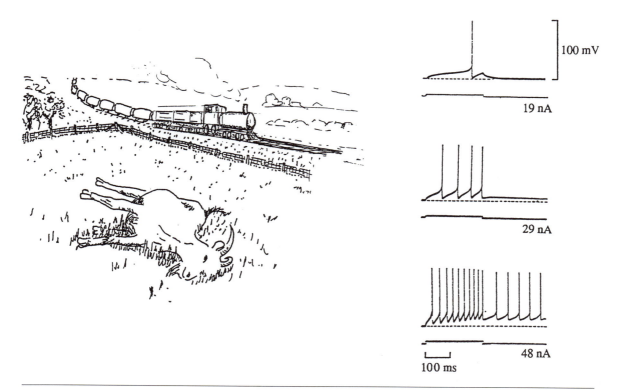

Figure 7.18   Responses of a muscle fiber from a myotonic goat to depolarizing pulses of current injected through an intracellular microelectrode. Unlike the responses of a normal fiber, there are repetitive discharges of action potentials as the stimulus intensity increases; this pattern is characteristic of myotonia.

Adapted from Adrian and Bryant (1974).

Chapter 8

Axoplasmic Transport

The function of the motoneuron axon is to act as a communication pathway between the spinal cord and muscle. Rapid signaling to the periphery is required if the muscle is to contract and is achieved by the nerve impulse; the ionic mechanisms underlying this event are considered in the next chapter. Equally important, however, is a slow signaling system that enables messages, coded in the form of chemicals, to be sent in both directions between the cell body of the motoneuron and the muscle fibers that it innervates. These messages are essential for the maintenance of the normal structure and cellular metabolism of both the muscle fiber and the motoneuron; the integrity of the Schwann cells and of the axis cylinder are also dependent on this chemical signaling system. These controlling effects are described as *trophic*, and they are analyzed in chapter 18. Of immediate concern is the nature of the axoplasmic transport that makes this chemical signaling system possible. It will be shown that a fast transport system operates in both directions between the motoneuron soma and muscle fiber; there is also a slow transport system which runs in a distal direction only (Table 8.1). In addition to their trophic roles, the transport systems serve other needs; thus, they carry enzymes and other special proteins to the nerve terminal, together with newly formed

mitochondria. In addition, aging mitochondria are returned to the soma for degradation or repair. Finally, there is a very slow movement of microtubules and neurofilaments down the axon, made possible by polymerization in the soma and depolymerization in the axon terminal.

Centrifugal Transport

The earliest intimation that there was a flow of cytoplasm from the cell body of the motoneuron outward along the axon came from a series of experiments by Paul Weiss and his colleagues in the 1940s (Weiss, 1944; Weiss & Davis, 1943; Weiss & Hiscoe, 1948). The results of these and later experiments have been summarized by Weiss (1969).

Slow Axoplasmic Flow Was Discovered by Constricting the Nerve

The technique employed in Weiss's laboratory was to apply a gentle constriction to the sciatic nerve of an

Table 8.1   Characteristics of Fast vs. Slow Axoplasmic Flow

Subtype	Fast			Slow	
	I	II	III	SCb	SCa
Velocity (mm/day)	240-410	34-68	4-8	2-4	0.1-2
Transported material	Organelles, mitochondria, vesicles			Cytoplasm, enzymes, microtubules, neurofilaments, actin filaments	
Transporting structure	Microtubules			Neurofilaments, microtubules, actin filaments	
Driving mechanism	Protein motors (kinesin-anterograde) (dynein-retrograde)			Extension of microtubules and filaments, bulk movement of cytoplasm, ? protein motor (dynamin)	
Blocking agents	Inhibitors of oxidative metabolism (DNP, NaCN, anoxia), colchicine			Inhibitors of protein synthesis (cycloheximide), colchicine	

animal by investing the nerve with a sleeve of artery. As the arterial cuff gradually contracted, due to its own elasticity, the nerve fibers underwent compression to varying extents, and the chronic effects of this were studied after several months. It was found that the segment of nerve abutting the proximal end of the constriction was swollen, partly because of edematous fluid between the nerve fibers but mainly on account of distortions and distensions of the axons themselves. With the light microscope, the axis cylinders appeared to be either beaded or else corkscrewed, ballooned, or telescoped (Figure 8.1). Within and beyond the constriction, the axis cylinders were narrowed.

To Weiss and his colleagues, the appearances of the axons suggested that there had been a ''damming up'' of axoplasm at the site of constriction. This conclusion implied that, under normal circumstances, axoplasm must be manufactured in the cell body and somehow propelled down the axon. Supporting evidence for Weiss's interpretation came from a subsidiary experiment in which the constriction was removed (Weiss & Cavanaugh, 1959). The dammed-up material was then released and could be observed to travel down the axon as a wave with a velocity between one and several millimeters per day. To Weiss these results indicated that the axons were perpetually growing from the cell bodies of the motoneurons and that the contents of the new axoplasm were then consumed distally as part of a continual replenishment process.

Although some of the conclusions reached by Weiss were challenged by Spencer (1972) on the basis of electron micrographs of ligated nerve, the existence of *slow* axoplasmic flow is not seriously doubted (Weiss & Cavanaugh, 1959). Much of the supporting evidence has come from the results of isotope experiments. For

example, Droz and Leblond (1963) showed that protein labeled with [3H]-leucine moved distally in rabbit sciatic nerve with a velocity of about 1 mm per day.

Slow Axoplasmic Flow Has Two Components

At present, two components of slow transport are recognized, and both are largely associated with movements of the three main cytoskeletal proteins—actin, tubulin, and neurofilament protein. The slower fraction, *slow component a* (SCa), travels at 0.1 to 2 mm/day and contains mainly tubulin and neurofilament protein. The faster phase, *slow component b* (SCb), moves at 2 to 4 mm/day; it contains actin and tubulin, together with cytoplasmic enzymes, clathrin, spectrin, and calmodulin (Lasek, Garner, & Brady, 1984).

Exactly how the cytoskeletal proteins move down the axon is not clear. Since the movement is slower than that of the axoplasm, it would appear that for much of the time the microtubules, neurofilaments, and actin filaments are stationary, possibly because of phosphorylation (see chapter 2). When the microtubules and filaments begin to move again, they may do so either as an interconnected lattice or as separate elements. Further, it is probable that molecules of the cytoskeletal proteins may be added from the axoplasm to the formed structures or else removed from the latter (for review, see Nixon & Sihag, 1991; Vallee & Bloom, 1991).

Fast Axoplasmic Flow Also Occurs and Has Been Studied With Radioactive Isotopes

After the clear demonstration of axoplasmic flow (transport) with the damming experiments described above,

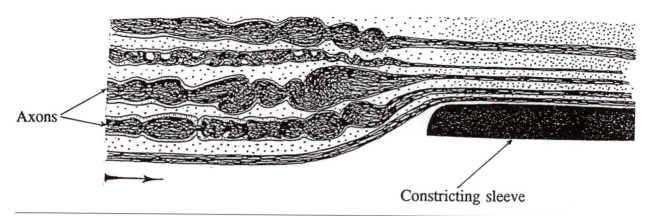

Axons

Constricting sleeve

Figure 8.1 Drawings of four axons showing characteristic deformations following chronic constriction. Weiss and Hiscoe (1948) described the deformations, from below upward, as "telescoping, ballooning, corkscrewing, and beading." From Weiss and Hiscoe (1948, p. 330).

attempts were made to explore the phenomenon more fully by labeling the axoplasm with radioactive isotopes and then studying their passage down the axon. The first experiments involved radioactive phosphorus, and later ones employed labeled carbon in glucose and amino acids; these methods have been superseded following the availability of tritiated amino acids. A commonly used strategy is to inject [³H]-leucine into the ventral gray matter of the lumbar region of the spinal cord; the amino acid is then taken up by the cell bodies of neighboring motoneurons and rapidly incorporated into newly synthesized proteins. At a given time after injection, the sciatic nerve is excised and cut into segments of uniform length. The amount of radioactivity in each segment is then measured with a scintillation counter and expressed as a function of distance from the spinal cord.

An example of fast axoplasmic transport is given in Figure 8.2, the lower curve of which shows the distribution of radioactivity in the sciatic nerve of the cat 6 hr after injection of [³H]-leucine into the L7 segment of the cord (Ochs & Ranish, 1969). Included for comparison, and displayed in the upper curve, are the results for sensory nerve fibers following an injection given at approximately the same time into the L7 dorsal root ganglion on the opposite side. The peak of radioactivity in the sensory fiber curve corresponds to isotope remaining in the ganglion; similarly, the high value at the left of the motor fiber curve denotes the activity left in, and around, the cell bodies of the motoneurons. The arrows at the bottoms of the two curves indicate the advancing fronts of the labeled axoplasm; the spatial disparity corresponds to the extra length of the motor pathway due to the inclusion of the L7 ventral root.

Studies such as these have demonstrated that, in addition to the slow movement of axoplasm already described, there is a very much faster flow with a maximum velocity, in mammals, of about 410 mm/day. Another technique is to produce a temporary arrest of axoplasmic flow by cooling a short length of nerve; the transported material then accumulates in, and proximal to, the cooled region. Upon rewarming the nerve, the movement of the material is resumed, and its velocity can be determined. This "stop-flow" method has been used by Brimijoin (1975) to study the passage of the enzyme dopamine-β-hydroxylase in rabbit sciatic nerve; a flow rate of 300 ± 17 mm/day was found.

Fast Axoplasmic Flow Has Three Components

The presence of *fast* axoplasmic flow is a property of all the motor and sensory axons that have been tested to date, and it has been demonstrated that the velocity of the flow is independent of fiber diameter. Isotope studies suggest that the fast-transported material travels at three different rates, approximately 240 to 410 mm/day, 34 to 68 mm/day, and 4 to 8 mm/day. However, the absolute values vary between one species and another and also between different nerves within the same species (Grafstein & Forman, 1980). Impulse activity has no effect on the rate of transport, although it may increase the amount of material leaving the cell soma (e.g., Lux, Schubert, Kreutzberg, & Globus, 1970) and possibly also that appearing in the synaptic cleft (Musick & Hubbard, 1972). Finally, it should be noted that the dendrites also exhibit fast and slow axoplasmic transport, though the phenomenon has only been studied in the olfactory nerve of the goldfish (Droz, Rambourg, & Koenig, 1975).

Newly Synthesized Proteins, Organelles, and Other Material Are Transported

By injecting inhibitors of protein synthesis, such as puromycin or cycloheximide, it has been shown that

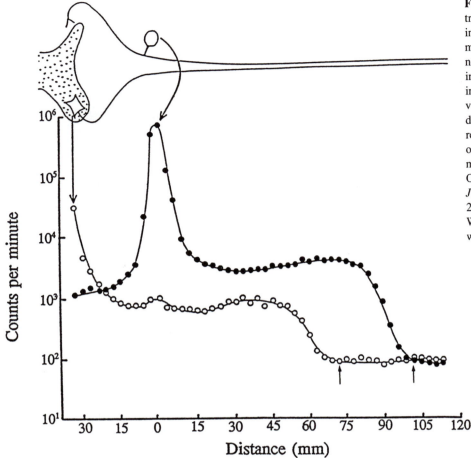

Figure 8.2 Centrifugal transport of radioactive leucine in sensory nerve fibers (●) and motor fibers (○) of cat sciatic nerve, measured 6 hr after injection of labeled material into dorsal root ganglion and ventral horn. See text for description of technique and results. From "Characteristics of the fast transport system in mammalian nerve fibres," by S. Ochs and M. Ranish, 1969, *Journal of Neurobiology*, **1**, p. 258. Copyright 1970 by John Wiley & Sons, Inc. Adapted with permission.

the motoneuron soma is able to complete its synthesis of protein within 10 to 20 min of the arrival of the labeled precursor, [³H]-leucine. Not all the amino acid is converted into protein, since polypeptides are also labeled, and there is a large fraction that is taken up by small axoplasmic particles.

Other studies have shown that lipids and polysaccharides are also rapidly conveyed down the axon. Included among the transported proteins are the enzymes acetylcholinesterase (AChE) and choline acetyltransferase; in addition Miledi and Slater (1970) have shown that a fast-traveling factor(s) is necessary for the maintenance of the axon terminal at the neuromuscular junction. Among a large number of proteins as yet unidentified are those that supervise the metabolic machinery of the axon, Schwann cells, and muscle fibers, in keeping with the trophic role of the motoneuron (see chapter 18). Of the particulate matter transported down the axon, the most prominent fraction is the mitochondria, and these can be observed in motion with phase-contrast and interference microscopes. The suggestion has been made that the mitochondria are transported to the axon terminal to supply energy needed for the synthesis of ACh, and that their aging components are continually replaced by

proteins conveyed by fast axoplasmic transport. Other transported particles include various kinds of vesicle, smooth endoplasmic reticulum, and plasma membrane.

Fast Axoplasmic Transport Requires a Local Supply of ATP

It has been suggested that the Golgi apparatus of the cell body acts as a "gate" by controlling the delivery of the material into the axon (Ochs, 1972). As far as the transport mechanism itself is concerned, a number of observations are relevant. First, axoplasmic flow is able to proceed in both directions at a normal rate even when the axon has been removed from the body; this finding indicates that the transport system obtains its energy from local sources in the axon (or the Schwann cell).

Other experiments show that fast transport is dependent on ATP supplied by oxidative metabolism. For example, transport can be blocked if a small segment of nerve is made anoxic. Alternatively, blocking can be produced by interfering with oxidative metabolism, by means of agents such as sodium cyanide or dinitrophenol. Glycolysis alone is not a sufficient source of ATP, for blocking will still occur, though slowly, in the presence of iodoacetic acid. Another requirement for axoplasmic transport is a critical concentration of Ca^{2+} in the interstitial fluid.

Kinesin Is a Motor Molecule Involved in Fast Axoplasmic Transport

How is the material conveyed along the axon by fast axoplasmic transport? In the case of slow transport, it was seen that entire neurofilaments and microtubules moved, either as relatively short segments or as extended linear structures. For fast transport, however, the microtubules serve as tracks, along which molecular *motors* carry the particulate matter. It is probable that in any living cell there are 50 or so distinct molecular motors that run on microtubules (and actin filaments) within the cytoplasm; additional motors move along strands of DNA and messenger RNA (Spudich, 1994). The picture which emerges is not unlike a complicated model railway layout, in which engines are constantly pulling wagons in different directions along crisscrossing tracks. Of these motors, *kinesin* is especially important in axoplasmic transport and was originally identified in the squid giant axon and in chick and calf brains (Brady, 1985; Vale, Reese, & Sheetz, 1985b). Kinesin's role as a motor was demonstrated by its capacity to move coated latex beads along an *in vitro* array of microtubules (Vale, Reese, & Sheetz, 1985a); for kinesin to transport an organelle *in vivo*, however, requires the additional presence of one or more accessory factors in the cytoplasm (Sheetz, Steuer, & Schroer, 1989; Figure 8.3). Since kinesin only moves particles toward the distal (+) end of a microtubule, it is obviously the main motor for anterograde transport. Analysis of the purified protein reveals that it has a molecular weight of 380 kD and is composed of two pairs of polypeptides. Within each pair there is a globular head connected to a feathered tail by a long stalk; it is probably the ability of the head to change its angle with the stalk, following interaction with ATP, that gives the protein its motor property.

Until recently it was not known whether the kinesin molecules remained attached to the microtubules or to the transported organelle. In the first case, the organelle would be passed along the microtubule from one kinesin molecule to another; in the second situation, a kinesin molecule would keep hold of an organelle and would carry it along the microtubular track. Svoboda, Schmidt, Schapp, and Black (1993) have now shown the second possibility to be the correct one. With an extraordinarily sensitive technique, combining optical "tweezers" with laser interferometry, these authors were able to study individual kinesin molecules. They found that each kinesin molecule moves along one of the protofilaments in a microtubule in a series of steps (see Figure 2.4, chapter 2). The length of one of the steps is 8 nm, exactly that required to span a tubulin dimer in the protofilament (Figure 8.4). By keeping one of its heads attached while the other head steps past, the kinesin molecule is prevented from straying onto adjacent protofilaments (Berliner, Young, Anderson, Mahtani, & Gelles, 1995).

Centripetal Transport

So far no consideration has been given to the axoplasmic flow that occurs in a retrograde direction, that is, from the muscle fiber to the motoneuron. Evidence for such a movement has come from several sources. For example, Lubinska and Niemierko (1970) found that AChE is transported proximally as well as distally; the rate of the centripetal flow is about one half of the centrifugal one. Other workers have observed particles, probably mitochondria, being carried in axons away from the periphery.

Transported Material Reaches Motoneuron Cell Bodies

Proof that some of the centripetally transported material arrives in the cell bodies of the motoneurons has been obtained by Glatt and Honegger (1973). These workers injected albumin coupled with the fluorescent dye, Evans blue, into triceps muscles of the rat forelimb and were able to detect dye in the ventral horns some 12 hr later. In the study by Kristensson and Olsson (1971), HRP was used as a marker instead. This substance had the advantage that it could produce an electron-dense reaction product, and with the electron microscope, its intracellular location was seen to be in small cytoplasmic granules around the nucleus. Just as the centrifugal axoplasmic flow largely mediates the trophic control of axon, Schwann cell, and muscle fiber by the motoneuron, so the centripetal flow exerts a trophic influence on the motoneuron soma by the periphery. Interruption of this trophic influence, as in cutting the axon, sets in motion the phenomenon of chromatolysis in the motoneuron soma and attempts at regeneration in the proximal nerve stump (see chapter 18).

Dynein Is the Molecular Motor for Retrograde Transport

While anterograde transport is made possible by the protein motor, kinesin, retrograde transport depends on

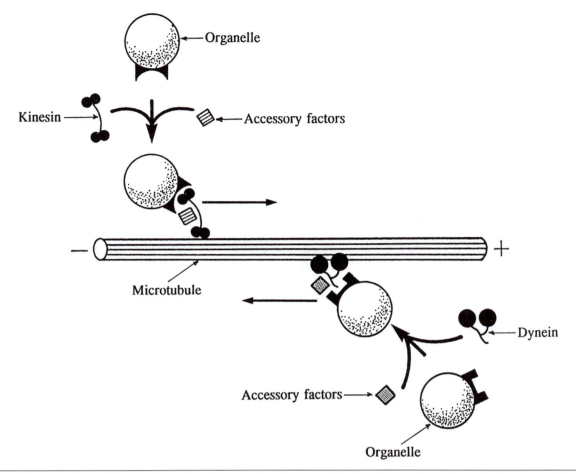

Figure 8.3   Assembly of ''organelle translocation complexes'' by combination of organelles with molecular motors (kinesin or dynein) and appropriate accessory factors. Adapted from Sheetz, Steuer, and Schroer (1989).

another motor, *dynein* (Vallee, Shpetner, & Paschal, 1989). The latter molecule had originally been identified as that responsible for the movements of cilia and flagella in other types of cell. It was differentiated from other *microtubule-associated proteins* (MAPs) by the ability to split ATP, a property shared with kinesin (and also myosin, see chapter 11). Proof of dynein's capacity to serve as a protein motor came from experiments in which a glass coverslip, coated with dynein, was able to propel microtubules in the presence of ATP (Paschal, Shpetner, & Vallee, 1987). In contrast to kinesin, dynein is attracted to the slow growing (–) end of the microtubule, which points to the motoneuron soma. Dynein is a rather larger protein than kinesin, being composed of two heads, each with its own stalk, and seven smaller subunits; the complete molecule has a molecular weight of 1200 kD (see Figure 8.5). At present it is not known if dynein and kinesin molecules can run along the same microtubule, carrying their respective organelles and other particles in opposite directions. An alternative possibility is that each organelle may have both types of molecular motor, but uses only one.

APPLIED PHYSIOLOGY

In this visit to the neuromuscular clinic, we encounter a second type of peripheral neuropathy. Whereas Charcot-Marie-Tooth disease, discussed in chapter 2, is a hereditary disorder, the second type is ''iatrogenic.'' This Greek-derived word means ''caused by the physician'' and sounds ominous; however, in most patients the benefits of the drug, which causes the neuropathy, outweigh the neurological complications.

Cancer Chemotherapy May Cause Peripheral Nerve Damage by Interfering With Tubulin

Vincristine and *vinblastine* are two alkaloids produced by the periwinkle plant (*Vinca rosea*), both of which are useful in the treatment of cancer, especially acute lymphoblastic leukemia, Hodgkin's disease, multiple myeloma, and lung cancer. Of the two compounds, vinblastine was the first to be used, in 1958, but vincristine was then found to be the more powerful and became the

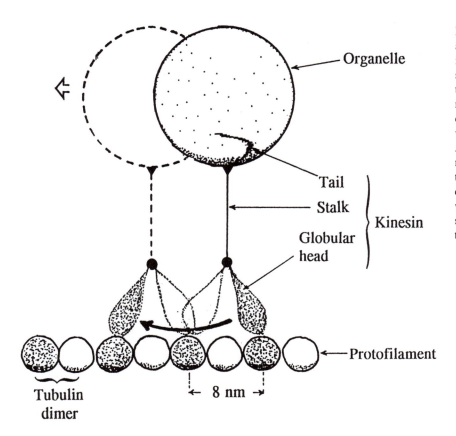

Organelle

Tail

Stalk

} Kinesin

Globular
head

Protofilament

Tubulin
dimer

← 8 nm →

Figure 8.4 Kinesin walking along a tubulin protofilament. It is probable that, in order to step, kinesin alters its shape in the same way that the myosin molecule does (see Figure 11.8, chapter 11). The kinesin head would flex on the stalk, once ATP had been inserted into the molecule and hydrolyzed. Note that one globular head retains contact with the protofilament while the other one (stippled) steps past and engages the next tubulin dimer.

Microtubule protofilament

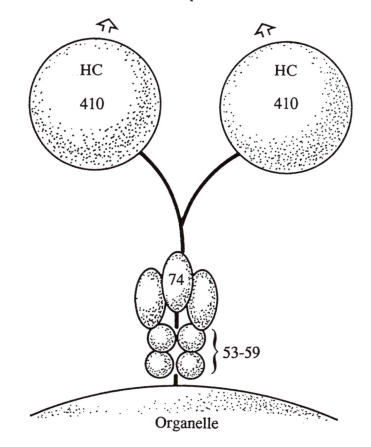

HC
410

HC
410

74

} 53-59

Organelle

Figure 8.5 Model of dynein. The two heavy chains (HCs) interact with the microtubule protofilament in the same way as kinesin, and the smaller subunits grip the organelle. Numbers refer to molecular weights (in kD) of subunits. Adapted from Valee, Shpetner, and Paschal (1989, p. 69).

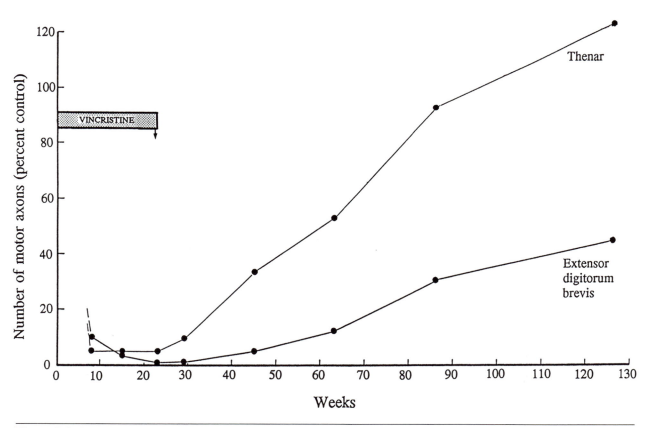

Figure 8.6   Recovery from a neuropathy induced by vincristine.

preferred therapeutic agent. Unfortunately, both drugs produce a peripheral neuropathy as a side-effect. By binding with the tubulin in the axis cylinders, they prevent microtubules from assembling and hence remove the track required for fast axoplasmic transport (Green, Donoso, Heller-Bettinger, & Samson, 1977); there is also an accumulation of neurofilaments in the peripheral nerve fibers (Shelanski & Wisniewski, 1969). These complications are unavoidable in that it is the combination of the vinca alkaloids with tubulin that also interferes with the formation of the mitotic spindle and, by preventing cancer cells from dividing, produces a beneficial therapeutic result. Through the reduction in axoplasmic transport, the motor and sensory neurons are no longer able to maintain the structural integrity of their respective axons, which then become inexcitable and degenerate.

In patients undergoing chemotherapy with vincristine, the first symptoms usually appear by 2 weeks and consist of numbness and tingling in the fingers and toes. Sensory loss is difficult to demonstrate at this stage, although an early rise in the thresholds of skin receptors can be demonstrated in experimental animals (Leon & McComas, 1984). Weakness appears rather later, and

is initially recognized by clumsiness of the fingers with some degree of wrist and foot drop; muscle cramps may also be prominent (Casey, Jellife, Le Quesne, & Millett, 1973). An example of a severe vincristine-induced neuropathy is shown in Figure 8.6. The patient, a 34-year-old plumber, was treated for embryonal carcinoma of the testis with surgery and multiple chemotherapy. When seen 8 weeks later in the EMG clinic, he had severe weakness with some wasting of distal limb muscles; there was also considerable numbness and diminished sensation. An EMG at this time revealed very few motor units in the intrinsic muscles of the hand and foot. Figure 8.6 shows that, after stopping vincristine, there was a slow recovery of motor nerve fiber function over the next 2 years, though this was still incomplete for the foot muscle studied.

In the next chapter, we will really begin to see nerve and muscle fibers in action, starting with the membrane mechanisms that are responsible for the phenomenon of excitability. We are entering the world of the neurophysiologist.

Chapter 9

Resting and Action Potentials

In this chapter, we will consider the nature of the electrical potential that exists across the surface membranes of nerve and muscle fibers, and how this potential may suddenly change so as to form a traveling impulse. An understanding of the membrane potential requires knowledge of the chemical composition of the fiber and its fluid environment, as well as some basic electricity.

Resting Membrane Potential

It has been known for more than a century that a difference in electrical potential exists between the inside of a cell and its fluid environment. It is possible to determine this potential difference directly by inserting a microelectrode through the cell membrane into the interior of the fiber and connecting the other terminal of the voltage measuring device to a reference electrode in the fluid bathing the cell (Figure 9.1A). In the case of inactive mammalian muscle, the inside of the cell has been found to be some 85 mV negative with respect to the outside; this potential difference is termed the *resting membrane potential* (Figure 9.1B). A similar potential is present across the membrane of a nerve fiber.

K⁺ Ions Are More Concentrated Inside the Cell Than Outside; the Reverse Is True for Na⁺ and Cl⁻

An understanding of the resting membrane potential requires knowledge of the ions present on the two sides of the membrane. Inside the muscle fiber, the most common cation is K^+; the anions are supplied by PO_4^{3-}, SO_4^{2-}, and HCO_3^-, together with amino acids, polypeptides, and proteins. Outside the cell, the interstitial fluid has a composition similar to plasma; Na^+ and Cl^- are the dominant cation and anion respectively, and smaller amounts of K^+, Ca^{2+}, Mg^{2+}, HCO_3^-, and PO_4^{3-} are also present.

The concentrations of the various ions in mammalian skeletal muscle and in its extracellular fluid are given in Table 9.1. It is difficult to determine the proportion of muscle fluid that is intracellular rather than extracellular, so the intracellular values are very approximate. Even if an intracellular value was accurately established, it would only reflect a gross average for the various organelles, myofibrils, tubular systems, and cytoplasm of the fiber. Also, the intracellular values would include those cations bound to amino acid residues in the various proteins of the cytosol and structural elements.

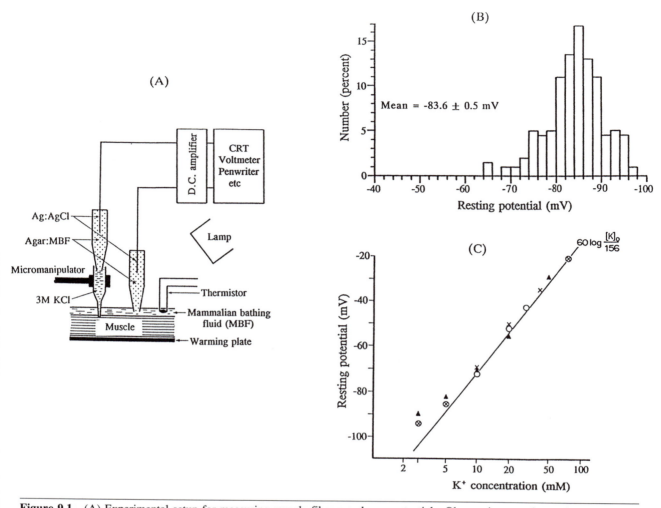

Figure 9.1 (A) Experimental setup for measuring muscle fiber membrane potentials. Observations can be made not only on an excised muscle, as shown here, but on fibers in tissue or organ culture and, of course, on muscles *in vivo*. (B) Resting potentials in biceps brachii and quadriceps of three healthy human subjects, recorded *in vivo*. Data of McComas, Mroźek, Gardner-Medwin, and Stanton (1968). (C) Resting potentials of human intercostal fibers; the muscles were excised and subjected to different concentrations of K+ in the bathing fluid. The straight line is a plot of the K+ equilibrium potential for an internal K+ concentration of 156 mM, at 37° C. The small deviation between the observed and predicted values for low K+ concentrations is due to the Na+ permeability of the membrane. Reprinted from Ludin (1970).

A more precise approach, which circumvents the problem of the extracellular space, is to use ion-sensitive microelectrodes to measure the ion concentrations. With this technique, a microelectrode is manufactured from a special glass known to attract ions of a particular species; alternatively, an ordinary glass microelectrode can be coated on the inside with an ion-binding resin. When an ion-sensitive electrode is dipped into a solution, the attraction between the selected ion and the electrode generates a potential at the electrode tip, which is proportional to the concentration of the ion. When inserted into a muscle fiber, however, the electrode will also record the potential across the cell membrane, so that a correction must be made; this is done by subtracting the resting potential, as measured with a standard microelectrode or, more commonly, with the untreated second channel of a double-barreled electrode.

In practice, the reliability of the determination of ion concentration is hampered by the appreciable junction potentials among dissimilar solutions in the microelectrode, bridge, and reference electrodes. Mean values obtained by the ion-sensitive electrode method range from approximately 120 to 180 mM for [K+]i and are around 10 mM for [Na+]i; slow-twitch muscle fibers tend to have a lower [K+]i and a higher [Na+]i than fast-twitch fibers (cf. Charlton, Silverman, & Atwood, 1981; Fong, Atwood, & Charlton, 1986; Juel, 1986).

The Resting Membrane Is Much More Permeable to K+ and Cl− Than to Na+

There are two reasons why the compositions of the intracellular and extracellular fluids are so different.

Table 9.1 Ionic Composition of Mammalian Skeletal Muscle Fibers and of the Interstitial Fluid

	Interstitial fluid mM	Intracellular fluid mM
Sodium	145	10
Potassium	4	160
Calcium	5	2
Magnesium	2	26
Total cations	156	198
Chloride	114	3
Bicarbonate	31	10
Phosphate	2	100
Sulfate	1	20
Organic acids	7	—
Proteins	1	65
Total anions	156	198

Note. From *Clinical Disorders of Fluid and Electrolyte Metabolism* (p. 28), by M.H. Maxwell and C.R. Kleeman, 1962, New York: McGraw-Hill. Copyright 1962 by McGraw-Hill.

First, a large proportion of the organic anions in the interior of the fiber are part of the cell structure and are therefore unable to migrate. Second, the resting membrane is semipermeable; it allows K+ and Cl- ions to diffuse through, but it impedes the passage of Na+. The internal anions exert an electrostatic attraction for cations that can only be satisfied by a high internal concentration of K+, because Na+ ions have difficulty in crossing the resting membrane. The external Na+ ions will, however, be effective in balancing the anions of the extracellular fluid, the most prevalent of which is Cl-. This unequal distribution of K+ and Cl- ions is termed a *Donnan equilibrium*, such that

$$\frac{[K]_o}{[K]_i} = \frac{[Cl]_i}{[Cl]_o} \qquad (1)$$

where the brackets signify the concentration of the ion inside (i) or outside (o) the cell.

The Resting Potential Is Mainly Due to the Differences in [K+] on Either Side of the Membrane

The potential difference across the membrane arises because the K+ ions, although attracted into the cell by the electrical force generated by the internal anions, are required to move uphill against their concentration gradient. Hence the internal K+ concentration does not quite balance the anions present within the cell; the slight excess of the latter is responsible for the internal

negativity of the resting membrane potential. The size of this electrical potential is related to the concentrations of K+ on the two sides of the membrane by the *Nernst equation*, in which

$$E_K = \frac{RT}{F} \log_e \frac{[K]_o}{[K]_i} \qquad (2)$$

where E_K is the K+ equilibrium potential, R is the universal gas constant, T is the absolute temperature, and F is the Faraday constant. For a mammalian muscle at 37° C, the equation can be simplified by inserting values for the various constants and transforming the logarithms so that

$$E_K = 61.5 \log_{10} \frac{[K]_o}{[K]_i} \qquad (3)$$

Assuming values of 4.5 and 160 mM respectively for external and internal concentrations of K+, E_K would be −95 mV, which is slightly more than the value actually observed for the resting membrane potential.

The Membrane Potential Can Be Altered by Changing [K+]o

An obvious way to test the postulated ionic basis of the resting membrane potential would be to observe the effect of altering [K+]o, the concentration of K+ in the fluid surrounding the muscle fiber. In keeping with the Nernst equation, an increase in the external K+ concentration causes the resting potential to fall (*depolarization*), while a reduction in concentration induces a rise in potential (*hyperpolarization*; Figure 9.1C).

Equilibrium potentials can also be calculated for both Cl- and Na+ ions. It is found that the equilibrium potential for Cl- has the same value and polarity as the K+ potential. Further, because the membrane is freely permeable to Cl-, one would expect the resting potential to be affected by changes in external concentration of the ion. Hodgkin and Horowicz (1959) were able to show that this was so but that the effect was short lived because diffusion of Cl-, K+, and water subsequently took place until the previous concentration gradient for Cl- was restored. In effect, the Cl- ion had been redistributed so as to form a new Donnan equilibrium with K+. In the case of Na+, the calculated equilibrium potential has a polarity opposite to the resting membrane potential. The discrepancy indicates that Na+ ions cannot be contributing significantly toward the resting potential and is consistent with the low permeability of the membrane to Na+ described previously. It follows that an alteration of the external Na^+ concentration should have little effect on the membrane potential, and experimentally this can be shown to be so.

The Membrane Potential Can Be Predicted From the GHK Equation

The theoretical basis for these observations can also be stated using the constant field concept, originally developed by Goldman (1943) and subsequently explored by Hodgkin and Katz (1949). This concept, which links together ion concentration, membrane permeability, and membrane potential, can be expressed as an equation (the *Goldman-Hodgkin-Katz*, or *GHK*, *equation*). It takes the form:

$$E_m = \frac{RT}{F} \log_e \frac{P_K[K]_o + P_{Na}[Na]_o + P_{Cl}[Cl]_i}{P_K[K]_i + P_{Na}[Na]_i + P_{Cl}[Cl]_o} \quad (4)$$

where P denotes the permeability of the membrane for an ion species and E_m is the resting potential. Since at rest P_{Na} is very low, and the effects of changes in Cl^- concentration are transient, the behavior of the resting membrane will be that of a K^+ electrode, and the equation will simplify to equations (2) and (3). Suppose, however, that the permeability of the membrane to Na^+ were to increase as a result of electrical or chemical stimulation. Under these circumstances, the potential would depolarize to reach a value between E_{Na}, on the one hand, and E_K and E_{Cl}, on the other. If the membrane permeability to Na^+ increased still further, so that the K^+ and Cl^- permeabilities were comparatively small, the membrane potential would approximate to E_{Na}, the Na^+ equilibrium potential. In fact, this is precisely what happens at the summit of the action potential, which will be considered later.

The Properties of the Membrane Can Be Simulated by Electrical Components

Another way of analyzing the behavior of the membranes is to depict it as an electrical analogue (Figure 9.2). The various equilibrium potentials can be represented as batteries, with the polarity of the Na^+ battery opposite to those of the K^+ and Cl^- cells. The permeability of the membrane for an ion may be symbolized as an electrical resistance, because resistance is a measure of the ease, or difficulty, with which that ion will traverse the membrane. A highly permeable membrane would therefore be shown as having a low electrical resistance. The arrow over each resistance in Figure 9.2 emphasizes that the resistances may vary, depending on whether the membrane is at rest or participating in impulse activity. From the figure it is obvious that if

any one resistance becomes very small (permeability very high) in comparison with the others, the potential appearing across the membrane will be that of the corresponding ionic equilibrium potential.

The net resistance of the membrane to the various ions may be determined experimentally by measuring the change in potential produced by passing current through the membrane from an intracellular stimulating electrode. According to Ohm's law, the fiber *resistance* will be equal to the induced *potential* divided by stimulating *current*. In Figure 9.3A, the smallest of the three currents, 2.5×10^{-8} A, produced a depolarization of 12 mV (trace a), which would give a value for the resistance of 0.48 MΩ. The resistance measured in this way is the "input" resistance of the fiber; it depends on the resistance of the fiber interior as well as on that of the membrane. The contribution of the various species of ion toward the membrane resistance may be determined by replacing each ion, in turn, with an impermeant substance. For example, by replacing Cl^- in the external bathing solution with methylsulfate, Hutter and Noble (1960) were able to show that the resting membranes of frog muscle fibers were twice as permeable to Cl^- as to K^+.

Membrane Capacitance Is Largely Due to the T-Tubules

It should be noted that the electrical analogue shown in Figure 9.2 includes the symbol for capacitance (C_m). This is required because the lipid membrane of the fiber acts as a dielectric between two conducting media, namely the interstitial fluid and the cytosol. Each time a depolarizing current is made to flow across the membrane, the capacitor has to be discharged; hence the change in membrane potential is never instantaneous but takes place exponentially, as in Figure 9.3A for the smallest current (response a). Similarly, there is an exponential decline in potential as the stimulus is terminated, caused by the recharging of the membrane capacity. The time constant of the membrane, τ_m, is a measure of the capacity of the membrane, being equal to the product of the capacity and the membrane resistance. It can be determined experimentally by measuring the time taken for the membrane potential to reach 84% of its final value following the onset of the rectangular pulse of current. The capacity of the mammalian muscle fiber membrane is 4 to 5 $\mu F/cm^2$ of surface membrane (Lipicky, Bryant, & Salmon, 1971) and is much larger than that estimated for nerve fiber membrane (about 1 $\mu F/cm^2$). This increased capacity of the muscle fiber is due to the substantial invaginations of membrane that form the transverse tubular system (T-system). Convincing proof of this supposition came from the experiments of Gage and Eisenberg (1969b) in which glycerol

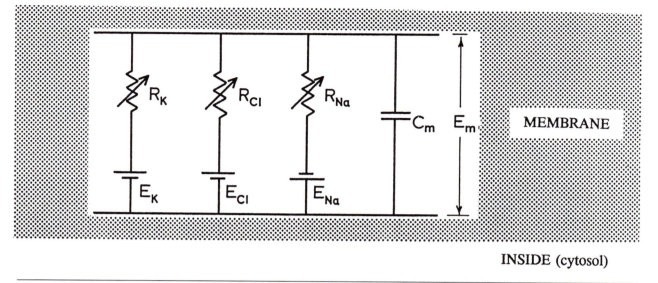

OUTSIDE (extracellular fluid)

MEMBRANE

INSIDE (cytosol)

Figure 9.2 Electrical analogue of membrane. E_m = membrane potential; C_m = membrane capacity. E_K, E_{Cl}, and E_{Na} are ionic equilibrium potentials, and R_K, R_{Cl}, and R_{Na} are the corresponding membrane resistances.

was used to disrupt the connections of the T-tubules to the surface plasmalemma; the membrane capacity was decreased by a factor of almost three.

Tubules Have High K⁺ Permeability

So far the membrane of the resting muscle fiber has been discussed as if it were a homogeneous structure, with the permeability of one region similar to that of another. It turns out that this is not so. One of the first indications of such a variation came from the experiments of Hodgkin and Horowicz (1959), already mentioned, in which the ionic composition of the solution bathing single frog muscle fibers was changed abruptly. These authors discovered that, whereas alterations in Cl⁻ concentration produced an immediate alteration in membrane potential, the effects of K⁺ were slower, particularly when the external concentration was being reduced. Hodgkin and Horowicz suggested that the reason for the slowness of the K⁺ effect was that the membrane permeability for this ion was restricted to the T-tubules and that diffusion into, and out of, these channels would be retarded because of their narrowness.

Eisenberg and Gage (1969) put this hypothesis to direct test by measuring the K⁺ and Cl⁻ permeabilities of muscle fibers after the necks of the T-tubules had been disconnected from the surface plasmalemma by treatment with glycerol. They found that the Cl⁻ permeability was restricted to the surface but that the K⁺ permeability was shared by the surface membrane and the T-system, the permeability of the latter being twice as high as that of the former.

The T-Tubules Are Responsible for Rectification and for Negative After-Potentials

The slow diffusion of K⁺ out of the T-tubules is responsible for two other effects. The first of these is the high resistance displayed by the muscle fiber to the movement of K⁺ ions *outward* through the fiber, as when an intracellular stimulating electrode is used to pass current. By discriminating between inward and outward K⁺ currents, the fiber is said to act as a *rectifier*. The second effect of the T-tubules is that, during impulse activity, there is an efflux of K⁺ from the muscle fiber into the T-system. Since this K⁺ is slow to diffuse away, the concentration gradient for K⁺ will be reduced across the tubular membrane, and so the K⁺ equilibrium potential, E_K, will be correspondingly smaller. This, in turn, will cause the membrane potential to fall provided that the permeability of the fiber to K⁺ is significantly larger than that to Cl⁻; normally the reverse is true. The condition of a high K⁺ permeability occurs during and immediately after the spike component of the action potential. Because of the accumulation of K⁺ in the tubules, the membrane potential will remain partially depolarized for a few milliseconds, producing a *negative after-potential*. This potential is more obvious in frog muscle fibers than in mammalian ones (see, however, Figure 9.3B), and it is enhanced by cooling or by repetitive impulse activity (because there is a greater accumulation of K⁺). Sometimes the negative after-potential becomes so large that it triggers an impulse.

These observations are also relevant to a consideration of the disease phenomenon of *peripheral myotonia*, in which the fibers are hyperexcitable, discharging

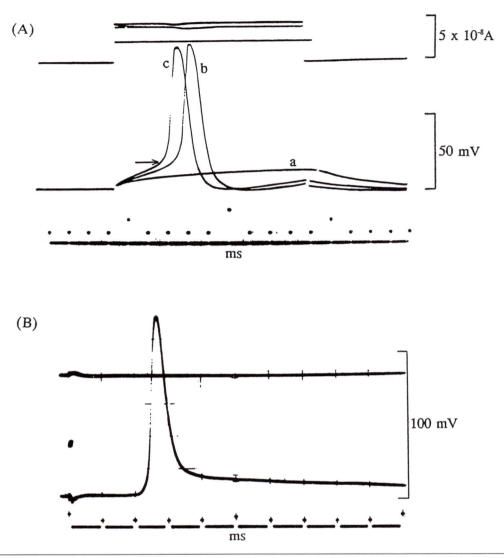

Figure 9.3 (A) Responses evoked in a normal human muscle fiber *in situ*, using intracellular stimulation through a Wheatstone bridge circuit. At the top are shown three rectangular current pulses of different intensities; the middle section displays the corresponding changes in membrane potential. Response a, produced by the smallest current, is subthreshold, but in b and c, the depolarizations are large enough to reach the firing level (*arrow*) for an action potential. Adapted from McComas, Mroźek, Gardner-Medwin, and Stanton (1968). (B) Response of a mouse tibialis anterior muscle fiber *in situ* to stimulation of the sciatic nerve. The action potential reached the recording point 2 ms after the stimulus had been delivered (artifact at left of trace) and had an overshoot of 39 mV; the resting potential was −80 mV. Note the negative after-potential following the spike.

long trains of impulses spontaneously and after transient mechanical or electrical stimulation of the fiber membrane. It has been shown that in the myotonic fiber the Cl⁻ permeability is greatly reduced (see chapter 7); hence K⁺ in the T-tubules is likely to be effective in reducing the resting potential and thereby bringing the membrane to the threshold for firing spontaneous impulses. A similar situation would be predicted to occur if all the Cl⁻ were removed from a solution bathing a normal muscle fiber; such fibers do indeed become hyperexcitable and exhibit myotonic-like behavior (Bryant & Morales-Aguilera, 1971).

The Na⁺-K⁺ Pump Expels Na⁺ Ions That Have Leaked Into the Fiber

Until now the membrane has been treated as if it were an entirely passive structure obeying physico-chemical laws. Suppose, however, the muscle fiber is allowed to become anoxic or is treated with a chemical, such as cyanide or iodoacetate, that interferes with oxidative metabolism. Under either of these conditions, the resting membrane potential will slowly fall to a low level, despite the large differences in ionic concentrations which existed across the membrane at the start of the experiment. The explanation for the slow depolarization involves the small permeability of the membrane to Na^+, noted earlier, and also the mechanism by which the cell normally counteracts the effects of the Na^+ permeability. These two factors may now be considered in more detail.

Since there is a small permeability of the membrane to Na^+ in the resting condition, Na^+ ions will leak into the muscle fiber down their concentration gradient and under the influence of the strong electrostatic attraction of the internal negativity of the fiber. As they accumulate in the cell, the Na^+ ions will tend to depolarize the membrane, because their positive charges will make the fiber interior less negative. This depolarizing tendency is normally balanced by a pump in the cell membrane that returns Na^+ to the interstitial fluid in exchange for K^+. Since the pump consumes ATP as its energy source, it can also be regarded as an enzyme, in this case Na^+K^+-ATPase (see chapter 7).

The Na^+-K^+ Pump Contributes to the Resting Potential

For each cycle of the pump, three Na^+ ions are expelled from the fiber and only two K^+ ions are admitted; thus, the pump not only maintains the normal chemical composition of the fiber but also contributes to the internal negativity. This "electrogenic" action of the pump, in turn, will promote the net inward diffusion of K^+ ions into the fiber and will thereby maintain the K^+ equilibrium potential.

How much potential does the Na^+-K^+ pump generate? Under resting, steady-state conditions, the contribution of the pump can be calculated if the ionic concentrations on either side of the membrane are known, together with the relative permeabilities of the membrane for Na^+ and K^+. Thus, according to Mullins and Noda (1964),

$$E_m = \frac{RT}{F} \ln \frac{r[K^+]_o + b[Na^+]_o}{r[K^+]_i + b[Na^+]_i} \qquad (5)$$

where E_m is the resting potential, r is 1.5 and is the Na^+:K^+ ion exchange ratio for the pump, and b is the ratio of membrane permeabilities for Na^+ and K^+. This value of E_m, which corresponds to the observed resting potential as measured with a microelectrode, will be rather higher than that calculated by the GHK equation using the same values of ion concentration and permeability. The difference in E_m will be the voltage contributed by the electrogenic Na^+-K^+ pump.

It is now appropriate to substitute real values of ion concentration and resting potential into equation (5), so that the permeability factor, b, can be calculated. With reference to Table 9.1, $[K^+]_o = 4$ mM, $[Na^+]_o = 145$ mM, $[K^+]_i = 160$ mM, $[Na^+]_i = 10$ mM, and E_m (the observed

resting potential) ≈ -85 mV for mammalian muscles of mixed fiber type. At 37° C, the terms

$$\frac{RT}{F} \ln$$

simplify to $61.5 \log_{10}$, so that equation (5) becomes

$$-85 = 61.5 \log_{10} \frac{1.5(4) + b(145)}{1.5(160) + b(10)} \qquad (6)$$

and b, the ratio of the membrane permeabilities for Na^+ and K^+, will be 0.03. This value for b can now be inserted into the GHK equation (4). Under steady-state conditions (in which the inward ion fluxes are equal to the outward ones), Cl^- can be ignored for the reason given earlier. For the same values of $[Na^+]$ and $[K^+]$ used in equation (6), the calculated resting potential would be -79.4 mV. The difference between the observed and calculated values of E_m is then -5.6 mV and is the voltage contributed by the electrogenic Na^+-K^+ pump to the resting membrane potential.

During muscle contraction, however, or following the application of depolarizing drugs, the pump may produce as much as -30 mV of internal negativity (Creese, Head, & Jenkinson, 1987; Hicks & McComas, 1989). During exercise the electrogenic contribution of the pump maintains the excitability of the muscle fiber membranes and prevents the onset of K^+-induced depolarization block (see chapter 15).

Action Potential

The ability of their membranes to develop transient changes in potential, which can be transmitted from one point to another, confers upon nerve and muscle fibers the special property of excitability. The propagated change of membrane potential is the *action potential*, or *impulse*.

The Action Potential Is a Brief Electrical Signal That Travels Along the Fiber

In order to record the full amplitude of the impulse, it is necessary to measure, during excitation, the potential developed between an electrode in the cytosol of the fiber and an external electrode. In Figure 9.3A, the same electrode was used simultaneously to stimulate and record from a human muscle fiber, and a Wheatstone bridge circuit was employed to nullify the stimulus artifact. It can be seen that the smallest rectangular pulse of current produced only a small maintained depolarization of the membrane (*a*). In contrast, each of the two larger pulses evoked reversals of membrane potential (*b, c*), such that the inside of the fiber became 13 mV positive with respect to the outside for about 1 ms.

One of the advantages of the intracellular stimulation technique is that it enables measurement to be made of the critical amount of membrane depolarization required to trigger the action potential mechanism. From Figure 9.3A it can be seen that the 12 mV depolarization induced by the weakest stimulus was insufficient, whereas depolarizations of 14 mV, produced by the two larger stimuli, were adequate. In one case, the onset of the action potential from the evoked depolarization has been indicated by an arrow. Figure 9.3B shows, for comparison, the action potential set up in a muscle fiber following electrical stimulation of its motor axon; the transient reversal of membrane potential and the negative afterpotential following the spike are both clearly shown.

Like the resting membrane potential, the action potential has an ionic basis, depending on the permeability of the membrane to the ions on its inner and outer surfaces, as well as upon the concentrations of those ions. Many years ago it was realized that Na^+ ions had an important role, because nerves and muscles bathed in Na^+-free solutions lost their excitability.

Voltage Clamp Experiments Reveal the Presence of an Inward Na^+ Current During the Action Potential

The key to our present understanding of the ionic mechanisms involved in the action potential is the *voltage clamp experiments* of Hodgkin and Huxley (1952a, 1952b). These authors chose to study the giant axon of the squid, for its large diameter, from 0.5 to 1.0 mm, permitted them to insert both a stimulating and a recording electrode into the axoplasm; the stimulating and recording circuits were completed by two external electrodes in the artificial seawater bathing the fiber (Figure 9.4A; see also Box 9.1). Their technique was to rapidly change the potential across the membrane to any desired level by passing current between the stimulating electrodes. By using a feedback circuit from the recording electrodes, they were able to keep the membrane potential steady at its new level. The amount of current necessary to clamp the membrane in this way was equal to, and opposite in sign to, the ionic current that would have normally been flowing through the membrane if the depolarization had been a naturally occurring one. Hodgkin and Huxley then determined the contribution of Na^+ ions to the total current by substituting choline, an impermeant cation, in the bathing solution. They were able to show that, as the membrane potential was reduced, there was an initial flow of Na^+ ions into the cell, and that this was followed by a flow of K^+ ions in the opposite (outward) direction.

It is obvious that the membrane permeability must have changed during the action potential, since under resting conditions, the membrane is largely impermeable to Na^+. However, during the action potential the Na^+ permeability not only increases, but the increase becomes self-regenerative. The explanation for this phenomenon is that, as the membrane becomes permeable to Na^+, these ions will diffuse into the fiber down their concentration gradient and also under the attraction of the internal negativity of the cell. However, the entry of Na^+ ions will depolarize the membrane further because, being positively charged, the ions make the inside of the cell less negative. In turn, the further depolarization causes the membrane to become even more permeable to Na^+, and the cycle is repeated (Figure 9.6; see also Figure 7.3 in chapter 7).

The Action Potential Is Terminated by a Rise in the K^+ Permeability of the Membrane

From a consideration of the electrical analogue of the membrane (Figure 9.2) and of the GHK field equation (see p. 130), it is apparent that a membrane that is much more permeable to Na^+ than to K^+ or Cl^- should have a potential across it equal to the Na^+ equilibrium potential, E_{Na}. The fact that the recorded potential is rather less than E_{Na} is due to two additional properties of the membrane; first, the Na^+ permeability mechanism is soon switched off (*inactivation*), and second, the membrane increases its permeability to K^+ as that to Na^+ declines. The time-courses of these changes are shown in Figure 9.4B, in which the Na^+ and K^+ permeabilities have been expressed as conductances. It is the increase in K^+ permeability that is responsible for the delayed outward current recorded in the voltage clamp experiment. In the normal unclamped fiber, the increased K^+ permeability serves to restore the membrane potential to its resting level.

At the end of an action potential, the fiber will have gained Na^+ and lost some K^+. The quantities are so small, however, that a motor nerve axon or muscle fiber can conduct several thousand action potentials without difficulty. In a thin muscle fiber or axon, the situation is rather different, for appreciable changes in internal Na^+ concentration occur after fewer impulses. During and following periods of impulse activity, the changes in ionic composition of the fiber are corrected by means of the Na^+-K^+ pump; the pump actively expels Na^+ and has been described in chapter 7.

Experiments With Squid Axon Sheaths Confirm the Ionic Basis of the Membrane Potentials

Following its formulation, the ionic hypothesis of the resting and action potentials was verified in a strikingly

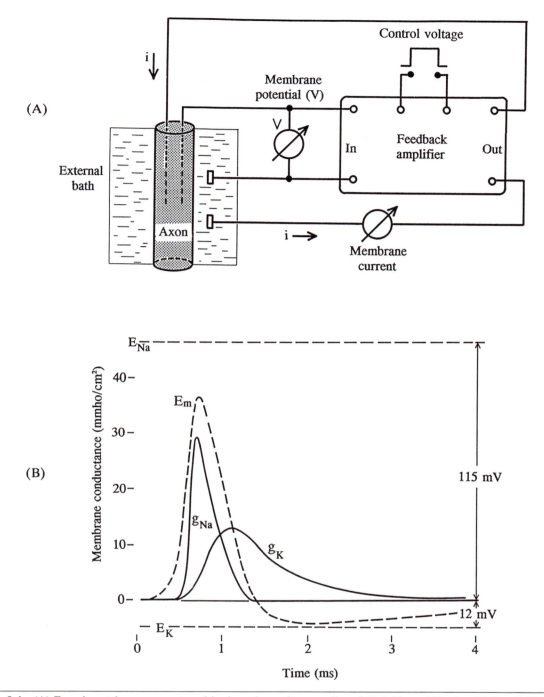

Figure 9.4 (A) Experimental arrangement used in the voltage clamp studies of Hodgkin and Huxley (1952a, 1952b) on the squid giant axon. The potential between an intracellular and an extracellular electrode is measured (*V*) and compared with a controlling voltage; any difference is automatically eliminated by the passage of a current (*i*) through another pair of electrodes situated on either side of the membrane. From *Nerve, Muscle and Synapse* (p. 82), by B. Katz, 1966, Whitby, ON: McGraw-Hill Ryerson Ltd. Copyright 1966 by McGraw-Hill, Inc. Adapted with permission. (B) Time-courses of the ionic permeability changes during an action potential, as determined by the voltage clamp experiment. The permeabilities of the membrane to Na$^+$ and K$^+$ are indicated by the corresponding conductances (*g*); E_m is the action potential recorded when the clamp was no longer applied. (Since the recordings were made from an axon, in which the T-system is not present, there is no negative after-potential; the hyperpolarization recorded instead is a consequence of the fact that the potassium equilibrium potential, E_K, may be larger than the resting membrane potential if the fiber is not in an optimal condition.) Adapted from Hodgkin and Huxley (1952b).

BOX 9.1   THE SQUID AND ITS GIANT AXON

It was the eminent British zoologist J.Z. Young (1936) who first drew attention to the possible usefulness of the squid (Figure 9.5A) in studies of the nerve impulse. These creatures, like certain other invertebrates, have a few unusually large axons that are essential for the rapid transmission of impulses in predatory or escape reactions. Figure 9.5D shows a longitudinal section through *Loligo pealei*, the common squid of the coast of North America. The animal is able to swim by undulating its tail fins but, if necessary, can move more rapidly by jet propulsion. In the latter mode, sea water is first drawn into the cavity of the mantle through the siphon and is then expelled rapidly through the same aperture by a sudden contraction of the mantle muscles. The synchronous timing of the contraction depends on the rapid transmission of impulses from the stellate ganglion along the giant axons on each side (Figure 9.5D). Each axon is a true neural syncytium, formed by the fusion of several hundred smaller axons as indicated in Figure 9.5C. Whereas the largest vertebrate axons have diameters of 20 μm or so, which include the myelin sheaths, the squid giant axons are 0.5 to 1.0 mm in width.

Using the squid while working at Woods Hole on Cape Cod, Curtis and Cole (1938) applied surface electrodes and an oscillating current to demonstrate the enormous increase in membrane permeability (conductance) during the action potential (Figure 9.5B). The usefulness of the squid axon was quickly appreciated by Alan Hodgkin during a visit to Woods Hole in 1938, and was later exploited with Andrew Huxley in a series of brilliant experiments at the Marine Biological Station, Plymouth, U.K. These experiments started in 1939, while Huxley was still an undergraduate student, and were interrupted by the Second World War. Hodgkin's account of this work (Hodgkin, 1977) describes the first insertion of an electrode down the inside of the axon and the significant early finding, that the action potential exceeded the resting potential.

After the war, Hodgkin and Huxley adopted the idea of Cole and Marmont for clamping the membrane of the squid axon at any given potential with a feedback amplifier and measuring the current required (Figure 9.4A). They were then able to plot the current-voltage relationship for the membrane and to determine its conductance. The results were fitted by a series of differential equations, which Huxley solved numerically with a hand-operated calculating machine. It is said that not a single error was made in the many thousand steps in the calculations! This pioneering work put the ionic basis of the resting and action potentials on a firm quantitative basis, and for the first time, encouraged ideas as to how ion channels might open and close. For their work on the squid giant axon, Hodgkin and Huxley were awarded the Nobel Prize in 1963, sharing it with John Eccles.

elegant series of experiments by Baker, Hodgkin, and Shaw (1962a, 1962b). These authors were able to extrude the axoplasm from a squid giant axon and to fill the bag of membrane left behind with solutions of any desired composition. By altering the constituents of the extracellular fluid as well, they could alter the resting potential at will. For example, if the external concentration of K^+ was higher than the internal one, the resting membrane potential reversed its sign and became positive inside the fiber, in agreement with predictions from the Nernst equation.

During the Action Potential, Na⁺ Ions Move Rapidly Through Special Channels in the Membrane

Once the changes in Na^+ and K^+ permeability during the action potential had been recognized, a major problem was to understand the nature of the underlying molecular events in the membrane. Initially it was thought that the Na^+ and K^+ ions might be transported by "carrier" molecules that would combine with the respective ions, transfer them across the membrane, and

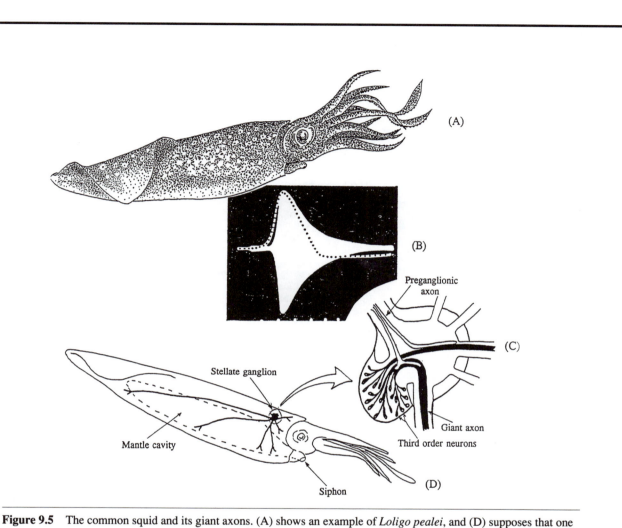

Figure 9.5   The common squid and its giant axons. (A) shows an example of *Loligo pealei*, and (D) supposes that one of these animals has been cut longitudinally so as to reveal the innervation of the muscles that supply the mantle cavity. An enlargement of the stellate ganglion is given in (C), which shows that each giant axon is formed by the fusion of many smaller fibers belonging to third-order neurons. (B) Change in membrane conductance (white envelope), measured by a bridge circuit during the passage of an action potential (dotted line). White dots at bottom are milliseconds. Reprinted from Cole and Curtis (1939).

then release them. This concept was challenged when estimates were made of the densities of the different types of channels in excitable membranes. In the case of the Na$^+$ channel, use was made of two biological toxins that combine with the channel and inactivate it, *tetrodotoxin* (TTX) and *saxitoxin* (STX; see chapter 7). By using very dilute solutions of TTX or STX, and also by labeling these compounds, it has been possible to estimate the density of the Na$^+$ channels in excitable membranes. It appears that the channels are remarkably sparse in muscle fibers and nonmyelinated nerve fibers; for example, in lobster axons, there are probably fewer than 13 Na$^+$ channels per μm^2 of membrane (Moore, Narahashi, & Shaw, 1967), while for rat diaphragm, the corresponding figure is about 21 channels (Colquhoun, Rang, & Ritchie, 1974). Such low densities imply that during the impulse Na$^+$ ions must flow through each channel at a very high rate, probably in the region of several million ions per second (Keynes, Ritchie, & Rojas, 1971). This flow is far too high to be accounted for by the activation of carrier molecules, and it was necessary to postulate instead that Na$^+$ ions enter through special water-filled pores in the membrane. Hodgkin and Huxley (1952a, 1952b) suggested that

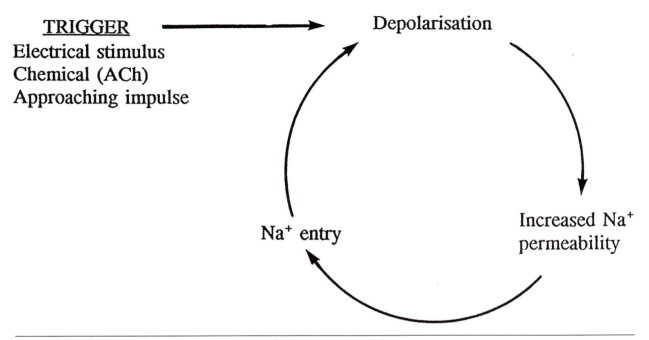

Figure 9.6   The regenerative nature of the increase in Na⁺ permeability of the membrane.

each channel would have two components; one part would sense the potential difference across the cell membrane and would control access to the second part, now known to be the critically narrowed pore. Hodgkin and Huxley also predicted that the conformational change in the sensing part of the channel would give rise to a small gating current.

Ion Channels Have Been Studied by Patch Clamping and by Molecular Biology Techniques

Further insights into the nature of ion channels came from the introduction of patch clamping by Neher and Sakmann (1976b), both of whom were awarded the Nobel Prize in 1991. These authors reported that it was possible to record the currents flowing through single ion channels, once a small piece of cell membrane had been tightly sealed to the tip of a glass microelectrode. The abrupt onsets and terminations of the currents were attributed to the openings and closings of the ion channels; the frequency and durations of the opening could be controlled by applying different voltages across the patch of membrane or by exposing the latter to selected drugs or neurotoxins (see Figure 7.6, chapter 7).

The latest approach to the understanding of ion channels has been to deduce their three-dimensional structures and amino acid sequences using molecular biology techniques. This approach was first applied successfully to the Na⁺ channels of the electroplax organ of the electric eel (Noda et al., 1984) and subsequently extended to Na⁺ channels in skeletal muscle and rat brain,

as well as to K⁺ and Ca²⁺ channels (see chapter 7). It is now known that each Na⁺ channel consists of four identical domains and that each of the domains contributes to the formation of the central pore. In turn, each domain is composed of six regions that span the membrane; one of these regions acts as the gating mechanism (see p. 95) and another forms part of the pore (see Figure 7.2, chapter 7). In such a model, the channel will open and admit Na⁺ ions when gating movements have taken place in all four domains. However, the same depolarization that opens the *m gate* and allows Na⁺ ions through the channel will also close the channel after a delay of 1 to 2 ms. This closing, *inactivation*, is brought about by a second type of gate, the *h gate*. Thus, Na⁺ ions can only flow across the membrane when the m and h gates are both open (Figure 9.7).

Inability of Sufficient Na⁺ Channels to Open Causes Refractoriness

The Na⁺ permeability system has added significance in that it underlies the phenomena of the *refractory period* and *accommodation*. The refractory period of a nerve or muscle fiber is the length of time following an impulse during which there is a depression of membrane excitability. At first the fiber is *absolutely* refractory and cannot respond with an impulse to a testing stimulus, no matter how large the stimulus is. This phase is succeeded by the *relatively* refractory period in which the fiber is able to fire an action potential provided that the stimulus intensity has been increased beyond its initial threshold value. In humans, the absolutely refractory

(A)   (B)   (C)

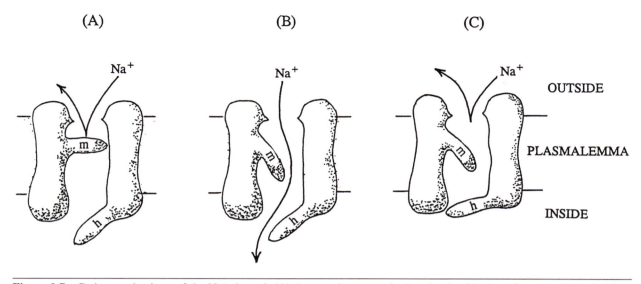

Figure 9.7 Gating mechanisms of the Na⁺ channel. (A) At rest, the m gate is shut, barring Na⁺ ions from the channel. (B) Depolarization of the membrane opens the m gate, allowing Na⁺ ions to pass through the central pore of the channel. (C) After a short interval (1-2 ms), the same depolarization inactivates the channel by closing the h gate.

periods of muscle fibers vary from 2.2 to 4.6 ms (Farmer, Buchthal, & Rosenfalck, 1960), and that for the largest motor axons is about 0.5 ms. The explanation of the absolutely refractory period is that the Na⁺ channels can no longer function because the h gates have closed; that is, inactivation has taken place. However, in the relatively refractory period some, but not all, of the Na⁺ channels have recovered and are available for participation in a further action potential.

A membrane is said to display *accommodation* if it fails to fire an action potential in response to a slowly rising current pulse, even though a rapidly increasing pulse is effective. In terms of the Na⁺ permeability mechanism, one can envisage that, during the slowly rising pulse, some channels are already inactivating while others have yet to be opened; in addition, the K⁺ channels will already be opening. During the rapidly rising pulse, more Na⁺ channels will be opened initially and, through the regenerative action of the resulting depolarization, will be more effective in causing others to open.

Ion Current Flows Between Adjacent Areas of Membrane During the Action Potential

So far only the events that take place locally in the nerve or muscle fiber membrane have been considered. The functional significance of the action potential mechanism is that, once an impulse has been initiated, excitatory changes automatically take place in neighboring regions of the membrane and so cause the impulse to travel the length of the fiber. This spread of excitation depends on the difference in potential that exists between the region of membrane at which the action potential is momentarily located and more distant regions that are still in the resting state. At the crest of the action potential, the interior of the fiber is some 30 mV positive with respect to the exterior, while farther along, the fiber will exhibit the normal resting membrane potential such that the inside of the fiber is some 85 mV negative to the outside. Because of this difference in potential, current will flow between the two regions of membrane. The resting membrane is said to act as a *source* of current for a *sink* at the site of the action potential. This current comes from the discharge of the membrane capacity in the resting membrane. It is usual to state the direction of a current by indicating the movement of the positive charges; in this situation, cations will flow inward through the membrane at the sink. As the resting (source) membrane discharges, the potential will fall to the critical depolarization necessary for local impulse generation; in this way, the excitatory disturbance is transmitted along the fiber.

The Action Potential Can Be Simulated by Electronic Components

In Figure 9.8, the situation has been displayed as an electronic circuit. The resting membrane potential at the source is shown as a battery of −85 mV, which is in series with a membrane resistance, R_m, and in parallel with a membrane capacity, C_m. Similarly, the action potential is depicted as a battery of +30 mV in series with another resistance, R'_m, and a second capacity, C_m.

Finally, R_i and R_e represent the resistances of the interior of the fiber and of the extracellular fluid respectively. For a muscle fiber, R_e is small because of the large volume of fluid surrounding the fiber; R_m is also small because of the huge increase in Na$^+$ permeability during the action potential. Therefore, the potential E_m will drop to a value intermediate between -85 mV and $+30$ mV, given approximately by $R_m(30 - 85)/(R_m + R_i)$ mV. The speed with which this depolarization takes place will depend on the product of the membrane capacity, C_m, and the series resistances $R_i + R_m + R_e$; of these, only R_i is significant.

Figure 9.8 illustrates that the velocity with which the impulse propagates will depend on the dimensions of the fiber, being lower for a small muscle fiber than for a larger one. The reason for this is that the smaller fiber will have a narrower core of cytosol and hence a higher electrical resistance, R_i. In turn, this means that the internal current, i, flowing between source and sink, will be relatively small and that it will take rather longer for the membrane capacity of the source to discharge to the threshold potential required for an impulse.

The Myelin Sheath Enables the Action Potential to Travel Quickly Down the Axon

In nerve fibers, a modification of structure enables the action potential to be conducted much more efficiently than in muscle fibers. This modification is the incorporation of a special lipid insulating layer around the nerve fiber membrane; the extra coat is the *myelin sheath* and is formed by the Schwann cells (see chapter 2). At regular intervals, the myelin sheath is interrupted by the *nodes of Ranvier* at which the nerve membrane, the *axolemma*, is exposed to the extracellular fluid and *gap*

substance. The effect of the myelin sheath is to prevent current from leaking across the axolemma; that is, R_e in Figure 9.8 becomes very large in relation to the other resistances in the circuit. Thus, as the action potential advances, only the axolemma at the nodes of Ranvier will be able to depolarize to the levels necessary for impulse initiation. In this way, the action potential jumps from node to node, skipping the intervening myelinated segments of axon.

This form of propagation is termed *saltatory conduction*, and it enables the impulse to travel much faster than in nonmyelinated structures. Saltatory conduction has the additional advantage of restricting the ionic exchange during the impulse to the nodes of Ranvier; as a result, the fiber expends less energy in pumping out Na$^+$ in exchange for K$^+$. Not all mammalian nerve fibers are myelinated; in fact, both cutaneous and muscle nerves contain more nonmyelinated fibers than myelinated ones. The nonmyelinated fibers (C-fibers) are thin, most less than 1 μm in diameter, and they are arranged in small colonies within the cytoplasm of individual Schwann cells. However, as far as muscle is concerned, all the nerve axons supplying the extrafusal muscle fibers have myelin sheaths. In humans, the largest of these axons have diameters of about 13 μm, including the thicknesses of the myelin sheaths (see Figure 9.9B); rather thicker axons are found in some of the lower mammals.

APPLIED PHYSIOLOGY

In this section, we will see how impulse conduction can be studied in human nerves and will look at the effects

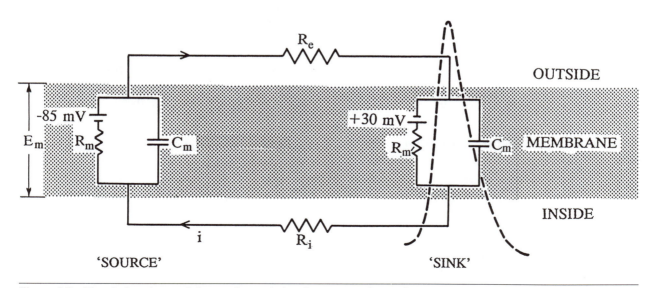

Figure 9.8   Electrical analogue of the membrane during the passage of an action potential.

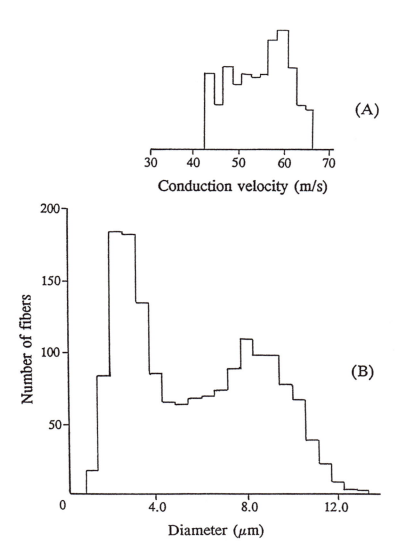

(A)

(B)

Figure 9.9 (A) Spectrum of impulse conduction velocities in motor axons of the median nerve in a single subject, as derived by mathematical decomposition. Adapted from Cummins, Dorfman, and Perkel (1979).(B) Diameters of myelinated sensory and motor axons in the recurrent branch of the median nerve to the thenar muscles. Adapted from Lee, Ashby, White, and Aguayo (1975).

of damage to the myelin sheath on conduction. Some of the methods for measuring impulse conduction in muscle fibers will also be described.

Impulse Conduction Can Be Easily Studied in Human Nerves

Impulse velocity in the motor axons can readily be determined by successively stimulating a nerve at two points and measuring the time elapsing before the muscle fibers are excited on each occasion. The difference between these times is then divided into the extra distance that the impulses had to travel when the stimulating electrodes were applied to the nerve farthest from the muscle (Figure 9.10). The impulse conduction velocities are rather higher in the upper limb than in the lower limb, about 50 to 65 m/s for the median and ulnar nerves in the forearm and 40 to 55 m/s for the tibial and peroneal nerves below the knee. These values are the

velocities in the fastest conducting axons only, since the time measurements are made from the earliest responses in the muscle.

To measure the impulse velocities in the slowest conducting fibers requires a special technique, such as the use of a second stimulus, given at a different point over the nerve and at varying time intervals, to produce a series of collisions between the two volleys of nerve impulses (Hopf, 1962, 1963). Other techniques for measuring the full range of axonal impulse conduction velocities are based on F-wave latencies (Doherty, Komori, Stashuk, Kassam, & Brown, 1994) and mathematical decomposition of compound action potentials (Cummins, Dorfman, & Perkel, 1979; Lee, Ashby, White, & Aguayo, 1975; Figure 9.9A). Since impulse velocities are proportional to nerve fiber diameters, the velocity measurements can be used to calculate the distribution of axon sizes in the motor nerve trunk. For example, in cat peripheral nerves,

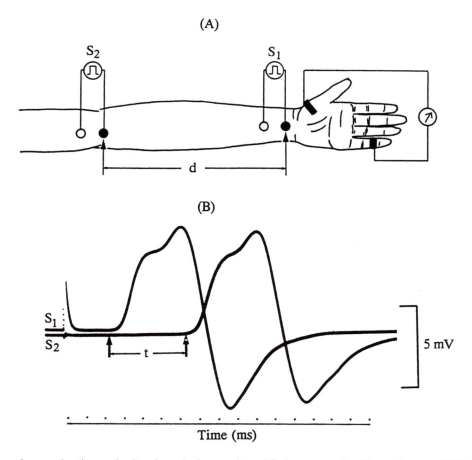

(A)

(B)

Time (ms)

Figure 9.10 Determination of the maximum impulse conduction velocities in motor axons of the median nerve. (A) The nerve is maximally stimulated at two sites by the electrode pairs S_1 (wrist) and S_2 (elbow). The corresponding muscle responses are recorded by a stigmatic electrode over the thenar muscles and an indifferent electrode on the little finger. A ground electrode is placed on the dorsum of the hand. (B) The distance, d, separating the cathodal electrodes (●) at the wrist and elbow respectively is divided by the difference, t, between the latencies of the two evoked muscle potentials; the result is the impulse velocity of the fastest conducting motor axons. In this example, $d = 25$ cm and $t = 4.2$ ms, giving a conduction velocity of approximately 60 m/s.

the conduction velocity, in m/s, is equal to 6.0 times the myelinated fiber diameter, in μm (Hursh, 1939); in the baboon, the corresponding factor is 5.3 (McLeod & Wray, 1967).

Reduced Impulse Conduction Velocities Are Found in Patients With Demyelinating Lesions

Measurements of impulse conduction velocity are invaluable in the EMG clinic, where they can be used to detect injury, degeneration, or inflammation of nerve fibers. Each of these agents interferes with impulse conduction by destroying part, or all, of the myelin sheaths covering the axis cylinders. The successive stages of myelin destruction are illustrated schematically in Figure 9.11 (top), and partially demyelinated motor axons in a human muscle biopsy are shown below.

The effect of myelin loss (demyelination) is to make impulse conduction less efficient (see the following section), and an example of this phenomenon in the EMG clinic is given in Figure 16.8A (chapter 16). This figure may be compared with the normal results in Figure 9.10; the former shows the greatly delayed response in the small muscles of the hand following stimulation of the median nerve at the elbow.

In Experimental Diphtheritic Neuropathy, the Time Taken for the Action Potential to Jump From One Node to Another Is Increased

Figure 9.12 illustrates the effect of demyelination in more detail and is taken from the work of Rasminsky and Sears (1972). These authors devised a special electrophysiological technique that enabled them to record action potential currents at successive points along the outside of single ventral root fibers (Figure 9.12A). In the normal animal, the latency of the recorded current increased in definite steps along the axon, with each step corresponding to the location of a node (Figure 9.12, B & C). These findings were, of course, to be expected if saltatory conduction was occurring. In nerve fibers treated with diphtheria toxin, discrete increments of latency were also observed with increasing distance along the axon. In contrast to the findings in the normal animal, in which the conduction time from one node to the next (approximately 1 mm) was about 20 μs, values as large as 600 μs were observed in the demyelinated preparation (Figure 9.12D).

Why is conduction slowed in these poorly myelinated axons? The obvious explanation depends on the fact that an affected axon becomes a less efficient cable. The loss of myelin insulation reduces the electrical resistance between the inside and the outside of

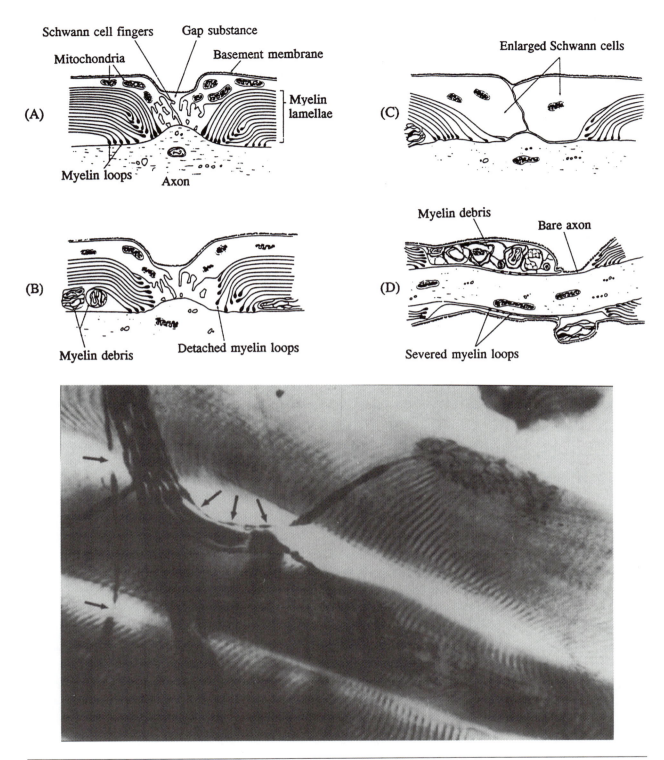

Figure 9.11   (Top) Morphological studies of nodes of Ranvier in the sciatic nerves of a normal rat (A) and of animals that had received intraneural injections of diphtheria toxin. An early change, detectable at 3 days, is disorganization of the myelin loops (B); by 7 days, the loops have become detached from the axolemma at the node and swollen Schwann cells take their places (C). At 10 days, much more of the axolemma has become denuded of myelin; myelin debris can be seen in the cyto-plasm of the Schwann cells (D). Reprinted from Allt and Cavanagh (1969). (Bottom) Demyelination in human nerve fibers. The preparation, a "squash," shows parts of three muscle fibers and a bundle of motor axons, most of which have various de-grees of demyelination (*arrows*). The neuromuscular junction in the uppermost fiber is clearly depicted. (Courtesy of Dr. D.F. Harriman.)

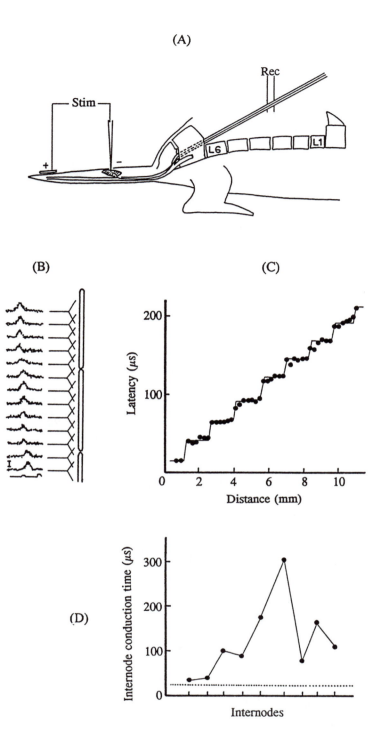

Figure 9.12 Studies of impulse propagation in the normal rat and in an animal treated with diphtheria toxin. (A) The stimulating electrodes (*Stim*) are shown applied to peripheral nerve fibers in the tail. The recording electrodes (*Rec*) consist of two fine wires mounted 400 to 600 μm apart, which are moved to different positions along the course of a ventral root fiber. (B) The external currents recorded during passage of an impulse, together with the positions of the recording electrodes in relation to the nodes of Ranvier. (C) The increments in the latency of the action current as the electrodes are moved; note their regularity. Each step corresponds to a node of Ranvier. (D) The results for an axon with partial demyelination. Notice the large and irregular values for internode conduction time compared with the average values for a normal axon (.......). Adapted from Rasminsky and Sears (1972).

the axon, and the thinner layer of myelin remaining causes the capacitance to increase. The decreased resistance will allow a larger part of the outwardly flowing action potential current to leak through the internodal region of the axolemma, and the increased capacitance will store more of the charge carried by the current. In both instances, current is expended in the internode instead of being concentrated at the next node. Hence the current must flow for a longer time before the critical depolarization is reached at the node and allows an impulse to be initiated.

What happens if an axon becomes completely demyelinated? In large diameter axons, impulse conduction is usually blocked, as is evident in Figure 16.7A in chapter 16. In smaller diameter axons, the density of Na⁺ channels in the internodal axon membrane, although much smaller than that at the node, may sometimes be sufficient to allow impulse conduction—at a greatly reduced

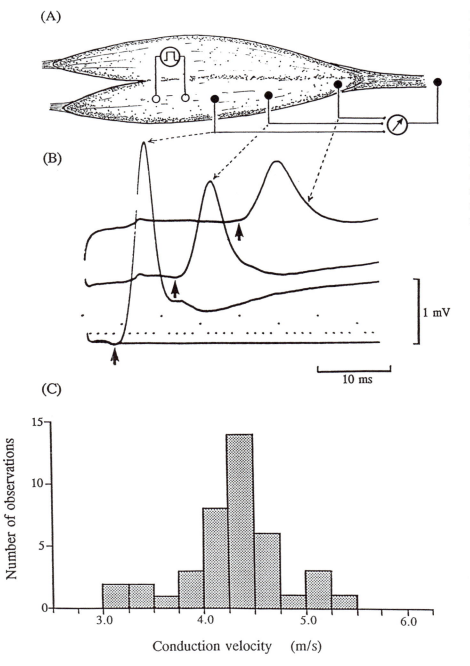

Figure 9.13 Measurement of impulse conduction velocities in muscle fibers of human biceps brachii. (A) Arrangement of stimulating and recording electrodes on skin overlying muscle. (B) Muscle compound action potential (M-wave) recorded at different distances from stimulating electrodes. Large arrows indicate response onsets of fastest conducting muscle fibers. (C) Values for fastest conducting fibers in 41 muscles.

rate, however (Bostock & Sears, 1978; Black, Kocsis, & Waxman, 1990).

Action Potentials Propagate More Slowly in Muscle Fibers Than in Motor Axons

Although not used in the EMG clinic, impulse conduction velocity measurements are easily made in skeletal muscle fibers. For example, the muscle can be stimulated directly or through its nerve supply, and the arrival of the action potentials can be timed at different points along the fibers (Buchthal, Guld, & Rosenfalck, 1955; Kereshi, Manzano, & McComas, 1983; Figure 9.13).

Alternatively, the subject is asked to make a small voluntary contraction, and the passage of an action potential down a muscle fiber can be detected at successive lead-off surfaces in a special recording needle (Stålberg, 1966). Finally, the same motor unit potential can be recognized at different sites in a muscle during weak effort (Nishizono, Saito, & Miyashita, 1979). All these methods indicate that there is a range of velocities, two of the mean values for the brachial biceps muscle being 4.02 ± 0.13 m/s (Buchthal et al., 1955) and 3.37 ± 0.67 m/s (Stålberg, 1966). In a large human muscle such as the tibialis anterior, the fibers with the highest impulse conduction velocities belong to the motor units that have the fastest twitches and develop the greatest forces (Andreassen & Arendt-Nielsen, 1987). In Duchenne muscular dystrophy, in which abnormally large and small muscle fibers are present, one would expect the range of conduction velocities to be increased. Surprisingly, no abnormality was found by Buchthal, Rosenfalck, and Erminio (1960), although previous work from the same laboratory has indicated that the atrophied fibers of denervated muscle may have 50% to 75% reductions in velocity (Buchthal & Rosenfalck, 1958).

In the next chapter, we will see how an action potential in a motor axon can provoke a similar signal in a muscle fiber. The task is a formidable one, for the nerve fiber is many times thinner than the muscle fiber and cannot provide enough current to excite the muscle fiber directly. In evolution the problem has been solved by the introduction of a chemical amplifier; the chemical is acetylcholine.

Chapter 10

Neuromuscular Transmission

The structure of the neuromuscular junction was described in chapter 3 (see Figure 3.1), and an account of the ionic mechanisms underlying the resting and action potentials was given in chapter 9. Using this framework and additional experimental observations on the junction itself, we can describe how excitation spreads from the motor axon to the muscle fiber (see Figure 10.1). In this excitation sequence, the impulse invades the motor nerve terminal and allows Ca^{2+} ions to enter and to release ACh from the synaptic vesicles. The ACh diffuses across the synaptic cleft and combines with receptors in the muscle fiber membrane; this combination alters the permeability of the membrane, causing it to depolarize and fire an impulse.

Acetylcholine Release

The role of ACh is to act as a chemical intermediary in the transfer of excitation from nerve to muscle. The need for such an intermediary is clear, for it serves to amplify the small impulse current in the motor nerve terminal into one that is sufficiently large to trigger an impulse in the muscle fiber. Each of the steps in the transmission process is described below in more detail; the account is particularly satisfying because the structural basis of most of the steps is understood.

ACh Is Synthesized in the Nerve Ending by the Acetylation of Choline

The synthesis of ACh is undertaken by the enzyme *choline acetyltransferase*, which, as its name implies, transfers an acetyl group (from acetyl-coenzyme A) to choline. This enzyme is found in the axon terminal, and at least some will have been carried there from the motoneuron soma (Jablecki & Brimijoin, 1974). The axon terminal obtains its supply of choline by uptake from the extracellular fluid; part is derived from the plasma and part from the hydrolysis of ACh by the enzyme acetylcholinesterase (AChE) in the basement membrane. The choline uptake can be inhibited by the substance *hemicholinium* (HC-3), and use has been made of this in the experimental analysis of ACh synthesis. The most widely held view is that the synthesis of ACh occurs in the axoplasm and that the transmitter is then taken up and concentrated by the synaptic vesicles. The rate of synthesis is geared to the amount of impulse

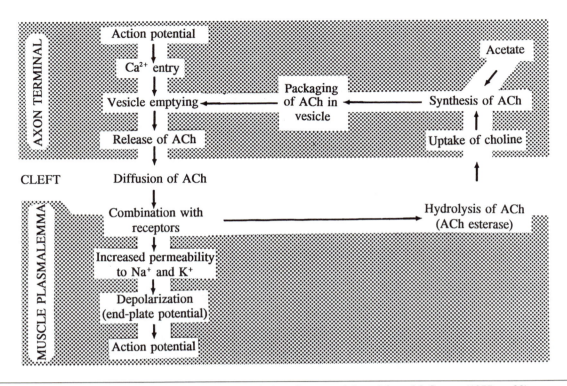

Figure 10.1   Summary of steps in neuromuscular transmission. See text. Adapted from McComas (1977, p. 30).

activity in the terminal, but a small amount of synthesis continues even in the resting state.

ACh Is Packaged Into Synaptic Vesicles for Release

Experiments using repetitive nerve stimulation have indicated that the ACh can be visualized as existing in three compartments. The first of these comprises the ACh that is *readily available* for release; electron micrographs suggest that this corresponds to the synaptic vesicles found bordering the synaptic region of the axolemma (Figure 3.1, chapter 3). The second compartment is the large remaining population of vesicles found more centrally within the terminal; these form the *main store* of ACh. Finally, there is an appreciable amount of ACh within the *cytoplasm* of the terminal; some of this ACh enters the synaptic cleft in the resting state and is responsible for the electrical ''noise'' that can be recorded at the end-plate. Isotope studies show that the cytoplasmic ACh exchanges freely with ACh in the synaptic vesicles. As already stated, ACh can be synthesized quickly and in surprisingly large amounts by the terminal during continuous impulse activity; it appears that the newly formed transmitter is largely directed into the ''readily available'' compartment.

Further information about the vesicular ACh has come from studies of the electric organ of the electric ray, *Torpedo*. This structure is especially rich in ACh

and is a good source of vesicles; in these vesicles the concentration of ACh is as high as 0.5 M. It has been established that the interiors of the vesicles are acidic and that it is the H^+ gradient across the vesicle membrane which attracts ACh into the vesicle (Anderson, King, & Parsons, 1982). Although ACh is the main component of the vesicles, the vesicles also contain an appreciable amount of *ATP* and a *proteoglycan*; the ATP is evidently released with ACh into the synaptic cleft (Silinsky & Hubbard, 1973), but its subsequent fate is unknown.

Electrical Impulses Travel to the Endings of the Motor Nerve Twigs

How is the ACh released from the synaptic vesicles? The first step, obviously, is the transmission of the action potential into the distal reaches of the motor axon. Katz and Miledi (1965) were able to show that the impulse travels all the way to the synapse by recording a propagated potential with an extracellular microelectrode positioned close to the axon terminal. As expected on theoretical grounds, the conduction velocity of the impulse is much lower in the fine terminal branches of the axon than in the main trunk. In the frog, the velocity distally is 0.3 m/s, and this value may be compared with the usual value of about 20 m/s for motor axons within a peripheral nerve of the same species.

When the impulse invades the axon terminal, it depolarizes the membrane in the usual way, through the opening of Na^+ channels (see chapter 9). This depolarization then activates Ca^{2+} channels, allowing Ca^{2+} ions to enter the axon terminal down their concentration gradient; this inward Ca^{2+} current starts rather slowly and lasts longer than the Na^+ current.

The Particles in the Active Zone Are the Ca^{2+} Channels

The entry of Ca^{2+} into the nerve terminal is essential for the release of ACh. Thus, neuromuscular transmission in *in vitro* preparations will cease if Ca^{2+} is omitted from the bathing fluid or if the concentration of Mg^{2+} is raised instead; Mg^{2+} ions appear to compete for the same receptors as Ca^{2+} and in this way can block the effect of the latter. The extent of the rise in Ca^{2+} concentration within the vertebrate motor nerve terminal is not known because of the difficulty in examining such a small structure. In the giant synapse of the squid, however, the axon terminal may measure 50 µm × 1,000 µm and is large enough to permit the introduction of one or more recording electrodes as well as the injection of Ca^{2+}-sensitive dyes such as Fura-2 and arsenazo III (Charlton, Smith, & Zucker, 1982; Miledi & Parker, 1981). It has been found that an impulse invading the axon terminal causes a 10-fold rise in Ca^{2+} concentration in the region of axoplasm adjacent to the plasmalemma. Freeze-fracture studies of the axolemma overlying the synaptic cleft reveal multiple arrays of particles, each array consisting of two sets of double rows (Figure 10.2, top). These structures comprise the *active zones*, for it is in their immediate vicinity that the ACh is discharged into the synaptic cleft; the particles are thought to be the Ca^{2+} channels themselves (Pumplin, Reese, & Llinas, 1981).

Vesicles Dock Against the Axolemma and Then Fuse With It

The events between the entry of Ca^{2+} and the release of ACh are still not fully understood, but at least two steps must be involved. The first of these is *docking* of the synaptic vesicles against the axolemma in the vicinity of the active zone. The second step is the *fusion* of the membrane of the vesicle with the axolemma, a process that instantly opens up the vesicle and allows the ACh to be released into the synaptic cleft. Since Ca^{2+} is necessary for transmitter release, it is evident that this ion participates in one or the other step; further, the steepness of the curve relating postsynaptic response to Ca^{2+} concentration suggests that from three to four Ca^{2+} ions cooperate in the emptying of a single vesicle (Dodge & Rahamimoff, 1967). It now seems likely that one pair of proteins is needed for docking and another pair for membrane fusion. A number of proteins have

been identified as likely to be involved in vesicle transactions (Bennett & Scheller, 1993; Warren, 1993); they include the following:

- Synaptobrevin
- SNAP-25
- Syntaxin
- Synaptophysin
- Synaptotagmin

Of these proteins, *synaptobrevin* and *syntaxin* probably mediate vesicle docking. Synaptobrevin, which appears to be identical with the vesicle associated membrane protein (VAMP), is to be found attached to the walls of the synaptic vesicles. Syntaxin is linked to the axolemma. In the presence of SNAPs (soluble accessory factors), synaptobrevin and syntaxin are able to interact (Söllner et al., 1993), a process that would thereby allow docking to occur. In molecular terms, both synaptobrevin and syntaxin are similar in that each has a single membrane-spanning domain with most of the protein extending into the cytoplasm (Kutay, Hartmann, & Rapoport, 1993).

In relation to the fusion step, an important experimental result was recently obtained by Bennett, Calakos, and Scheller (1992). By blocking synaptotagmin-acceptor sites with an engineered peptide fragment, these authors showed that *synaptotagmin* was essential for transmitter release, though not for vesicle docking. They therefore proposed that the combination of this protein, in the wall of the vesicle, with its acceptor on the axolemma, enables membrane fusion to take place. It is possible that the acceptor is syntaxin again (Bennett et al., 1992); further, since synaptotagmin binds Ca^{2+}, the fusion process could be the Ca^{2+}-dependent step in the release of ACh. The sites of fusion appear to be just to the side of the active zones in the nerve terminal, since a series of depressions can be seen in this region immediately after impulse activity using the technique of freeze-fracture electron microscopy (Heuser, Reese, & Landis, 1974; Figure 10.2, bottom).

At Each Neuromuscular Junction, 20 to 300 Vesicles Are Emptied Following a Single Impulse

How many synaptic vesicles release ACh after a single impulse invades the nerve terminal? A comparison of the sizes of miniature end-plate potentials (MEPPs; see chapter 3) with those of end-plate potentials in curarized nerve-muscle preparations would indicate that at least 100 vesicles are emptied at each neuromuscular junction. Another estimate, obtained by a different technique that avoided the use of curare, is that 200 to 300 quanta (vesicles) of ACh are released per impulse (Hubbard &

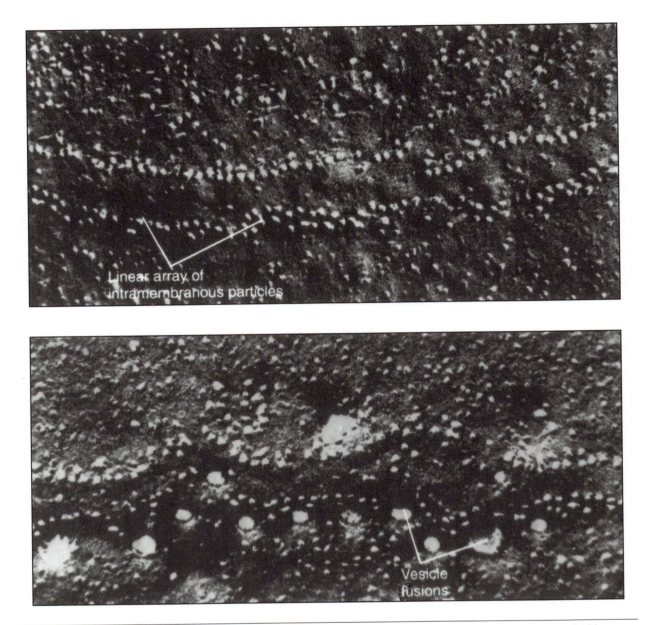

Figure 10.2   (Top) Freeze-fracture electron micrograph of the axolemma of the motor nerve terminal. The two double rows of intramembranous particles identify an active zone and are thought to be Ca^{2+} channels. (Bottom) Following an impulse, synaptic vesicles dock and fuse with the axolemma close to the Ca^{2+} channels before emptying. The fusion can be seen as the pale circular structures in the electron micrograph. Reprinted from Heuser et al. (1979).

Wilson, 1973). However, both these values, which were obtained in small laboratory animals, are considerably higher than those estimated by Slater, Lyons, Walls, Fawcett, and Young (1992) in biopsy specimens of human quadriceps muscles. In this preparation, the estimated mean number of ACh quanta at 37°C was 28 and was similar to the corresponding value for human intercostal muscle (41 quanta; Engel, Walls, Nagel, & Uchitel, 1990). Since there are many active zones in a given neuromuscular junction, only a few vesicles need

to be emptied at each of them, even if individual zones respond intermittently to successive impulses (see p. 152).

A Period of Facilitation Follows the Discharge of ACh

Most of the *synaptic delay* between the excitation in the nerve and its subsequent appearance in the muscle fiber is due to the Ca^{2+}-mediated release of transmitter. For the frog neuromuscular junction this delay is of the order of 0.5 ms, but at the mammalian junction the value is rather smaller (about 0.2 ms; Eccles & Liley, 1959; Hubbard & Schmidt, 1963). Experiments with double or multiple stimuli have shown that for the first 100 ms or so following an impulse, the axon terminal is in a hyperexcitable state. Thus, although some 30 to 100 vesicles will have been emptied of ACh, the transmitter remaining is more available for release by a second impulse. This *facilitation* is mainly due to residual Ca^{2+} in the nerve terminal following the first impulse; this Ca^{2+} is added to that admitted by the second impulse, to produce a larger-than-normal Ca^{2+} concentration and hence a greater release of ACh (Katz & Miledi, 1968).

The period of facilitation is succeeded by a *depression* of synaptic transmission, which typically lasts for several seconds. During this time, a second stimulus releases a smaller-than-normal number of ACh quanta, due to delays in the movement of synaptic vesicles to the active zones from more remote parts of the axon terminal. A detailed analysis of the postexcitation changes in the nerve terminal has been made by Magleby and Zengel (1982), who have identified four components.

Membrane Is Removed From the Nerve Terminal to Form New Synaptic Vesicles

One interesting question concerns the fate of the vesicles after they have discharged their ACh into the synaptic cleft. If the membranes of the vesicles fuse with the axolemma, as electron micrographs indicate, the area of the axolemma would be increased to such an extent that some later mechanism would have to be available for removing the excess membrane. By using extracellular markers, Heuser and Reese (1973) have shown that recycling takes place, the presynaptic membrane being reformed into vesicles toward the periphery of the synapse (*endocytosis*). In the frog neuromuscular junction, endocytosis can be detected within 1 s of ACh release and is maximal at 30 s; the process is almost over by 90 s (Miller & Heuser, 1984).

Combination of Acetylcholine With Receptor

When ACh is released from a micropipette near an endplate, the electrical response of the muscle fiber can be detected almost instantaneously. It seems, therefore, that once ACh is discharged from the synaptic vesicles into the synaptic cleft, it diffuses very quickly to combine with receptors in the muscle fiber membrane. That the ACh-binding sites are on the outer surface of the membrane was shown by early experiments in which ACh was released iontophoretically from a micropipette. When the tip of the pipette was inside the fiber, no response to injected ACh could be detected, whereas external application of the substance produced a rapid fall in muscle fiber membrane potential.

The Combination of ACh With Its Receptor Allows Na⁺ and K⁺ to Flow Through the Muscle Fiber Plasmalemma

The ACh receptors (AChRs) have already been described in chapter 3. The receptors are attached to the muscle fiber plasmalemma where they line the crests of the junctional folds and the upper parts of the secondary synaptic clefts (Figure 3.1C, chapter 3). The AChRs, which can be seen with high-powered electron microscopy, each consist of five subunits, two of which (the α subunits) carry the ACh-binding sites (see Figure 3.4, chapter 3). Whenever two molecules of ACh combine with the α subunits, the channel in the center of the receptor opens sufficiently to allow K⁺ and Na⁺ ions to flow through. The ions travel in opposite directions, moving down their respective concentration gradients; thus, K⁺ ions leave the fiber while Na⁺ ions, under the additional influence of the electrical gradient across the membrane, enter the fiber. The membrane potential will tend to fall to a value between the respective equilibrium potentials for Na⁺ and K⁺ (see chapter 9). This depolarization is termed the *end-plate potential* (EPP) and, if fully developed, would bring the membrane potential from its resting value of approximately −85 mV to about −15 mV (Fatt & Katz, 1951).

The End-Plate Potential Triggers an Action Potential in the Muscle Fiber

Under normal circumstances, the rising end-plate potential is overtaken by an action potential, which arises in the membrane immediately adjacent to the AChRs and is due to the opening of voltage-gated Na⁺ channels; these channels are especially plentiful in the end-plate region, and large numbers line the deeper regions of the secondary synaptic clefts (Flucher & Daniels, 1989). Once initiated, the action potential propagates in both directions toward the end of the muscle fiber. It has been calculated that, during the end-plate potential, each of the activated AChRs allows 17,000 Na⁺ ions to flow into the fiber and a smaller number of K⁺ ions to leave it.

Analysis of end-plate "noise" indicates that the opening of one ACh channel depolarizes the membrane by 0.3 µV. Since the fully developed end-plate potential would be approximately 70 mV (see the preceding section), it follows that rather more than 200,000 AChRs must be activated following a single impulse in the motor nerve terminal. At human neuromuscular junctions, where relatively few ACh quanta are released by a single impulse, the EPP is smaller and probably not much larger than the threshold potential for action potential generation.

Slater et al. (1992) have suggested that the Na^+ channels in the deeper parts of the secondary synaptic clefts provide a safety margin by acting as an amplifying system.

AChR Channels Open for 1 ms or Longer

One way to examine the full time-course of the EPP is to block a proportion of the AChRs with curare, so that the EPP becomes too small to trigger an action potential. The EPP is then seen to have a sharply rising onset and an exponential decay lasting several milliseconds. In view of the time needed to discharge the capacitance of the muscle plasmalemma (see chapter 9), the current flowing at the end-plate is briefer than the EPP. Since this current is the aggregate of those of the AChR channels, it is important to determine the periods of time that individual ACh channels are open. This task can be accomplished by patch-clamp recording, and the AChR was the first ion channel to be studied by this technique (Neher & Sakmann, 1976a). It was found that individual channels opened for periods of time that could be as short as 1 ms or as long as 50 ms (Figure 10.3).

Acetylcholinesterase Terminates Synaptic Transmission

The enzyme acetylcholinesterase (AChE) is one of the proteins in the basement membrane and is therefore positioned in the synaptic cleft between the ACh-release sites in the axolemma and the AChRs in the muscle fiber plasmalemma. By virtue of its location, the AChE will capture a fraction of the ACh discharged into the synaptic cleft, before the transmitter can reach the AChRs. However, because there are many more ACh molecules released than there are AChE molecules to catch them, most of the ACh nevertheless reaches the receptors. After opening the pores in the receptors, the ACh molecules are detached and diffuse toward the basement membrane where they are hydrolyzed by AChE. The hydrolysis of ACh brings the synaptic transmission process to a conclusion.

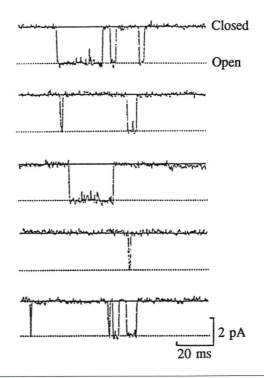

Figure 10.3 Recordings of the synaptic currents flowing through a single AChR channel in the presence of ACh. The recording was made with a patch-clamp electrode (see Figure 7.6). Reprinted from Kandel and Siegelbaum (1991). (Courtesy of Dr. B. Sakmann.).

Each Active Zone and Its Opposing AChRs Constitute a "Saturating Disc"

The active zones releasing ACh from the motor nerve terminal respond differently from one impulse to the next. It is therefore appropriate to consider the neuromuscular junction as an array of *saturating discs*, in which each part of the postsynaptic membrane is able to respond maximally to a corresponding region of the motor nerve terminal (Fertuck & Salpeter, 1976; Engel, 1987).

In each disc, the AChRs in a junctional fold are close to an active zone in the nerve terminal. Since the density of AChRs ($10,000-20,000/\mu m^2$) is much greater than that of the AChE molecules in the basement membrane ($2,000-3,000/\mu m^2$), most of the ACh will combine with its receptor. Further, a quantal packet of ACh needs to diffuse only 0.3 µm along the top, or down the side, of a synaptic fold in order to saturate the receptors and become completely bound. The small sizes of the saturating discs prevent them from overlapping and ensure that only a small fraction of the AChRs are saturated when a number of ACh quanta are released by a nerve impulse. Because successive impulses will activate different active zones and hence different saturating discs, the AChRs are prevented from becoming desensitized by continued exposure to ACh.

APPLIED PHYSIOLOGY

In view of the complex sequence of events involved in neuromuscular transmission, it is hardly surprising that there are many ways in which the process can be disrupted. In this section, we will study first a number of drugs, including some that are used in the operating theater, then a bacterium that can produce a fatal paralysis, and last, two immunological disorders.

Neuromuscular Transmission Can Be Blocked to Produce Muscle Paralysis During Surgery

During surgery it is usually important to have the muscles of the patient relaxed, so that the trachea can be intubated by the anesthetist and so that, at the site of the operation, the muscles can be more easily incised and pulled to one side. The favored drug for producing relaxation is *suxamethonium*. It is a quaternary ammonium compound that blocks neuromuscular transmission by combining with AChRs and producing a depolarization. As the depolarizing action commences, the muscle fibers begin to twitch spontaneously, indicating that their membrane potentials have fallen to the critical level for action potential initiation. As depolarization proceeds, the muscle fibers become refractory to motor nerve excitation, and paralysis ensues. In comparison with the natural transmitter, ACh, the action of the drug in prolonged, and this enables it to be used as an effective muscle relaxant during surgery. Suxamethonium can be administered by intravenous drip in the form of its chloride (*succinylcholine*). *Decamethonium* is a similar type of compound.

Tubocurarine is a different type of drug and is less frequently used. It is an alkaloid that can be prepared from crude extracts of the plant toxin curare. By competing for the same postsynaptic receptors as ACh, tubocurarine diminishes the action of the transmitter. Since the drug does not produce depolarization of the muscle fiber membrane itself, it is classified as a nondepolarizing blocking agent. Experimentally, microelectrode recordings from single muscle fibers reveal progressive reductions in the sizes of the end-plate potentials as they continue to fall below the level required for impulse initiation. In humans, the action of tubocurarine usually lasts about 30 min. *Gallamine* is another drug with a similar type of action.

Anticholinesterase drugs have opposite effects; they potentiate neuromuscular transmission by preventing the hydrolysis of ACh by the AChEs attached to the basement membrane. Consequently, ACh released from a nerve terminal exerts a larger and more prolonged depolarization of the muscle fiber. Examples of anticholinesterases used in medicine are *neostigmine, pyridostigmine bromide,* and *edrophonium chloride.*

Botulism

The toxins produced by the bacterium, *Clostridium botulinum*, remain the most powerful yet known to man; as little as 0.5 µg of toxin A is fatal. The bacterium is anaerobic and is to be found in the soil and in animal feces. Poisoning is fortunately uncommon and usually results from the improper canning or bottling of food; if heat sterlization has been insufficient, the spores persist in the food and then germinate in the anaerobic environment.

Very Small Amounts of Botulinum Neurotoxin Can Produce a Fatal Paralysis

Symptoms of poisoning appear within 48 hr of ingestion of the toxin; initially they consist of double vision (diplopia) and unsteadiness on standing. Subsequently, the lower cranial nerves are affected, producing paralysis of speech and swallowing. Unless treatment is available, death occurs from respiratory paralysis within a few days. Until the end, the patient is fully conscious and is without any disorder of sensation.

The Paralysis Is Due to the Inability of Impulses to Release ACh

Electromyography reveals small evoked muscle responses following nerve stimulation, although the muscle can be shown to respond normally to direct stimulation. Repeating the nerve stimulation at low rates (e.g., 2 Hz) causes further decrement in the muscle responses, whereas rapid stimulation (at 50 Hz) produces potentiation; this pattern of behavior resembles that found in the Lambert-Eaton syndrome (see p. 158). Together, these findings indicate that the paralysis results from a failure of neuromuscular transmission, and this has been confirmed by microelectrode investigations in frog muscle. Thus, the finding of spontaneously occurring MEPPs indicates that transmitter is available within the axon terminal and that there is no loss of postsynaptic responsiveness. However, although the nerve impulse still invades the axon terminal during the phase of paralysis, none of the available transmitter can be released (Harris & Miledi, 1971).

The Various Botulin Neurotoxins Attack Synaptic Vesicle Proteins

There are seven types of botulinum neurotoxins, each of which can block neuromuscular transmission. It turns out that each type is a zinc-containing *endopeptidase*, capable of degrading one of the proteins necessary for the release of ACh from the nerve terminal. As an example, Schiavo et al. (1992) have shown that botulinum B neurotoxin splits the protein synaptobrevin (as does tetanus toxin), while Blasi et al. (1993) found that neurotoxin A cleaves SNAP-25. Not only do observations of this type explain the actions of this deadly family of neurotoxins, but they also demonstrate that the targeted proteins are essential for either vesicle docking or membrane fusion.

Myasthenia Gravis

The diseases that affect the neuromuscular junction primarily are relatively infrequent but are nevertheless well understood as a result of studies employing electron microscopy, immunology, and microelectrode recordings. In the least common conditions, all of which are genetic in nature, the defects range from faulty ACh packaging (*familial myasthenia gravis*) to the impaired synthesis of AChRs and AChE.

The Neuromuscular Junction May Be Subjected to Immunological Attack

Two other transmission disorders are immunologically determined; in *myasthenia gravis*, the antigen is the AChR, while in the *Lambert-Eaton myasthenic syndrome*, it appears to be the voltage-gated Ca^{2+} channels in the motor nerve terminal.

Did the 17th-Century Physician Thomas Willis Describe a Patient With Myasthenia?

In 1672 Willis wrote:

> Nevertheless those labouring with a want of spirits, will use these spirits for local motions as well as they can; in the morning they are able to walk firmly, to fling their arms about hither and thither or take up any heavy thing; before noon the stock of spirits being spent, which had flowed into the muscles, they are scarcely able to move hand or foot. At this time I have under my charge a prudent and honest woman who for many years has been subject to this sort of spurious palsy, not only in her members, but also in her tongue. She for some time can speak hastily or eagerly, she is then not able to speak a word, but becomes suddenly as mute as a fish, nor can she recover the use of her voice for an hour or two.

Myasthenia Usually Affects Ocular Muscles First

Myasthenia attacks about 1 in 10,000 persons, mainly young women and older men. The presenting symptom is often drooping of an eyelid or double vision, and is most noticeable in the late afternoon and evening. In more severe cases, there is difficulty in swallowing, talking, and chewing, and the arms and legs also fatigue easily. In the most seriously affected patients, even the respiratory muscles are involved, and artificial ventilation may be needed.

The Diagnosis of Myasthenia Can Be Confirmed by Pharmacological, Immunological, and Electrical Testing

The intravenous injection of a small dose of edrophonium chloride can improve muscle strength within a minute, as in the woman shown in Figure 10.4, who was then able to open her eyes and smile. Edrophonium blocks AChE, allowing ACh to linger in the synaptic cleft and to combine repeatedly with AChRs. Since myasthenia is an autoimmune disorder, it is not surprising that antibodies to the AChR can be detected in the plasma of 90% of patients with the disorder; in general, the more ill the patient, the higher is the antibody titer.

Electrical testing is based on the observation that repeated excitations cause fluctuations in the efficacy of neuromuscular transmission. For example, if the motor nerve is stimulated repetitively, the evoked muscle responses decline, as in Figure 10.5. A more accurate and less painful electrical test is *single fiber EMG*, devised by Ekstedt and Stålberg (1967). The principle of the test is to record with a needle electrode from two muscle fibers belonging to the same motor unit. In normal subjects, the time interval between the respective fiber action potentials fluctuates slightly as the person carries out a weak contraction; in myasthenic patients, however, the fluctuations, or "jitter," are more pronounced, as in Figure 10.6.

Myasthenia Can Be Induced in Experimental Animals

While attempting to prepare antibodies to the AChR, Patrick and Lindstrom (1973) were puzzled to find that rabbits injected with receptor from the electric organs of the electric eel became weak and floppy. Vanda

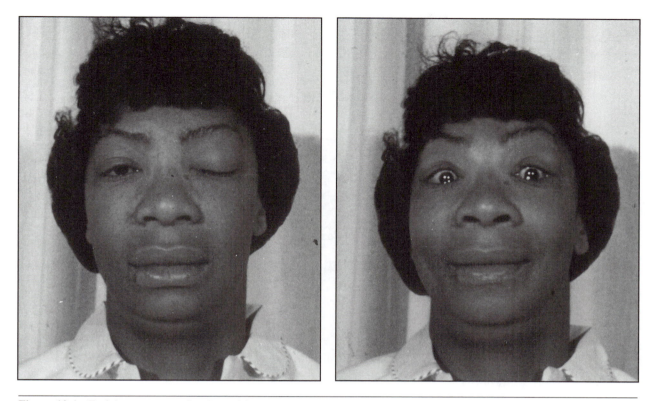

Figure 10.4 Facial appearance of patient with myasthenia gravis before (left) and approximately one minute after (right) an injection of an anticholinesterase drug. Notice how the drooping of the eyelids (ptosis) disappears. (Courtesy of Dr. H.S. Barrows.)

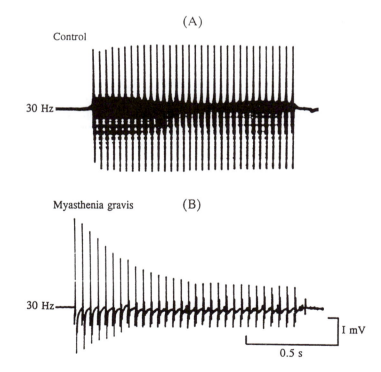

Figure 10.5 (A) Normal response (absence of decrement) in the extensor digitorum brevis muscle of a healthy subject; the peroneal nerve was stimulated at 30 Hz. (B) Decrementing response in the corresponding muscle of a 35-year-old woman with generalized myasthenia gravis.

Lennon, who had earlier suggested that the AChR might act as an antigen in myasthenia (Lennon & Carnegie, 1971), recognized that the injected animals had become myasthenic and was later able to show that the disease takes place in two stages (Lennon, Lindstrom, & Seybold, 1976). In the first stage, thymus-derived lymphocytes (T-cells) become sensitized to AChR at 4 to 5 days after inoculation of AChR. The sensitized T-cells

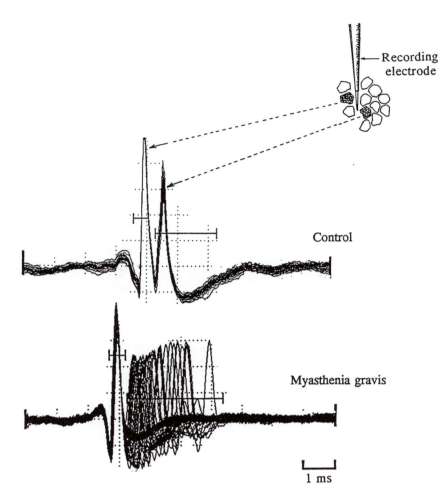

Recording electrode

Control

Myasthenia gravis

1 ms

Figure 10.6 Superimposed recordings made from two muscle fibers belonging to the same motor unit in a healthy subject and in a patient with myasthenia gravis. Notice how the interval between each pair of impulses fluctuates markedly in the patient (jitter phenomenon) but not in the healthy subject. These recordings are examples of single fiber EMG.

then react with the animal's own AChRs and release pharmacologically active *lymphokines*; these increase vascular permeability and cause masses of inflammatory cells to accumulate at the neuromuscular junction (Figure 10.7). Destruction of the synapses then occurs and is responsible for the first episode of weakness; during this phase, the axon terminals become separated from the postsynaptic membranes.

After the acute inflammation response subsides, the neuromuscular junctions are partially repaired but now assume a simpler structure with fewer secondary clefts. Further degeneration of the synapses takes place following the appearance of circulating antibody to the AChR; this stage is responsible for the second phase of muscle weakness. Electron microscope studies have revealed a strong similarity between the structures of the neuromuscular junctions in this chronic phase of experimental myasthenia and in that of the human disease (Engel, Tsujihata, Lindstrom, & Lennon, 1976).

AChR Antibodies Have Several Myasthenic Effects

How do the AChR antibodies produce their effects in myasthenia? It is now clear that there is more than one type of AChR antibody but that most types bind to the α subunit (Tzartos, Langeberg, Hochschwender, & Lindstrom, 1983), though not usually at the ACh-binding site. A more serious effect of the receptor antibody is to cause complement fixation, which, in turn, results in destruction of the synaptic folds. Thus, not only are AChRs lost as pieces of membrane are shed into the synaptic cleft, but there is less membrane available for the insertion of newly synthesized receptors (Engel, 1987). A further action of the antibodies is that, after combination with receptors, they hasten the internalization and degradation of the receptors (Stanley & Drachman, 1978).

The End-Plate Potentials Are Diminished in Myasthenia

In myasthenia, neuromuscular transmission is jeopardized for two reasons. First, there are too few receptors for the ACh to combine with, and second, the widened primary synaptic cleft encourages the diffusion of ACh away from the receptors.

Since the ACh-AChR combinations are diminished, following an impulse in the nerve terminal, the EPP

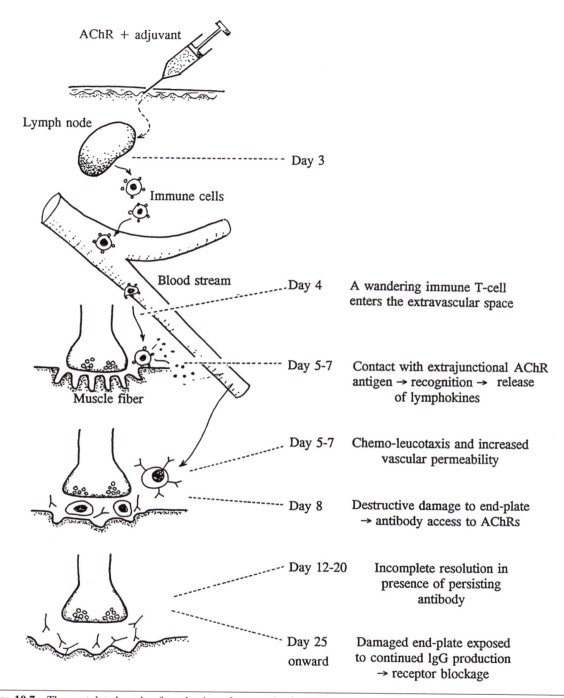

AChR + adjuvant

Lymph node

Day 3

Immune cells

Blood stream

Day 4 — A wandering immune T-cell enters the extravascular space

Muscle fiber

Day 5-7 — Contact with extrajunctional AChR antigen → recognition → release of lymphokines

Day 5-7 — Chemo-leucotaxis and increased vascular permeability

Day 8 — Destructive damage to end-plate → antibody access to AChRs

Day 12-20 — Incomplete resolution in presence of persisting antibody

Day 25 onward — Damaged end-plate exposed to continued IgG production → receptor blockage

Figure 10.7   The postulated mode of production of a myasthenia gravis-like illness in animals following the injection of AChR and adjuvant. Adapted from Lennon, Lindstrom, and Seybold (1976, p. 295).

develops more slowly than usual and may not reach the threshold for generating an action potential. Such impulse failures can be seen in Figure 10.8, in which a microelectrode was used to record from the end-plate region of a human intercostal muscle fiber. On those occasions when the threshold for excitation is reached, the timing of the action potential varies considerably; this variation is the basis for the increased neuromuscular jitter, which can be recorded with a single fiber EMG

electrode (see Figure 10.6). As would be expected, the deficiency of AChRs also reduces the sizes of the MEPPs (Elmqvist, Hofmann, Kugelberg, & Quastel, 1964). It must be emphasized, however, that there is nothing wrong with the synthesis and release of ACh from the motor nerve terminal in myasthenia; indeed, measurements with gas chromatography and mass spectrometry indicate that the amount of ACh may be increased in myasthenia (Ito et al., 1976).

The Lambert-Eaton Myasthenic Syndrome (LEMS)

The eponym given to this disorder is well deserved. Lee Eaton was a senior neurologist at the Mayo Clinic who had under his care a group of patients with an unusual form of muscle weakness, while Ed Lambert not only recognized the unusual nature of the disorder but continued for many years as the most distinguished electromyographer in North America.

Earlier cases of the condition had likely been misattributed to myasthenia gravis, such as that described by Anderson, Churchill-Davidson, and Richardson (1953). The paper by Eaton and Lambert (1957) was definitive, however, for it emphasized that the myasthenic-like weakness is often associated with a malignancy, usually small cell carcinoma of the lung. Further, the condition can be differentiated from myasthenia gravis by the

response of the muscles to high-frequency motor nerve stimulation. In myasthenia, the muscle response is of normal amplitude initially and then declines, whereas in LEMS, the response is at first small and then potentiates markedly as stimulation is continued (Figure 10.9); potentiation of the evoked responses can also be achieved by voluntary contraction.

In LEMS, Insufficient ACh Is Released From the Nerve Endings

The fact that the muscle responses grow larger as stimulation continues strongly suggests a presynaptic disorder, and microelectrode studies on intercostal muscle biopsies have confirmed this prediction (Lambert & Elmqvist, 1971; see also Hofmann, Kundin, & Farrell, 1967). The MEPPs are found to be of normal size;

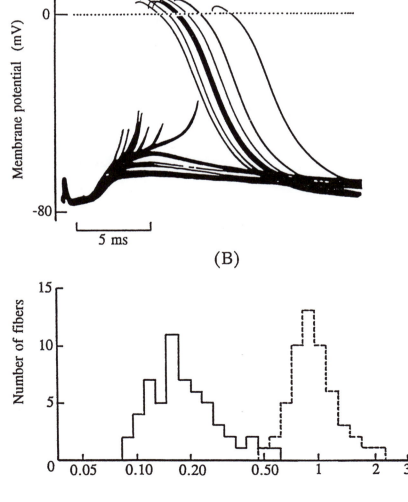

Figure 10.8 (A) Intracellular recordings from an intercostal muscle fiber of a patient with generalized myasthenia gravis. When the nerve was stimulated at 5 Hz, the evoked responses from the fibers often consisted of subthreshold end-plate potentials instead of action potentials. (B) Amplitudes of miniature end-plate potentials (MEPPs) in intercostal muscle fibers of myasthenic patients (solid line) and control subjects (dashed line). Adapted from Elmqvist, Hofmann, Kugelberg, and Quastel (1964).

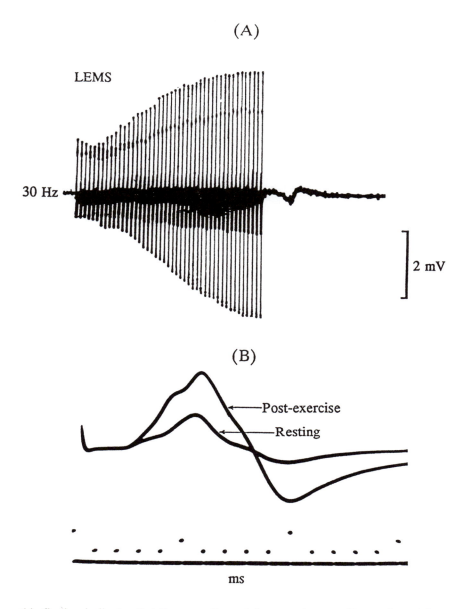

(A)

LEMS

30 Hz

2 mV

(B)

Post-exercise

Resting

ms

Figure 10.9 (A) Incrementing muscle action potentials during repetitive nerve stimulation at 30 Hz in a patient with LEMS. The progressive enlargement of the electrical responses should be compared with the situation in the healthy subject of Figure 10.5A. (B) A simple, and less painful, test for LEMS is to contrast the muscle responses to single shocks delivered before, and immediately after, a brief voluntary contraction. The enlargement is obvious in a LEMS patient, as in the example here.

this finding indicates that the synaptic vesicles contain normal amounts of ACh and that the AChRs are available and responsive. Direct measurements of ACh contents in LEMS muscle also yield normal results (Molenaar, Newsom-Davis, Polak, & Vincent, 1982). When the motor nerve is stimulated, however, too few quanta are released to depolarize the muscle fiber membrane to the threshold for generating an action potential. The quantal release can be enhanced by raising the Ca^{2+} concentration in the fluid bathing the muscle fibers, or by adding guanidine. All of these findings, then, point to defective release of ACh in LEMS.

The possibility that the voltage-gated Ca^{2+} channels might be abnormal in the motor nerve terminal has been investigated by subjecting this membrane to freeze-fracture. The parallel rows of particles, which constitute the active zones for ACh release (Figure 10.2, top), are seen to be markedly decreased, although other randomly disposed particles are normal in number (Fukunaga, Engel, Osame, & Lambert, 1982).

The Presynaptic Ca^{2+} Channels Are Attacked by Antibodies Raised Against Tumor Cells

Convincing evidence for an autoimmune basis for LEMS is the ability to induce the electrophysiological features of the condition in mice by injecting them with immunoglobulin from affected patients (Lang, Newsom-Davis, Prior, & Wray, 1983; Vincent, Lang, & Newsom-Davis, 1989). The animals develop the same type of disruption of the active zones in the motor nerve membrane as do LEMS patients. Supporting evidence favoring an autoimmune etiology in LEMS is that removing immunoglobulin from patients by the technique of plasmapheresis is often beneficial.

The most plausible explanation for the etiology and pathogenesis of LEMS is that the voltage-gated Ca^{2+} channels in the motor nerve become the targets of antibodies. In patients with cancer, the antibodies are probably raised against voltage-gated Ca^{2+} channels in the tumor cells and then react with similar channels in the motor nerve endings. In the minority of LEMS patients without tumors, the origin of the antibodies is unknown. The combination of antibodies with Ca^{2+} channels causes the latter to become cross-linked, destroying the normal structure of the active zones and preventing the influx of Ca^{2+} during impulse invasion of the motor nerve terminal.

The last step in muscle activation is, of course, the contraction itself and the production of force or movement. In the next chapter, we will learn that muscle contraction, like neuromuscular transmission, is a complex process involving a number of steps.

Chapter 11

Muscle Contraction

Even the simplest unicellular creatures move about. Further, individual cells in multicellular organisms may change their sizes and shapes, depending on their functions and stages of growth. These movements involve different types of protein filament—actin, myosin, intermediate filaments, and microtubules. A further type of movement, considered in chapter 8, is the mechanism responsible for axoplasmic transport. However, the most specialized type of movement, and that most completely studied and understood, is the contraction and relaxation of skeletal muscle fibers. It will be seen that, like axoplasmic transport, the muscle movements are brought about by molecular motors interacting with filaments; in this case, the motors are myosin molecules, and the filaments are actin.

Sliding Filament Mechanism

The way in which skeletal muscle fibers contract was deduced by two quite different types of microscopic study in the early 1950s, and in each case a Huxley was involved. Andrew Huxley, of the voltage clamp experiments (see chapter 9), made his observations with an interference microscope of his own design, while Hugh Huxley (no relation) used an electron microscope to examine what were, in those days, exquisitely thin sections of muscle.

Microscopic Studies Indicate a Sliding Filament Mechanism

With the electron microscope, it was seen that, within a muscle fiber, each myofibril was composed of many *myofilaments* and that the latter were of two types, thick and thin. In cross sections of muscle fibers, each thick filament was seen to be surrounded by a *hexagonal array* of thin filaments (Figure 11.1, top); in longitudinal sections the thick filaments of a myofibril were found to be in register with each other (Figure 11.1, bottom). For a long time it had been known that the contractile proteins of muscle were *myosin* and *actin*, and next it was shown that these proteins corresponded to the thick and thin filaments respectively. This step was achieved by dissolving the myosin in KCl solution and demonstrating that the thick filaments were no longer visible with electron microscopy.

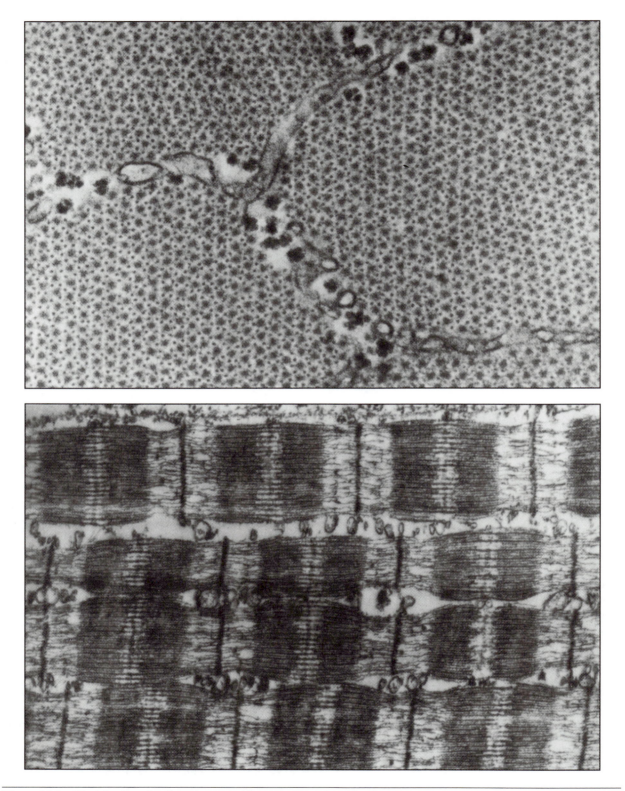

Figure 11.1   (Top) Lattice of actin and myosin filaments. Six actin (thin) filaments surround each myosin (thick) filament in a hexagonal array, as shown diagramatically in Figure 11.2F. The electron micrograph shows part of a frog sartorius myofiber in cross section. Reprinted from Huxley (1972). 123,000×.

(Bottom) Thin, longitudinal sections through four myofibrils showing overlapping thick and thin filaments. See also Figure 1.7 in chapter 1 for identification of features. Reprinted from Bourne (1972). 53,000×.

At the same time that Hugh Huxley was undertaking these important studies, Andrew Huxley was using the interference microscope to examine the muscle striations of living frog muscle fibers during contraction and relaxation. He observed that during contraction the light *I-band* became shorter while the dark *A-band* remained the same length; within the A-band, however, the pale *H-zone* narrowed and might disappear completely. Quite independently, the Huxleys proposed that their respective findings could be explained by a sliding movement of the actin and myosin filaments past each other (A.F. Huxley & Niedergerke, 1954; H.E. Huxley & Hanson, 1954; see also A.F. Huxley, 1959, 1974). This *sliding filament hypothesis* is now accepted (see Figure 11.2, D & E).

Figure 11.2 shows that the I-band is the region of fiber where only actin filaments are present; the A-band corresponds to the position of the myosin filaments. In the relaxed state (Figure 11.2D), although there is some overlap between the actin and myosin filaments, opposing actin filaments are separated from each other along the myosin filament. The gap between the actin filaments is responsible for a pale region (H-zone) at the center of the A-band.

When the muscle contracts (Figure 11.2E), the opposing actin filaments are propelled toward each other and slide along the intervening myosin filament. As they near each other, they cause the H-zone to become narrower; similarly, as more of each actin filament is drawn into the space between the myosin filaments, the I-band becomes shorter. Since the myosin filaments do not alter their shape, the length of the A-band stays unchanged.

The function of the *Z-line*, or *Z-disc*, is to tether the actin filaments together, while the *M-line* in the center of the A-band corresponds to links between the myosin filaments; both types of structure maintain the orderly geometrical relationship of the filaments to one another (see chapter 1).

The Sliding Mechanism Is Produced by Cross-Bridges

The next step in understanding the sliding filament mechanism was the demonstration, using electron micrographs at high magnification, of small projections from the myosin filaments (H.E. Huxley, 1958; Figure 11.3). These projections were termed *cross-bridges*, and it was proposed that they could momentarily attach themselves to the actin filaments and propel the latter to new positions. The cross-bridges are approximately 13 nm long and lie in six rows along the myosin filament; each row can engage one of the actin filaments in the hexagonal array surrounding the myosin filament.

As shown in Figure 11.2C, the cross-bridges are arranged in pairs separated by 180°; the next pair is always 14.3 nm away, with its axis rotated through 60°.

Isolated Cross-Bridges Can Be Made to Work *In Vitro*

Further insights into the mechanism of cross-bridge action came from studies of the alga *Nitella*; the advantage of using this preparation is that the cells are relatively long and contain parallel bundles of actin filaments. If the cells are cut open and spread out, the cytoplasm can be washed away, leaving the bundles of actin filaments in place. Sheetz and Spudich (1983) applied tiny plastic beads coated with myosin to the treated *Nitella* cells and observed that, in the presence of ATP, the beads were moved along the actin filaments, through the action of the myosin cross-bridges. An extension of this type of experiment was to replace *Nitella* with an artificial membrane to which longitudinally oriented actin filaments had been attached (Spudich, Kron, & Sheetz, 1985). In subsequent experiments from the same laboratory, the need for beads was eliminated by coating a glass surface with myosin filaments, applying ATP, and observing the movement of fluorescently labeled actin filaments under the microscope. All three types of experiment indicated that the cross-bridge movements occur with a constant velocity and that the direction of movement is determined by the polarity of the actin filaments (see p. 170). In a single "*working stroke*," a cross-bridge moves an actin filament through 10 nm in the space of 2 ms, while hydrolyzing one molecule of ATP (see H.E. Huxley, 1990).

Contractile Force Depends on the Number of Cross-Bridge Interactions

One of the predictions of the sliding filament hypothesis was that each cross-bridge would act as an independent force generator, and that the force developed in a contraction would therefore depend on the number of simultaneous interactions between the cross-bridges and the actin filaments. Experimentally, the number of cross-bridge–actin interactions can be varied by stretching the muscle fiber and so altering the amount of overlap between the actin and myosin filaments. This type of experiment is easily carried out in humans, for example, by recording the isometric twitches developed by the calf muscles after they have been stretched by dorsiflexing the ankle (Figure 11.4). It can be seen that the force increases, with muscle lengthening, up to a maximum and then, with further stretching, starts to decline.

Although the preceding result is consistent with the sliding filament hypothesis, the degree of overlap between the actin and myosin filaments cannot be known

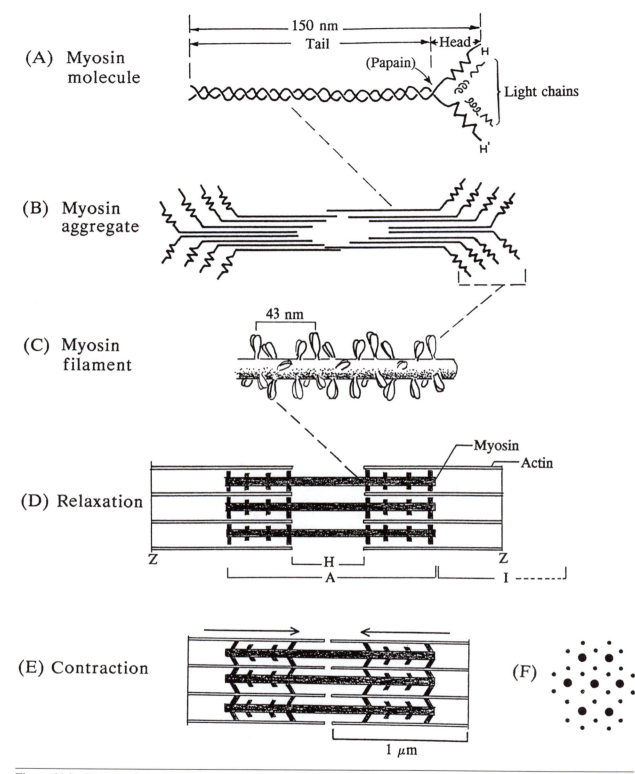

Figure 11.2 The myosin molecule and its arrangement in the thick filaments of the myofibrils. (A) A single molecule, consisting of a double-helical rod terminating in two globular heads, each of which has two light chains attached. The site of enzymatic cleavage, with papain, is shown. (B) In solution, myosin molecules spontaneously aggregate to form filaments with heads at both ends. (C) In a thick filament, the two globular heads of a myosin molecule project to form a cross-bridge. The next bridge is separated by 14.3 nm and 60° (see text). (D) The overlap of actin and myosin filaments in a relaxed myofibril and the various refractive bands that are created. (E) Contraction is produced by actin filaments sliding over the myosin filament, causing approximation of the Z-lines (discs) and narrowing of the H-region. (F) A cross section through a myofibril to display the hexagonal disposition of the actin filaments around the myosin filaments. The diagram can be compared with the electron micrograph in Figure 11.1 (top).

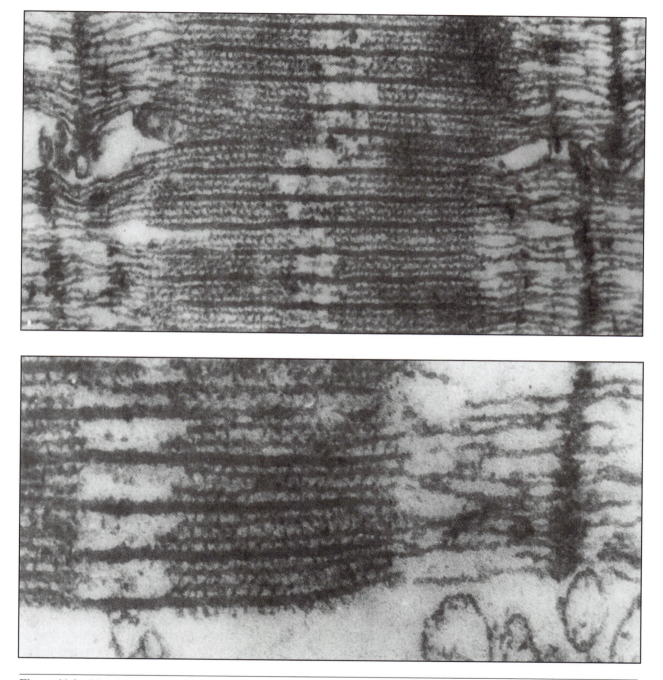

Figure 11.3 Myosin cross-bridges. (Top) Two thin (actin) filaments separate adjacent thick (myosin) filaments. (Inspection of Figures 11.1, top, and 11.2F shows how this appearance arises.) The small projections, perpendicular to the overlapping filaments, are the cross-bridges. (148,000×). (Bottom) The cross-bridges are seen more clearly. (144,000×). Reprinted from Huxley (1972).

precisely, because stretching affects the central regions of the fibers more than the ends. For this reason Gordon, Huxley, and Julian (1966a) devised an elegant experiment in which they examined short lengths of single muscle fibers in the frog. These authors fixed two small pieces of gold leaf to the fiber and used these to interrupt

two beams of light passing from oscilloscope tube spots to two photocells (Figure 11.5). The output from the photocells was fed into a control circuit that operated a servomotor attached to one end of the muscle fiber. The other end of the fiber was attached to a strain gauge. By using negative feedback, the system enabled the

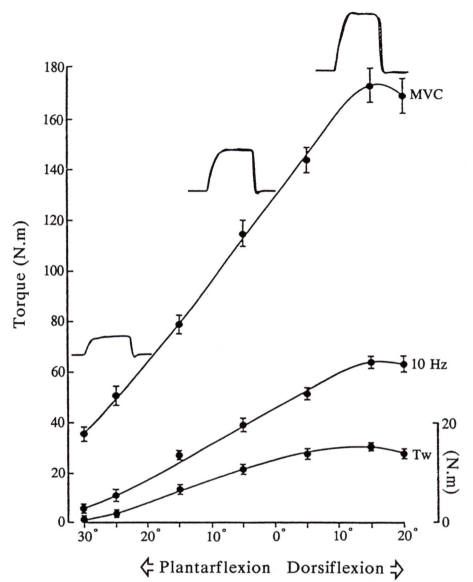

Figure 11.4 Effect of altering the lengths of human muscles on force, and hence torque, produced during single twitches (*Tw*), 10 Hz stimulation (*10 Hz*), and maximal voluntary contraction (*MVC*). The lengths of the calf muscles were increased by moving the ankle from plantarflexion to dorsiflexion. Sample MVCs are shown beside top curve. Data of Sale, Quinlan, Marsh, McComas, and Bélanger (1982).

length of the fiber between the gold-leaf markers to be kept at a preset value while the tension developed during tetanic stimulation was measured.

Gordon, Huxley, and Julian (1966b) observed that, over a certain range (traces 1 and 2 in Figure 11.6), the tension developed is indeed proportional to the degree of overlap between the actin and myosin filaments and hence to the number of active cross-bridges on the latter. If further shortening is allowed, such that opposing actin filaments now overlap each other as well as the underlying myosin filament, the isometric tension declines (traces 4-6 in Figure 11.6). It is probable that in some way the overlapping actin filaments interfere with the cross-bridge mechanism at these short sarcomere lengths. In other experiments, Gordon and colleagues stretched the muscle fiber so far that there was no overlap between the myosin and actin filaments. Because the myosin cross-bridges were unable to reach the actin

filaments, no tension could be developed (beyond 3.65 μm in Figure 11.6). This is exactly the situation that is thought to obtain in the cardiac failure associated with ventricular distension. Whether excessive lengthening can ever take place in normal skeletal muscles is doubtful, for the permissible ranges of joint movement are possibly too small. In diseased muscles, however, the possibility of hyperextension seems more likely, since the partial replacement of muscle fibers by relatively inelastic fibrous tissue could well allow surviving segments of fibers to be stretched excessively.

The Heads of the Myosin Molecule

Much more has been learned about the cross-bridge through the use of high-powered electron microscopy

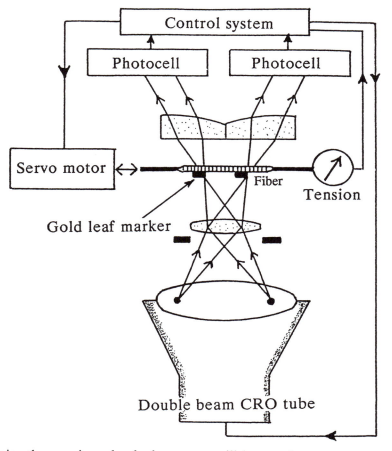

Figure 11.5 Experimental system to examine the length-tension relationship in short lengths of single muscle fibers. See text. Adapted from Gordon, Huxley, and Julian (1966a).

to examine the myosin molecule, because as will be seen, it is the heads of the molecule that form the cross-bridges. The complete myosin molecules appear as structures some 150 nm long, each consisting of two globular heads and a single tail (Figure 11.2A). The heads can be snapped apart from the tail by the proteolytic enzyme papain. The relatively long *tail* is formed by two α-helices that coil round each other; the function of the tails is to cause the myosin molecules to associate together so that several hundred can form a single myosin filament. This aggregation occurs spontaneously and can be demonstrated in myosin precipitated from solution. The structure of the tail also ensures that, within each filament, half the myosin molecules within a filament face away from the other half; the point of attachment of two tails is within a base area devoid of cross-bridges (Figure 11.2B).

The Cross-Bridges Are the Globular Heads of the Myosin Molecules

The two *globular heads* of each myosin molecule are often referred to as S1 fragments (Figure 11.7A). Each head consists of a *heavy chain* of about 2,000 amino acids and has two *light chains* attached to it. The globular heads, and the necks to which they are attached, form the cross-bridges. Biochemical studies have identified

three components of the globular head, with molecular weights of 50 kD, 25 kD, and 20 kD respectively (Figure 11.7B). The two light chains are attached to the 20 kD moiety, while the 50 kD segment contains a *pocket* for binding and hydrolyzing ATP, as well as sites for attachment to actin (cf. Vibert & Cohen, 1988). A recent advance, by Rayment et al. (1993), has been the use of computerized image analysis to study the results of X-ray crystallography and electron microscopy of myosin. The major discovery was a *cleft* through the 50 kD segment, dividing the latter into upper and lower domains (Figure 11.7B). The width of the cleft is controlled by the interactions of ATP with the pocket; the opening of the cleft weakens the binding of the 50 kD segment to actin, while the closure of the cleft strengthens the binding. The ATP pocket can also open, tilting the myosin head and thereby altering the angle of the cross-bridge (discussed next).

The Myosin Head Hydrolyzes ATP and Tilts to Produce the Power Stroke

Since the description of the sliding filament mechanism 40 years ago, evidence has gradually accumulated to show how the myosin cross-bridge might alter its angle, so as to give a power stroke to the actin filament, and how this movement could be related to the hydrolysis

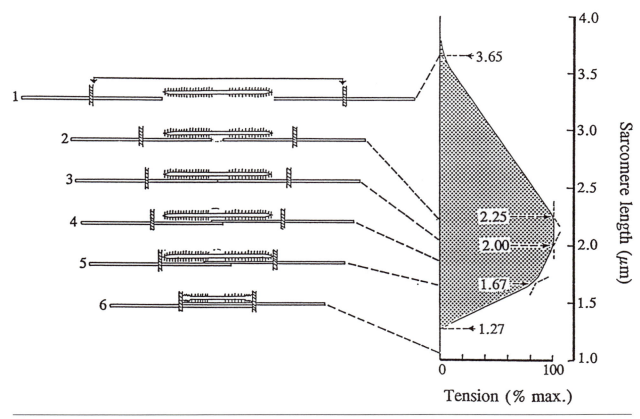

Figure 11.6　Length-tension results of experiment depicted in Figure 11.5. At the left is shown a single sarcomere, the arrows in trace 1 indicating the two Z-lines. The amount of overlap of the actin and myosin filament at different sarcomere lengths is shown in traces 2 to 6, and the corresponding tensions developed are displayed at the right. Critical sarcomere lengths are indicated by arrows. Adapted from Gordon, Huxley, and Julian (1966b).

of ATP. It was long known that the combination of ATP with the myosin head was necessary to free the latter from the actin filament. Thus, if ATP is unavailable or incapable of being hydrolyzed in the fiber, the two types of filament remain locked together in the *rigor* state, as after death. Lymn and Taylor (1971) proposed that the hydrolysis of ATP, to ADP and P_i, led to the formation of an *intermediate complex* with myosin (the myosin products complex). A further feature of this model was that the binding of actin to myosin took place in two stages, being weak initially and then strong. Another concept, suggested by several authors (e.g., A.F. Huxley & Simmons, 1971), was that the movement of the cross-bridge might arise from rotation of the globular myosin head. With the improved definition of the myosin head by Rayment et al. (1993), it has been possible to provide a more complete account of the probable structure and function of the myosin head. This account, which incorporates several of the previous ideas, has as its main feature the opening and closing of a newly discovered cleft in the 50 kD segment of the myosin head.

In this scheme, the first stage is the *rigor complex*, formed by the strong binding of actin to myosin in the absence of ATP (Figure 11.8A). ATP now enters the pocket in the 50 kD myosin segment phosphate first but with the adenine ring left protruding. This incomplete entry is sufficient to open the narrow cleft between the upper and lower domains of the 50 kD myosin segment, weakening its binding to actin (Figure 11.8B).

The remainder of the ATP molecule is then enclosed in the pocket, and this causes the 50 kD segment to detach itself from the actin filament and to move 5 nm farther on (Figure 11.8C). ATP is then hydrolyzed, but the products (ADP and P_i) remain within the myosin head, so as to form an *intermediate complex*. Next, the 50 kD segment reattaches itself to actin (5 nm farther down the filament than in Figure 11.8A). The reattachment is initially to the lower domain of the 50 kD segment and is weak, but it becomes stronger when the upper domain is involved. The reattachment allows the cleft between the upper and lower domains to close, and this permits P_i to be expelled from the pocket. In turn, the loss of P_i causes the ATP pocket to open and, as it does so, to make the lower part of the myosin head swivel (Figure 11.8E); this is the *power stroke*, during which force is produced and the actin filament is moved 5 nm onward. In addition,

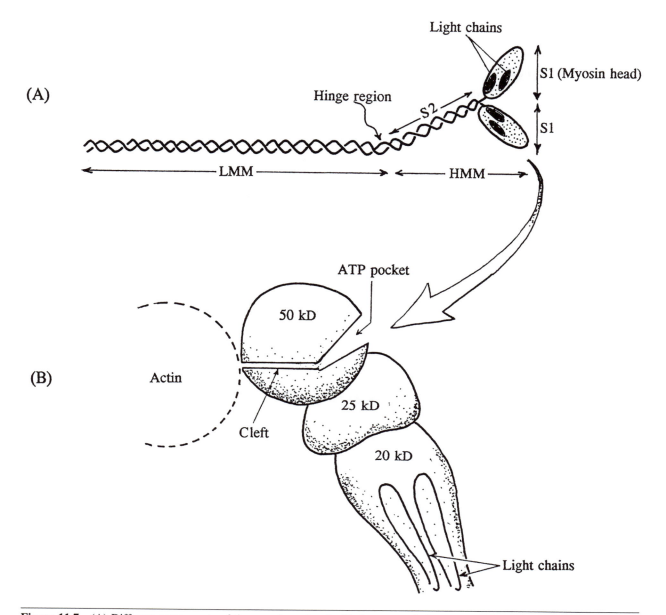

Figure 11.7 (A) Different components of the myosin molecule. Various proteolytic enzymes cleave the molecule into heavy meromyosin (HMM) and light meromyosin (LMM). The point of cleavage is flexible in the intact molecule (*hinge region*) and serves to bring the cross-bridge to the surface of the myosin filament. The HMM moiety comprises the remainder of the α-helical rod (S2 fragment) and the two globular heads (S1 fragments), to which the light chains are attached. The two globular heads form a cross-bridge. (B) Enlargement of one globular myosin head to show the three components (segments). The 50-kD segment is divided into upper and lower domains by a cleft, the size of which is regulated by the interaction of ATP with its ''pocket.'' Based on Vibert & Cohen (1988) and Rayment et al. (1993).

ADP is released from the ATP pocket as it opens. The actin and myosin molecules remain tightly bound until another molecule of ATP enters the pocket and the cross-bridge cycle is repeated.

A complete cross-bridge cycle lasts approximately 50 ms; of this time, the myosin head is attached to the actin filament for only 2 ms, and the power stroke is therefore a very brief event. In extraordinarily delicate

experiments, in which single myosin molecules and actin filaments were allowed to interact, it has been possible to estimate the forces and movements developed by individual myosin cross-bridges (Finer, Simmons, & Spudich, 1994). The movements, measured without any opposing force, average 11 nm, and are rather larger than those predicted by Rayment et al. (1993; see above). When feedback is applied with a laser system,

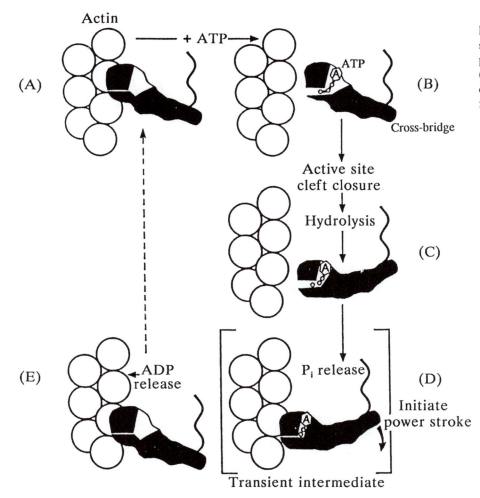

Figure 11.8 Hypothetical scheme of cross-bridge cycling proposed by Rayment et al. (1993); the dark structure is the cross-bridge. See text. Adapted from Rayment et al. (1993).

so as to prevent movement of the actin filament, a single myosin head generates a force of 3 to 4 pN.

Myosin Light Chains Regulate the Myosin Heads

As noted previously, each myosin head contains two *light chains* and each has a molecular weight of approximately 20 kD. The two light chains appear to be attached to the neck region of the myosin head (Figure 11.7A) and are termed *regulatory* and *essential* respectively. In some way the light chains are thought to affect the actions of the myosin heads, possibly by modifying the rate of movement in each power stroke (Lowey, Waller, & Trybus, 1993). Phosphorylation of the light chains has been proposed as the mechanism responsible for potentiating isometric twitch tension (Moore & Stull, 1984).

Actin Filaments

In chapters 1 and 2, actin filaments were encountered as part of the cytoskeletons of muscle and motoneurons,

while in the present chapter, the filaments, through their attachments to the cross-bridges, have been viewed as one half of the contractile mechanism. In the present section, we will learn a little more about their structure and about the special proteins, troponin and tropomyosin, which are associated with them in the thin filaments of the muscle fiber.

Actin Filaments Have a Polarity

With the electron microscope, an actin filament is seen as a thread about 1 μm long and 8 nm wide. The filament contains between 300 and 400 actin molecules, and each of these, in turn, is made up of 375 amino acids. Although the actin molecules are spherical, they have a distinct polarity, and they line up facing in the same direction, giving the appearance of a double-helical strand twisted about its own axis (Figure 11.9). This union of the actin molecules is brought about by polymerization, a process that requires the presence of a nucleotide (either ADP or ATP). An actin filament can be ''decorated'' by exposing it to myosin heads, which then appear as arrowhead-like structures along the filament. Since the actin molecules in each filament face the same way, the arrowheads behave similarly, pointing to the minus end of the filament.

Actin Filaments Are Associated With Two Accessory Proteins, Tropomyosin and Troponin

Tropomyosin is rather like the tail of the myosin molecule. It consists of a double helix that lies on the surface of the actin filament and stiffens it; each tropomyosin molecule spans about seven of the actin molecules. *Troponin* is a complex of three polypeptides, one of which (*troponin T*) attaches it to tropomyosin. *Troponin I*, in contrast, binds to actin and indirectly prevents the latter from interacting with the myosin head. During a muscle contraction, this inhibitory effect of troponin I is overcome by a sudden rise in cytosolic [Ca^{2+}]; each *troponin C* molecule captures four Ca^{2+} ions (see chapter 7) and, as it does so, undergoes a conformational change that lifts the tropomyosin molecule away from the actin filament. This movement of tropomyosin exposes sites on the actin filament to which the myosin heads can become attached, allowing the contraction to proceed.

Excitation-Contraction Coupling

As previously noted, the signal for a contraction to begin is a sudden rise in the concentration of Ca^{2+} ions in the vicinity of the myosin and actin filaments. In this way Ca^{2+} acts as a second messenger, acting as an intermediary between the action potential and the contractile apparatus. The cellular mechanism by which the Ca^{2+} concentration is increased is called excitation-contraction coupling and takes place in two steps: (1) depolarization of the T-tubules, and (2) diffusion of Ca^{2+} ions

from the sarcoplasmic reticulum to the myofilaments (Figure 11.10).

The T-Tubules Conduct the Electrical Signal Into the Fiber

The detailed structure of the T-system has been determined with the electron microscope. The tubules are situated at regular intervals along the muscle fiber and run inward toward its center, meeting with each other to form a continuous structure (see Peachey, 1965, and Figure 1.11, in chapter 1). In the frog, the tubules are situated in the vicinity of the Z-disc and thus occur once in each sarcomere. In mammals, however, the tubules are to be found at the junctions of the A- and I-bands, giving two tubular arrays per sarcomere. Electron micrographs have revealed that the tubules open onto the surface of the muscle fiber so that the tubular membrane becomes continuous with the sarcolemma. The existence of such openings has been confirmed by the ability of relatively large molecules to pass into the T-system from solutions bathing the muscle fiber. The substances chosen for this type of experiment have included ferritin (H.E. Huxley, 1964) and fluorescent dyes (Endo, 1966), as well as albumin, thorium dioxide, and horseradish peroxidase.

The key role of the T-system in excitation-contraction coupling was shown by two contrasting experimental approaches. First, A.F. Huxley and Taylor (1958) applied weak stimulating currents at different points along the surface of frog muscle fibers. They found that local contractions of myofibrils only occurred when the electrode tip was over the I-band, in the center of which the T-tubules are situated. In the second type of experiment, Gage and Eisenberg (1969a) demonstrated that action potentials were unable to elicit twitches in muscle fibers treated with glycerol. In these experiments, glycerol was

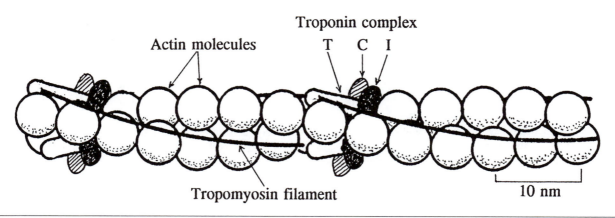

Figure 11.9 Schematic drawing of part of an actin filament showing the relationship between successive pairs of actin molecules, thin filaments of tropomyosin, and the three types of troponin. Another depiction of a (nonmuscle) actin filament is given in Figure 2.4A.

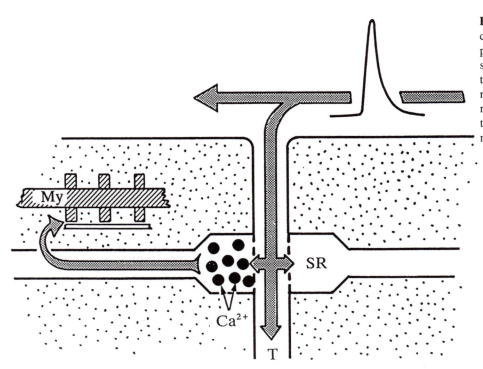

Figure 11.10 Excitation-contraction coupling. The action potential propagates along the surface plasmalemma and down the transverse tubule (*T*). Ca²⁺ is released from the sarcoplasmic reticulum (*SR*) and diffuses toward the cross-bridges on the myosin filaments (*My*).

added to the physiological bathing fluid and penetrated the muscle fibers. When the fibers were returned to a glycerol-free solution, the T-tubules swelled and broke off from the surface plasmalemma.

The speed with which the signal to contract travels into the fiber has been measured by González-Serratos (1971). The technique used was to embed single frog semitendinosus fibers in gelatin-Ringer and to compress them longitudinally; the myofibrils were photographed with a high-speed ciné camera; in shortening, they became straight instead of wavy. It was found that the evoked action potential caused the superficial myofibrils to contract first and the central ones last. At 20° C the velocity of inward spread of activation was 7 cm/s.

The T-Tubules Conduct Impulses Into the Interior of the Fiber

By themselves, the measurements of inward activation velocity and of its change with temperature do not enable the nature of the inward spread to be determined. In theory two such mechanisms are possible. One is a passive electrotonic spread of depolarization along the T-system, and the other is that the tubules propagate action potentials into the center of the fiber. One way of deciding between these possibilities has been to abolish any propagated action potentials with tetrodotoxin and to depolarize the surface plasmalemma using an intracellular stimulating electrode. In this way, only the effects of electrotonic spread are observed. Adrian, Costantin, and Peachey (1969) found that when brief stimuli were used, only the superficial myofibrils would

contract when the fiber membrane had been depolarized to 0 mV; larger surface depolarizations were necessary to make the central myofibrils shorten. In an ingenious extension of their work, the same authors imposed an artificial action potential on the membrane by using a voltage clamp circuit driven by an action potential in another fiber. It appears that the reversal of membrane polarity during the spike is just sufficient to activate the central myofibrils by electrotonic spread along the T-system. If, however, the same type of experiment is carried out on toad muscle at 20° C, it is found that the surface action potential only produces about 30% of the normal twitch tension, the remaining 70% presumably requiring a propagated impulse in the T-tubules (Bastian & Nakajima, 1974).

Other indirect evidence for a T-impulse has come from the work of Costantin (1970), who investigated the responses of single frog muscle fibers to focal stimulation through a microelectrode. By reducing the concentration of Na⁺ in the solution bathing the fiber, the initiation of an action potential in the surface plasmalemma was prevented. Costantin found that the central and superficial myofibrils had very similar thresholds for contraction, suggesting that an active mechanism was present and still effective in the T-system even though it was absent at the fiber surface. This conclusion was strengthened by the results of experiments involving tetrodotoxin, in which the superficial myofibrils now had lower thresholds than the central ones and were the first to shorten. Under these last conditions, only passive electrotonic depolarization of the fiber interior could have taken place.

The most convincing evidence for an inwardly propagated action potential would be to record it. Although there have been several attempts to do this (Natori, 1975; Strickholm, 1974), there is doubt as to the significance of the recorded potentials.

Ca²⁺ Ions Are Needed for Contractions to Take Place

It has been known that Ca^{2+} ions play an important role in muscle contraction ever since Ringer (1883) showed that the frog heart stops beating if this element is omitted from the bathing solution. Mines (1913) was able to demonstrate that the quiescent heart still generates action potentials, indicating that Ca^{2+} ions must be required for some step beyond the excitation of the fibers. Similarly, in the case of skeletal muscle, Frank (1958) showed that frog muscle fibers will not develop contractures when placed in K^+-rich solutions unless Ca^{2+} is also present. Further evidence for the important role of Ca^{2+} is that injections of this ion into a muscle fiber through a micropipette evoke local contractions; Na^+, K^+, and Mg^{2+} ions are ineffective (Heilbrunn & Wiercinski, 1947). Another experiment has been to desheath a muscle fiber, by removing the plasmalemma, and then to induce a contraction by applying calcium ions directly to myofibrils with a micropipette (the Natori preparation; see Podolsky, 1964).

Ca²⁺ Is Released Into the Cytosol From the Sarcoplasmic Reticulum

The Ca^{2+} ions, which are necessary for contraction to proceed, are released from the sarcoplasmic reticulum; the onset and duration of the efflux can be measured with Ca^{2+}-sensitive dyes. In the first such study, Ashley and Ridgway (1968, 1970) injected large, single muscle fibers of barnacles with *aqueorin*, a protein that luminesces in the presence of Ca^{2+} ions. Following direct stimulation, the light emitted by the fiber was detected by a photomultiplier tube, and the electrical output from the latter was amplified and displayed on an oscilloscope; the membrane depolarization and the force generated by the fiber were also recorded (Figure 11.11).

Ashley and Ridgway were able to show that the Ca^{2+} ions are released between the onset of the membrane response and the start of the muscle contraction, as would be expected for a coupling mechanism. The concentration of Ca^{2+} ions increases rapidly during the membrane response and begins to fall on its completion. The change in muscle tension runs a much slower course, with relaxation beginning at a time when the Ca^{2+} ion concentration has already returned to the resting value (Figure 11.11). This last observation is of considerable interest, for it might indicate that relaxation does not result simply from the removal of calcium but, at

least in barnacle muscle, may also involve an intermediary step. Other experiments, showing Ca^{2+} transients occurring between excitation and contraction, were subsequently reported for frog muscle (Rüdel & Taylor, 1973) and mouse muscle (Westerblad & Allen, 1991).

The Ca^{2+} transients have also been measured when, instead of a single stimulus, a series of shocks are given to the muscle fiber. The Ca^{2+} concentration is seen to rise to a higher level and then to plateau (see Figure 15.7, chapter 15). The rise in Ca^{2+} concentration in the cytosol following excitation is impressive; the resting concentration is only 10^{-7} M, but following successive stimuli, it rises to 10^{-5} M, a 100-fold change.

The Ca²⁺ Release Mechanism Involves Two Types of Ca²⁺ Channel

The electrical signal traveling down the T-tubules uses two types of Ca^{2+} channel in order to release Ca^{2+} from the sarcoplasmic reticulum (SR). The first type is termed the *DHP channel* because it can be blocked by dihydropyridine; it is an example of an L (large or long-lasting) Ca^{2+} channel and is described more fully in chapter 7. This type of channel is found in high concentration in the membrane of the T-tubules. According to Fosset, Jaimovich, Delpont, and Lazdunski (1983), the density of DHP receptors found in the T-tubules of skeletal muscle is from 50 to 100 times greater than that of any other tissue. Unlike L-channels in other membranes, however, the function of the DHP channel is not to admit Ca^{2+} ions, because vertebrate muscle fibers can still contract if this ion is omitted from the bathing fluid (Armstrong, Bezanilla, & Horowicz, 1972). Rather, the DHP channel appears to act primarily as a voltage sensor that transmits a signal to a second type of Ca^{2+} channel, found in the membrane of the SR. This second type of channel is the *ryanodine receptor* (RYR); it is a huge protein, consisting of four subunits, each of which has a molecular weight of 564 kD (see chapter 7).

Low-power electron micrographs of muscle fibers show that the T-tubules and SR have a special relationship; a T-tubule and the SR terminal cistern on either side form a *triad* (Figure 11.12). Higher power electron micrographs can identify the DHP channels in the T-tubule and the ryanodine channels in the SR. The two types of channel are then seen to have a precise relationship to each other; thus, the DHP channels occur in groups of four, and each of these *tetrads* faces every second ryanodine channel (Figure 11.13). Even in some of the earliest electron microscopic studies of the triads, it was possible to identify additional structures, the *feet* (Franzini-Armstrong, 1970), which spanned the 15 nm gap separating the T-tubular and SR membranes, and are now known to comprise part of the ryanodine channel.

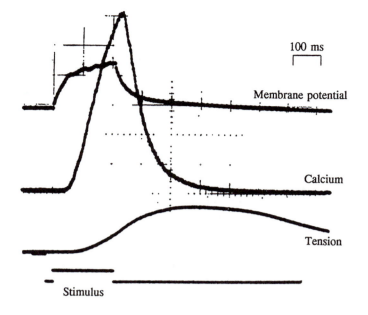

100 ms

Membrane potential

Calcium

Tension

Stimulus

Figure 11.11 Transient rise in Ca^{2+} concentration within the cytosol of a barnacle muscle fiber. Note that the rise follows the excitation of the plasmalemma and precedes the development of force. Adapted from Ashley and Ridgway (1970).

Three-dimensional reconstruction from multiple electron micrographs has revealed further details of the ryanodine channel (Wagenknecht et al., 1989). There appears to be a central pore which runs through that part of the molecule which lies in the SR membrane. This pore is connected to four radial outlets in the foot part of the molecule, and these, in turn, open into the gap between the SR and T-tubule (see Figure 11.14B). The central pore and the four radial outlets presumably provide the pathway for Ca^{2+} ions to exit from the SR into the cytosol.

The DHP and Ryanodine Channels May Be Connected Mechanically

How can a signal get from the DHP channel to the ryanodine channel? Well before the structures of the two channels were known, Chandler, Rakowski, and Schneider (1976) suggested that they were physically coupled to each other. In this model the DHP channel, after activation, would be able to unplug the ryanodine channel by means of a connecting rod (Figure 11.14A). The new information concerning the ryanodine receptor, previously described, would be quite consistent with such a model. For example, the top of the ryanodine channel foot could be attached to the DHP channel; the conformational change in the DHP channel, brought about by a voltage signal in the T-tubule, could then lift the top of the ryanodine channel and allow Ca^{2+} to flow through the four radial outlets (see Ríos, Ma, & González, 1991; Figure 11.14B).

Chemical Links Between the DHP and Ryanodine Channels Have Not Been Excluded

The "plunger" mechanism proposed by Chandler et al. (1976) may not be the only means by which the DHP

channel can activate the ryanodine channel. Additional mechanisms may involve Ca^{2+} ions and *inositol triphosphate* (IP$_3$). Such a role for Ca^{2+} would be consistent with the Ca^{2+} transients recorded in the cytosol following excitation; also, Ca^{2+} release from the SR does appear to be positively affected by the Ca^{2+} level in the cytosol.

The evidence favoring IP$_3$ is that this messenger increases rapidly in the cytosol after stimulation of the muscle fiber (Nosek, Guo, Ginsburg, & Kolbeck, 1990) and that, even in very low concentrations, it can release Ca^{2+} ions from the SR. One obvious role for Ca^{2+} or IP$_3$ would be to turn on the alternate ryanodine channels that are not opposed to the DHP tetrads.

Ca^{2+} Ions Are Pumped Into the SR From the Cytosol

Muscle contractions must be terminated at the appropriate moment, and the muscle fiber achieves this by pumping Ca^{2+} ions from the cytosol back into the SR. Since the pump has to move Ca^{2+} ions against a large concentration gradient, it requires ATP to provide the energy necessary. With each cycle of the pump, two Ca^{2+} ions are transported into the SR in return for two K^+ ions; this unequal exchange of positive charges causes the cytosol near the SR to become increasingly negative. On gaining entry to the SR, the Ca^{2+} ions are detached from the pump and taken up by Ca^{2+}-binding proteins, such as a high-affinity ATPase enzyme as well as the low-affinity *calsequestrin*. Since individual pumps have a low capacity for transporting ions (e.g., 20/s), it is necessary for the SR to have a high density of pumps in its membrane in order to bring down the Ca^{2+} concentration in the cytosol quickly. Freeze-fracture electron micrographs reveal rounded particles, almost certainly the Ca^{2+} pumps, that are packed tightly

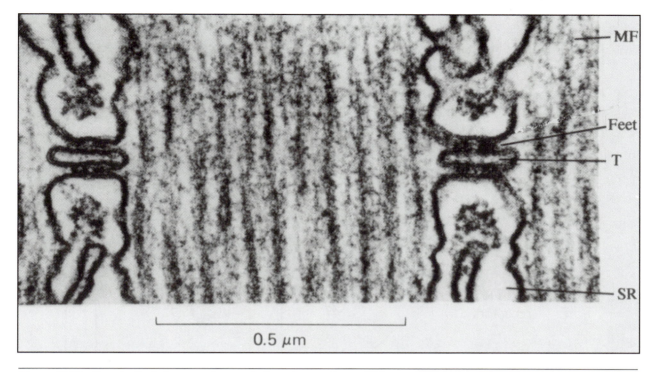

Figure 11.12 Longitudinal electron micrograph through part of a toadfish muscle fiber showing two triads. Note the flattening of the T-tubules (*T*) and the junctional feet linking the tubules to the adjacent SR membranes (*SR*). *MF* = myofibril. Reprinted from Franzini-Armstrong (1980).

against each other in the SR membrane. Further details of the Ca^{2+} pump were considered in chapter 7.

Active State

When the myosin cross-bridges are attaching themselves to the actin filaments, the muscle fiber is said to have entered its *active state*; the duration and the intensity of the active state depend on the concentration of Ca^{2+} around the contractile filaments.

The Duration of the Active State Is Shorter Than That of the Twitch

The isometrically recorded muscle twitch gives a poor indication of the intensity and duration of the active state because of the elastic component in series with the contractile elements. Most of this elasticity is in the connective tissue attachments at the ends of the muscle fibers, but some may also reside in the myosin cross-bridges. Since the series elastic component has to be stretched before tension can be recorded, the active state has an onset that is earlier than the start of the isometric twitch and a duration that is less than the *contraction time* (see Figure 11.15); the tension generated by the cross-bridges will be greater than that recorded.

The Active State Can Be Prolonged by Repetitive Stimulation

A simple way of prolonging the active state is to give a rapid sequence of stimuli to the fiber. The active state following each excitation then merges with the next, and because there is sufficient time for all the series elasticity to be taken up by the contractile elements, the tension recorded is considerably greater than that of a single twitch (Figure 11.15). As the stimulus frequency is increased, the tension (force) gradually rises; eventually a frequency is reached beyond which there is no greater tension. The latter frequency is the *optimal* value and is higher in fast-twitch muscles than in slow-twitch ones; an example of the *force-frequency curve* for human muscles is given in Figure 11.16. A further property of contracting muscles is that after the tetanus has been stopped, the twitch responses to single shocks, given during the next few seconds, remain enlarged; the same phenomenon can be demonstrated following a voluntary contraction, particularly in fast-twitch muscles, as is evident in Figure 11.17. This behavior has been termed *posttetanic potentiation* and is again probably due to a temporary alteration in the intensity or duration of the active state. Such potentiation has been attributed to phosphorylation of the light chains on the myosin head (Moore & Stull, 1984), but another possible mechanism is elevation of Ca^{2+} in the cytosol

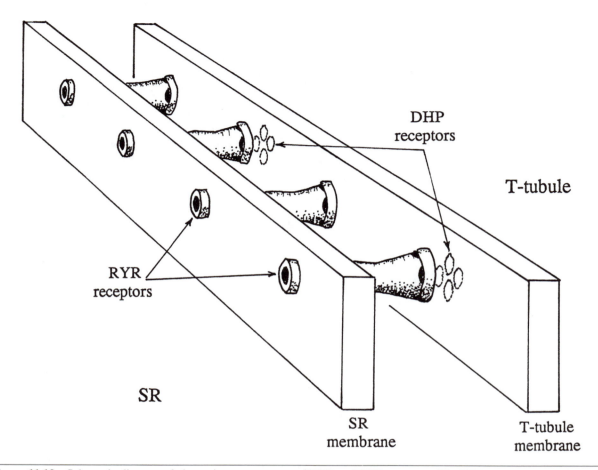

Figure 11.13 Schematic diagram of alternating arrangement of RYR and DHP receptors. The DHP receptors occur in tetrads (groups of four) and face every second RYR receptor.

for a short while after the end of the tetanus or voluntary activation (Allen, Lee, & Westerblad, 1989). Another phenomenon with a similar underlying basis is the *positive staircase*, which was originally observed in the frog gastrocnemius muscle but has also been demonstrated in the human adductor pollicis muscle (Slomíc, Rosenfalck, & Buchthal, 1968). It consists of a progressive enlargement of isometric twitches when stimuli are given at a low repetition rate, such as 2 Hz.

Chemicals and Drugs Can Modify the Active State

The active state may also be extended by cooling the muscle fiber or by applying a variety of chemical agents (see Sandow, 1965, for review). *Lyotropic anions* such as bromide, nitrate, and methylsulfate seem to act by reducing the mechanical threshold of the fiber. Thus, the active state is prolonged because the contractile machinery is switched on at a lower membrane depolarization during the action potential. *Divalent metal ions*

such as Zn^{2+} and uranyl can augment the twitch tension by a factor of two to three times; they do so by increasing the duration of the action potential and hence of the active state. *Caffeine* in low concentrations (above 5 mM for frog muscle) will produce a maintained shortening of the muscle (the caffeine contracture). This alkaloid produces its effect by releasing Ca^{2+} from the sarcoplasmic reticulum and preventing its subsequent recapture by the Ca^{2+} pump.

In contrast, the intensity of the active state process can be *diminished* by reducing the availability of Ca^{2+} ions to the myofibrils. One method is to bind the Ca^{2+} as an organic salt by injecting potassium citrate or potassium oxalate into the fiber. Another technique is to immerse the fiber in *hypertonic saline*; although the fiber shrinks, the T-tubules and the lateral sacs become dilated. This behavior would be expected of the T-system, because this is really an extracellular space that will draw water from the fiber by osmosis; the SR appears to behave as an intermediate compartment between the intra- and extracellular fluids. Evidently, the end result of the distension of the triads is to reduce

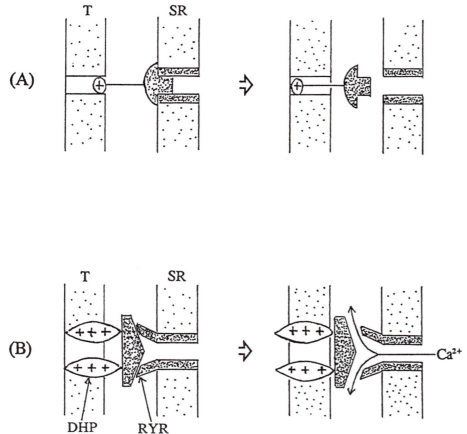

Figure 11.14 Possible connections between Ca^{2+} channels in the T-tubular and SR membranes. (A) is the mechanical "plunger" system proposed by Chandler, Rakowski, and Schneider (1976), while (B) incorporates subsequent knowledge of the molecular structure of the ryanodine channel (RYR). Adapted from Ríos, Ma, & González (1991, p. 129).

the release of Ca^{2+} from the lateral sacs (Ashley & Ridgway, 1970). The effect of *epinephrine* is to shorten the active state and thereby to speed up the isometric twitch; in theory, the tension developed should be rather smaller, but Marsden and Meadows (1970) observed little change in the human soleus muscle. These authors interpret their findings in terms of an action of epinephrine on β-receptors within the muscle fibers.

Dantrolene sodium also reduces the intensity of the active state and does so by interfering with the release of Ca^{2+} from the SR (Desmedt & Hainaut, 1977). Patch-clamp experiments have demonstrated a direct effect of this drug on SR Ca^{2+} channels (Suarez-Isla, Orozco, Heller, & Froehlich, 1986). Dantrolene is an important drug in the treatment of patients with malignant hyperthermia (see p. 179).

Effects of Muscle Length and Load

In experimental situations, it is usual to study the contractile behavior of muscle fibers under either isotonic or isometric conditions, because the analysis of the results is made considerably easier.

Contractions Can Take Place at Different Muscle Lengths

In an *isotonic* contraction, the muscle is able to shorten and to lift whatever load has been attached to its tendon. Under these circumstances, external work is performed, equal to the product of the load and the distance through which it has been moved (i.e., Work = Load × Distance). In an *isometric* contraction, the muscle is prevented from shortening by fixing both its ends. Instead of performing external work, the muscle develops tension at its points of attachment; the energy expended in the contraction is released as heat. In everyday life there are many examples of contractions that are purely isotonic or isometric. Isotonic contractions include cycling, combing the hair, placing objects on shelves, and loading a grocery cart. An isometric contraction would be an unsuccessful attempt to lift a heavy object, providing a counterforce to prevent an object from toppling over, or simply maintaining one's posture against gravity. Other activities are more complicated. For example, in walking the forward swing of the leg is checked by the action of the gluteal muscles, which thus develop tension while being *lengthened*.

In human subjects, the contractions in which the muscles are allowed to shorten are usually referred to as

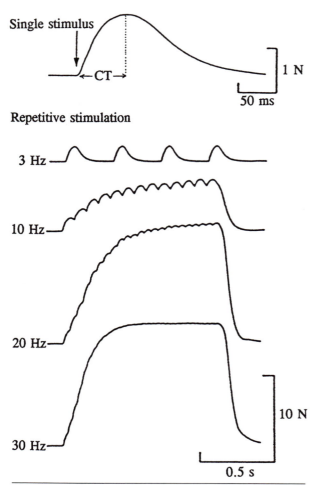

Figure 11.15 Tension developed by a human extensor hallucis brevis muscle following a single stimulus (top) and trains of shocks given at progressively higher frequencies (lower four traces). The twitch contraction time (*CT*) is the interval between the onset of tension and the peak value.

concentric, and contractions during muscle lengthening are termed *eccentric*.

The Velocity and Extent of Muscle Shortening Depend on the Load

Consider a situation in which progressively larger loads need to be lifted. It is a well-known experience that the lift becomes increasingly slow as the load is made larger. In the laboratory, this phenomenon may be investigated by stimulating the muscle repetitively (tetanically) and measuring the velocity of muscle shortening for different loads (Figure 11.18). It can be seen that, not only is the velocity of shortening reduced when a relatively large load is applied, but there is a latent period before the object is moved at all. During this latent period, the muscle is contracting isometrically until sufficient tension has been produced to equal the load. Figure 11.18 also shows that, not only is the rate of shortening

reduced for heavy loads, but the amount of shortening is also decreased.

A point to be emphasized is that, irrespective of the circumstances of the contraction—isotonic, isometric, or lengthening—the interactions between the cross-bridges and the actin filaments remain the same, in that the cross-bridges engage the actin filaments and attempt to slide them along. In an isotonic (concentric) contraction with a small load, the sliding movement allows the myosin filaments to become completely overlapped by actin. In an isometric contraction, the amount of filament overlap depends on the length at which the muscle is held prior to activation. During activation the cross-bridges repeatedly make and break connections with the actin filaments, producing tension equal to the external load but no movement of the actin filaments. In a lengthening (eccentric) contraction, the cross-bridges generate less tension than the external stretching force applied to the muscle, and the opposing actin filaments are pulled away from each other in the sarcomeres.

Muscle Power Is Determined by the Velocity of Shortening and the Load

By definition, the *power* developed by a muscle in a contraction is equal to the product of the load and the velocity with which it is moved. Thus,

$$\text{Power} = \text{Load} \times \text{Velocity}$$

Inspection of the force-velocity curve (Figure 11.18) indicates that power will be zero both when there is no load on the muscle and when the load is so heavy that it cannot be moved at all; at some intermediate load, the power will be greatest. For many activities, such as running, jumping, rowing, and cycling, success will be governed by the power that can be developed. As an example of research in this field, the relationship between the velocity of pedaling and power output during sprint cycling has been examined by Sargeant, Dolan, and Young (1984). The fiber-type compositions of the muscles will also be important in deciding how much power can be developed. For example, whereas a slow-twitch muscle (see chapter 12) may not be able to shorten sufficiently quickly to generate a given power, a fast-twitch muscle may be able to do so.

APPLIED PHYSIOLOGY

In this section, we consider the curious condition of muscle contracture, in which the fibers stay shortened even though no longer firing impulses. We will discover

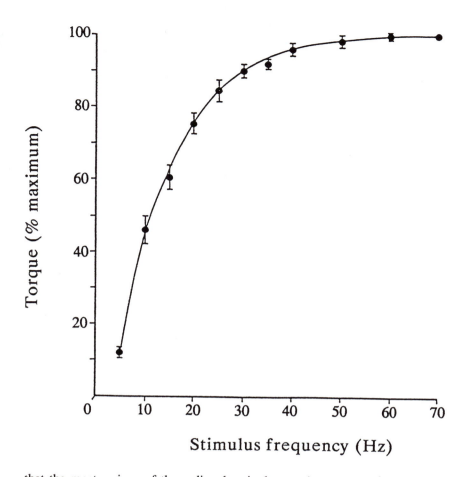

Figure 11.16 Force-frequency curve for human plantarflexor muscles. As the stimulus frequency to the tibial nerve is raised, the force developed by the muscle (expressed here as torque) becomes larger. However, beyond a certain optimal frequency, in this case 60 Hz, no further increase in force can be generated. Reprinted from Sale, Quinlan, Marsh, McComas, and Bélanger (1982).

that the most serious of these disorders is due to abnormal Ca^{2+} fluxes and can cause death during anesthesia. There is a strain of mouse, however, in which a Ca^{2+} defect results not in contracture, but in paralysis.

Muscle Fibers Can Remain Shortened Without Being Excited

A *contracture* differs from a contraction in that the shortening of the myofibrils is prolonged in the absence of action potential activity in the plasmalemma. One method of producing a contracture is to immerse the fiber in a solution rich in K^+ ions. By reducing the K^+ equilibrium potential, these ions maintain the membrane in a depolarized state and thereby cause a lasting release of Ca^{2+} ions from the SR. Another technique is to add caffeine to the bathing solution; the nature of this drug action has been described earlier in the chapter. Contracture may also occur as part of a disease process; in *McArdle's disease* (myophosphorylase deficiency), the muscle fibers stay shortened and swell after physical exertion, at a time when impulse activity in the membrane has ceased (see chapter 14).

It should be noted that the term "contracture" is used here in a physiological sense. Unfortunately, the term is also applied by clinicians to permanent shortening of

muscle produced by excessive growth of fibrous tissue within the muscle belly and possibly by resorption of sarcomeres as well (Tabary, Tabary, Tardieu, Tardieu, & Goldspink, 1972). One of the best examples of this type of contracture is the severe shortening of the calf muscles that affects boys with advanced Duchenne muscular dystrophy (see chapter 1). In such cases, it is best to refer to the phenomenon as a "pathological" contracture to avoid confusion with the "physiological" type considered in this section.

In *rigor* there is also a maintained shortening that is independent of action potential activity. In this condition, unlike a contracture, the fiber is depleted of ATP. Because of this deficiency, the cross-bridges on the myosin molecules cannot be detached from the actin filaments, and the myofibrils are unable to relax. A rigor state supervenes after death once the ATP in the fiber has been consumed by the various autolytic reactions that take place during the early degenerative processes.

A Hereditary Disorder, Malignant Hyperthermia, Can Cause Sudden Anesthetic Deaths

In 1960, Denborough and Lovell discovered an Australian family in which no fewer than 10 members had

Tibialis anterior   Plantarflexors

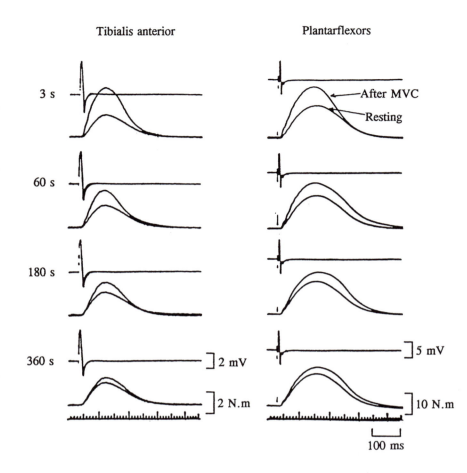

Figure 11.17 Twitch potentiation at different times after a brief maximal voluntary contraction (*MVC*). In each pair of traces, the upper is the M-wave (compound action potential), and the lower shows two twitches, the smaller of which is the control (resting) and the larger is the potentiated one (*after MVC*). Note that potentiation is greater for the fast-twitch tibialis anterior than for the slow-twitch plantarflexors and that it is not associated with any change in the M-wave. Adapted from Vandervoort, Quinlan, and McComas (1983).

died unexpectedly after having been given a general anesthetic. The manner of the deaths was alarming. Within minutes of starting an anesthetic, the patient would develop a rapid pulse, become cyanosed, and have increasing stiffness of his or her skeletal muscles. Since the body temperature could rise as high as 43° C (109° F), the condition was termed *malignant hyperthermia* (MH). The Australian cases were identical to other reported instances of unexpected anesthetic deaths, and it was clear that the condition was a hereditary one, transmitted by an autosomal dominant gene.

The incidence of the MH gene in the population is much higher than the incidence of MH reactions during surgery (1:15,000 in children), for many patients with the susceptibility can tolerate one or more general anesthetics without complications. The anesthetic agents which have been incriminated are halothane, ether, ethyl chloride, methoxyflurane, trichloroethylene, ethylene, and cyclopropane. Another triggering agent is succinylcholine, a drug that depolarizes skeletal muscle fibers by combining with the ACh receptors and is commonly employed as a muscle relaxant. With the increased awareness of MH, the routine monitoring of body temperature during surgery, the availability of dantrolene (an excitation-contraction

blocking agent, see p. 177) and of body-cooling apparatus, the incidence of anesthetic deaths from MH has declined from 80% to 7%.

Malignant Hyperthermia Is Due to Abnormal Ca²⁺ Levels in the Muscle Fibers

It was suspected that MH might be due to abnormal handling of Ca^{2+} inside the muscle fibers. For one thing, muscle contraction is controlled by the release of Ca^{2+} from the SR into the cytosol, and, were the latter to be prolonged for any reason, the muscle fibers would stay contracted as they do in MH. At the same time, Ca^{2+} would stimulate a number of muscle enzymes, including phosphorylase kinase; the latter would increase the breakdown of glycogen in the muscle fibers, producing excessive amounts of lactate and heat—two other features of the MH reaction. Suggestive evidence for a Ca^{2+}-based mechanism is that muscle fibers from MH-susceptible patients are unusually sensitive to caffeine, a compound that increases Ca^{2+} release from the SR and can cause both potentiation of the twitch and muscle contractures. Indeed, the most reliable test of genetic susceptibility until now has been to expose a biopsy specimen of muscle to a bathing fluid containing halothane and caffeine, and to find the concentrations of

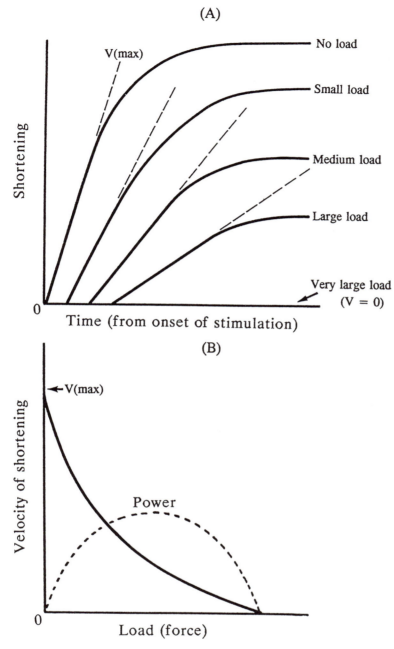

(A)

(B)

Figure 11.18 (A) The amount of muscle shortening that takes place during contraction depends on the load that has to be lifted. (B) Relationships among velocity of shortening, load, and power.

drug which, singly and in combination, produce contractures of the fibers (Kalow, Britt, Terreau, & Haist, 1970).

In Pigs and in Some Human MH Subjects, the Ryanodine Receptor Is the Culprit

The first clue to the location of the abnormal gene came from studies on strains of pigs susceptible to MH. These pigs were bred for their heavy muscling and leanness, but it was also noted that occasionally during stress (e.g., prior to slaughter) animals might die spontaneously and also develop large, pale, soft areas in their muscles. Genetic studies showed tight linkage between the genes

for MH and for the enzyme glucose phosphate isomerase (GPI). In humans, the GPI gene occurs on the long arm of chromosome 19 and is close to the gene for the ryanodine receptor. The next step was to show, in MH-susceptible pigs, that the ryanodine receptor was the abnormal protein due to the substitution of a single amino acid (cystine for arginine; Fujii et al., 1991). This small change in protein structure is apparently sufficient to prevent the ryanodine channel from closing normally (Fill et al., 1990).

Although the ryanodine receptor is the culprit in all breeds of pig susceptible to MH, the situation in humans is more complicated. In a minority of patients, the ryanodine receptor is indeed abnormal, but in others the

MH trait is due to some other deficit in the cell or is part of another recognized neuromuscular disorder, such as central core disease or Duchenne muscular dystrophy. Evidently, any lesion that can produce abnormally large or prolonged Ca^{2+} levels in the muscle fiber cytosol is capable of initiating a MH reaction (MacLennan & Phillips, 1992).

Mice With Muscular Dysgenesis Have Defective DHP Ca^{2+} Channels

There is a very interesting disorder of mice, caused by a mutant gene, in which the animals are almost totally paralyzed. In this condition, named *muscular dysgenesis*, physiological recordings show that, despite the absence of contractile activity, action potentials propagate normally along the myotubes (Powell & Fambrough, 1973). This dissociation of events strongly suggests that the defect is due to a disorder of excitation-contraction coupling. Furthermore, in affected animals, the triads are disorganized in their microscopic structure; there is a 5- to 10-fold reduction in DHP binding sites (Pinçon-Raymond, Rieger, Fosset, & Lazdunski, 1985), and the slow Ca^{2+} current, normally mediated by the DHP chan-

nel, is absent (Beam, Knudson, & Powell, 1986). This combination of findings points to the DHP channel as the primary abnormality in muscular dysgenesis, and indeed, injection of an expression plasmid containing the cDNA sequence for the DHP channel can restore excitation-contraction coupling in the myotubes (Tanabe, Beam, Powell, & Numa, 1988). It is not yet known, however, how the affected myotubes can be rescued by co-culture with normal spinal cord (Rieger et al., 1987). Contractions are also restored if dysgenic myoblasts are cultured with normal fibroblasts (Courbin, Koenig, Ressouches, Beam, & Powell, 1989); in this case it is probable that the two types of cell fuse together and that the fibroblasts provide the DHP receptors needed by the myoblasts.

In the last three chapters, we have seen how impulses, discharged from motoneurons in the spinal cord, result in contraction of the muscle fibers. In the next chapter we will find that the colonies of muscle fibers, controlled by single motoneurons, differ from each other in various ways, including the speeds of their contractions.

Chapter 12

Motor Units

So far the muscle and nerve fibers have been treated as single cells. It is time to consider the anatomical and physiological relationships between these two types of cells and then to study how the body makes use of the neuromuscular system. As Sherrington (1929) appreciated, all the varied reflex and voluntary contractions of a muscle are achieved by different combinations of active motor units. This term, *motor unit*, describes a single motoneuron and the many muscle fibers to which its axon runs (Figure 12.1). Expressed differently, all movements are planned by the central nervous system in terms of motor units rather than individual muscle fibers. Some of the more important aspects of motor unit function can now be discussed.

Organization of Motor Units

One of the problems that has repeatedly interested anatomists concerns the *number* of motor units in individual muscles. In this chapter, we will examine how these numbers can be determined in animal muscles, and even in human ones. Of equal interest, perhaps, are the *sizes* of the motor units, that is, the numbers of muscle fibers

supplied by single motoneurons through their axons. Last, there is the *architecture* of the motor units to consider, by which is meant the way that the muscle fibers of any one motor unit are distributed within the muscle belly.

Some Motor Nerve Fibers Innervate the Muscle Spindles

To find the numbers of motor units in individual muscles, the usual approach in animals has been to cut dorsal roots distal to their ganglia (i.e., dorsal rhizotomy) and to allow time for degeneration of the severed sensory nerve fibers. If the nerve to a muscle is then examined, the surviving nerve fibers are those that left the cord in the ventral roots and are motor in function. Until the important study of Leksell (1945), it was not realized that the smallest motor axons in the ventral roots, the *gamma (γ) axons*, are those passing from fusimotor neurons to the small muscle fibers inside spindles (see chapter 4). In the first studies, the numbers of motor units estimated were therefore too high, in most cases by 20 to 40% (see, for example, Eccles & Sherrington, 1930). Even in these early studies, however, it was clear

(A)

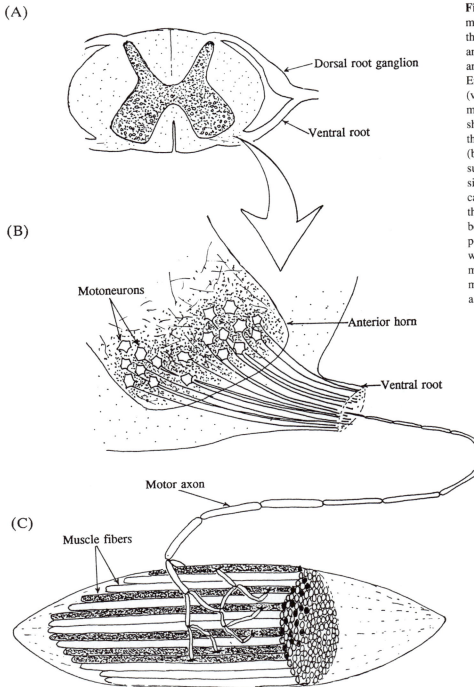

Figure 12.1 The anatomy of a motor unit. (A) Cross section of the spinal cord to show grey and white matter, with dorsal and ventral nerve roots. (B) Enlargement of anterior (ventral) horn with motoneurons, one of which is shown sending its axon through the ventral root to the muscle (below). (C) The muscle fibers supplied by the branches of the single axon are darkened and can be seen on the surface of the muscle belly and where the belly has been transected. This population of fibers, together with the motor axon and motoneuron, constitute a single motor unit. Most muscles have a hundred or more such units.

that there were considerable differences in the numbers of motor units between muscles within the same animal.

In Animals, Motoneurons Can Be Counted After Retrograde Labeling

The numbers of motor units can also be determined in animals by the retrograde labeling of motoneurons. The substance originally used for this purpose was *horseradish peroxidase* (HRP); when applied to a muscle belly as a conjugate with wheat germ agglutinin, HRP is taken up by the motor nerve terminals and transported in the axons to the motoneuron cell bodies. After 24 hr or longer, depending on the length of the motor axons, the spinal cord is fixed, removed, and sectioned serially. The HRP-containing neurons are stained with tetramethyl benzidine and can then be recognized under the light microscope and counted (Mesulam, 1982). Since HRP may sometimes spread from the injected muscle to neighboring ones, more reliable results may be obtained by cutting the motor nerve and placing the proximal stump in a solution containing HRP. As an alternative to HRP, new types of tracers have been developed; these include fluorescent markers such as the dextran-amines (Glover, Petursdottir, & Jansen, 1986).

In Human Muscles, Anatomical Estimates of Motor Units Have Depended on Comparisons With Pathological Material

For obvious reasons, it has not been possible to determine numbers of motor units in human muscles with the same precision as in the animal studies. The best that can be done in humans is to count the number of large myelinated fibers in a muscle nerve and to assume that a certain proportion of these are *alpha (α) motor* and that the remainder are sensory fibers from spindles and tendon organs. The uncertainty introduced by this assumption is evident from the fact that the proportion of large fibers that are motor varies between 40% and 70% in different muscle nerves in the cat hindlimb (Boyd & Davey, 1968). In the study of Feinstein, Lindegård, Nyman, and Wohlfart (1955), 60% of the large diameter fibers were regarded as α-motor. This value was chosen on the basis of observations of the muscle nerves of a single patient who had died from poliomyelitis; no motor axons were thought likely to have been left. Leaving aside the questionable validity of this assumption, there is no doubt that Feinstein and colleagues (1955) have provided the best anatomical data for the numbers and sizes of human motor units.

Small Muscles of the Eyes and Face Have Large Numbers of Small Motor Units

Table 12.1, which includes most of the results obtained by Feinstein et al. (1955), emphasizes the considerable differences between limb muscles. Note that the intrinsic muscles of the foot and hand (lumbrical, first dorsal interosseous) have fewer motor units than the larger muscles (brachioradialis, tibialis anterior, and medial gastrocnemius) situated nearer the trunk. However, because the motor units in the distal muscles are relatively small, these muscles have greater numbers of units in relation to their masses. Table 12.1 also draws attention to the remarkable properties of the external eye muscles (external rectus) and the superficial muscles of the face and neck (platysma), which contain large numbers of very small units.

Physiological Estimates of Human Motor Units Agree With Anatomically Derived Numbers

Although the anatomical data are valuable, axon counting is tedious and depends on the availability of postmortem tissue. A very different approach enables motor

Table 12.1 Comparison of Numbers and Sizes of Motor Units in Various Human Muscles

Muscle	No. of motor units	Muscle fibers/unit
External rectus	2,970	9
Platysma	1,096	25
First lumbrical	96	108
First dorsal interosseous	119	340
Thenar group	203	—
Brachioradialis	333	>410
Tibialis anterior	445	562
Gastrocnemius (medial)	579	1,934

Note. Thenar values from Lee, Ashby, White, & Aguayo (1975); other data from Feinstein et al. (1955).

unit numbers to be estimated during life and can be performed rapidly; this electrophysiological method is based on the comparison of an average motor unit potential with the response evoked from the whole muscle (see Box 12.1).

Table 12.2 gives some of the results obtained by this method. The computer-based version produces rather lower estimates than the original "manual" technique, partly because corrections can be made for "alternation" (Box 12.1). How well do the numbers of units estimated by the computer-based method compare with anatomical determinations? The sparse anatomical data are not particularly reliable, for reasons given previously; nevertheless, comparison of Tables 12.1 and 12.2 shows that there is fair agreement for the tibialis anterior muscle and a better match for the median-innervated thenar muscle group. Since the median nerve normally supplies two of the thenar muscles completely (abductor pollicis brevis and opponens pollicis) and part of a third muscle (flexor pollicis brevis), it would appear that each of the intrinsic muscles of the hand has about 100 motor units. This conclusion would be consistent with Feinstein et al.'s values of 119 units for the first dorsal interosseous muscle and of 96 units for the first lumbrical (Table 12.1). A surprising result is that the biceps brachii, a large muscle, appears to have a number of motor units similar to one of the small muscles in the hand. Further, if the values for humans are compared with those of smaller mammals (Table 12.3), it would seem that in the course of evolution the motoneuron populations supplying the small muscles of the hand increased, in relative terms, much more than those innervating the larger limb muscles. Presumably this trend allowed the extraordinary manual dexterity of human beings to develop.

Motor Unit Populations Vary Among Subjects

A further surprise from motor unit estimations in humans is that there is so much variation in the results for the same muscle between healthy individuals of the same sex, size, and age. Part of this variation is undoubtedly due to methodological factors, especially in the sampling of motor units of different sizes. However, it is quite clear that some individuals have significantly more motor units than others, and this phenomenon tends to be generalized throughout the body musculature. It is not yet known whether this is a genetic endowment and, if so, whether it is due to greater proliferation of motoneurons during embryogenesis or to reduced cell death (see chapter 6). Evidence of variation among individual muscles has also come from animal experiments using HRP to label motoneurons; for example, Wasserschaff (1990) obtained a coefficient of variation of 21% for the mouse tibialis anterior muscle.

Table 12.2   Numbers of Motor Units in Control Subjects Below the Age of 60 Years, Estimated by Automated and Manual Incremental Methods[a]

Muscle	Automated method Mean ± SD (n)	Manual method Mean ± SD (n)
Thenar	230 ± 90 (90)	342 ± 89 (115)
Hypothenar	411 ± 174 (41)	390 ± 94 (109)
Biceps brachii	109 ± 43 (80)	—
Extensor digitorum brevis	143 ± 73 (86)	210 ± 65 (151)
Tibialis anterior	256 ± 107 (22)	—
Soleus	—	957 ± 254 (41)
Vastus lateralis	224 ± 112 (24)	—

[a]For information about the methods of estimation, see Box 12.1.

Bulky Muscles Contain the Largest Motor Units

Various methods have been employed to determine the sizes of motor units, that is, the number of muscle fibers innervated by individual motor axons. For a given muscle, the average size of the motor units may be estimated simply by dividing the total number of muscle fibers by the number of motor axons. It has been found, for example, that a relatively large muscle, the human medial gastrocnemius, has an average of about 2,000 fibers in each unit, whereas the much smaller external rectus muscle has only 9 (Feinstein et al., 1955; Table 12.1).

Even in the Same Limb Muscle, Motor Units May Vary 10 to 100 Times in Their Sizes

Electrophysiological studies show that, even within the same muscle of an individual animal or human subject, the sizes of motor units vary so greatly that the estimation of mean size has limited significance. For example, Burke (1967) stimulated single motoneurons and measured the tensions produced in the cat triceps surae muscle. He found that the twitch tensions ranged from 2 to 950 mN and noted that the scatter of results was much greater in the gastrocnemius muscle than in the soleus (see Figure 12.8). Although these results indicate that the motor unit sizes vary considerably, they do not permit the absolute numbers of muscle fibers to be determined. These determinations may be performed using a technique developed independently by Edström and Kugelberg (1968) and by Brandstater and Lambert (1969). In this method, a single motor axon is stimulated repetitively, and the muscle is then excised and stained with periodic acid-Schiff (PAS) reagent for glycogen. The glycogen-depleted fibers are those that have been stimulated and therefore correspond to a single motor

Table 12.3   Number of Motor Units in Muscles of Small Mammals

	Mouse	Author	Rat	Author	Cat	Author
Extensor digitorum longus	20	Bateson & Perry (1983)	40	Close (1967)	130	Boyd & Davey (1968)
Tibialis anterior	223	Harris & Wilson (1971)	—	—	200	Ibid.
Soleus	22	Lewis & Parry (1979)	30	Close (1967)	155	Ibid.
Lumbrical	—	—	11	Betz, Caldwell, & Ribchester (1979)	—	—
Gastrocnemius	—	—	41 (lat[a])	Gillespie, Gordon, & Murphy (1986)	280 (med[b])	Ibid.

[a]lat = lateral head.

[b]med = medial head.

unit (Figure 12.2A). In the rat tibialis anterior muscle, the sizes of the motor units are within the 50- to 200-fiber range (Brandstater & Lambert, 1973; Edström & Kugelberg, 1968). As might be expected, the comparatively large tibialis anterior and extensor digitorum longus muscles of the cat have large units, the range observed by Mayer and Doyle (1970) being 43 to 1,099 fibers. In the cat triceps surae complex, the sizes of the motor units in the soleus range from less than 50 to more than 400 fibers, while in the medial gastrocnemius, an average unit contains between 400 and 800 fibers (Burke, Levine, Salcman, & Tsairis, 1974; Burke & Tsairis, 1973).

In Human Muscles, the Relative Sizes of Motor Units Can Be Determined by Comparing Their Electrical and Mechanical Responses

In humans, it is not possible to determine the full sizes of motor units by the glycogen-depletion technique. However, the *relative* sizes of motor units can be assessed by measuring the sizes of their electrical and mechanical responses when motor axons are excited by threshold stimuli (McComas, Fawcett, Campbell, & Sica, 1971; Garnett, O'Donovan, Stephens, & Taylor, 1979).

In the extensor digitorum brevis, Sica and McComas (1971) found that the twitch tensions of 122 single units in the extensor hallucis brevis (EHB) varied from 20 to 140 mN (mean 54 ± 22 mN); in the medial gastrocnemius, the range was 15 to 2,000 mN (Garnett et al., 1979). A recent stimulation method has been to excite motor axons by fine tungsten wires inserted through the skin and into the nerve trunk (Westling, Johansson, Thomas, & Bigland-Ritchie, 1990). With this technique, the forces developed by thenar motor units, following stimulation of median nerve axons above the elbow, range from 3 to 34 mN (Thomas, Johansson, Westling, & Bigland-Ritchie, 1990).

Spike-Triggered Averaging May Underestimate the Sizes and Durations of Twitches

Twitch tensions of human muscles have also been studied by the technique of spike-triggered averaging during voluntary contractions (see Figure 13.4, chapter 13). With this approach, single motor units in the first dorsal interosseous muscle of the hand are found to have a 100-fold range in twitch tension (Milner-Brown, Stein, & Yemm, 1973b). Other results for this and different human muscles are given in Table 12.4.

One of the problems with the spike-triggered averaging technique is that, even during the weakest contractions, the frequency of motor unit discharge (8-12 Hz; Monster & Chan, 1977) is still sufficiently high to cause the twitches to become partially fused. This fusion, in turn, would be expected to reduce the amplitudes of the twitches by approximately half and to shorten the contraction times to an even greater extent (Thomas, Bigland-Ritchie, Westling, & Johansson, 1990). The fact that such reductions are not observed, when the results of spike-triggered averaging and motor axon stimulation are compared for the same muscle, suggests that the averaging technique may not recognize synchronous firing by more than one unit or that potentiation and compliance effects may be present (Thomas, Bigland-Ritchie, et al., 1990).

Comparisons of Potentials Also Point to a Large Range in Human Motor Unit Sizes

Rather less satisfactory is to measure the sizes of the motor unit potentials recorded with large surface electrodes. Although the size of each potential will be proportional to the number of muscle fibers in that unit, the potential will also be influenced by the diameters of the fibers and their proximity to the stigmatic recording electrode. Figure 12.3 should be viewed with these qualifications in mind; it suggests that there is a greater than 30-fold variation in motor unit size in the extensor digitorum brevis muscle but that the number of large units is small.

Glycogen-Depletion Studies Reveal a Random, Scattered Arrangement of Motor Unit Fibers

It used to be thought that the muscle fibers supplied by a single axon were distributed in clusters, or subunits, within the muscle belly. The validity of this concept was questioned by Ekstedt (1964) on the basis of recordings with a special electrode that had small recording surfaces 180° apart (the "Janus" electrode). He observed that when a motor unit fired, usually only one of the two surfaces was next to an active fiber; had subunits been present, the electrode should have been surrounded by discharging fibers. It is now recognized that the subunit arrangement is only true of motor units in partially denervated muscles, in which collateral reinnervation of muscle fibers has taken place.

The correct architecture of the motor unit has been revealed by the glycogen-depletion method. Within a given volume of muscle, the fibers in a single motor unit are arranged randomly among those of other units, with only a few fibers in the same unit being contiguous

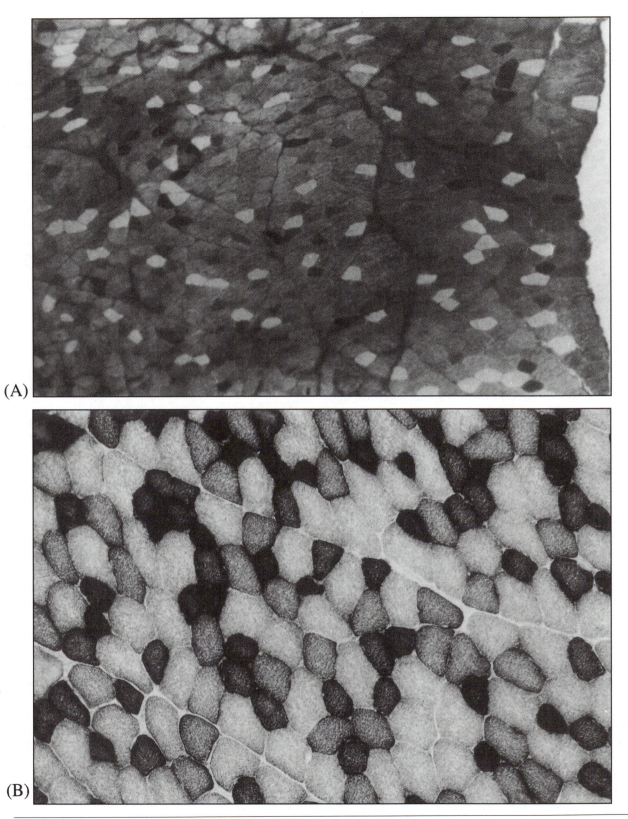

Figure 12.2 (A) Part of a motor unit in the tibialis anterior muscle of a rat. A single motor axon has been stimulated repetitively; the activated muscle fibers became depleted of glycogen and failed to stain with PAS, appearing white in the photograph. (Courtesy of Dr. M.E. Brandstater.) (B) Transverse section of tibialis anterior muscle from a 3-month-old mouse, stained for succinic dehydrogenase activity. Three types of fiber can easily be differentiated on the basis of their staining reactions; the darkest fibers are type I and the lightest are type IIB. (Courtesy of Dr. Ethel Cosmos.) Magnifications 150 × (A) and 320 × (B).

Table 12.4   Contractile Properties of Human Motor Units

Muscle	CT[a] (ms) Mean ± SD	CT[a] (ms) Range	Force[b] (mN) Mean ± SD	Force[b] (mN) Range	Method[c]	Authors
Masseter	57 ± 14	24-91	58 ± 58	1-329	STA	Yemm (1977)
	35 ± 10	20-72	—	20-763	STA	Nordstrom & Miles (1990)
Temporalis	49 ± 11	30-76	76 ± 59	3-297	STA	Yemm (1977)
Abductor pollicis brevis	74 ± 20	37-102	28 ± 25	2-163	STA	Thomas, Ross, & Calancie (1987)
Thenar	50 ± 9	35-80	11 ± 8	3-34	INS	Thomas, Johansson, Westling, & Bigland-Ritchie (1990)
First dorsal interosseous	56 ± 13	30-100	15 ± 18	1-122	STA	Milner-Brown, Stein, & Yemm (1973b)
	—	32-122	—	2-294	STA	Stephens & Usherwood (1977)
	65 ± 18	34-140	34 ± 47	2-423	IMS	Young & Mayer (1981)
	68	40-99	52	3-215	STA	Thomas, Ross, & Stein (1986)
	—	—	—	1-100	STA	Dengler, Stein, Thomas (1988)
Flexor carpi radialis	—	—	—	0.2-98	STA	Calancie & Bawa (1985)
Extensor digitorum communis	—	40-70	—	5-49	STA	Monster & Chan (1977)
Extensor hallucis brevis	63	35-96	54 ± 22	20-140	SS	Sica & McComas (1971)
Medial gastrocnemius	—	40-110	200 ± 304	15-2000	IMS	Garnett, O'Donovan, Stephens, & Taylor (1979)

[a]CT = contraction time, i.e., the time elapsing between the start of the contraction and the moment of peak tension during an isometrically recorded twitch.

[b]Published values expressed in grams were converted to mN.

[c]STA = spike-triggered averaging; INS = intraneural stimulation; IMS = intramuscular stimulation; SS = surface stimulation.

(Figure 12.2A; Brandstater & Lambert, 1969; Edström & Kugelberg, 1968).

Individual Motor Unit Territories May Extend Across Much of the Muscle Belly

Glycogen-depletion studies also show that fibers belonging to the same motor unit can be distributed over a surprisingly large volume of muscle, for example, up to a quarter of the cross-sectional area in the rat tibialis anterior and up to three quarters of the cat soleus (Edström & Kugelberg, 1968; Bodine-Fowler, Garfinkel, Roy, & Edgerton, 1990; see also Burke & Tsairis, 1973). When seen in muscle cross sections, the motor unit territories tend to be elliptical rather than circular, and to contain regions in which the fibers may be present in high densities or else altogether lacking, as in the FR (fast-contracting, fatigue-resistant) unit in Figure 12.4; the uneven distributions have been attributed to the presence or absence of major axonal branches (Bodine-Fowler et al., 1990). In Figure 12.5, the fiber distributions in muscle cross sections have been displayed graphically, and their unevenness is reflected in the varying heights of the peaks. Although the fibers tend to be randomly disposed within a dense region, remarkably few fibers touch (Kelly & Schotland, 1972; Willison, 1978), with the implication that the elimination of polyneuronal innervation during embryogenesis involves a mechanism that discriminates against the clustering of fibers belonging to the same unit (see chapter 6).

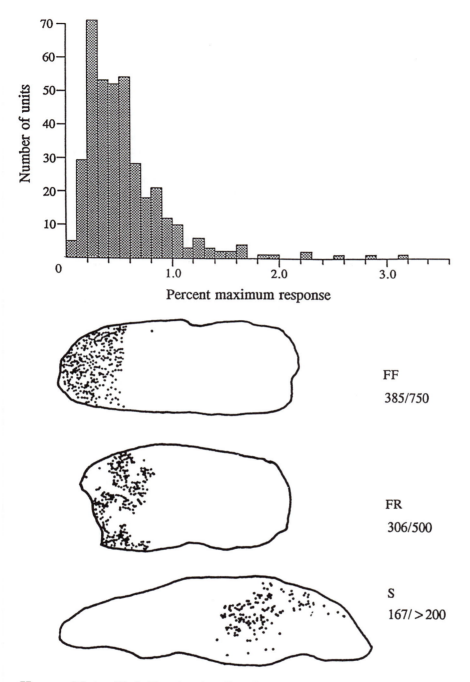

Figure 12.3 The sizes of 380 motor unit potentials in the extensor digitorum brevis muscles of 41 control subjects below the age of 60. Each value has been expressed as a percentage of the maximum response in order to compensate for variations in electrode placement and in the thickness of the tissues overlying the muscle. Reprinted from McComas, Fawcett, Campbell, and Sica (1971).

FF
385/750

FR
306/500

S
167/>200

Figure 12.4 Territories of three different motor units in the cat medial gastrocnemius muscle, as determined by the glycogen-depletion method. Since, in this muscle, the fibers do not run the full length of the belly, only about half of the respective fiber populations can be seen in a single cross section. The numbers of fibers counted in these sections and the total numbers of fibers estimated to be present in the three units are given at the right, together with the designations of the motor units. (The designations FF, FR, and S are discussed later in this chapter.) Adapted from Burke and Tsairis (1973).

Human Motor Unit Territories Can Be Surveyed With Special Needle Electrodes

In humans, the information available on motor unit territories has come from electrophysiological recordings, using either a *multilead* electrode (Buchthal, Guld, & Rosenfalck, 1957) or a *scanning* electrode that is slowly drawn through the muscle belly. In the scanning technique, a second electrode is maintained in a constant position and is used to ensure that the potentials picked up by the scanning electrode come from the same motor unit (Stålberg & Antoni, 1980). Both techniques show that the territories of individual human motor units have

cross-sectional diameters of 5 to 10 mm, with 10 to 25 units overlapping with each other. Even greater degrees of overlap, involving at least 40 to 50 motor units, have been found by Burke and Tsairis (1973) in the cat medial gastrocnemius muscle.

Finally, it is important to recognize not only that motor units may extend over a considerable part of the muscle cross-sectional area, but that, in long muscles, they may include muscle fibers in different regions of the longitudinal axis. For example, the hamstring muscles of the thigh are composed of two or more serial arrays of muscle fibers separated from each other by fibrous *inscriptions* (Loeb, Pratt, Chanaud, &

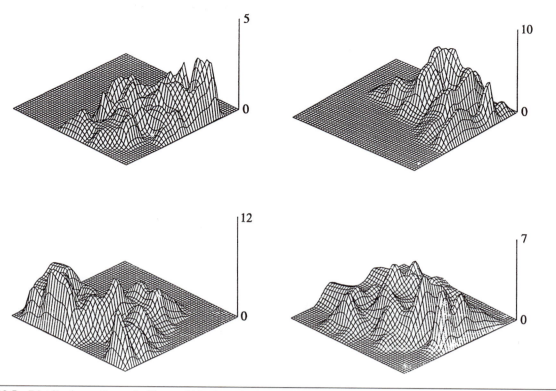

Figure 12.5 Distributions of muscle fibers within four single motor unit territories of the rat tibialis anterior. The muscle cross sections have been stained and divided into ''quadrats,'' based on the territorial boundary of a single motor unit; the number of fibers in each quadrat has been counted and displayed on the vertical axis. From ''Spatial distribution of muscle fibers within the territory of a motor unit,'' by S. Bodine-Fowler, A. Garfinkel, R.R. Roy, and V.R. Edgerton, 1990, *Muscle and Nerve*, **13**, p. 1138. Copyright 1990 by John Wiley & Sons, Inc. Reprinted with permission.

Richmond, 1987). A single motor axon may innervate fibers belonging to different arrays (Manzano & McComas, 1988; Wines & Hall-Craggs, 1986).

Physiological and Biochemical Properties of Motor Units

Not only do motor units differ from each other in size but also in the biochemical and physiological properties of their muscle fibers. Ranvier (1873) provided evidence of differences, observing in a comparative study of several species that some muscles contracted more slowly than others and had a more pronounced red coloration. The red color of the muscle is due to the increased amounts of the oxygen-bearing pigment *myoglobin* in the muscle fibers. These fibers also have a rich capillary network. Denny-Brown (1929) noted that the slowly contracting red muscles are those that are used continuously in the maintenance of posture. In contrast, the fast-contracting pale muscles are employed intermittently, being reserved for movements of a nonrepetitive nature.

Pale Fibers Are Most Plentiful in Superficial Parts of Muscle Bellies and Fiber Fascicles

Denny-Brown (1929) pointed out that some mammalian muscles contain both pale and red parts; for example, the deeper region of the cat tibialis anterior is red, but the superficial part is pale. In other muscles, and this is particularly true of humans, red and pale muscle fibers are generally intermingled so as to form a *mosaic*. Within fascicles there is a tendency for pale fibers to form a ring under the perimysial connective tissue border, with a layer of dark fibers immediately underneath (Pernuš & Erzen, 1991).

With the development of methods for staining muscles histochemically, further differences between muscle fibers have emerged. Some fibers are found to stain strongly for glycolytic enzymes such as phosphorylase and for myosin ATPase. Other fibers, because of their numerous mitochondria, are rich in oxidative enzymes such as malate and succinate dehydrogenase, which are associated with these organelles (Figure 12.2B). Some of the differences between muscle fibers will now be considered in more detail.

Marked Differences Exist in the Twitch Contraction Times of Mammalian Motor Units

Following the lead of Henneman and his colleagues (McPhedran, Wuerker, & Henneman, 1965; Wuerker, McPhedran, & Henneman, 1965), it has become routine to record the twitch and tetanic forces developed by single motor units in mammalian muscles. The techniques employed have been either to divide ventral root filaments until stimulation yields an all-or-nothing contractile response, or to excite single motoneurons through intracellular microelectrodes (Burke, Levine, Tsairis, & Zajac, 1973). Although it is not possible to review the results for the various species examined, it can be said that, for most muscles, there are considerable ranges in the times taken for motor units to contract and relax. One conventional measure is the *contraction time* of a twitch; this is the interval elapsing between the onset of tension and the peak value (see Figure 11.15, chapter 11). In the cat triceps surae, the contraction times are as short as 20 ms in the medial gastrocnemius and as long as 130 ms in the soleus (Burke et al., 1974). In the corresponding muscles of smaller mammals, such as the rat and mouse, the motor unit contraction times also display considerable variations, but are shifted toward briefer values. Thus, the smaller the animal, the more rapidly can it move. In all species investigated, the soleus has the motor units with the longest contraction times, while other hindlimb muscles, such as the extensor digitorum longus and plantaris, are mainly composed of motor units with fast twitches.

Human Motor Units Also Differ in Their Contraction Times

In human muscles, some motor units have contraction times as short as 20 ms, but others are as long as 140 ms. An example of the spectrum of contraction times is given in Figure 12.6 for the extensor hallucis brevis muscle, and Table 12.4 contains values for other human muscles. Note that the masseter has mostly fast-twitch motor units, even though the histochemical profile of this specialized muscle does not reflect a predominance of "fast" muscle fibers (Butler-Browne, Eriksson, Laurent, & Thornell, 1988; see Table 12.6). In the extensor digitorum communis and medial gastrocnemius, there are fairly even mixtures of fast- and slow-twitch units (Garnett et al., 1979; Monster & Chan, 1977); in all of these muscles, the ranges of contraction times are similar to those of the very much smaller muscles in the hand and foot (see Table 12.4).

Contraction Times of Whole Muscles Reflect Motor Unit Composition

It is obvious that the shape of the whole muscle twitch will depend on the proportions of motor units with different contraction times. Figure 12.7 contrasts the twitches of four limb muscles in the same individual, recorded with a piezoelectric device pressed over the muscle bellies. It can be seen that in the facial muscles the units are nearly all of the fast-twitch type, because the mean contraction time is only 43 ms; in contrast, the relatively long contraction time in the human calf muscles indicates the presence of a substantial slow-twitch motor unit population. Last, in the small muscles of the hand and foot, the contraction times are intermediate between those of the calf and facial muscles, indicating similar numbers of fast- and slow-twitch units (see also Table 12.5).

Different Types of Muscle Fiber Can Be Recognized With Enzyme Histochemistry

Many different staining methods have been devised to study the enzyme activity of muscle fibers (see, for example, Dubowitz, 1985), but the most useful one for many purposes is that for myosin ATPase (Engel, 1962). The fibers that stain strongly at pH 9.4 are referred to as *type II*, in contrast to the poorly reacting *type I* fibers. Brooke and Kaiser (1970) have divided the type II fibers into A, B, and C subtypes on the basis of differences in staining, which can be brought out by preincubating the tissue at various pH values. Brooke and Kaiser (1974) consider the IIC fibers to be precursors that have the ability to develop into type IIA or IIB fibers. It is now known that the myosin ATPase activities of the fibers are related to their rates of shortening, as measured from force-velocity studies (see chapter 11). The reason for this is that the velocity with which the actin and myosin filaments can slide over each other depends in part on the speed with which the head of the myosin molecule can break down ATP. It is possible that the sites for this enzymatic reaction differ in the type I and type II fibers.

Histochemical Studies Show Consistent Differences Between Some Human Muscles

The myosin ATPase staining reaction has been employed extensively in the study of normal and diseased muscles and has proved a useful technique for detecting denervation. Prior to the 1970s, however, little was known concerning the distributions of fiber types among normal human muscles. In an autopsy study of 36 human muscles, Johnson, Polgar, Weightman, and Appleton

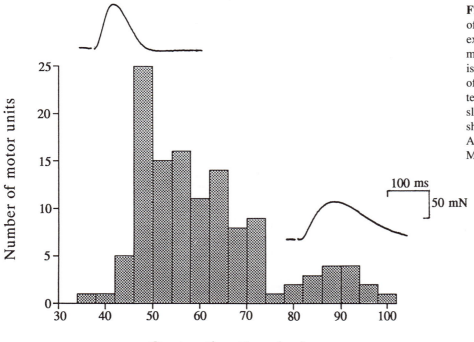

Figure 12.6 Contraction times of 122 motor units in human extensor hallucis brevis muscles. (The contraction time is the interval between the onset of the twitch and the peak tension). Examples of fast and slow motor unit twitches are shown above the histogram. Adapted from Sica and McComas (1971).

(1973) showed that in most muscles there is a considerable variation in the proportions of type I and type II fibers. Exceptions include the soleus and adductor pollicis muscles, which are largely composed of type I fibers, and the orbicularis oculi muscle, in which type II fibers predominate. Some of the reports of type I and type II fiber proportions in human muscles have been tabulated in Table 12.6. Other histochemical studies, also included in Table 12.6, make use of the fact that some of the muscle fibers are specially equipped for anaerobic metabolism. These fibers are able to produce ATP by breaking glycogen down to glucose and thence to lactic acid; they contain correspondingly large amounts of glycogen phosphorylase and lactate dehydrogenase that can be detected histochemically. Other fibers have numerous mitochondria and therefore react strongly in histochemical tests for reaction products of the various enzymes associated with these organelles, for example, succinate and malate dehydrogenase.

Immunocytochemistry Confirms the Existence of Multiple Molecular Forms of Muscle Proteins

A more recent refinement of fiber typing is the use of immunocytochemistry to identify different molecular forms (*isoforms*) of key muscle proteins. The protein most extensively studied in this way has been myosin itself, which consists of two heavy chains and four light chains (see chapter 11). The heavy chains determine the rate of cross-bridge reactions with the actin filaments

and hence the speed of muscle shortening (Reiser, Moss, Giulian, & Geaser, 1985). The myosin heavy chains also control the pH sensitivity of the ATP-splitting reaction and are therefore responsible for the depth of histochemical staining at the various pHs in the Brooke and Kaiser (1970) procedure. It is now established that there are at least five major heavy chain isoforms and seven light chain isoforms in adult mammalian muscle (Staron & Pette, 1990). Four of the myosin heavy chain isoforms correspond to the I, IIA, IIB, and IIC fiber types respectively, and the other isoform may be responsible for distinguishing a further subtype of the type II fibers (IIX or IID; see Bär & Pette, 1988; Schiaffino et al., 1990).

In addition, developing muscle has its own unique isoforms, and so far embryonic, fetal, and perinatal types have been described (for example, Draeger, Weeds, & Fitzsimons, 1987). During muscle development, there are transitions from one of the early isoforms to another and eventually to the appropriate adult isoform. At any given stage, however, there may be more than one isoform expressed in the same muscle fiber; in later life this situation is especially likely to occur when muscles are undergoing a change of usage, as the result of training programs or a period of disuse (see chapter 20). Table 12.7, adapted from Billeter, Heizmann, Howald, and Jenny (1981), summarizes the myosin heavy and light chain composition of human muscle fibers and is based on the use of monoclonal or polyclonal antibodies.

Successful searches have been made for variations in the structures of other key molecules in the muscle fiber. For example, the regulatory protein, troponin-T,

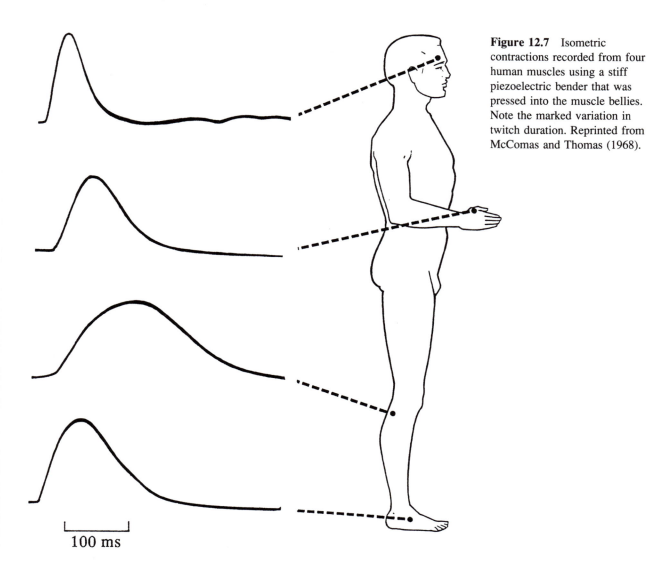

Figure 12.7 Isometric contractions recorded from four human muscles using a stiff piezoelectric bender that was pressed into the muscle bellies. Note the marked variation in twitch duration. Reprinted from McComas and Thomas (1968).

100 ms

Table 12.5 Varying Twitch Speeds of Human Muscles (Selected Values)

Muscle	CT[a] (ms) Mean ± SD	Authors
Frontalis/orbicularis oculi	43 ± 4.3	McComas & Thomas (1968)
Abductor digiti minimi	63	Burke, Levine, Tsairis, & Zajac (1973)
Adductor pollicis	65 ± 7	Slomíc, Rosenfalck, & Buchthal (1968)
First dorsal interosseous	65 ± 3.3	McComas & Thomas (1968)
Vastus lateralis	66 ± 4	Round et al. (1984)
Biceps brachii	66 ± 8.9	Bellemare, Woods, Johansson, & Bigland-Ritchie (1983)
Extensor digitorum brevis	67 ± 3.3	McComas & Thomas (1968)
Tibialis anterior	81 ± 7.4	Marsh, Sale, McComas, & Quinlan (1981)
Diaphragm	88 ± 7	Faulkner, Maxwell, Ruff, & White (1979)
Triceps surae	112 ± 11.1	Sale, Quinlan, Marsh, McComas, & Bélanger (1982)
Lateral gastrocnemius	118 ± 6.5	McComas & Thomas (1968)
Soleus	120	Buller, Dornhorst, Edwards, Kerr, & Whelan (1959)

[a]CT = contraction time, i.e., the time elapsing between the start of the contraction and the moment of peak tension.

Table 12.6 Type I Fiber Compositions of Selected Human Muscles

Muscle	%Type I	Authors	Muscle	%Type I	Authors
Orbicularis oculi	15	Johnson, Polgar, Weightman, & Appleton (1973)	Quadriceps	52	Round et al. (1984)
Triceps brachii	33	Johnson et al. (1973)	First dorsal interosseous	57	Johnson et al. (1973)
	50	Schantz, Randall-Fox, & Hutchison (1983)			
Biceps brachii	38	Brooke & Kaiser (1970)	Masseter	60-70	Butler-Browne, Eriksson, Laurent, & Thornell (1988)
	42	Johnson et al. (1973)			
EDB	45	Johnson et al. (1973)	Abductor pollicis brevis	63	Round et al. (1984)
Vastus lateralis	46	Green et al. (1981)	Tibialis anterior	73	Johnson et al. (1973)
Gastrocnemius (lateral)	49	Green et al. (1981)	Adductor pollicis	80	Johnson et al. (1973)
Diaphragm	50	Bellemare et al. (1986)	Soleus	80	Gollnick, Sjödin, Karlsson, Jansson, & Saltin (1974)

Table 12.7 Myosin Heavy Chain and Myosin Light Chain Content of Human Muscle Fibers

Fiber type	Myosin heavy chains	Myosin light chains
Type I	Slow	LC_{1s}, LC_{1f}, LC_{2s}
Type IIA	Fast IIA	LC_{1f}, LC_{2f}, LC_{3f}
Type IIB	Fast IIB	LC_{1f}, LC_{2f}, LC_{3f}
Type IIC	Fast + slow	LC_{1f}, LC_{2f}, LC_{1s}, LC_{2s}, LC_{3f}
Neonatal	Neonatal	LC_{1f}, LC_{2f}, LC_{3f}

Note. Adapted from Billeter, Heizmann, Howald, and Jenny (1981).

has two slow and four fast isoforms (Schachat, Bronson, & McDonald, 1985). Similarly there are at least two isoforms of Ca^{2+}-ATPase in the sarcoplasmic reticulum, corresponding to fast and slow fibers respectively, and the enzyme creatine kinase has long been known to exist in fetal and adult forms.

Fibers Can Also Be Distinguished From Each Other by Their Ultrastructure

Attempts have been made to differentiate mammalian muscle fibers on the basis of their ultrastructural appearances (Gauthier, 1986). The feature that can be correlated best with other properties of the fiber is the thickness of the Z-line. For example, in the guinea pig, the Z-line is more than twice as wide in the red fibers of the soleus as in the white fibers of the vastus muscle

(Eisenberg, 1974; see also Padykula & Gauthier, 1967). In addition, the width of the M-line tends to follow that of the Z-line, being greater in the red fibers (Schiaffino, Hanzlíková, & Pierobon, 1970). The mitochondria also exhibit differences; in the red fibers they are large and spherical, with abundant cristae, and they tend to be grouped under the fiber plasmalemma as well as between myofibrils. In white fibers the mitochondria are thinner and longer, and fewer are found under the plasmalemma. Finally, the T-tubular system and the cisterns of the SR are more plentiful in red fibers (Eisenberg & Kuda, 1976).

Classifications of Motor Units

In the preceding sections, it emerged that there were many physiological and biochemical differences between muscle fibers, and some evidence was presented to suggest that their differences might be linked together. Indeed, many attempts have been made to correlate the various characteristics of muscle fibers and to use such associations as a basis for classifying motor units.

The Histochemical and Contractile Properties of Motor Units Are Linked Together

In one of the most comprehensive correlative studies, Burke, Levine, and Zajac (1971; see also Burke et al., 1973) stimulated spinal motoneurons through intracellular microelectrodes and recorded the twitch and tetanic contractions of the corresponding motor units. At the end of an experiment, they demonstrated the architecture of the motor unit that they had been investigating using the glycogen-depletion technique (see p. 186). By staining consecutive sections for various enzymes, they were able to establish other biochemical characteristics of the previously activated fibers. As part of this study, they were able to confirm that all the muscle fibers belonging to the same motor unit were histochemically similar in that they stained strongly for the same enzyme activities.

Burke and colleagues (1971) concluded that certain properties of the motor units are closely linked together and that three types of units can be recognized. One type (*type S*) has a slow twitch, develops relatively small tension, and is resistant to fatigue (Figure 12.8); the muscle fibers have high contents of mitochondrial enzymes, are poor in glycogen, stain weakly for alkali-stable myosin ATPase, and have rich capillary networks. These fibers appear to be well equipped for aerobic metabolism and hence for prolonged activity.

A second type (*type FF*), in complete contrast, has a fast twitch, commonly develops large tensions, and is susceptible to fatigue. These fibers have low contents of mitochondrial enzymes, are rich in glycogen, stain strongly for alkali-stable myosin ATPase, and have poor capillary networks. They are suited for anaerobic metabolism and can engage in brief contractions.

A third type (*type FR*) has intermediate properties. It has a fast twitch, develops moderate tensions, and is resistant to fatigue. The muscle fibers have high contents of glycogen and mitochondrial enzymes and have considerable myosin ATPase activity; they are well supplied with capillaries. These fibers would be expected to work under both aerobic and anaerobic conditions and are therefore likely to be involved in prolonged work as well as in intense effort.

An attempt to summarize the contrasting histochemical and physiological characteristics of the three main motor unit types is given in Figure 12.9; the different terminologies employed are described below.

Different Classifications Exist for the Motor Unit (and Muscle Fiber) Types

As already noted, Burke and colleagues (1971) termed the three types of motor unit S (slow-contracting), FF (fast-contracting, fast-fatigue) and FR (fast-contracting, fatigue-resistant) respectively. A fourth type of unit, *FInt*, with properties in between those of FF and FR units, was subsequently recognized by Burke (1975); the existence of such intermediate units has been emphasized by McDonagh, Binder, Reinking, and Stuart (1980).

In a detailed biochemical investigation, involving hindlimb muscles of the rabbit and guinea pig, Peter, Barnard, Edgerton, Gillespie, and Stempel (1972) have recognized three types of fibers. These authors have proposed a different classification, and the FF, FR, and S units become *fast-glycolytic* (FG), *fast-oxidative-glycolytic* (FOG), and *slow-oxidative* (SO) types respectively. In Table 12.8, the classification of Burke and colleagues has been reconciled with that of Peter et al. (1972) and with the widely used histochemical scheme of Brooke and Kaiser (1970), discussed earlier.

Motor Units Tend to Overlap in Their Properties

In reviewing Table 12.8, it must be emphasized that, even though certain contractile properties tend to be associated in a particular motor unit, there are considerable overlaps between units. For example, although the motor units generating the largest tensions always have fast twitches, other fast-twitch units may develop tensions similar to those of slow-twitch units. This overlap is not surprising, since parameters such as twitch and tetanic tensions, and contraction and half-relaxation times, are distributed unimodally for a given muscle. One contractile parameter that does *not* show overlap is *fatigability*, which enables fatigue-resistant units (S, FR) to be distinguished from fatigue-sensitive ones (FF). The other reliable feature is the presence or absence of the "*sag*" *phenomenon* in short, unfused tetani, which differentiates between fast-twitch and slow-twitch units (Burke et al., 1973).

Most Properties of Human Motor Units Can Also Be Correlated

In the case of human muscle, Buchthal and Schmalbruch (1970) were the first to attempt a systematic correlation between histochemical properties and contractile behavior. In their studies, small bundles of muscle fibers were stimulated, and the time-courses of the twitches were recorded by a needle-mounted strain gauge inserted into the tendon. This technique did not enable the responses of single motor units to be analyzed, but the results were nevertheless important. Buchthal and Schmalbruch (1970) found that long contraction times predominate in muscles that have a large proportion of fibers rich in mitochondria (soleus, gastrocnemius), but short contraction times are common in muscles that consist mainly of motor units poor in mitochondrial enzymes (lateral head of triceps brachii and platysma).

A more complete examination of human motor unit and muscle fiber types has been undertaken by Garnett et al. (1979) in the medial gastrocnemius. The axons

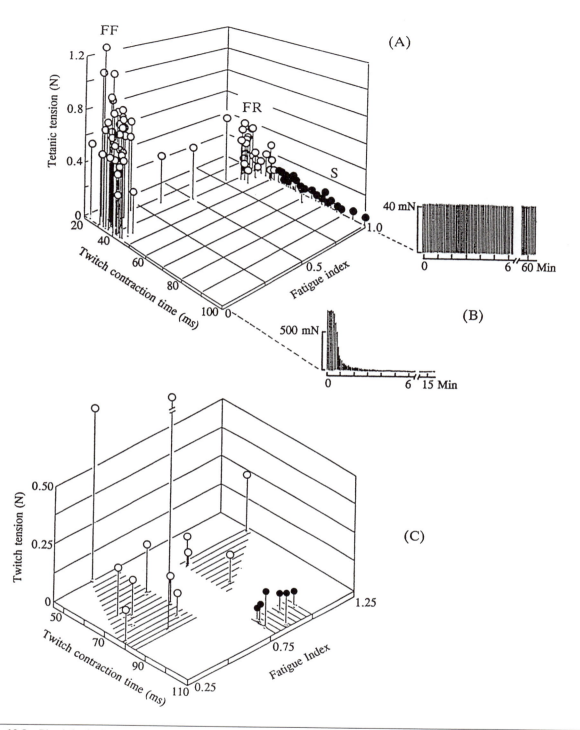

Figure 12.8 Physiological properties of different motor unit types. (A) shows a simulated three-dimensional display of the results for 81 cat medial gastrocnemius units. It can be seen that all but two of the units fall into one of three clusters (see text). (B) shows examples of the responses of a type S unit (upper trace) and a type FF unit (lower trace) to intermittent tetanic stimulation. Note that force is sustained in the type S unit but not in the type FF unit. (C) displays the results of a similar type of study undertaken in the human gastrocnemius; although fewer units were examined than in (A), there is a tendency for the values to fall into three groups. Both (A) and (B) are reprinted from Burke, Levine, Tsairis, and Zajac (1973). (C) is reprinted from Garnett, O'Donovan, Stephens, and Taylor (1979).

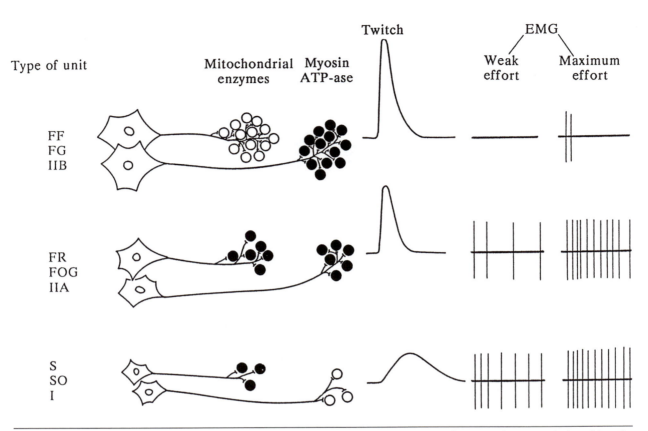

Figure 12.9   Summary of the histochemical and physiological properties of the three main types of motor unit, based largely on the animal studies of Burke, Levine, Tsairis, and Zajac (1973).

Table 12.8   Fiber-Type Classification

	Motor unit type		
	I	IIA	IIB
	(S)	(FR)	(FF)
Property	(SO)	(FOG)	(FG)
Twitch speed	Slow	Fast	Fast
Twitch force	Small	Intermediate	Large
Fatigability	Low	Low	High
Red color	Dark	Dark	Pale
Myoglobin	High	High	Low
Capillary supply	Rich	Rich	Poor
Mitochondria	Many	Many	Few
Z-line	Intermediate	Wide	Narrow
Glycogen	Low	High	High
Alkaline ATPase	Low	High	High
Acid ATPase	High	Low	Moderate
Oxidative enzymes	High	Medium - High	Low

of single motor units were stimulated through needle electrodes in the innervation zone, and recordings were made of the twitch and tetanic responses. At the end of the experiment, the same motor unit was tetanized to exhaustion, and a needle biopsy was taken from the region of the muscle belly that had previously shown the greatest visible contractions. By staining alternate sections for glycogen and for myosin ATPase activity, the authors were able to combine the histochemical and physiological properties of the stimulated units. The results obtained, although fewer in number, were remarkably similar to those of Burke et al. (1973) for the medial gastrocnemius in the cat (compare A and C in Figure 12.8). Another large human muscle, the tibialis anterior, also shows correlations between the contraction times and the tensions developed in single motor unit twitches (Andreassen & Arendt-Nielsen, 1987).

In Some Human Muscles, Correlations Are Absent

Not all human muscles lend themselves to clear demarcations of motor unit properties. In small muscles, such as the first dorsal interosseous and abductor pollicis brevis muscles of the hand, and the extensor digitorum brevis muscle of the foot, there are no significant correlations between contraction time and twitch force (Sica & McComas, 1971; Milner-Brown et al., 1973b; Thomas, Johansson, et al., 1990; Young & Mayer, 1981).

Further, in the adductor pollicis, there is an unusually high proportion of type I fibers, and yet the contraction time of the muscle is brief (Round et al., 1984). It should be noted, however, that the contraction time is determined by the availability of cytosolic Ca^{2+} ions to the myofilaments in addition to the rate of ATP hydrolysis and the speed of shortening (see chapter 11).

The Electrical Properties of Motoneurons Differ According to Motor Unit Type

In animal experiments, the use of intracellular stimulation and recording techniques in motoneurons provides the opportunity to establish whether or not there are features of the motoneuron itself that can be correlated with the type of muscle fiber that it innervates. The motoneurons of S motor units are those most readily excited by weak currents and have the highest *input resistances* (change in membrane potential ÷ applied current). Injections of horseradish peroxidase confirm that the S motoneurons are the smallest (Burke et al., 1982; Ulfhake & Kellerth, 1982). In addition, the action potentials of the S motoneurons are followed by *after-hyperpolarizations* (AHPs); because the motoneurons are refractory to excitation during the AHPs, the AHPs impose a limit on the discharge rate of a motoneuron and its motor unit (Eccles, Eccles, & Lundberg, 1958).

In humans, the only accessible information concerning the physiological properties of a motoneuron is the axonal conduction velocity. On the basis of animal studies, the motoneurons with the largest cell bodies would be predicted to be those with axons having the highest conduction velocities (Kernell, 1966; Burke, 1967) and, therefore, the greatest diameters (Hursh, 1939). Dengler, Stein, and Thomas (1988) found that the conduction velocities of axons supplying the first dorsal interosseous muscle of the hand ranged from 40 to 63 m/s and were highly correlated with the twitch forces of the respective motor units. Thus, it would appear that, in humans as in other mammals, the largest motoneurons supply the greatest numbers of muscle fibers (see, however, Thomas, Johansson, et al., 1990).

APPLIED PHYSIOLOGY

Motor units may undergo striking alterations in disease or following nerve injury. For example, a motoneuron disorder or nerve damage results in very large motor units, while muscle diseases have the opposite effect. Further information on some of these changes is given in chapter 17, and in the next chapter we see how the abnormal motor units can be recognized in the EMG laboratory. Before closing this chapter, though, we will discuss how the number of motor units in a human muscle can be determined automatically within a few minutes (see Box 12.1).

BOX 12.1 ESTIMATION OF HUMAN MOTOR UNITS

There is a simple way to estimate the number of motor units in a living human muscle. All that is required is a stimulator, amplifier, and storage oscilloscope. A recording electrode is applied to the skin overlying the motor innervation zone of the muscle, and a reference is attached elsewhere. The motor nerve is then stimulated with weak electrical shocks, starting with an intensity that is just subthreshold. As the stimulus intensity is gradually raised, the amplified compound action potential recorded from the muscle (M-wave) is seen to grow in discrete steps on the oscilloscope screen (Figure 12.10). The assumption is made that each of these steps is due to the excitation of an additional motor unit. If the average amplitude or area of the putative motor unit potentials is then compared with the maximum M-wave of the muscle, an estimate of the number of motor units in the muscle is obtained.

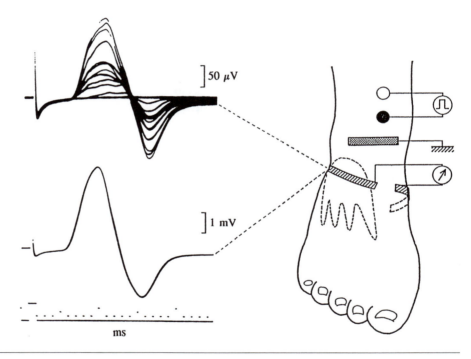

Figure 12.10 Original, "manual" method of motor unit estimation as applied to the human extensor digitorum brevis muscle. The placement of the stimulating and recording electrodes is shown on the right, and on the left are the superimposed, incremental responses to increasing stimulus intensities, together with the maximum M-wave. This subject was estimated to have just over 200 motor units in the muscle.

Obviously, there are many assumptions in the method, not the least of which is that each increment in the graded response results from the discharge of a single motor unit. Nevertheless, the method does seem to work and has recently been made easier by entrusting the stimulation and response analysis procedures to a computer program (Galea, De Bruin, Cavasin, & McComas, 1991). A subroutine deals with the vexatious problem of "alternation," that is, fictitious motor unit responses caused by axons with similar thresholds discharging in different combinations as the stimulus is repeated. Another subroutine arranges the evoked potentials in order of increasing area and subtracts each response from the next largest, so as to yield the potentials generated by single putative motor units. Figure 12.11 shows an example of the automated procedure as applied to a medial vastus muscle; the estimate of 286 units was well within the normal range for this muscle—clearly, motoneuron aging had not started in this healthy 68-year-old man!

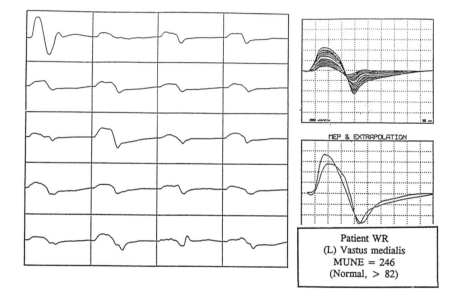

Figure 12.11   Automated motor unit estimation in the medial vastus muscle of a 68-year-old man. A computer software program increases the stimulus intensity and analyzes the progressively larger muscle responses (top, right). The potentials of individual motor units can be derived by successive subtractions of the incremental responses and are shown at the left. The two traces at bottom right compare the shapes of the sample of motor unit potentials and of the maximum M-wave. Further descriptions of this methodology can be found in Galea, DeBruin, Cavasin, and McComas (1991).

Chapter 13

Exercise

Having dealt with the various characteristics of motor units, it is opportune to consider how the units are used in voluntary and reflex contractions. For want of a better term, we have described this use of motor units as "exercise." In the course of this chapter, we will find out about the impulse firing rates of motor units, and the sequence in which the different types of units are brought into play.

Motor Unit Activation

The very fact that the units exhibit such striking differences among themselves in terms of size, speed of contraction, and biochemistry suggests that their involvement is unlikely to be random but suited to the task demanded of them. Indeed, to some extent the task will shape the unit (see chapter 18).

Motor Units Have Different Thresholds for Recruitment

The earliest study of motor unit function in humans appears to be that of Adrian and Bronk (1929), who were able to record the activity of single motor units by means of a fine wire positioned within, but insulated from, a hypodermic needle (see Box 13.1). Adrian and Bronk found that their subjects were able to increase the force of their muscle contractions by raising the firing frequency of the motoneurons and by calling additional motor units into activity (*recruitment*). Since this important study, there have been many investigations of firing rate in which a variety of fine electrodes have been used to distinguish the activities of single motor units (see, for example, Bigland & Lippold, 1954; Lindsley, 1935; Petajan & Philip, 1969). Because of the large movements of the muscle fibers that take place during muscle contraction, some workers have preferred to introduce the fine recording wires through a hypodermic needle and then to withdraw the needle, allowing the wires to flex with the distortions of the muscle belly. From all these studies, it is evident that the motor units have different thresholds for recruitment, with some units coming in during weak contractions and others only involved in forceful effort. In addition, however, each unit has the ability to modulate its firing frequency.

BOX 13.1   NEEDLE EMG—FROM RESEARCH LAB TO BEDSIDE

Needle electromyography (EMG) had its origin in 1929 when Edgar Adrian and Detlev Bronk published a report in the *Journal of Physiology* on "The Discharge of Impulses in Motor Nerve Fibres." In order to obtain their results, Adrian and Bronk made a special recording electrode from a hypodermic needle into which they had inserted a fine insulated wire, which was then glued in place (Figure 13.1). This electrode proved capable of detecting the discharges of individual motor units within a muscle, and hence provided information as to the frequency of impulse firing in motor nerve fibers. The recordings were displayed on a capillary-electrometer, a much less satisfactory instrument than the cathode ray tube, which started to be used in electrophysiology laboratories in the United States at about this time. Although Adrian is said to have doubted whether his simple electrode would have any clinical applications, it subsequently proved capable of detecting action potentials that were reduced in number but prolonged in duration in patients with chronic denervation—though the significance of this finding was initially misinterpreted (Buchthal & Clemmesen, 1941). The large amplitudes and complex waveforms of the potentials in denervation were recognized afterward. An example of this kind of activity, in a patient with spinal muscular atrophy, is shown in the bottom trace of Figure 13.1. Rather later, Kugelberg (1949) described potentials that were unusually brief and "polycyclic" in patients with muscular dystrophy, and such potentials are now known to characterize all degenerative muscle disorders. In the middle trace of Figure 13.1, such activity is shown from a patient with limb-girdle muscular dystrophy. It was not until the 1960s, however, that needle EMG became widely practiced as a diagnostic test for the investigation of neuromuscular disorders. Since then the field has expanded to such an extent that one or more international meetings are held each year, and in North America, several training courses are offered annually. The American Association of Electrodiagnostic Medicine, to which many American and Canadian electromyographers belong, has over 3,000 members.

Impulses Are Discharged at High Frequencies at the Start of a Strong Contraction

There is agreement among investigators that the lowest frequency at which motor units can discharge steadily is from 8 to 12 Hz and that this is found during weak contractions (Figure 13.2; Freund, 1983; Monster & Chan, 1977). Lower frequencies can only be generated in repetitive voluntary contractions of a muscle or in certain movement disorders (e.g., Parkinsonian tremor). It has been much more difficult to determine the maximum firing frequencies, partly because of the considerable overlap of motor unit territories in any region of the muscle (see chapter 12). During a maximum contraction, as many as 20 motor units may be discharging in the vicinity of the recording electrode tip, and it is virtually impossible to recognize the firing pattern of an individual unit. Even if a tungsten wire microelectrode is employed, so as to improve the signal-to-noise ratio, the movement of the muscle belly during the contraction will invariably dislodge the tip from its proximity to the active fiber.

For the reasons given above, it was therefore not surprising that considerable disagreement existed among investigators as to the maximum firing frequency. On the one hand, Norris and Gasteiger (1955) reported rates as high as 140 Hz, but others, such as Dasgupta and Simpson (1962), have insisted on much lower values. Another factor that may have contributed to the differences in experimental results is the speed of contraction. Tanji and Kato (1972) have found that a motoneuron will discharge at a much higher rate if a given tension is to be reached rapidly rather than slowly. They observed many units that could fire initially at rates of 70 to 90 Hz. Similar high rates were also noted by Marsden, Meadows, and Merton (1971), who made

What of the EMG pioneers? Edgar Adrian continued his distinguished career at Cambridge University and was awarded the Nobel Prize in 1932 (with Charles Sherrington) for demonstrating how perception and movement depended on patterns of electrical impulses conducted in nerve fibers. He was elevated to the peerage in 1955. Detlev Bronk returned to the United States and pursued wide-ranging studies of nervous system function, and Kugelberg and Buchthal were leaders in the EMG field for many years. At the time of writing (1995), Fritz Buchthal, now in his mid-80s, is still practicing EMG, but in California rather than in his native Denmark.

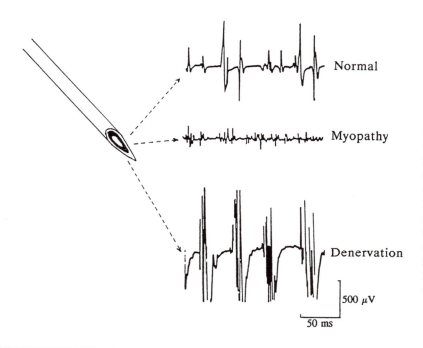

Normal

Myopathy

Denervation

500 μV

50 ms

Figure 13.1 A concentric (coaxial) needle electrode. The central wire is usually 100 μm in diameter, and the outside diameter of the needle varies but is typically about 0.6 mm. At a given site in a normal muscle, the central wire can detect activity in some of the fibers belonging to 20 or so motor units. During a strong contraction, however, the impulse patterns are so dense and complex that the discharges of individual units cannot be distinguished. The three recordings displayed in the figure were obtained during weak effort (see text).

use of the anomalous innervation of some units within the adductor pollicis by median nerve fibers. On blocking the major innervation (ulnar nerve) with local anesthesia, the activity of single (median nerve) units could be followed even during maximum contractions. Rates of 100 Hz or more were observed, but only at the start of a contraction.

Steady Impulse Firing Rates Depend on the Tension Required of the Muscle and on the Contraction Speed of the Muscle

Within a few seconds of activation, the maximal motor unit firing frequencies fall to values that depend both on the muscle and on the individual. Bellemare, Woods, Johansson, and Bigland-Ritchie (1983) observed that the mean firing rates for units in the adductor pollicis and biceps brachii were approximately 30 Hz, but for the soleus they were only 11 Hz (Figure 13.3). They pointed out that the soleus has a much slower twitch than the other two muscles and that the reduced firing rate would still have been sufficient to achieve a fused contraction.

Leaving aside the very high frequencies that can be generated at the start of a maximal contraction, it would appear that there is a three- to fourfold range (i.e., 8-30 Hz) within which units can modulate their *steady* firing rates, depending on the force required (Figure 13.2). As the effort continues and mechanical fatigue sets in, the maximum firing rate declines further (e.g., to 8-15 Hz, see Figure 13.3); nevertheless, the rate still remains appropriate for ensuring a fused contraction, because the fatiguing muscle relaxes ever more slowly. It is thought that an inhibitory reflex from the fatiguing muscle is responsible for decreasing the motoneuron discharge rate (see chapter 15).

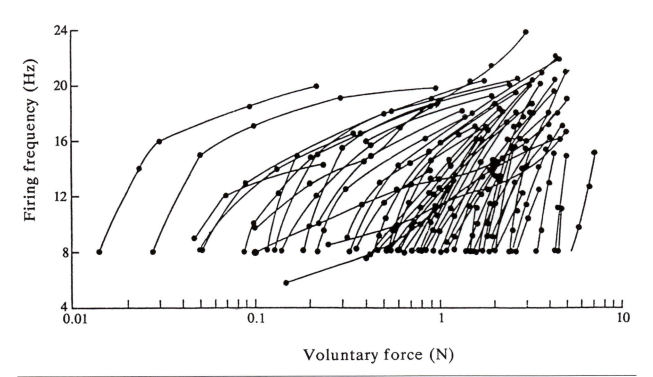

Figure 13.2 Firing frequencies of motor units in the human extensor digitorum communis muscle during steadily increasing contractions. It can be seen that most units commence discharging at approximately 8 Hz and that the frequency rises to 16 to 24 Hz as the contractions become stronger. Unlike the high-threshold units, the lowest threshold units increase their firing rates over a relatively large range of forces. Reprinted from Monster and Chan (1977).

Impulse "Doublets" Boost the Initial Tension Generated by a Muscle

There is an interesting refinement of the motoneuron discharge pattern that has been described by Zajac and Young (1975) in decerebrate cats made to walk on a treadmill. The authors observed that during each step the flexor and extensor motoneurons commenced firing with *doublets*; that is, the second impulse of each train followed quickly upon the first. The functional advantage of this maneuver is that the first two twitches summate, bringing the motor unit tension rapidly toward the tetanic level, where it stays for the remainder of the discharge.

Muscles Differ in Their Ranges of Relative Forces Over Which Motor Units Are Recruited

Adrian and Bronk (1929) noted that increased force was produced by increasing the impulse firing rate and by recruiting extra motor units. But which of the two mechanisms is the more powerful? One approach is to examine the relationship between impulse firing frequency (or stimulus repetition rate) and force for individual motor units or whole muscles. At the minimum voluntary firing frequency of 8 Hz, the twitches of a muscle such as tibialis anterior are only partly fused, so that the mean tension is approximately half the peak tension of the twitch. At the maximum steady firing frequency of 35 Hz, the tension developed is about five times larger than the twitch (see Figures 11.15 & 11.16 in chapter 11). Thus, the motor units in a muscle can increase their force 10-fold simply by adjusting their firing rates.

On the other hand, recruitment must also be a very effective way of increasing force, since some motor units can develop up to 100 times as much tension as others within the same muscle belly (see chapter 12). It was Adrian and Bronk's opinion that, at relatively low forces, recruitment was the more important mechanism but that increased firing rates were increasingly employed to bring contractile force closer to the maximum possible. These conclusions were confirmed by the spike-triggered averaging study of Milner-Brown, Stein, and Yemm (1973a, 1973b); in the first dorsal interosseous muscle of the hand, they observed that half of the motor units had already been recruited when only 10% of the maximal force had been developed (Figure 13.4). These authors conceded, however, that the highest threshold units might not have been detected by their

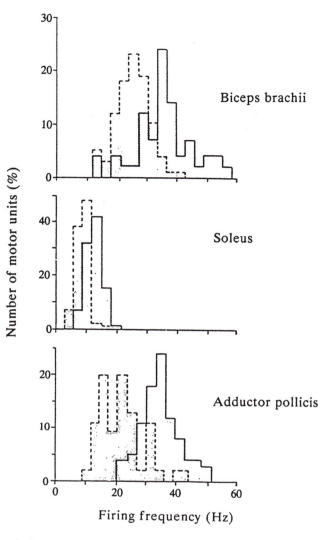

Figure 13.3 Firing rates of motor units in the biceps brachii, soleus, and adductor pollicis muscles of two subjects (solid and dashed lines, respectively), soon after the onset of maximum voluntary contractions; the recordings were made with tungsten microelectrodes. Note that the firing frequencies are consistently higher in one subject than in the other, and that the values are significantly lower for soleus than for the other muscles. Reprinted from Bellemare, Woods, Johansson, and Bigland-Ritche (1983).

averaging technique, and other investigators have argued that recruitment is an important factor at all force levels (for example, Bigland & Lippold, 1954).

Small Muscles Depend Heavily on Firing Rate to Achieve Maximal Force

The explanation for the apparent confusion between experimental results is that the two strategies, recruitment and firing rate, are employed to different extents from one muscle to another. This point was nicely demonstrated by Kukulka and Clamann (1981), using fine wires inserted into the muscle through a hypodermic needle. In the biceps brachii, recruitment was observed from 0 to 88% of maximum voluntary force, while in the adductor pollicis, no additional motor units were recruited at forces greater than 50% of maximal. Similar differences between the first dorsal interosseous muscle of the hand and the deltoid muscle (DeLuca, LeFever, McCue, & Xenakis, 1982) suggest that small, distal muscles rely more on increases in firing rate for the

development of large forces, whereas large, proximal muscles continue to recruit additional motor units.

An interesting refinement of motor unit firing rate has been found by Broman, DeLuca, and Mambrito (1985), who observed that the recruitment of an additional motor unit was associated with a transient drop in the firing rates of those units previously active. Such an effect would help to ensure that the increase in force following recruitment was smooth rather than abrupt.

In a Steady Contraction, Small Motor Units Are Recruited Before Large Ones

While investigating the responses to stretch of cat hindlimb muscles, Henneman, Somjen, and Carpenter (1965) pointed out that the recruitment of motor units was orderly; that is, the smaller motor units had lower thresholds than the larger units. Since then numerous experiments have been conducted on human muscles, during both voluntary and reflex contractions, to ascertain whether the same principle applies. One

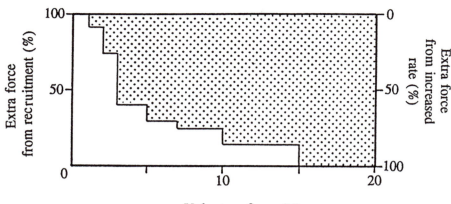

Figure 13.4 The relative importance of recruitment of additional motor units and of increase in firing frequency, during increasingly strong contractions of the first dorsal interosseous muscle of one subject. Based on experimental data of Milner-Brown, Stein, and Yemm (1973a).

method of estimating motor unit sizes is to measure the amplitudes of the motor unit action potentials, since the amplitudes are proportional to the numbers of muscle fibers in each unit. Two such studies showed that the majority of units are, in fact, recruited voluntarily in order of increasing size (Monster & Chan, 1977; Tanji & Kato, 1973).

A more accurate indication of motor unit size is the twitch tension, and this may be determined by the technique of spike-triggered averaging during steady voluntary contractions (Figure 13.5A). Figure 13.5B shows the results of multiple observations in the first dorsal interosseous muscle of the same subject, plotted on a log-log scale; it can be seen that there is a nearly linear relationship between the twitch tensions of the units and the voluntary forces at which they are recruited. Similar results, indicative of linear or curvilinear relationships, have been obtained by Monster and Chan (1977) in the extensor digitorum communis and by Yemm (1977) and Goldberg and Derfler (1977) in human jaw muscles. The impulse conduction velocities of the motor axons are also related to recruitment order, the most rapidly conducting axons belonging to the units with the highest thresholds (Freund, Buedingen, & Dietz, 1975; Grimby, 1984). If the voluntary contractions are made more quickly, motor units appear to become active at lower forces. As Freund (1983) points out, this is illusory because the background force, increasing rapidly, will have reached the same level by the time that the newly recruited unit would have developed its maximal twitch tension.

Orderly recruitment is also a feature of stretch reflexes, as observed originally by Henneman et al. (1965) in animals; Calancie and Bawa (1985) have found this to be true in the human flexor carpi radialis also.

The Order of Motor Unit Recruitment Changes Only if the Motor Task Is Modified

Is the recruitment order truly invariable? Basmajian (1963) reported that when human volunteers were given

feedback from the oscilloscope screen and loudspeaker, some learned to alter the order of recruitment among motor units in the thumb muscles. This challenging finding has been the source of much controversy and has received some support from experiments in which recruitment order was altered by sensory stimuli applied simultaneously to the skin (Garnett & Stephens, 1980; Grimby & Hannerz, 1968). It is also quite clear that recruitment order can vary if the same muscle is used for different purposes or in slightly different ways. For example, in the extensor digitorum communis, the recruitment order will depend on which of the four fingers is to be extended (Schmidt & Thomas, 1981). Similarly, in the biceps brachii, the threshold of a motor unit depends on whether the muscle is being used to flex the elbow, supinate the forearm, or externally rotate the humerus (Gielen & Denier van der Gon, 1990). Passive factors, such as joint position, may also affect recruitment order, as was noted in the rectus femoris by Person (Person, 1974; Person & Kudina, 1972).

Although the preceding observations must be regarded as genuine examples of altered recruitment order, the point must be emphasized that, when the same motor task is undertaken in exactly the same way, the order in which motor units are recruited remains fixed (see also Henneman, Shahani, & Young, 1976).

In Steady or Repetitive Contractions, the Type I (Slow-Twitch) Motor Units Are Recruited First

In chapter 12, it was seen that the smallest motor units in a human or mammalian muscle are those with the slowest twitches and the greatest resistance to fatigue; depending on the classification employed, they were described as type I (Brooke & Kaiser, 1970), type S (Burke, Levine, & Zajac, 1971), or type SO (Peter, Barnard, Edgerton, Gillespie, & Stempel, 1972). In view of the ordered recruitment of motor units during voluntary contractions, described earlier, the type I units are

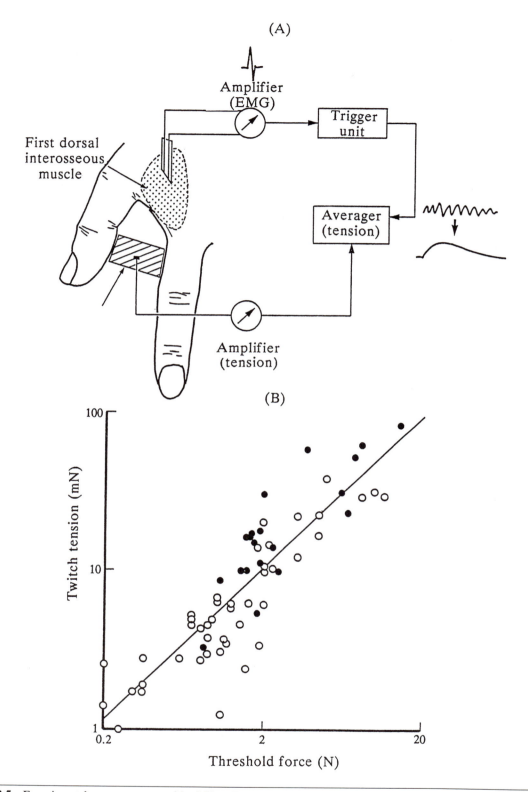

Figure 13.5 Experimental arrangement used by Milner-Brown, Stein, and Yemm (1973a, 1973b) for detecting the force developed by a single motor unit during voluntary contraction (A). The action potential of the motor unit is distinguished and used to trigger an averager, which then samples the muscle tension. By repeating the process many times, the contribution of the unit can be recognized from among the activity of all the other units. (B) Proportional relationship between the forces developed by motor units and their thresholds for recruitment. Reprinted from Milner-Brown, Stein, and Yemm (1973b).

those that should be called into action first. To examine this supposition, Gollnick, Karlsson, Piehl, and Saltin (1974) and Gollnick, Piehl, and Saltin (1974) exercised volunteers on a bicycle or had them sustain isometric contractions of the quadriceps muscles. Specimens of muscle were taken at intervals using a needle biopsy technique (Bergström, 1962) and were stained histochemically. By correlating the presence of glycogen depletion with the staining reactions of the fibers for myosin ATPase, it was possible to determine whether type I (slow-twitch) or type II (fast-twitch) fibers had been employed in the exercise. In the cycling exercise, it was found that slow-twitch fibers were the first to become depleted of glycogen at all work loads requiring less than the maximal oxygen consumption. At supramaximal levels, when the oxygen uptake was insufficient and anaerobic metabolism also took place, the fast-twitch fibers were also involved.

In the static exercise experiments, the findings were similar. At isometric contractions of less than 20% of maximum force, there was preferential use of the slow-twitch fibers. At higher forces, however, the fast-twitch fibers were the first to lose glycogen, suggesting that these had been most active. Thus, in both types of exercise, pedaling and sustained contraction, the slow-twitch fibers were used for weak contraction, the fast-twitch ones being reserved for greater effort. In the glycogen-depletion experiments of Garnett, O'Donovan, Stephens, and Taylor (1979), it was possible to differentiate between type IIA and type IIB motor units in the human medial gastrocnemius muscle and to show that the former had lower thresholds for voluntary contraction than the latter.

The preferential involvement of type I units has also been demonstrated in animal experiments. For example, Gillespie, Simpson, and Edgerton (1974) showed that these units were those used most heavily by the bush baby, *Galago senegalensis*, in steady running of moderate intensity. Similarly, in reflex contractions of cat plantaris muscles evoked by muscle stretch, Zajac and Faden (1985) found that recruitment invariably followed the order:

$$S \rightarrow FR \rightarrow FInt \rightarrow FF \text{ units}$$
$$(\text{or } I \rightarrow IIA \rightarrow IIint \rightarrow IIB).$$

In Sudden Movements, the Type II (Fast-Twitch) Units May Have the Lowest Thresholds

What about rapidly executed movements? Under these circumstances, there is evidence from both human and animal experiments that type II units may sometimes have the lowest thresholds. For example, in the bush baby, Gillespie et al. (1974) found that jumping made

greatest demands on the type IIB units. Again, when cats shake their hindpaws, the gastrocnemius muscles, containing both type I and type II units, are active, while the soleus muscles, comprising only type I units, are silent (Smith, Betts, Edgerton, & Zernicke, 1980). The results of Grimby (1984), in the human short extensor muscle of the toes, are of particular interest, since he was able to follow the activities of single motor units during walking and running as well as in controlled toe extension. He found that, as expected, the lowest threshold units have the lowest impulse conduction velocities. However, there are some high-threshold units that do not participate in walking or even in running, but only in rapid corrective movements—such as accelerations or sudden changes in direction. Even then, these units fire very few impulses in a single high-frequency burst.

In conclusion, there is good evidence that the type I motor units are mostly employed during weak or moderately strong contractions of a sustained or repetitive nature. As the contractions become stronger, type IIA units are recruited and, finally, the type IIB units also. In certain very rapid or sudden corrective movements, however, the type II units may have lower thresholds than type I units.

The Low Thresholds of the Type I Motor Units Depend on the Small Sizes of the Motoneuron Cell Bodies and on the Densities of Synapses

It is functionally advantageous for the largest units to be reserved for extreme effort, for they would impart marked unevenness to a weak contraction (as in an unfused tetanus). But how is the order of recruitment established? Henneman et al. (1965) suggested a simple explanation based on a *size principle*. Thus, the smallest α-motoneurons in the ventral horn are likely to correspond to the smallest motor units, because the metabolic activity of a motoneuron is probably proportional to the number of muscle fibers and nerve endings that it is required to maintain through its trophic action (see chapter 18). Suppose that the motoneuron pool receives an excitatory input from the motor cortex and that roughly the same number of synapses are activated on all the motoneurons. The density of the synaptic current flowing between the excitatory synapses on a motoneuron and the axon hillock will differ among the motoneurons. The smallest cells will have the highest densities because the current is concentrated in a smaller membrane area. A large current density will, in turn, produce a correspondingly large depolarization, and so the smallest cells should have the lowest thresholds for excitation.

The size principle hypothesis can be tested experimentally by impaling motoneurons with microelectrodes and measuring their input resistances during current pulses; the input resistance of a motoneuron will be inversely proportional to its size. It turns out that, although the largest axons arise from the largest motoneurons, as predicted (Burke, 1967), they do not necessarily supply the greatest numbers of muscle fibers, at least in the large muscles of the cat hindlimb (Stephens & Stuart, 1975; but see Bagust, 1974); in the small distal muscles a better correlation exists. On the basis of these observations, it would appear that motoneuron size cannot be the sole factor determining excitation threshold during voluntary or reflex contractions. Rather, it would appear that there is some specialization of the synaptic input among the motoneurons, with those neurons belonging to the type I and smaller IIA motor units receiving the heaviest projections from the muscle spindles and from the corticospinal pathway (Burke, 1986).

In Most Muscles, All the Motor Units Can Be Maximally Activated

As previously described, Grimby (1984) found that there were some motor units in the short extensor muscle of the toes that discharged during sudden corrective movements but not during sustained contractions. What of other muscles? One experimental approach has been to compare the force developed during a maximal voluntary contraction with that generated by tetanic stimulation of the same muscle. In the small muscles of the hand, the results obtained to date suggest that the forces are similar and imply that all motor units can be recruited voluntarily and made to discharge at optimal rates for tension development (Bigland-Ritchie, Johansson, Lippold, Smith, & Woods, 1983; Merton, 1954).

There are, however, technical problems that make such experiments difficult and cloud the interpretation of the results. For one thing, the muscles are usually stimulated through their nerves. Not only is tetanic stimulation painful, but there are few accessible nerves that innervate a single muscle or even a group of muscles having only the one action at the same joint. Also, some muscles have rather complex actions, and it is difficult to measure all the resultant forces. Rather than use tetanic stimulation, an alternative approach has been to inject a single (or double) maximal stimulus to the motor nerve in the course of a supposedly maximal voluntary contraction. If all the motor units have been recruited and are firing at optimal frequencies, no additional tension will be detected by an appropriately mounted strain gauge. If some motor units are silent or are firing at low frequencies, the interpolated stimulus will produce

a twitch on top of the voluntary force record (see Figure 13.6). This relatively painless technique, originally described by Merton (1954), has shown that during isometric contractions, motor units can be fully activated in some muscles but not in others. Activation is complete in the small muscles of the hand, the dorsiflexors of the ankle, the quadriceps, and the diaphragm (Bélanger & McComas, 1981; Bellemare & Bigland-Ritchie, 1984; Edwards, Hill, & Jones, 1975; Merton, 1954). In some subjects, however, there are motor units in the triceps surae that cannot be fully recruited in steady contractions (Bélanger & McComas, 1981), and as already noted, the same is true of the short extensor of the toes.

Changes During Exercise in the Respiratory and Cardiovascular Systems

Although the muscles and their innervation play the central role in exercise, important adaptive changes take place in other body systems—the pulse quickens, blood pressure rises, muscle blood flow increases, and breathing becomes deeper and more rapid. All of these adaptations are beneficial because they result in a greater delivery of oxygen, glucose, and lipid to the contracting muscles, while removing unwanted metabolites—H^+, K^+, and CO_2 and lactate. These effects are, in themselves, the subjects of textbooks, and only the blood flow changes will be considered here.

During Exercise Muscle Blood Flow May Increase 10-Fold

The flow of blood through a muscle can be estimated by various methods, including the time-honored one of venous occlusion plethysmography. In this method, the venous return is prevented by inflating a cuff around the limb; the initial rate of increase in limb volume is equal to the arterial inflow. At rest the flow is 2 to 3 ml per 100 ml muscle per min, but there is an immediate increase on beginning exercise, and the value may then rise 10-fold. A further increase in blood flow occurs when the effort ceases (*reactive hyperemia*; see Figure 13.7) and may reach 70 ml per 100 ml muscle per min (Barcroft & Dornhorst, 1949). Much depends, however, on the nature of the contractions, because if they are strong, the intramuscular pressure may exceed the systolic blood pressure (Barcroft & Millen, 1939; Humphreys & Lind, 1963).

On the other hand, the pressure required to close the arteries will be affected by the elevation of the blood pressure in exercise and will also depend on the size

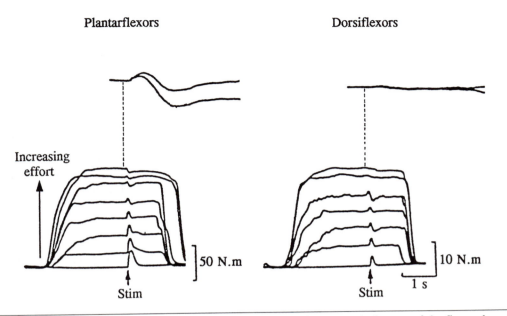

Figure 13.6 Interpolated twitch technique. The two sets of superimposed traces at the bottom of the figure show the responses of the dorsiflexor and plantarflexor muscles to single stimuli (*stim*) delivered during relaxation and during the course of increasingly forceful contractions. In the strongest contractions (uppermost traces), no interpolated twitch can be detected in the dorsiflexors, but a small one appears to be present in the plantarflexors. The paired traces at the top of the figure were made at higher amplification and on a faster sweep; they confirm the persistence of a small twitch in the plantarflexors and the absence of one in the dorsiflexors. Adapted from Bélanger and McComas (1981, p. 1133).

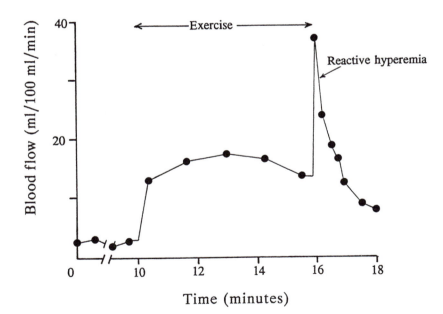

Figure 13.7 Blood flow, measured by occlusion plethysmography, during and after calf muscle exercise. The exercise consisted of plantarflexion against a 15-kg weight, repeated every second. The blood flow through the muscles during this 6-minute period is an average value, since the flow would have been higher during the relaxations than the contractions. At the end of the exercise, there is a pronounced increase in blood flow (reactive hyperemia). From Barcroft and Dornhorst (1949, p. 407).

and architecture of the muscle belly. Much larger forces will be generated in the centers of bulky muscles, such as those of the upper arm and calf, than in thin, pliant bellies such as the platysma and diaphragm. When strong isometric contraction of a large muscle is performed, the rise in blood flow is delayed until the effort ceases; there is then a reactive hyperemia that continues for several minutes. The same phenomenon can be demonstrated when repetitive contractions are performed

with an arterial cuff inflated around the limb; deflation of the cuff results in prompt hyperemia.

Multiple Factors Are Responsible for Muscle Hyperemia

The changes in blood flow depend to only a small extent on the increase in cardiac output; much more important is the relaxation of smooth muscle in the walls of the arterioles and small arteries within the muscle belly. It is probable that several factors are responsible for this relaxation, one of them being a rise in interstitial $[K^+]$. Thus, the injection of K^+ into an artery increases blood flow (Dawes, 1941), and there is a close correspondence between the relative magnitudes and time-courses of the changes in $[K^+]$ and blood flow (Kiens, Saltin, Wallye, & Wesche (1989).

Another factor causing vasodilatation is reduced O_2 tension in the blood perfusing the small arteries and arterioles of the muscle. In most mammals, there is also a reflex component mediated by the efferent sympathetic nerve supply to the muscle. In humans, in whom the sympathetic muscle vasodilator fibers appear to be absent, the ability of the adrenal medulla to secrete epinephrine provides the sympathetic nervous system with a different method for increasing muscle blood flow. Last, prostaglandins and a number of neuropeptides have been shown to produce vasodilatation in skeletal muscle; it is probable that many effects involve the release of nitric oxide from the endothelial wall (Persson, Hedqvist, & Gustafsson, 1991).

APPLIED PHYSIOLOGY

Motor unit potentials, as recorded with an intramuscular needle electrode, can be greatly altered by nerve and muscle diseases. Indeed, these changes can be used to help diagnose patients complaining of weakness or increased fatigability. More information about this aspect of motor unit function has been given in Box 13.1.

The physiology of muscle excitation and contraction has now been explored, both at the level of the muscle fiber and at that of the motor unit. One more feature of the working muscle remains to be examined. This feature is the biochemical reactions in the muscle fiber that produce the energy required for muscle contraction. These reactions are discussed in the next chapter.

Chapter 14

Muscle Metabolism

This chapter is largely devoted to adenosine triphosphate (ATP), the source of energy for many enzyme reactions in the muscle fiber, including the contractile process itself. We will see how the energy associated with the molecular structure of glucose and lipid is transferred to ATP, and how some ATP synthesis continues even when the fiber is no longer supplied with adequate oxygen. The special role of the mitochondrion in cell metabolism will also be considered.

ATP Synthesis

Muscles are designed for work, and work requires energy. A muscle fiber may be called upon to transform itself from a state of quiescence into one developing maximum power or force in less than a second. Therefore, the energy must be plentiful, readily available, and—because many efforts must be maintained for minutes or even hours—renewable.

ATP Provides the Immediate Source of Energy for Contraction and Pumping Ions

Like other cells in the body, the muscle fibers obtain their immediate energy by splitting *adenosine triphosphate* (ATP; see Figure 14.1).

$$ATP \Leftrightarrow \underset{\substack{\text{(adenosine} \\ \text{diphosphate)}}}{ADP} + \underset{\substack{\text{(inorganic} \\ \text{phosphate)}}}{P_i} + \underset{\text{(energy)}}{e}$$

How is ATP consumed during muscle activity? Most of the ATP is used in the contractile process itself, because ATP is required to detach the myosin cross-bridges from the actin filaments, thereby allowing the bridges to move to new positions on the filaments (see chapter 11). However, ATP is also needed for two important metabolic pumps. One of these, the Na^+-K^+ pump, maintains the muscle fiber plasmalemma in an excitable state during muscle activity. The other pump returns Ca^{2+} to the sarcoplasmic reticulum at the conclusion of the period of excitation. Together, these two cation pumps probably consume at least a third of the ATP used during mammalian muscle activity and, in avian muscle, as much as 70% (Fambrough, Wolitzky, Tamkun, & Takeyasu, 1987; Homsher, 1987).

ATP is also required for many of the normal "housekeeping" functions of the muscle fiber during and between periods of activity. One of these functions is to build up the store of *phosphocreatine* (PCr) in the resting fiber. Furthermore, ATP is needed for the phosphorylation of enzymes by *protein kinases* and for conversion to *cyclic*

215

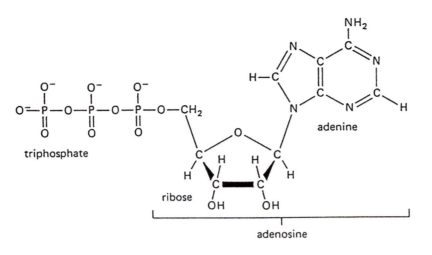

Figure 14.1 The ATP molecule. Like other nucleotides, ATP consists of a base (adenosine), a sugar (ribose), and phosphate (in this case, three phosphates). Reprinted from Alberts et al. (1983).

adenosine monophosphate (cAMP) at the fiber plasmalemma (see p. 219).

Phosphocreatine Replenishes ATP Quickly

Even for relatively brief periods of activity, the muscle must be able to renew its supply of ATP. In this respect, the muscle fiber differs from other cells in having a substantial energy reserve in the form of PCr. As ATP is expended, it is instantly replenished from PCr by the action of the enzyme *creatine kinase*:

$$ADP + PCr \Leftrightarrow ATP + \underset{\text{(creatine)}}{Cr}$$

So effective is PCr in maintaining ATP levels that, when a muscle fiber is exercised to fatigue, substantial amounts of ATP remain, even though the supply of PCr is exhausted. In a rested muscle, this store of PCr is 3 to 4 times larger than that of ATP.

ATP can be synthesized from ADP in another way. Since the enzyme *myoadenylate kinase* can convert two molecules of ADP to one of ATP and one of AMP,

$$2ADP \Leftrightarrow ATP + \underset{\text{(adenosine monophosphate)}}{AMP}$$

AMP can then be deaminated:

$$AMP \Leftrightarrow \underset{\text{(inosine monophosphate)}}{IMP} + \underset{\text{(ammonia)}}{NH_3}$$

This pathway for the production of ATP probably plays a relatively minor role, though in exercise of moderate or high intensity, the concentration of ammonia in the arterial plasma rises severalfold (Banister & Cameron, 1990).

The Oxidation of Glucose and Fat Provides Additional ATP

As muscles continue to contract, the production of ATP from PCr becomes inadequate. The fibers then become totally dependent on the oxidation of fat and also of glucose (see Figure 14.2), the latter being largely derived from the storage polysaccharide, *glycogen*. Even in resting muscle, some fat is consumed for most of the housekeeping expenditures of energy, and in low-intensity exercise, it remains the major source of ATP production. Although part of the glycogen store is broken down to glucose in low-intensity exercise as well, this source of energy becomes increasingly important as the effort increases and the type II fibers start to be recruited. The enzyme pathways that consume fat and glucose convert both fuels to *acetyl coenzyme A* (acetyl CoA) and then break the latter down to CO_2 and H_2O, consuming oxygen as they do so.

Glucose Can Be Partially Processed Even in Anaerobic Conditions

The breakdown of glucose (*glycolysis*) commences with its phosphorylation, and proceeds with the conversion of glucose to fructose, further phosphorylation, cleavage of the six-carbon fructose molecule into two three-carbon aldehyde molecules, and finally, oxidation of the

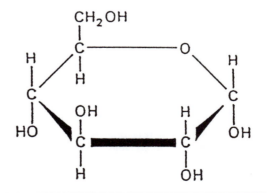

Figure 14.2 The glucose molecule. Containing three types of atom (C, H, and O), glucose can be completely broken down to CO_2 and H_2O by oxidative metabolism in the muscle fibers.

aldehyde molecule to a carboxylic acid (*pyruvic acid*). Figure 14.3 shows the nine reaction steps involved in glycolysis. It can be seen that, although two ATP molecules are required for successive phosphorylations (steps 1 and 3), four ATP molecules are produced from ADP and inorganic phosphate (steps 6 and 9). For each glucose molecule that is converted to pyruvic acid, there is therefore a net gain of two ATP molecules. Additional molecules of ATP are created after the *hydride* ion (one proton and two electrons) is passed from NADH along the electron transport chain in the mitochondria (see p. 218).

One of the important features of the glycolytic pathway is that it can still operate and supply ATP under anaerobic conditions, that is, when sufficient oxygen is no longer available to the muscle fiber. Such a situation is liable to occur during sustained strong contractions, because the rise in pressure inside the muscle belly then becomes greater than the systolic blood pressure, and perfusion of the muscle with oxygenated arterial blood ceases. In anaerobic conditions, the pyruvate that is formed from glycolysis is converted to lactic acid:

$$CH_3 \cdot CO \cdot COOH + H_2 = CH_3 \cdot CHOH \cdot COOH$$

(pyruvic acid)   (lactic acid)

Glucose Breakdown Is Completed by the Citric Acid Cycle

The reactions described in the previous section take place in the muscle fiber cytosol; once pyruvic acid has been formed, however, it is quickly taken up by the mitochondria and oxidized to CO_2 and H_2O. This degradation process produces much more ATP than the glycolytic pathway provided. The oxidation of pyruvic acid begins with its decarboxylation and combination with

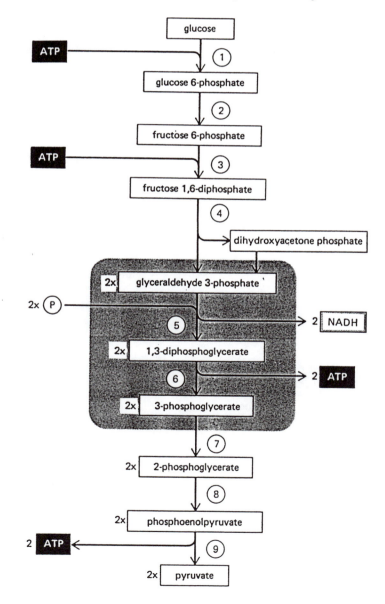

Figure 14.3 Steps in the breakdown of glucose to pyruvic acid, as described in the text. At step 4, the six-carbon sugar molecule is split, and steps 5 and 6 are those responsible for the net synthesis of ATP and NADH molecules. Reprinted from Alberts et al. (1983).

coenzyme A (CoA) to create the key intermediary metabolite, *acetyl CoA*. Once formed, acetyl CoA combines with *oxaloacetic acid* to make *citric acid*, a six-carbon molecule. Through a series of reactions, two of the carbon atoms are oxidized and removed as CO_2, resulting in the eventual formation of oxaloacetic acid again. Thus, the metabolic steps form a cycle, the *citric acid cycle* (Figure 14.4).

Energy Is Trapped in the Respiratory Chain

The energy from the oxidation of acetyl .CoA is harnessed in two ways. One of these is through the creation of a high-energy phosphate bond, in *guanosine triphosphate* (GTP). In turn, GTP donates its high-energy phosphate bond to ADP, forming ATP. However, most of the energy captured in the citric acid cycle comes from the transfer of electrons. The electrons are first received by the hydrogen receptor molecules, *nicotinamide adenine dinucleotide* (NAD) and *flavin adenine dinucleotide* (FAD), and then passed along a special series of molecules (the *electron transport chain* or *respiratory chain*) in the mitochondria. As the electrons travel along the respiratory chain, their energy levels become progressively lower; the energy released is used to pump protons across the inner mitochondrial membrane (the *chemi-osmotic process*; Mitchell, 1961). Since the protons become more concentrated on the outside of the membrane, and since the inside of the mitochondrion will have a surplus of negative charges, the protons will tend to diffuse back across the membrane down their electrochemical gradient. By a mechanism that is not fully understood, the reentry of protons stimulates *ATP synthetase*, a membrane-bound enzyme that converts ADP and inorganic phosphate to ATP. At the end of the respiratory chain, the electrons are accepted by oxygen. For each molecule of acetyl CoA oxidized in the citric acid cycle, 12 molecules of ATP are formed. Since two molecules of acetyl CoA are produced from each glucose molecule, the citric acid cycle creates 24 ATP molecules. The glycolytic pathway (glucose to pyruvate) adds another 12 molecules of ATP, making a total of 36 for each molecule of glucose consumed.

In the Muscle Fiber, Glucose Is Produced From Glycogen

Although muscle fibers can oxidize glucose entering from the blood stream via the interstitial fluid, during

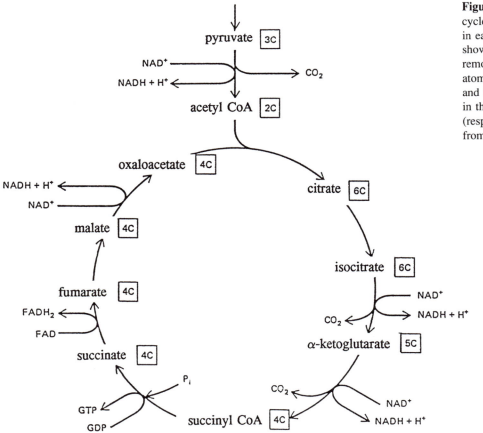

Figure 14.4 The citric acid cycle. The number of C atoms in each intermediate molecule is shown. The C atoms are removed as CO_2, while the H atoms are taken up by NAD^+ and FAD, two carrier molecules in the electron transfer (respiratory) chain. Reprinted from Alberts et al. (1983).

contractile activity most glucose is provided by the hydrolysis of *glycogen*. As the storage form of glucose, glycogen can be seen in electron micrographs of muscle fibers as granules, 10 to 40 nm in diameter, in the cytosol. Bound to the surfaces of the granules are the enzymes necessary to synthesize glycogen from glucose (*glycogen synthetase*) and also those required to degrade it back to glucose (*glycogen phosphorylase*). Each glycogen molecule is formed from the linking together of 10,000 to 30,000 glucose units in a branching pattern, giving the molecule a molecular weight of 2,600 kD (see Figure 14.5A). The various enzymatic reactions involved in the synthesis and degradation of glycogen are shown in Figure 14.5B; note that glucose must be phosphorylated before it can be used to manufacture glycogen and also that some of the glycogen is hydrolyzed in the lysosomes.

Glycogen Hydrolysis Is Stimulated by Epinephrine

For glycogen to be broken down to glucose, it must first be phosphorylated by glycogen phosphorylase; however, this enzyme must itself be phosphorylated by another enzyme, *protein kinase*. The latter, in turn, is switched on by cAMP, which is released from the inner face of the muscle fiber plasmalemma, following the combination of epinephrine (adrenaline) with its receptor; *G protein* acts as an intermediary. The complete sequence of events, culminating in the degradation of glycogen, is set out in Figure 14.6. Cyclic AMP is an example of a *second messenger*, that is, a molecular signal which works in the interior of the cell, having been generated by an event at the cell membrane. In the case of cAMP, such an event is the combination of epinephrine with its receptor; however, the opening of an ion channel allows Ca^{2+}, another second messenger, to invade the muscle fiber directly. Other known second messengers are *arachidonic acid* and *inositol triphosphate*.

Fat Oxidation Proceeds With the Conversion of Fatty Acids to Acetyl CoA

As already noted, fat (lipid) metabolism is used to supply the energy requirements of resting muscle; it also provides much of the energy in exercise, especially when the exercise is of low intensity. In prolonged exercise, such as distance running, all the muscle glycogen will be consumed, and continued effort will then depend largely on lipid metabolism (Åstrand, 1967). Most of the fat is available to the muscle fiber as fatty acids, which are transported across the plasmalemma from the blood stream; some fat, however, is in the form of fine lipid droplets in the cytosol. Within the droplets, each fat molecule is composed of three fatty acid molecules linked to glycerol and hence is termed a *triglyceride*. The lengths of the fatty acid chains (tails) vary considerably; palmitic acid (see Figure 14.7) and stearic acid are considered to have "long" chains with 16 and 18 carbon atoms respectively.

The first step in the breakdown of a triglyceride molecule is for it to be split, with an enzyme, into glycerol and the three component fatty acids. The fatty acids are then taken into the matrix space of the mitochondrion for metabolic processing. While medium- and short-chain fatty acids can pass directly through the inner mitochondrial membrane, the long-chain fatty acids need a special transport system. The long-chain fatty acids are first *esterified* at the outer mitochondrial membrane with CoA; the latter is in turn converted to *acetyl-carnitine*, by the enzyme *carnitine palmityltransferase*, at the outer surface of the inner mitochondrial membrane. Acetylcarnitine is then passed through the inner membrane by the enzyme *acetylcarnitine translocase* in exchange for *carnitine* from the matrix space (see Figure 14.8).

The long-chain acetylcarnitine, like the short- and medium-chain fatty acids that have passed directly into the matrix space, is next converted to *fatty acyl CoA* (Figure 14.9). The latter enters a sequence of reactions that cuts off two carbon atoms from the fatty acid tail and produces one molecule of *acetyl CoA*; the cycle is repeated with two carbons being removed during each turn. It can be seen from Figure 14.8 that hydrogen atoms, each with an additional electron, are given to the carrier molecules, NAD and FAD. These electrons are then passed down the respiratory chain and enable ATP to be produced. Meanwhile, the acetyl CoA enters the citric acid cycle (Figure 14.4) and is oxidized to CO_2 and H_2O, with the creation of further ATP.

The Mitochondrion

The mitochondrion was first encountered in chapter 1, in which the structure of the muscle fiber was described, and its role in lipid oxidation has just been stressed. It is now time to consider this important organelle in rather more detail.

Mitochondria Are Elongated Organelles With Double Membranes

In most cells, the mitochondria are long, mobile structures, from 1.0 to 2.0 μm in diameter. Since the muscle fiber is a compact structure with tightly packed myofibrils, the mitochondria are rather shorter than in other

(A)

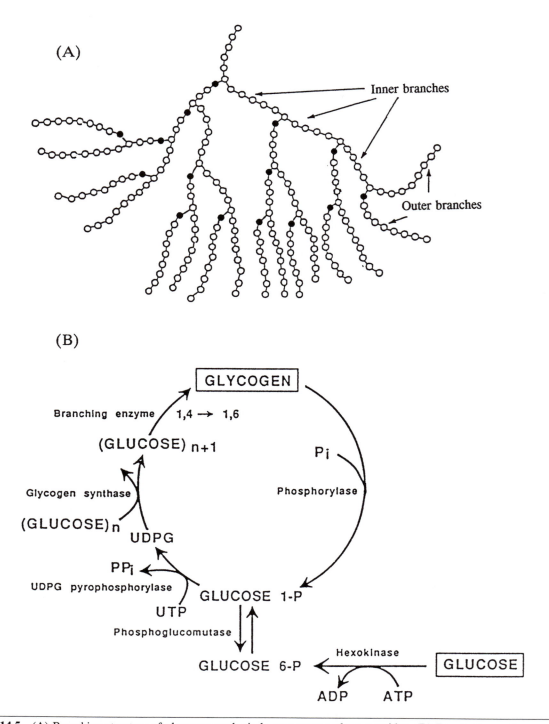

(B)

Figure 14.5 (A) Branching structure of glycogen; each circle represents a glucose residue. (B) Enzymatic reactions involved in the synthesis and degradation of glycogen. (A) and (B) adapted from DeBarsy and Hers (1990).

cells and are more stable in their positions. The mitochondria are found where ATP is needed most; thus, they tend to lie alongside, or to be wrapped around, the myofibrils and sarcoplasmic reticulum, and others are situated beneath the plasmalemma.

Each mitochondrion has two membranes, the outer one of which is richly embedded with transport pro-

teins. These proteins permit molecules of 10 kD or less to pass freely between the cytosol and the *intermembrane space* (see Figure 14.10). The inner membrane has a much larger surface area than the outer membrane, due to the presence of numerous folds, or *cristae*. The inner membrane also contains transport proteins that control the movement of metabolites between the inter-

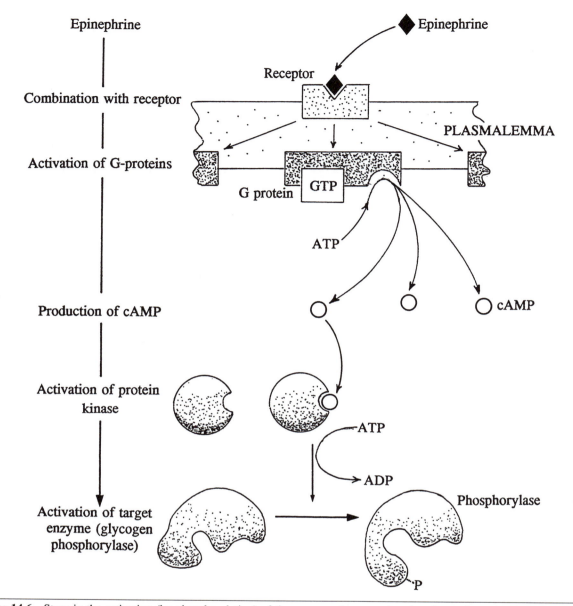

Epinephrine

Combination with receptor

Activation of G-proteins

Production of cAMP

Activation of protein kinase

Activation of target enzyme (glycogen phosphorylase)

Epinephrine

Receptor

PLASMALEMMA

G protein   GTP

ATP

cAMP

ATP

ADP

Phosphorylase

P

Figure 14.6 Steps in the activation (by phosphorylation) of the enzyme glycogen phosphorylase, after epinephrine reaches the outer face of the muscle fiber plasmalemma.

membrane space and the *matrix space*; unlike the outer membrane, the inner membrane is permeable to most small ions.

The Respiratory Chain Is Found in the Inner Membrane

In addition to the transport proteins, the inner membrane houses the 15 different molecules that pass electrons from one to the next in line (the respiratory chain). Included in the respiratory chain are five different *cytochromes* and a *NADH dehydrogenase complex*. Finally,

the inner membrane incorporates the enzyme *ATP synthetase*, which is driven by the passive inward flux of protons to synthesize ATP from ADP and inorganic phosphate (see p. 218). The innermost cavity of the mitochondrion, the matrix space, is rich in enzymes, including those of the citric acid cycle and those needed to form acetyl CoA from pyruvate and fatty acids. The matrix space also contains DNA unique to the mitochondrion; it is necessary for multiplication of the organelles and for the formation of the many types of mitochondrial enzymes.

Figure 14.11 summarizes the way in which the mitochondrion is able to transform the energy in pyruvate

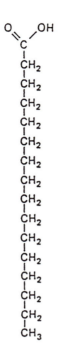

Figure 14.7 Palmitic acid. This example of a fatty acid has 16 C atoms.

and fatty acids into ATP, with the simultaneous consumption of O_2 and production of CO_2.

APPLIED PHYSIOLOGY

Many inherited diseases of muscle metabolism are known, in most of which a particular enzyme is missing or malfunctioning. Fortunately, all of these disorders are rare. In this section we will learn about two types of metabolic defect, including the first such condition to be recognized. The section concludes by examining the proposition that glycogen loading of muscle can improve athletic prowess.

Lack of Myophosphorylase Causes Exercise Intolerance

In 1947 Dr. Bryan McArdle investigated a 30-year-old man who had been admitted to Guy's Hospital in London with a lifetime history of pain, weakness, and stiffness of his muscles during exertion (McArdle, 1951). Even mild exercise, such as walking 100 m, was sufficient to induce the symptoms, which were always relieved by rest. When the muscles were stiff and swollen, it was impossible to record any EMG activity with a needle electrode—hence the stiffness was a physiological *contracture*. McArdle carried out a number of biochemical

tests on this patient and showed that the lactate concentration *fell* in the venous blood during ischemic exercise, whereas in a normal subject a substantial rise occurred. On the basis of this observation, McArdle postulated that this patient was unable to break glycogen down to pyruvate and thence to lactate. Since glycogen breakdown depends successively on a debranching enzyme, muscle glycogen phosphorylase (*myophosphorylase*), and phospho-glucomutase (Figure 14.5B), a hereditary deficiency of either of these enzymes might have been responsible.

The mystery was solved 10 years later when a group of investigators in Los Angeles had the opportunity to study a 19-year-old man who gave a similar history and also showed no rise in blood lactate during ischemic muscle contractions (Figure 14.12). Pearson, Rimer, and Mommaerts (1961) further showed that the ability to exercise was greatly improved if glucose was given intravenously—another pointer to the presence of a glycogen breakdown disorder. Finally, by histochemical staining of a muscle biopsy, the investigators were able to demonstrate that the enzyme that was deficient was myophosphorylase. If patients with McArdle's disease are made to keep on exercising, despite their pain, something very interesting happens—the pain disappears, as does the fatigue. This "second wind" phenomenon, which is present to a lesser extent in normal subjects, is due to the muscle fibers switching from glycogen to fatty acids as a source of fuel, and to the arrival of glucose from the liver.

Further light on the pathophysiology of McArdle's disease has been obtained by Richard Edwards and his collaborators at the University of Liverpool (Cooper, Stokes, & Edwards, 1989). They have shown that not only does force decline more rapidly than normal in patients with this disorder, but so does the muscle compound action potential (Figure 14.13), the latter indicating a loss of muscle fiber excitability. It would be tempting to ascribe these various changes to insufficient ATP for the myosin-actin interactions and for Na^+-K^+ pumping. This explanation is too simple, however, for in McArdle's patients, as in normal subjects, there is only a modest decline in muscle ATP during exercise. An alternative hypothesis is that the fatigue is due to the accumulation of contraction metabolites, such as ADP and inorganic phosphate.

After McArdle's disease had been described and named after its discoverer, a number of other enzyme deficiencies were recognized, in which the formation or utilization of glycogen was impaired; all of these conditions are extremely rare and appear to be inherited through autosomal recessive genes (see Figure 2 of DeBarsy & Hers, 1990).

Lipid Disorders Can Also Cause Fatigability and Muscle Cramps

Twenty years after McArdle's pioneering study, reports began to appear of occasional patients in whom muscle

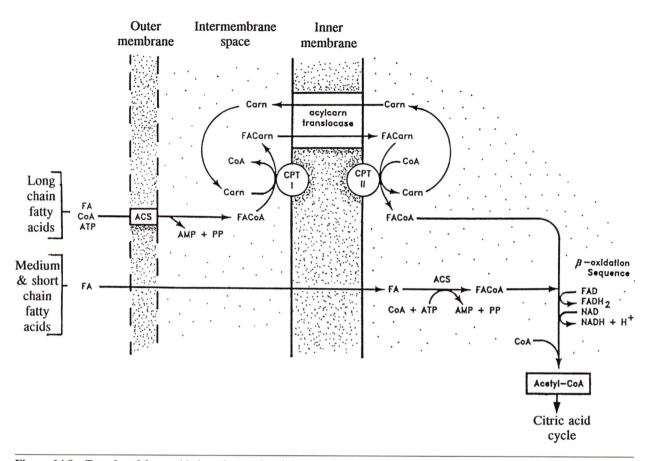

Figure 14.8 Transfer of fatty acids into the matrix of the mitochondrion. See text. ACS = acyl-CoA synthetase; Carn = carnitine; CPT = carnitine palmityltransferase; FA = fatty acid. From *Myology-Basic and Clinical* (p. 704), by A.G. Engel and B.Q. Banker, 1986, New York: McGraw-Hill, Inc. Copyright 1986 by McGraw-Hill, Inc. Adapted with permission.

cramps, weakness, and fatigability were associated with excessive amounts of triglyceride fat in the muscle fibers. Engel and Angelini (1973) showed that the inability of such a patient to oxidize long-chain fatty acids could be corrected by adding carnitine to her diet; indeed, her disorder was one of hereditary *carnitine deficiency*. In the same year, DiMauro and DiMauro (1973) reported a patient in whom there was a deficiency of the lipid-handling enzyme, *carnitine palmityltransferase* (see Figure 14.8). In such patients, prolonged exercise causes aching and excessive fatigability of muscles; in severe episodes, the muscle fibers become necrotic and release myoglobin into the bloodstream and thence into the urine.

Like carnitine deficiency, the absence of carnitine palmityltransferase is an inherited mitochondrial disorder. However, mitochondria can be abnormal in other respects; for example, very rarely patients are found in whom the energy from glucose and lipid oxidation, instead of being used for the synthesis of ATP, is released as heat—causing perspiration, heat intolerance, and increased thirst and hunger. In other mitochondrial disorders, there may be abnormal function in one of the molecules in the respiratory chain or in one of the enzymes in the citric acid cycle. Some of these conditions are associated with mitochondria that appear abnormal in electron micrographs; for example, they may be greater in size or number, or distorted in shape and internal structure.

Does Building Up Glycogen Stores in Muscles Improve Athletic Performance?

In view of the importance of glycogen as a source of energy, a muscle that has its glycogen store increased would be expected to fatigue more slowly than before. This proposition was first examined by Christensen and Hansen (1939), who showed that a carbohydrate-rich diet was associated with better performance in prolonged exercise. Some 30 years later, Bergström, Hermansen, Hultman, and Saltin (1967) repeated this work, with similar results. In addition, they analyzed samples of muscle taken with a biopsy needle and showed that muscle glycogen

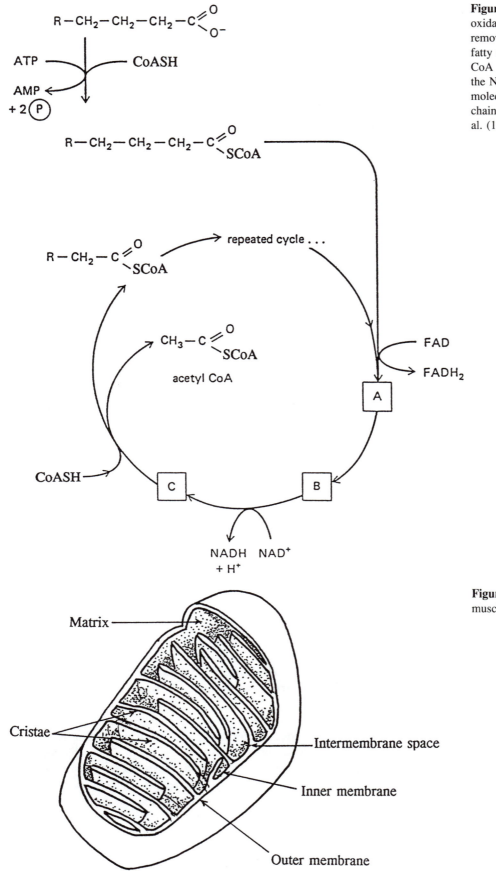

Figure 14.9 The fatty acid oxidation cycle. Each cycle removes two C atoms from the fatty acid in the form of acetyl-CoA and donates H atoms to the NAD and FAD carrier molecules of the respiratory chain. Reprinted from Alberts et al. (1983).

Figure 14.10 Structure of a muscle mitochondrion.

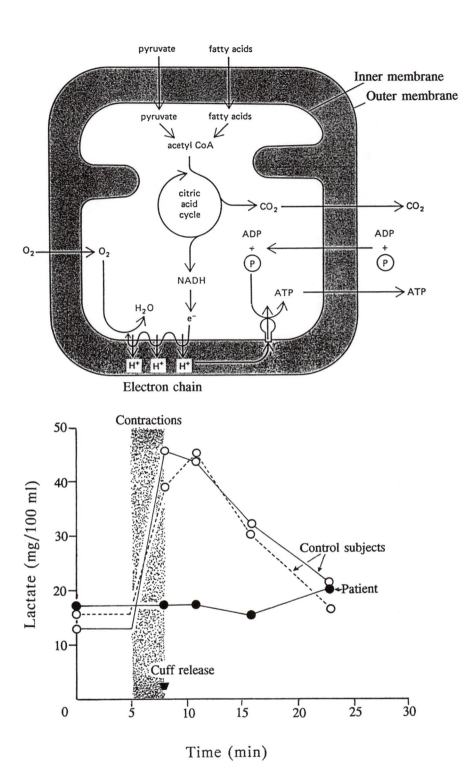

Figure 14.11 Summary of the oxidative reactions in the mitochondrion that convert pyruvic and fatty acids to CO_2 and H_2O, while producing energy in the form of ATP. Note the electron (respiratory) chain in the inner membrane. Reprinted from Alberts et al. (1983).

Figure 14.12 Lactate concentrations in venous blood taken from the forearm after repetitive handgrips, performed under ischemic conditions. Adapted from Pearson, Rimer, and Mommaerts (1961, p. 506).

had been increased by the carbohydrate-enriched diet. The principle of glycogen loading has now been adopted by many athletes, especially those in endurance events. A common strategy is to exercise the muscles to exhaustion 1 week before the competition, so as to deplete the muscle of glycogen. Over the next 3 days, the athlete prevents glycogen synthesis by eating a carbohydrate-*free* diet; he or she then consumes a carbohydrate-*rich* diet, so as to stimulate glycogen synthesis. Such a procedure can increase muscle glycogen from the normal 1% of muscle weight to as much as 3 to 4%.

But does this strategy work? Doubts of its effectiveness have come from a recent study by Bangsbo, Graham, Kiens, and Saltin (1992), in which it was found that glycogen breakdown was *not* enhanced in quadriceps muscles with double their normal amounts of glycogen.

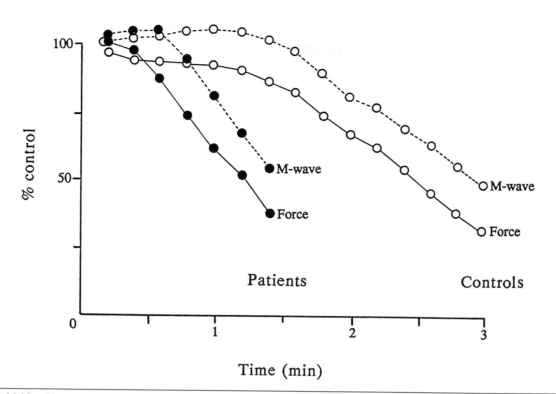

Figure 14.13   Changes in mean force and muscle excitation (*M-wave*) in 7 patients with McArdle's disease and in 9 control subjects, during intermittent repetitive stimulation at 20 Hz under ischemic conditions. Note the much greater declines in force and excitation in the patients. Adapted from Cooper, Stokes, and Edwards (1989, p. 5).

Meanwhile athletes, ever desperate to seize any suspected advantage, will probably continue to "glycogen load."

This chapter has completed our study of the electrical and chemical processes that underlie the maintenance of resting muscle and its transformation into a device for generating force or movement. Our study has ranged from the survey of individual enzyme reactions to the performance of muscle fibers, motor units, and entire muscles. In the third part of the book, we will see how the muscle fibers and their motoneurons can adjust to a variety of stresses.

Part III

The Adaptable Neuromuscular System

One of the remarkable features of skeletal muscle is its adaptability. This adaptability is seen in the short term, in those situations in which muscles must continue to develop force for as long as possible. The adaptations are diverse, for they include changes in blood flow, ion pumping, and motoneuron impulse firing. Were it not for these adaptations, muscle fatigue, considered in chapter 15, would set in much sooner than it usually does. In the longer term, adaptations take place when muscle fibers lose their nerve supply, as discussed in chapter 16. Even in these adverse circumstances, many of the adaptations appear advantageous, and some of them, such as the presumptive release of chemical attractants by the muscle fibers, are designed by evolution to restore the nerve supply. We will discover in chapter 17 that this restoration can be achieved by regeneration of the injured nerve fibers or by sprouting of those healthy nerve fibers remaining.

The study of denervation leads naturally to consideration, in chapter 18, of the trophic effects of muscle and nerve. Thus, the nerve not only excites the muscle fiber, but also sustains it; were it not so, the muscle fiber would not shrink and degenerate after the nerve supply is lost. And the sustaining influence works in the reverse direction also; the motoneuron is dependent on trophic signals, in the form of chemicals, secreted by the muscle fibers.

Next, in chapter 19, is the topic of disuse. Although the effects on the muscle fibers are not as profound as those resulting from denervation, they can be severe nonetheless, affecting not only the sizes and functional properties of the muscle fibers, but the impulse firing patterns of the motoneurons too.

The subject of chapter 20 is training, as intended for the development of muscle strength or endurance. The huge gains in muscle mass following strength training in humans are remarkable, and have become the inspiration for body-building competitions and fitness magazines. In contrast, muscles trained for endurance usually appear normal, and it is only during continuous activity that the improved function is evident. We are now begin-

ning to understand the way in which the different adaptations are achieved, and to appreciate the importance of muscle fiber stretch as a stimulus.

Chapter 21 deals with the adaptations of muscle fibers in response to injury, and emphasizes their effectiveness in being able to repair themselves rapidly. The injury is not necessarily from some external agent, for it will be shown that untrained muscles can be damaged by unaccustomed contractions, especially those taking place as the muscle is lengthened.

Finally, and perhaps fittingly, we conclude the survey of skeletal muscle with a consideration of aging. Even though an older adult may not be able to compete as effectively as a younger one, force generation is usually maintained until the 6th decade. Moreover, the elderly muscles retain some degree of plasticity and can respond to training programs and even compensate for partial loss of innervation.

Chapter 15

Fatigue

In this chapter, we begin a survey of situations in which muscles and motoneurons exhibit adaptations. The choice of fatigue is perhaps surprising, but, as will be seen, the adaptations can involve processes as disparate as Na⁺-K⁺ pumping and motoneuron firing. Even the loss of force that characterizes fatigue can be viewed as an adaptation, for without it, serious muscle damage would undoubtedly occur. Fatigue will be shown to have both central and peripheral components; although more is known about the factors involved in peripheral fatigue, there is still uncertainty over their relative contributions. The chapter closes by considering the way in which muscle recovers from fatigue.

Concepts of Fatigue

In everyday language, the word "fatigue" is used to describe any reduction in physical or mental performance. For the physiologist, however, the word has a more restricted meaning, *muscle fatigue* being defined as the "failure to maintain the required or expected force" (Edwards, 1981). The rate of fatigue depends on the muscles employed and whether or not the contractions are continuous or intermittent. Figure 15.1 shows, as an example of fatigue, the decline in force of ankle muscle during a steady contraction. In general, up to half the force can be lost in the first minute of a maximal contraction. The simple definition of fatigue, given above, is adequate to distinguish *fatigue* from *weakness*, in which there is an inability to develop an initial force appropriate to the circumstances. If necessary, the definition of muscle fatigue could be broadened to include an inability to sustain rapidly executed movements as, for example, when the fingers are tapped as quickly as possible. This kind of fatigue has been poorly studied, and it is possible that the most important component is a reduction in the "intensity" of the commands developed in the central nervous system.

Most of our knowledge of fatigue has come from the study of prolonged voluntary or stimulated contractions, made at a constant muscle length (i.e., *isometric*), and it is this body of work that will be reviewed. An adequate understanding of muscle fatigue requires determination of both the site of fatigue and the cellular factors involved. In relation to the site, fatigue may affect either *central* or *peripheral* elements in the motor system.

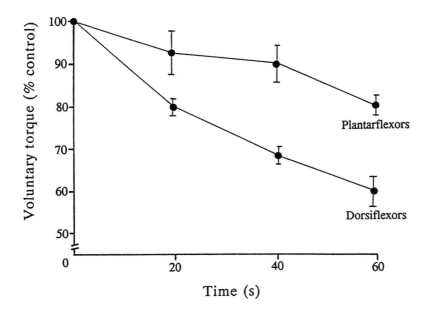

Figure 15.1 Rates of fatigue during continuous maximal isometric contractions of ankle dorsiflexor and plantarflexor muscles in healthy adults; values are means ± SEMs. Note that the slow-twitch plantarflexors fatigue less rapidly than the faster twitch dorsiflexors. Adapted from Bélanger and McComas (1983, p. 629).

Central Fatigue

The elements in the central nervous system that may be affected in fatigue encompass the emotions and other psychological factors responsible for the sense of *effort*, as well as the various descending motor pathways and the interneurons and motoneurons in the brainstem and spinal cord (Figure 15.2). For many years, the descending motor pathways have been known to comprise the corticospinal, rubrospinal, tectospinal, vestibulospinal, and reticulospinal tracts. Similarly, the topographic representation of movements in the motor cortex is well understood. In contrast, almost nothing is known about the identities of the neurons involved in the desire to move or in the generation and assessment of effort.

Central Fatigue Can Be Estimated by Stimulus Interpolation

Recordings of brain activity using scalp electrodes have shown that a surprisingly large area of cerebral cortex is involved in the initiation of a movement. It has long been known that the basal ganglia and cerebellum also participate in the preparatory events. Once the movement starts, however, it is probable that the brain activity becomes more spatially focused. Although the central mechanisms are complex (and beyond the scope of this review), it is relatively easy to determine whether central fatigue is present during a voluntary contraction. The simplest method is to interject a maximal electrical stimulus into the contracting muscle and to look for a twitch superimposed on the recording of voluntary force (see Figure 13.6 in chapter 13; also Bélanger & McComas, 1981; Merton, 1954). If a twitch *is* observed, then either not all motoneurons have been recruited or else some

are not firing impulses at the optimal frequency for force generation. The sensitivity of the technique can be improved by applying two or more stimuli close together, rather than a single one, so as to make the twitch force larger.

An alternative strategy for detecting central fatigue is to compare the voluntary force with that developed by tetanic stimulation of the same muscle. Unfortunately, tetanic stimulation of human peripheral nerves is invariably painful; further, the same nerve may innervate antagonist as well as agonist muscles, and some synergistic muscles may be supplied by other nerves.

Central Fatigue Becomes Increasingly Prominent as the Exercise Continues

The first observation, on applying the twitch interpolation technique, is that, even at rest, some subjects appear unable to activate all their motor units fully. Bélanger and McComas (1981) found this to be true of the ankle plantarflexor muscles in half of their subjects, although the ankle dorsiflexors, in contrast, could always be fully activated. An inability to activate some motor units fully during an isometric contraction is perhaps not surprising, since some motoneurons may only participate in rapid, brief (phasic) types of movement, as demonstrated by means of single unit recordings in the extensor digitorum brevis muscle (Grimby, Hannerz, & Hedman, 1981).

Regardless of whether or not all motor units are optimally recruited at the start of a contraction, central fatigue becomes an increasingly important factor the longer that the contraction is maintained (McKenzie, Bigland-Ritchie, Gorman, & Gandevia, 1992; Thomas, Woods, & Bigland-Ritchie, 1989). The most simple

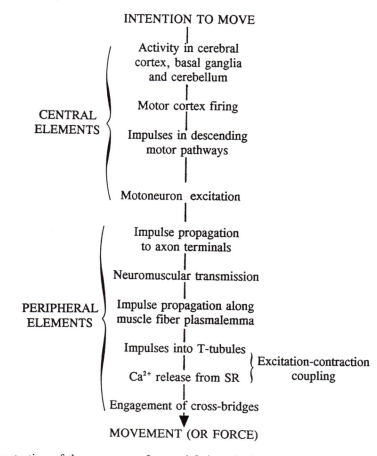

INTENTION TO MOVE

CENTRAL ELEMENTS

Activity in cerebral cortex, basal ganglia and cerebellum

Motor cortex firing

Impulses in descending motor pathways

Motoneuron excitation

PERIPHERAL ELEMENTS

Impulse propagation to axon terminals

Neuromuscular transmission

Impulse propagation along muscle fiber plasmalemma

Impulses into T-tubules

Ca^{2+} release from SR

Excitation-contraction coupling

Engagement of cross-bridges

MOVEMENT (OR FORCE)

Figure 15.2 Successive stages in the production of a voluntary movement.

demonstration of the presence of central fatigue is that force can be momentarily enhanced by giving encouragement or a sudden loud command to the subject.

Peripheral Fatigue

The peripheral elements in the motor system include impulse conduction in the motor axons and their terminals, neuromuscular transmission, conduction of impulses in the muscle fibers, excitation-contraction coupling, and the contractile process itself. In well-motivated people, the major contributions to fatigue come from these peripheral mechanisms. Several sites can be involved, singly or together (Figure 15.2), and these will be considered in the sequence in which they are normally involved during muscle contractions. As will be seen, the susceptibility of a site is affected by the nature of the experiment.

Impulse Conduction Can Fail at the Branch Points or Endings of Motor Axons

Sometimes, when stimulating a motor nerve repetitively, the muscle compound action potentials (M-waves) undergo step-like fluctuations in amplitude, suggesting that action potentials fail to propagate beyond

major *bifurcations* in the axons of the largest motor units.

A more vulnerable site for impulse failure is the *motor nerve terminal* as Krnjević and Miledi (1958) showed in the rat diaphragm. By stimulating single motor axons in the phrenic nerve and recording end-plate potentials with microelectrodes, they were able to compare the responses in two fibers belonging to the same motor unit. On some occasions they found that a normal-sized end-plate potential in one fiber was not associated with any detectable response in the other fiber. This type of defect could best be explained by a failure of the impulse to invade the axon terminal. A similar defect is known to occur in mice with *hereditary motor end-plate disease* (Duchen & Stefani, 1971).

In Krnjević and Miledi's (1958) experiments, it was found that the conduction failure was very sensitive to anoxia, but it is difficult to escape the conclusion that this susceptibility may have been induced by the *in vitro* conditions of the experiment. For example, in human single fiber EMG (see chapter 11) in which impulse propagation can be studied simultaneously in two or more muscle fibers belonging to the same motor unit, it is unusual to see one of the fibers "dropping out." Again, when single cat motoneurons are stimulated *in vivo*, the summated muscle fiber action potentials may show no significant decrement, even after several thousand stimuli (Burke, Levine, Tsairis, & Zajac, 1973).

Presumably the intact blood supply is able to maintain the preterminal axon in satisfactory condition.

Insufficient ACh Release May Be a Factor in Fatigue

A number of investigators have shown that the liberation of acetylcholine (ACh) from motor nerve terminals is markedly reduced during the course of repetitive stimulation. In the experiments of Brooks and Thies (1962), for example, the number of ACh quanta was depressed by about one third within seconds of starting stimulation at a frequency as low as 2 Hz. This experiment was carried out on the serratus anterior muscle of the guinea pig, but, as in the case of the presynaptic failure considered previously, the function of the neuromuscular junction may have been affected adversely by the *in vitro* nature of the experiment.

Muscle Action Potential Changes

It is known that the giant axon of the squid can continue to propagate action potentials for long periods even when the axoplasm has been extruded, and it would be surprising if the membrane of an intact muscle fiber was any less efficient. Yet there is strong evidence that, under certain circumstances, the muscle fiber membrane may become inexcitable because of the associated contractile activity.

In Frog Muscle Fibers *In Vitro*, Action Potentials Are Diminished by Contractions

Lüttgau (1965) isolated single fibers of the frog and stimulated them *in vitro*. Lüttgau found that, at a stimulation rate of 100 Hz, some action potentials began to drop out within 2 s (see also Ramsay & Street, 1942). Rather unexpectedly, Lüttgau found that if stimulation was given at *lower* rates until the fiber became exhausted and could twitch no longer, the action potential mechanism recovered. Membrane excitability was also maintained if the contractile responses were abolished by metabolic inhibitors such as sodium cyanide and iodoacetate, or by immersing the fiber in a hypertonic bathing solution. Lüttgau concluded that the fatigue of the action potential mechanism was caused by metabolic reactions connected with contraction of the muscle fiber.

Muscle Action Potentials May Also Diminish in Mammalian Muscle *In Vivo*

Interesting through Lüttgau's findings are, caution must be exercised before translating these results into the human *in vivo* situation. Not only is there a species difference, but the relatively large volume of fluid bathing the muscle in Lüttgau's experiments would eliminate any significant changes in the composition of the extracellular milieu. On the other hand, the findings of Burke et al. (1973; previously discussed) that mammalian motor units are capable of firing many thousands of impulses when stimulated individually, may give a misleadingly favorable impression of the action potential generating mechanism during fatigue. During strong effort, most or all of the motor units will be recruited rather than a single unit, and the rise in intramuscular pressure will be sufficient to occlude the arterial circulation, rendering the muscle ischemic (Barcroft & Millen, 1939). Also, there will be a large impulse-mediated efflux of K^+ from the muscle fibers, causing the concentration of K^+ to rise in the interstitial fluid and to affect the excitability of the muscle fiber plasmalemma (see Box 15.1).

Leaving theory aside, there is surprising disagreement over the results of experiments in which the M-wave is evoked by interpolated stimuli during the course of maximal voluntary contractions. For example, Stephens and Taylor (1972), employing a double-pulse technique, found that there was an early decline in the M-wave of the first dorsal interosseous muscle of the hand during isometric contractions. However, this result was at variance with the earlier one of Merton (1954) in the adductor pollicis muscle (see the next section). More recently, Bigland-Ritchie, Kukulka, Lippold, and Woods (1982) have reexamined this problem and have concluded that the M-wave is well preserved at a time when fatigue has set in and there is already a decline in voluntary EMG activity.

The Action Potential Decline Begins After Force Is Reduced

The general consensus at present is that, although the muscle fiber action potential may ultimately begin to fail during voluntary contraction, the attenuation is relatively modest and only occurs after there has been a significant reduction in the ability to develop force. This point was well made in the classic study of Merton (1954) on the adductor pollicis muscle. After repeated maximal voluntary contractions of this muscle (usually his own), performed with the arm ischemic, Merton found that the twitch could be completely abolished at a time when the M-wave had barely altered. Similar findings are obtained if, instead of using voluntary contractions, muscles are fatigued by motor nerve stimulation at modest rates (10-30 Hz), as shown in Figure 15.3. It is found that approximately 1,000 impulses can be conducted without decrement in the muscle fibers, even in the presence of ischemia. By this time the twitch is greatly reduced, although tetanic tension is better preserved. In the experiment illustrated in Figure 15.3, the depression of the twitch, together with marked slowing of the relaxation following the tetanus, is clearly seen after 2,000 stimuli. If higher rates of stimulation are employed (e.g., 80-100 Hz), the decline in the M-wave is much more rapid and does indeed become the limiting factor in force development (*high frequency fatigue*), since reducing the stimulation frequency leads to an improvement in force (Edwards, Hill, Jones, & Merton 1977).

The conclusion from the various studies is that when muscle is maximally activated, either voluntarily or following physiological rates of stimulation, the reduction in force begins when there is failure of a cellular mechanism subsequent to the muscle fiber action potential; that is, the failure must involve either excitation-contraction coupling or the contractile machinery. Both possibilities will now be considered.

Excitation-Contraction Failure

Before considering changes in excitation-contraction (E-C) coupling, it is necessary to review briefly the normal physiology of the coupling process. The first step in the coupling process is the propagation of an action potential into the interior of the muscle fiber from

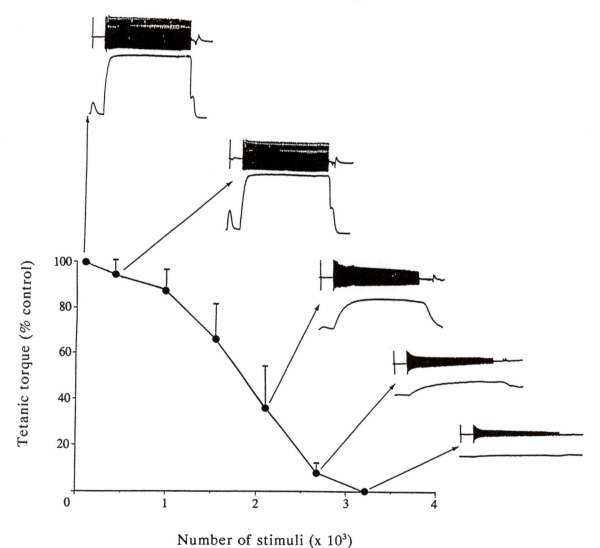

Figure 15.3 Mean fatigue (+ *SEM*) of ankle dorsiflexor muscles in healthy subjects during intermittent 30 Hz stimulation under ischemia. Each pair of sample traces shows M-waves above and torque below; a single stimulus, for eliciting a twitch, is delivered before each tetanus. Note the slowing of torque development and of relaxation as fatigue sets in. In the final pair of traces, M-waves are still present, though much reduced, although no torque can be detected. Adapted from Garland, Garner, and McComas (1988b, pp. 90-91).

the surface membrane (see chapter 11), the *transverse (T) tubules* providing this pathway. The *dihydropyridine* (DHP) receptors in the wall of the T-tubules act as voltage sensors and, in response to the action potential, transfer a signal to the *ryanodine receptors* linking the T-tubule to the sarcoplasmic reticulum. The affected ryanodine receptors permit the rapid escape of Ca^{2+} from the lumen of the sarcoplasmic reticulum into the cytosol, where the Ca^{2+} combines with the troponin molecules on the actin filaments. The Ca^{2+} is then pumped back into the sarcoplasmic reticulum, by a membrane-bound Ca^{2+}-ATPase, in anticipation of the next excitation of the fiber.

Fatigued Muscle Can Still Develop Force in Contractures

Perhaps the clearest demonstration of E-C coupling failure in fatigued muscle has come from experiments in which part of the coupling mechanism has been bypassed. In the first of these, Eberstein and Sandow (1963) treated isolated frog muscle fibers, which had been stimulated to exhaustion, with either 0.1 M KCl solution or with caffeine. Under both circumstances, sustained shortening (*contractures*) of the muscle fibers resulted, with the production of force (Figure 15.4). The presumed effect of the K^+ would be to produce a maintained depolarization of the T-tubules, whereas caffeine is known to promote the release of Ca^{2+} from the sarcoplasmic reticulum. This experiment showed that, in the fatigued muscle fibers, the contractile machinery was still capable of generating force but lacked the signal to do so. Eberstein and Sandow's important observations have been confirmed in amphibian muscle (e.g., Grabowski, Lobsiger, & Lüttgau, 1972) and more recently in mouse muscle fibers also (Lännergren & Westerblad, 1991).

Impulse Propagation Is Likely to Fail in the Transverse Tubules

If the major factor in fatigue is indeed failure of the E-C coupling mechanism, which component is likely to be at fault? There are good reasons for supposing that the inwardly propagated action potential would be vulnerable because of pooling of K^+ in the T-tubules during impulse activity. Hodgkin and Horowicz (1959) produced rapid changes in the chemical composition of the fluid bathing single frog muscle fibers and found that the effects of Cl^- were more rapid than those of K^+. They reasoned that a sizable fraction of the K^+ channels were less accessible and were likely to lie in the T-tubular membrane; diffusion of K^+ from the T-tubules

would be delayed because of their small lumens. Calculations of K^+ efflux during repetitive activity show that large changes in K^+ concentration would be expected in the extracellular fluid (see Box 15.1). Whereas the surface membrane can prevent K^+-induced depolarization by enhancing the activity of the electrogenic sodium pump, the density of pump sites is relatively low in the T-tubular membrane (Fambrough, Wolitzky, Tamkun, & Takeyasu, 1987). Thus, the combination of a high K^+ concentration in the T-tubules, due to slow diffusion, and only modest electrogenic Na^+-K^+ pumping would be expected to result in sustained depolarization of the tubular membrane and block of the local action potential mechanism. Failure of inward spread of the action potential is suggested by the experiments of Edman and Lou (1992). These authors rapidly froze single fibers that had been fatigued by tetanic stimulation, and examined the appearance of the myofibrils with the electron microscope. They found that the myofibrils in the periphery of a fiber had shortened (myofibrils straight), whereas the inner myofibrils had not (myofibrils wavy).

Also relevant to this issue are the voltage-clamp experiments of Adrian, Costantin, and Peachey (1969), which showed that, if inwardly propagated action potentials are abolished by tetrodotoxin (TTX), there is just sufficient electrotonic spread of the action potential from the surface of the fiber to the interior to allow the innermost myofibrils to shorten. In fatigue conditions, however, it is possible that the safety margin for electrotonic spread may no longer be sufficient.

Ca^{2+} Release From the SR Is Reduced in Fatigue

As already noted, the immediate consequence of a diminished signal entering the fiber from the T-tubules is that less Ca^{2+} would be released from the SR by the ryanodine receptors. It is now possible to test this prediction by measuring Ca^{2+} levels within the muscle fibers. The principle of this method is to inject a Ca^{2+}-sensitive luminescent compound, such as aqueorin or Fura-2, into a single muscle fiber and to measure the emitted light with a photomultiplier. This approach was first used in fatigue by Blinks, Rüdel, and Taylor (1978); the fact that these workers found little correlation between tension and intracellular Ca^{2+} concentration may have been due to the fact that the Ca^{2+} concentration was higher than that needed to saturate the troponin binding sites. In more recent studies, employing aqueorin or Fura-2 in single toad muscle fibers but with lesser amounts of fatigue, a close relationship between Ca^{2+} concentration and tetanic force has been found (Allen, Lee, & Westerblad, 1989; Lee, Westerblad, &

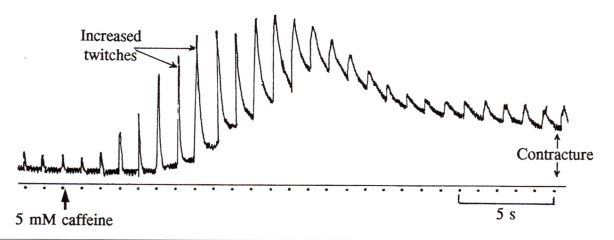

Figure 15.4 Effect of adding caffeine to the fluid bathing a highly fatigued frog muscle fiber. Not only do the twitches become much larger, but a steady tension (contracture) develops. Adapted from Eberstein and Sandow (1963).

Allen, 1991). Figure 15.7 shows one of the sets of recordings made from a single fiber, injected with Fura-2, during fatigue and recovery; it illustrates the parallel changes in Ca^{2+} release and muscle tension.

Lee et al. (1991) also observed a second factor that would contribute to E-C uncoupling; this was a reduced sensitivity of the contractile elements (troponin-C) to Ca^{2+}. These important and revealing experiments are considered again in relation to changes in the background concentration of Ca^{2+} within the cytoplasm during fatigue. Taken with the results of the contracture experiments previously discussed, they show conclusively that the main peripheral factor in the development of muscle fatigue is impairment of E-C coupling.

Fatigued Muscle Fibers Develop Vacuoles

A logical question would be to ask if the T-tubules show any structural changes in fatigued muscle fibers. In single frog muscle fibers, González-Serratos et al. (1978) observed vacuoles in *longitudinal* tubules, and by means of electron probes, these were shown to have higher Na^+ concentrations than the muscle fiber cytoplasm. Transient vacuolation, disappearing within seconds, has also been seen in tetanized rat muscle fibers, but the vacuoles are associated more with the sarcoplasmic reticulum than the T-tubules (Landon, 1982). If the vacuoles do indeed contain Na^+, admitted to the fiber during impulse activity (see Box 15.1), then this response on the part of the fiber may be an example of a nonspecific sequestering role by the sarcoplasmic reticulum; thus, the presentation of sugars such as sucrose and mannitol in the extracellular fluid appears to be dealt with by vacuolation also. As far as the T-tubules are concerned, however, the general consensus is that they do not change in appearance or diameter, despite changes in the ionic composition of the luminal fluid.

Biochemical Changes In Muscle Fibers

In addition to generating less force, the contractile machinery of the fatigued muscle fiber develops tetanic tension more slowly than normal and prolongs the relaxation phase following twitch and tetanic contractions (Figure 15.3). At the same time, there are marked changes taking place in the chemical composition of the muscle fiber cytoplasm (Figure 15.8). These include

- an accumulation of H^+ and lactate from the breakdown of muscle glycogen (see chapter 14);
- a rise in inorganic phosphate (P_i) and in ADP from the splitting of ATP by myosin and membrane ATPases, and an increase in diprotonated phosphate from the buffering of lactic acid by the bicarbonate/phosphate system;
- a fall in phosphocreatine, though with little change in ATP;
- a rise in Ca^{2+} concentration, due to impaired pumping of Ca^{2+} back into the sarcoplasmic reticulum; and
- a gain of water, some of it in the form of vacuoles.

Various Techniques Have Been Used to Measure the Chemical Changes

In animal muscles, it has long been possible to study the chemical events in fatigued muscles by analysis of whole muscles following stimulated or natural contractions. As long ago as 1807, Berzelius detected increased amounts of lactic acid in fatigued muscle (Lehman, 1850). In the case of Ca^{2+} measurements, a recent refinement has been to measure the appearance of free Ca^{2+} in the cytosol of single muscle fibers,

BOX 15.1   THE POTASSIUM CHALLENGE

With each impulse, there is a gain of Na^+ by the muscle fiber and a loss of K^+. The K^+ efflux into the interstitial fluid has been measured with isotopes for single frog muscle fibers (Hodgkin & Horowicz, 1959) and by chemical analysis of the fluid bathing the stimulated rat diaphragm (Creese, Hashish, & Scholes, 1958). Both studies yielded a value of approximately 10 pmol K^+/cm² membrane/impulse. Suppose that a strong isometric contraction is performed; the intramuscular pressure will rise above the arterial systolic level and will prevent the capillary circulation from removing K^+ from the interstitial spaces of the muscle (Barcroft & Millen, 1939). Also, ignore for the moment any effects of the Na^+-K^+ pump. Then if the mean muscle fiber diameter is taken as 50 μm, the interfiber distance as 1 μm, and the excitation frequency as 25 impulses/s, it can be calculated that the interstitial K^+ concentration will rise by 5 mM in only 1 second. Such a rise in K^+ concentration would be large enough to reduce the resting potential of the muscle fiber by approximately 15 mV and to block impulse conduction. Using ion-sensitive microelectrodes, it is now possible to measure the K^+ concentration in the interstitial spaces during or following muscle contractions; although values of 8 to 10 mM are usually found, concentrations as high as 15 mM have been observed in human forearm muscles (Vyskočil, Hník, Rehfeldt, Vejapada, & Ujec, 1983; Figure 15.5).

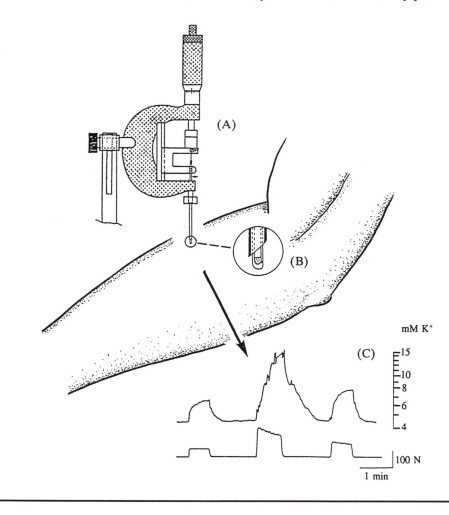

Figure 15.5 Measurement of interstitial [K^+] in intact human muscle. A sharp metal cannula is inserted through a small skin incision into the brachioradialis muscle. Within the cannula is an ion-sensitive glass electrode, the enlarged tip of which is shown in (B); the tip of the electrode can be gently driven out of the cannula into the muscle tissue by a micrometer screw, seen in (A). During voluntary contractions, the interstitial [K^+] rises, the highest value corresponding to the strongest contraction [middle section in (C)].
Adapted from Vyskočil, Hník, Rehfeldt, Vejapada, and Ujec (1983).

Why is it that despite such high K⁺ concentrations, a maximally stimulated muscle, even under ischemic conditions, can conduct a thousand or so impulses in each fiber? Similarly, how is the M-wave (muscle compound action potential) so well maintained during isometric voluntary contractions?

The answer lies in the Na⁺-K⁺ pump. This important molecule quickly enhances its activity when a muscle starts to contract and, because of its electrogenic nature, is able to maintain the resting potential at, or even above, the control level (Figure 15.6). Among its effects, the hyperpolarizing action of the pump is responsible for the enlargement of the M-wave that is seen in the first 30 s of a voluntary or stimulated contraction (*pseudofacilitation*; see Figure 15.10). Since, under the influence of the pump, the resting potential may become -30 mV or so greater than E_K (the potassium equilibrium potential), it follows that there will be a strong electrical gradient across the plasmalemma, causing K⁺ to diffuse back into the fiber from the interstitial space. This passive flux will become greater if, in the contracting muscle, the calcium-gated K⁺ channels and also the ATP-sensitive K⁺ channels open (see chapter 7; also Rudy, 1988).

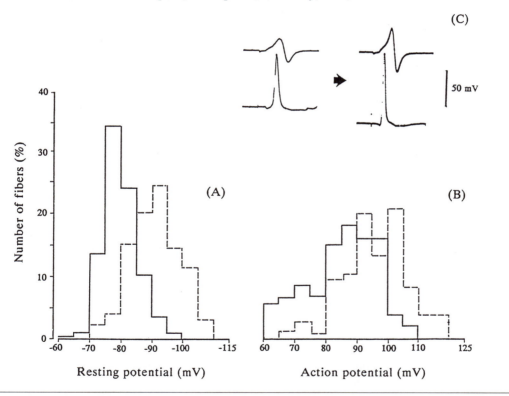

Figure 15.6 Hyperpolarization of rat soleus muscle fibers due to enhanced Na⁺-K⁺ pump activity. In (A), the resting potentials of the fibers are shown before and after tetanic stimulation (solid and dashed lines respectively). Because of the hyperpolarization, the action potentials are enlarged in the stimulated fibers, compared with the resting ones [dashed and solid lines respectively in (B)]. An example of action potential enlargement is shown in (C); the upper trace in each pair of records shows the muscle M-wave, and the distance separating the upper and lower traces equals the resting potential.

Reprinted from Hicks and McComas (1989).

continued

THE POTASSIUM CHALLENGE   *(continued)*

Suppose now that the contraction is submaximal. What would be the effect of the K^+ efflux from the active fibers on the noncontracting ones? It might be thought that the latter would become depolarized in accordance with the GHK equation (see chapter 9). In fact, studies of mammalian muscle show that Na^+-K^+-pump activity is increased in the noncontracting fibers also, and can cause them to hyperpolarize (Kuiack & McComas, 1992). The advantages of such a muscle strategy are obvious, since the previously quiescent fibers will remain available for recruitment if more force is required. Also, the quiescent fibers, by increasing their Na^+-K^+ pumping, will moderate the rise in interstitial K^+ concentration.

What is the stimulus responsible for facilitating pump activity? Probably the most powerful one for the contracting fibers is the impulse-mediated rise in Na^+ concentration in the fibers, because there are many studies, in a variety of cells, that show the importance of intracellular Na^+ in regulating pump activity (see Thomas, 1972). However, it also appears that epinephrine and norepinephrine promote pump activity, and are largely responsible for boosting the pump in the noncontracting fibers (Kuiack & McComas, 1992). It is intriguing that noradrenergic nerve fibers have been observed not only in the walls of the muscle arterioles but also terminating on some of the muscle fibers (Barker & Saito, 1981).

There is a different type of physiological challenge in situations in which a large fraction of the body muscle mass is contracting, as in vigorous cycling or treadmill running. Because the muscle contractions are intermittent, K^+ can enter the capillary circulation from the interstitial spaces. The K^+ efflux may be so large that the arterial concentration may double the resting value (Medbø & Sejersted, 1990); there is then the danger that cardiac excitability will be affected, causing death from ventricular fibrillation or atrial paralysis. Once again, it would seem that the Na^+-K^+ pump comes to the rescue by increasing its activity in muscle fibers throughout the body and assisting in K^+ homeostasis (Clausen, 1990). Evidence for such a response has come from the work of Sjøgaard (1986), who showed that, if only one leg is exercised, the K^+ level in the arterial blood of the other leg may be higher than the venous level, suggesting that the muscle fibers in the resting leg have been extracting K^+ from the circulation. In keeping with the results of Kuiack and McComas (1992), it is probable that catecholamines are implicated in stimulating the pump, especially since epinephrine and norepinephrine are released from the suprarenal gland by elevations in plasma K^+ concentration (Popham, Band, & Linton, 1990).

---

using one of the calcium-luminescent dyes (Lee et al., 1991; see p. 234).

For human muscles a variety of techniques have been employed for sampling metabolites, including the withdrawal of *venous blood* during ischemic contractions and, in some laboratories, the taking of small specimens of muscle (50-100 mg) with a biopsy needle at different times during the fatiguing procedure. The former technique was useful in demonstrating the efflux of H^+, K^+, and lactate during repeated muscle contractions and remains an important diagnostic approach for patients suspected of having myophosphorylase deficiency (McArdle's syndrome). One of the main contributions

of the needle biopsy technique was to show, in fatigued muscle, that there is a great decrease in the amount of phosphocreatine, the substrate necessary for phosphorylating ADP to ATP (see chapter 14). Although there was some controversy initially, the results of biopsy sampling indicate that, even in pronounced fatigue, there is only a modest reduction in the level of ATP. With the advent of *NMR spectroscopy*, it has become possible to investigate changes in the phosphorus compounds of the muscle fibers during fatigue, as well as alterations in H^+ concentration (see the next section). Some of the chemical changes, and their possible effects on the contractile machinery, will now be considered in more detail.

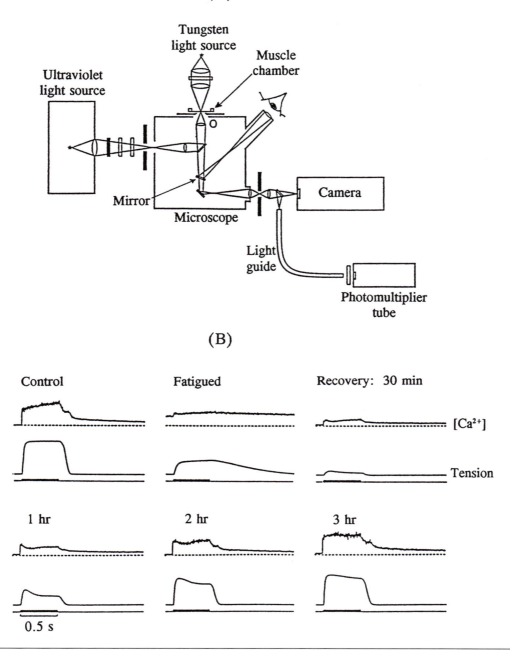

Figure 15.7 Measurement of Ca^{2+} transients in a single amphibian muscle fiber injected with Fura-2 and subjected to tetanic stimulation. (A) The fiber in the muscle chamber is transilluminated by the ultraviolet lamp via the dichroic mirror; the mirror reflects light of one wavelength and transmits light of another. In the muscle fiber, Fura-2 fluoresces in the presence of Ca^{2+}, and the intensity of the fluorescence is measured with the photomultiplier tube. The tungsten lamp is used to view the fiber by conventional microscopy. (B) Comparison of tension and intracellular $[Ca^{2+}]$ in control period, after fatiguing stimulation, and at various times in the recovery period. Note that although the background $[Ca^{2+}]$ is raised at the end of fatigue, only a small amount of Ca^{2+} is released by the tetanus. Both $[Ca^{2+}]$ and tension are depressed for the first 2 hr of the recovery process. (A) and (B) adapted from Lee, Westerblad, and Allen (1991).

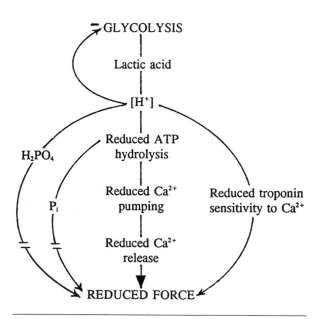

Figure 15.8 Interrelationship of main biochemical changes during development of fatigue.

The Concentration of H⁺ Rises in Fatigue and Can Depress Force Generation

Glycolysis, especially under ischemic conditions, results in the formation of lactic acid. The fall in cytoplasmic pH, due to the accumulation of H^+, can now be detected in contracting muscle fibers *in situ* by means of NMR spectroscopy (see Figure 15.9). The lowest pHs observed in human muscle fatigue are around 6.5. If rabbit muscle fibers are skinned and placed in a fluid medium with a similarly low pH, there is a 30% reduction in the force generated by fast-twitch fibers and a 10% loss in slow-twitch fibers (Donaldson & Hermansen, 1978). The force-reducing effects of H^+ could be produced in several ways:

- By interfering with the actin-myosin cross-bridge cycling and detachments
- Independently of the preceding action, by reducing the sensitivity of troponin for Ca^{2+}
- By inhibiting the enzyme phosphofructokinase and thereby slowing glycolysis

The inherited disorder, myophosphorylase deficiency (McArdle's syndrome; see chapter 14), is relevant to the effects of H^+. In this disorder, the breakdown of muscle glycogen is prevented, and hence there is minimal production of lactic acid and no drop in muscle pH. Nevertheless, the muscles of affected patients fatigue much more rapidly than normal. Does this observation mean that pH changes are of little significance in the fatigue of healthy muscles? Or are other metabolic changes in the myophosphorylase-deficient patient,

such as the unavailability of ATP, so powerful that they anticipate any effects of H^+ concentration?

Inorganic Phosphate and Diprotonated Phosphate Increase in Fatigue

There are two types of change in phosphate concentration in muscle fatigue. First there is a rise in *inorganic phosphate* (P_i) through the splitting of ATP by myosin ATPase; one molecule of ATP is hydrolyzed during each cycle of a myosin cross-bridge (see chapter 11). Additional P_i will, of course, come from the consumption of ATP by ionic pumps in the surface membrane (Na^+,K^+-ATPase) and in the sarcoplasmic reticulum (Ca^{2+}-ATPase). If skinned muscle fibers are exposed to different concentrations of P_i in the bathing fluid, force is reduced by·one third at a value of 10 mM P_i (Cooke & Pate, 1985). Although P_i levels as high as 40 mM can be seen in muscle fatigue, most of the increase occurs relatively early in contraction (within 30 s), by which time there is only a modest loss of force, during either voluntary or stimulated contraction (cf. Cady, Jones, Lynn, & Newham, 1989).

The second type of change in phosphate involves the buffering reaction: $HPO_4^{2-} + H^+ \Leftrightarrow H_2PO_4^-$. At the normal muscle pH of 7.0, the amount of monoprotonated phosphate (HPO_4^{2-}) is approximately twice that of the *diprotonated form* ($H_2PO_4^-$). As the pH falls, however, the proportions reverse so that, at pH 6.5, there is twice as much of the diprotonated form. An increase in diprotonated phosphate to 20 mM was found to reduce force by half (Nosek, Fender, & Godt, 1987). Since this concentration is similar to that observed in fatigue, it is quite possible that the production of diprotonated phosphate is an important factor in the loss of force. Further, there is a satisfactory temporal correspondence between the change in protonated phosphate and the loss of force in human adductor pollicis muscles examined by NMR spectroscopy (Miller, Boska, Moussavi, Carson, & Weiner, 1988).

Phosphocreatine Declines Markedly in Fatigue, but ATP Does Not

One of the earliest results in the biochemical investigation of muscle fatigue was the demonstration of a loss of phosphocreatine (PCr). This finding, initially made on whole animal muscles and human muscle biopsy specimens, has now been confirmed by NMR spectroscopy (Dawson, Gadian, & Wilkie, 1978; Miller et al., 1987; see also Figure 15.9). Since PCr provides the phosphate for the phosphorylation of ADP to ATP, it might be anticipated that the concentration of ATP would become similarly reduced. A loss of ATP would

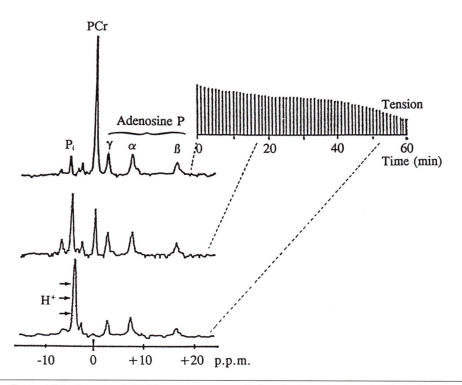

Figure 15.9 The first published report of NMR spectroscopy applied to the investigation of fatigue, by Dawson, Gadian, and Wilkie (1978). Frog gastrocnemius muscles were made anaerobic and were stimulated for 1 s every minute. The trace spectra show the changes in the various phosphorus compounds in the muscle as fatigue sets in. There is a complete loss of phosphocreatine (*PCr*), but substantial amounts of ATP remain (α, β, and γ in figure). Inorganic phosphate (*P$_i$*) increases, due to hydrolysis of ATP, and the corresponding peak is shifted to the right under the influence of the increasing [H⁺]. Adapted from Dawson, Gadian, and Wilkie (1978).

have important consequences for the cross-bridge cycling mechanism, because it would prevent the detachment of the actin filaments from the myosin cross-bridges, and would cause a contracture to develop. However, although there is a slowing of muscle relaxation during fatigue, the affected muscles are still able to relax completely.

Another consequence of ATP deficiency would be the loss of energy for ion pumping, both for Na⁺-K⁺ exchange at the surface membrane and for the return of Ca²⁺ to the sarcoplasmic reticulum. In fatigued muscle, there is certainly evidence of defective Ca²⁺ pumping, since the cytosolic concentration of Ca²⁺ remains elevated (see Figure 15.7B). Indeed, defective pumping could be the major cause of the prolongation of the relaxation period that is so characteristic of fatigued muscle.

The Rate of ATP Hydrolysis Is Reduced in Fatigue

Given these theoretical considerations, the results of ATP measurements are, at first sight, disappointingly normal. Thus, estimates made on muscle biopsies or by NMR spectroscopy show little, if any, change in ATP

concentration. However, Dawson et al. (1978) have stressed that it is not so much the total amount of ATP that is critical, but rather the availability of free energy through the hydrolysis of ATP. Since the reaction products of ATP hydrolysis (ADP and P$_i$) increase in fatiguing muscle, the splitting of further ATP should be reduced, causing a reduction in the phosphorylation potential

$$\frac{[ATP]}{[ADP] \cdot [P_i]}$$

and a shortage of free energy. Dawson et al. (1978) calculated the relative availability of free energy during a fatiguing contraction and found that, although it did not correlate well with the reduction in force, there was good correspondence with the slowing of muscle relaxation. Thus, the latter might result from defective Ca²⁺ pumping brought on by reduced hydrolysis of ATP.

The Background (Resting) Concentration of Ca²⁺ Rises in Fatigue

The fluorometric measurements of intracellular Ca²⁺ have already been considered in relation to the failure of E-C coupling. However, Lee et al. (1991), using Fura-2, made two other significant observations in fatigue. The first was that the slowing of relaxation, which is a characteristic property of fatigued muscle, is associated with, and presumably largely due to, a similar slow fall in Ca^{2+} concentration at the end of each activation (Figure 15.7B). The slowness of the decline in Ca^{2+} concentration could be due to reduced Ca^{2+} pumping by the SR or to saturation of the parvalbumin receptors.

The other finding was that the background concentration of Ca^{2+} in the cytoplasm rises steadily during fatigue (Figure 15.7B). The main factor in producing the rise is not K^+-induced depolarization, but is either increased leakiness of the SR or the reduced uptake of Ca^{2+} by the membrane pumps in the SR. Lee et al. (1991) have pointed out that an increased concentration of cytosolic Ca^{2+} could have important consequences for enzyme reactions in the fibers, in addition to those involved in cross-bridge attachment and release. For example, Ca^{2+} increases muscle metabolism (Solandt, 1936), stimulates protease activity (Turner, Westwood, Regen, & Steinhardt, 1988), and uncouples oxidative reactions in the mitochondria from ATP production (Chapman & Tunstall, 1987). Further, Ca^{2+} deposits in the mitochondria are one of the first signs of fiber degeneration (Wrogemann & Pena, 1976).

Water Content of Muscle Increases in Fatigue

It has been known for a long time that exercising muscles take up water from the plasma (Fenn, 1936). The loss of water from the plasma can be detected by rises in the concentrations of red blood cells and plasma proteins (Sejersted, Vøllestad, & Medbø, 1986). Sjøgaard, Adams, and Saltin (1985) have shown, by chemical analysis of human muscle biopsies, that during submaximal contractions, most of the increase in water content occurs in the interstitial spaces of the muscle. As the effort intensifies, however, water also moves into the fibers. The uptake of water is responsible for the swelling of the muscle during strenuous exercise and, in the short term, is also the cause of the intracellular vacuoles that can be transiently observed under the microscope (see p. 235).

Impaired E-C Coupling Likely Is the Key Peripheral Factor in Fatigue

As we have observed, muscle fatigue is an extremely complex state, and it is therefore useful to summarize the main features before proceeding to consider recovery from fatigue. Leaving aside central causes of fatigue, there are a number of peripheral factors that have been

incriminated. In some cases, the evidence is purely circumstantial and comes from the finding of a change in concentration of a particular metabolite. In other instances, the likelihood of a causal relationship has been strengthened by showing that muscle force is diminished if muscle fibers, especially skinned ones, are exposed to concentrations of metabolites similar to those found in fatigue. Nevertheless, it is possible to produce fatigue, through the use of repeated submaximal contractions, in such a way that there are only minor changes in metabolite concentration (Vøllestad, Sejersted, Bahr, Woods, & Bigland-Ritchie, 1988). This observation and the close correspondence between force and the intracellular Ca^{2+} concentration in fatigue suggest that *impaired E-C coupling* is likely to be the most important of the peripheral factors. This conclusion is reinforced by Eberstein and Sandow's (1963) finding that force can be restored in a fatigued muscle by the application of caffeine. Table 15.1 summarizes the physiological and biochemical changes known to take place in muscle fibers during fatigue.

Recovery From Fatigue

For human subjects and animals, the recovery from fatigue is as important as the development of fatigue. Further, it is possible to learn something more of the interrelationships of factors in producing fatigue by studying their associations as recovery takes place.

Ischemia Prolongs Muscle Fatigue

The recovery from fatigue can be investigated in various ways, but there is an attractive simplicity in studying the aftermath of repetitive stimulation of human muscle performed under ischemic conditions. The first point to make is that, even though the repetitive stimulation has stopped, the manifestations of fatigue remain for as long as the circulation is occluded. That is, there continues to be a diminution in force, a slowing of twitch and tetanic relaxation, a fall in the size of the M-wave (muscle compound action potential), and a reduction in voluntary EMG activity; at the same time, the pain associated with fatigue persists and is felt in the muscle (see p. 245).

Muscle Excitability Recovers Rapidly After Restoration of the Circulation

As soon as the arterial cuff is released, the pain is relieved and is succeeded by a pleasant feeling of warmth in the previously ischemic limb; this sensation is accompanied by a large blood flow well above that

Table 15.1   Summary of Changes in Muscle Fibers During Fatigue

Mechanical	Electrical	Biochemical	
Decline in force Slowed force development Slowed relaxation	Early hyperpolarization Late depolarization Slowed impulse conduction Reduced EMG activity	Increased [Na$^+$] Reduced [K$^+$]	Increased P$_i$ Increased ADP
		Increased [H$^+$] Increased lactate Increased H$_2$PO$_4$	Reduced Ca^{2+} fluxes Reduced Ca^{2+} sensitivity Increased background Ca^{2+}
		Reduced PCr Reduced ATP hydrolysis	Increased H$_2$O

in the resting state (*reactive hyperemia*), due to dilatation of blood vessels in the skin and muscles. The M-wave enlarges rapidly, usually returning to its control amplitude within seconds of cuff release. This increase has been attributed to the "washing out" of K$^+$ from the interstitial spaces of the muscle, producing a rise in E_K, the potassium equilibrium potential, and hence a recovery of the resting and action potential amplitudes

(see Box 15.1). In some muscles, and especially in the human brachial biceps, the M-wave continues to enlarge under the influence of electrogenic Na$^+$-K$^+$ pumping (Figure 15.10), returning to control values after about 15 min. After fatigue produced by tetanic stimulation, there may then be a late depression of the M-wave, due either to reduced Na$^+$-K$^+$ pumping or to a failure of Na$^+$ channel inactivation (Galea, McFadden, Cupido, &

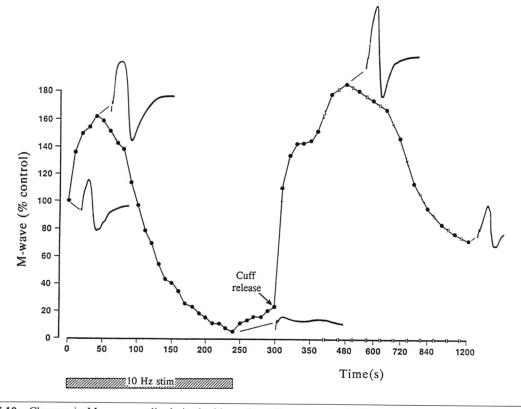

Figure 15.10   Changes in M-wave amplitude in the biceps brachii muscle of a healthy subject, during and following 10-Hz stimulation under ischemic conditions. Immediately after the arterial cuff is deflated, at 300 s, the M-wave rapidly increases, as K$^+$ is flushed out of the interstitial spaces in the muscle. The later increase in M-wave amplitude, at 480 s, is due to enhanced Na$^+$-K$^+$ pumping. Adapted from McComas, Galea, Einhorn, Hicks, and Kuiack (1993).

McComas, 1993). The significance of the late depression is not clear; similar changes of the M-wave have also been observed in single motor units of the rat (Lännergren, Larsson, & Westerblad, 1989). Voluntary EMG activity also increases quickly after cuff release, due to recovery of the individual muscle fiber action potentials and to removal of the reflex inhibition of motoneurons (Box 15.2).

Twitch Tension Recovers More Slowly and Then Declines Again

The changes in muscle force lag behind those in electrical excitability, suggesting the presence of defective E-C coupling during the early part of recovery from fatigue. As the twitch becomes larger, the relaxation phase becomes faster; the rise in tetanic force also becomes steeper. In some individuals, the twitch may become temporarily larger than the control contraction, a finding that has been attributed to persistence of the twitch potentiating mechanism (Garner, Hicks, & McComas, 1989). Following this, however, there is a depression of the twitch, and of the contractile responses to low-frequency stimulation, which may last for 24 to 48 hr (Edwards et al., 1977). This *low-frequency fatigue* is evidently caused by impaired E-C coupling, because higher frequencies of stimulation (50-100 Hz) evoke normal forces.

In single frog muscle fibers, depression of the twitch can also be demonstrated in the recovery period. Fluorometric measurements with Fura-2 show that the reduced force is associated with a smaller delivery of Ca^{2+} from

---

BOX 15.2   AN INHIBITORY REFLEX?

As a steady voluntary contraction is continued, there is a reduction in the electrical impulse (EMG) activity generated by the muscle. To investigate this observation further, recordings have been made from single motor units using fine wires or tungsten microelectrodes or, in specially favorable circumstances, surface electrodes.

Such recordings reveal that the impulse firing frequencies fall from initial values that may be as high as 100 impulses/s to values as low as 10 to 15 impulses/s (see chapter 13). As pointed out by various workers (Bigland-Ritchie, Johansson, Lippold, & Woods, 1983; Marsden, Meadows, & Merton, 1969), this decline in frequency does not, by itself, reduce the forces generated by the motor units. The reason is that the relaxation of the contractile mechanism becomes considerably prolonged during fatigue, such that the twitch half-relaxation time may be from 2 to 3 times the starting value. This prolongation enables a steady force to be maintained with fewer excitations of the muscle fiber than in the rested state. In this respect, the reduction in firing frequency improves the efficiency of the fatiguing motor units; the phenomenon has been aptly described as an example of muscle "wisdom." In addition to the fall in firing frequency, it is probable that some motoneurons stop discharging altogether as fatigue sets in; these are likely to be the ones that have already lost the ability to generate force, due to failure in E-C coupling or in the contractile machinery.

How is the reduction in EMG activity brought about? Brenda Bigland-Ritchie has suggested that there is an inhibitory reflex, such that afferents from the fatiguing muscle depress the excitabilities of the motoneurons supplying the same muscle, thereby rendering them less responsive to motor commands from the brain (Bigland-Ritchie, Dawson, Johansson, & Lippold, 1986).

There are a number of observations which are certainly compatible with this proposal. For example, Bigland-Ritchie et al. (1986) found that, if muscles are fatigued by voluntary contractions under ischemic conditions, the reduction in motor unit firing rate will persist for as long as the ischemic cuff is maintained, suggesting that metabolites are accumulating in the muscle belly and exciting inhibitory afferent fibers. Further, Garland, Garner, and McComas (1988a) showed that voluntary EMG activity was depressed if fatigue had been achieved by electrical stimulation of the muscle, rather than

the SR (Figure 15.7B; see also Lee et al., 1991). Further, normal force can be restored by applying caffeine to the fluid bathing the fibers (Allen et al., 1989), and this substance is known to release Ca^{2+} from the SR. It therefore appears that the low-frequency fatigue during the recovery period resembles the fatigue due to prolonged muscle activity in that both are due to impaired E-C coupling.

APPLIED PHYSIOLOGY

In this section, we will first consider some of the sensations associated with fatigue, not only during the performance of the fatiguing activity but in the next few days as well. Then, in Box 15.2, there is a discussion of an inhibitory reflex which may come into play during fatiguing contractions and may serve to regulate motoneuron firing frequency.

Muscle Pain and Tenderness Are Associated With Severe or Unaccustomed Exercise

During the development of muscle fatigue, especially under ischemic conditions, pain becomes an increasingly prominent distraction and, indeed, may contribute to central fatigue. Part of the pain can be described as "dull" or "hard" and is felt diffusely in the muscle belly; there is another component, with a burning quality, that is localized to the myotendinous region and is exacerbated with each relaxation. During repeated

by voluntary contraction. In these circumstances, the loss of EMG activity could not have been produced by exhaustion or refractoriness of the descending motor pathways but must have required some other mechanism, possibly a reflex. Finally, the Hoffmann (H) reflex testing technique has been used to demonstrate that there is a depression of reflex excitability in human soleus motoneurons after fatigue has been induced by muscle stimulation (Garland & McComas, 1990).

Which afferent fibers might participate in such a reflex? One possibility would be mechanoreceptors in the muscle belly with increased sensitivities to stretch during fatigue. These receptors include not only the muscle spindles and Golgi tendon organs, but also the free nerve endings supplied by the slowly conducting (type III and type IV) nerve fibers (see chapter 4). Although some mechanoreceptors do indeed discharge more frequently during fatigue, others do not, and overall there is a decline in impulse activity (Hayward, Wesselmann, & Rymer, 1991). More plausible candidates for the afferent limb of the reflex would be those metaboreceptors responsive to muscle hypoxia or to increased concentrations of H^+, K^+, bradykinins, and prostaglandins (Mense, 1986; Rotto & Kaufman, 1988).

Last, a note of caution: Although there is appeal in the idea of an inhibitory reflex that unites motoneuron firing rates to the slowing of muscle relaxation during fatigue, the problem with all the studies that have been considered is that none of them provides direct evidence for such a mechanism. The best that can be said is that the experimental data are consistent with the presence of a reflex, and this is perhaps as far as the argument can be carried, at least in human subjects. It is important to recognize that there may be other, equally plausible, interpretations of the evidence. For example, Bongiovanni and Hagbarth (1990) have shown that motoneuron excitation can be enhanced during fatigue if a vibratory stimulus is applied to the muscle tendon. They suggest that the effect of the vibration is to restore some of the reflex activation of α-motoneurons that had been lost through fatigue of the intrafusal muscle fibers in the muscle spindles. Unfortunately, it seems that a clear answer for or against the reflex hypothesis is unlikely to come from human experiments and must await carefully designed animal studies.

contractions without ischemia, pain builds up in those muscles that move joints and are allowed to shorten (*concentric contractions*). In some situations, however, muscles may contract while being lengthened (*eccentric contractions*); examples are the gluteal and quadriceps muscles that support the body weight when walking downstairs, the calf muscle when descending a ladder backwards, and the elbow flexors when a heavy weight is being lowered. In muscles that have been used strenuously in such ways, there is commonly pain and a feeling of stiffness when similar activities are performed on the next day (*delayed-onset muscle soreness* or DOMS). The muscles are also tender to palpation and may feel swollen and rather firmer than usual, due to tissue edema. These symptoms may last several days and, apart from the influx of water in the muscle, have been attributed to stretching of the connective tissue round the muscle belly and in the myotendinous junctions (Jones & Round, 1990). In addition, there may have been damage to the muscle fibers (see chapter 21).

Our study of muscle fatigue is complete. Despite the large number of experimental observations, there are still some uncertainties, and, in particular, it is not clear to what extents the various biochemical perturbations contribute to the loss of force. The answers may come from the combination of results obtained by increasingly refined *in vitro* techniques and further examination of human subjects by NMR spectroscopy. Now we must leave fatigue and consider other adaptations of muscle.

Chapter 16

Loss of Muscle Innervation

Injury to a limb may result in a motor nerve being crushed or divided. Alternatively, nerves may degenerate or become inflamed as part of a disease process. Regardless of the cause, nerve dysfunction has serious consequences not only for muscle fibers but also for the motoneurons and their axons.

Changes in Motor Axons and Neuromuscular Junctions

We will start by considering the changes that take place in an axon following section or crush; it will be shown that the main trunk of a motor axon differs rather sharply from the terminal arborization on the muscle fiber in both the onset and the rapidity of response. Partly for this reason and partly on account of their contrasting functions, the trunk and termination of the axon will be analyzed separately in relation to their degenerative processes. Chapter 17 deals with the remarkable regenerative capacity of the motor axons which, in successful instances, enables them to make new synaptic connections with the muscle fibers.

In the Distal Stump, the Myelin Sheath Retracts and Then Breaks Up

Following nerve section or severe crush, the distal nerve stump remains capable of propagating action potentials for many hours. In rats, usually 24 hr or so elapse before electrical activity is impaired and about 80 hr before all the axons become inexcitable. In humans and in baboons, impulse conduction in some fibers may persist for as long as 200 hr (Gilliatt & Hjorth, 1972; see also Table 16.1). When the axon commences to break up, the process is termed *Wallerian degeneration*, in recognition of Waller (1850), who was the first to describe the changes visible under the light microscope. The observations of Waller have been confirmed on many occasions, and the nature of the degenerative changes has been studied with the electron microscope (a good account is given by Williams & Hall, 1971a, 1971b).

Retraction of myelin at the nodes of Ranvier is one of the earliest changes. This process spreads along the distal stump from the site of the injury and by 1 hr may have involved 20 mm or so of axon. Within the next few hours, the nodes show other evidence of increased metabolic activity in the accumulation of mitochondria and lysosomes. At 24 hr the degenerative changes within

Table 16.1   Time to Conduction Failure Distal to Nerve Section

Species	Nerve	Time to failure (hr)	Authors
Nerve trunk recording			
Rabbit	Peroneal	71-78	Gutmann & Holubár (1950)
Rat	Peroneal	79-81	Gutmann & Holubár (1950)
Guinea pig	Peroneal	72-82	Gutmann & Holubár (1950)
Cat	Sciatic	72-101	Rosenblueth & Dempsey (1939)
Dog	Phrenic	96	Erlanger & Schloepfle (1946)
Baboon	Peroneal	120-216	Gilliatt & Hjorth (1972)
Muscle recording			
Rabbit	Peroneal	30-32	Gutmann & Holubár (1952)
Rat	Sciatic	24-36	Miledi & Slater (1970)
Guinea pig	Sciatic	40-45	Kaeser & Lambert (1962)
Cat	Sciatic	69-79	Lissák, Dempsey, & Rosenbleuth (1939)
Man	Median, Ulnar*	85-128	Landau (1953)
Baboon	Peroneal	96-144	Gilliatt & Hjorth (1972)
Man	Facial	120-192	Gilliatt & Taylor (1959)

*From observation of muscle twitch—no electrical recording.
Note. Adapted from Gilliatt and Hjorth (1972).

the *axon cylinder* are well advanced, with disruption of the microtubules, endoplasmic reticulum, and neurofilaments. Another early change, already visible within a few minutes of the nerve lesion, is extreme dilatation of the Schmidt-Lanterman incisures.

The myelin sheath, having already withdrawn slightly at the nodes of Ranvier, now begins to break up, with the lamellae peeling off at the nodes. Both at the nodes and at the dilated Schmidt-Lanterman incisures, the axons undergo progressive constriction; eventually a series of large *ellipsoids* are formed, each being bounded by the degenerating myelin sheath and cut off from the remainder of the axon. In time the ellipsoids become partitioned internally by unfolding of the myelin sheath, while at the ends of the ellipsoids, small globules of degenerating myelin become pinched off.

The Schwann Cells Multiply and Help to Digest the Myelin Sheath

The *Schwann cells* become highly active during the degenerative process, first spreading over the denuded nodes of Ranvier and then undergoing mitotic division so as to form a maximum number of cells at about 15 to 25 days. The Schwann cells hasten the disintegration of the myelin sheath and engulf some of the lipid droplets. It is probable that a portion of these cells are transformed into *macrophages*; however, autoradiographic studies have shown that the majority of macrophages are carried to the degenerating nerve in the bloodstream (Olsson & Sjöstrand, 1969). These hematogenous cells are first seen at about 3 days following the lesion, and they are especially active in removing the degenerating myelin. The macrophages, and the myelin debris that they ingest, also make the Schwann cells multiply (Baichwal, Bigbee, & De Vríes, 1988). With the passage of time, the lipid-laden macrophages appear to migrate from the endoneurial spaces toward the periphery of the nerve trunk.

If the axon has been crushed rather than cut, the surrounding basement membrane will remain intact and act as a scaffolding for the columns of dividing Schwann cells (bands of Bungner).

If the macrophage invasion is delayed, as it is in a mutant strain of mouse described by Lunn, Perry, Brown, Rosen, and Gordon (1989), the degeneration of the fibers is slowed considerably, so that action potentials can be conducted for up to 14 days, compared with the 3-day period for transected axons in normal mice.

In the Proximal Nerve Stump, the Axis Cylinders Shrink

While the distal nerve stump is degenerating, dramatic changes start to take place in the proximal stump. Gordon, Gillespie, Orozco, and Davis (1991) investigated these changes by implanting stimulating and recording electrodes around the sciatic and common peroneal nerves of cats, and then cutting the nerves distally. They found that the amplitude of the compound action potential in the proximal stump begins to decrease within a few days of nerve section, and by 80 days is less than 20% of the initial value (Figure 16.1). Similar reductions have been observed in the ulnar nerve stumps of patients who had suffered amputation of the hand months or years previously (McComas, Sica, & Banerjee, 1978). In the study by Gordon et al. (1991), there was also considerable atrophy of the axons (from a mean fiber diameter of 6.6 μm to one of 4.4 μm at 227 days); neither the atrophy nor the reduced action potential could be prevented by chronic intermittent stimulation of the proximal nerve stump. Since the atrophy is greatest in the largest axons, it has the effect of converting the normal bimodal distribution of fiber diameters into a unimodal one (Figure 16.2). Not only are the nerve fibers smaller, but many appear to have collapsed after distal nerve section, assuming a flattened rather than a circular profile on cross section; in some the axis cylinder is detached from the myelin sheath (Gillespie & Stein, 1983). In view of the proportionality between impulse conduction velocity and fiber diameter (see chapter 9), it would be expected that the impulse conduction velocities would be lower following nerve section, and a number of studies have shown that this is indeed the case (for example, Milner & Stein, 1981).

The Onset of Degeneration in the Motor Nerve Terminals Depends on Stump Length and Species

It is an interesting observation that, after an axon has been divided, the first severe degenerative changes are seen at the neuromuscular junction rather than in the separated stump of axon. The onset of the changes in the axon terminal depends on two factors: the length of the distal stump of axon and the species of animal affected. Miledi and Slater (1970) studied the effect of stump length in the rat by cutting the phrenic nerve either as close as possible to the diaphragm or else in the neck. They found that when the nerve is divided distally, about 8 hr elapse before there is any evidence of neuromuscular degeneration in the diaphragm. When the greater length of nerve is left, the onset of neuromuscular failure is delayed by approximately 1 hr for each 1.5 cm of axon remaining. This intriguing relationship between the length of the nerve stump and the onset of denervation phenomena is of significance for an understanding of the nature of the neurotrophic controlling system (see chapter 18).

As already stated, the onset of end-plate failure also depends on the species studied, being much later in humans and baboons than in smaller mammals such as rats (see Table 16.1). Once the degenerative changes start, however, they proceed rapidly, and only 3 to 5 hr are required for complete disruption of the end-plate. During this time, the synaptic vesicles form clumps; the mitochondria swell, and their cristae break up into small vesicular pieces; membrane whorls and glycogen bags appear in the axoplasm, and lysosomes become evident. At this stage the whole end-plate may become fragmented; meanwhile, the Schwann cell sends a cytoplasmic process into the primary synaptic cleft, and this completely envelops the degenerating end-plate. It is probable that the Schwann cell assists in the final dissolution of the end-plate. Subsequently the Schwann cell withdraws, and after a month or so, only the sole-plate is left of the original neuromuscular junction.

The electrophysiological behavior of the end-plate parallels the morphological changes described above. For the first 8 hr or more, depending on the nerve stump length, there is the usual spontaneous discharge of MEPPs, and electrical stimulation of the axon results in normal neuromuscular transmission. From these observations it would appear that the synaptic vesicles contain normal numbers of ACh molecules and that there is no impediment to the mobilization and release of transmitter from the vesicles. After this latent period is over, neuromuscular function fails abruptly; the MEPPs cease, and no postsynaptic response can be detected following nerve stimulation.

Intact Neuromuscular Junctions May Cease Transmission (Silent Synapses)

Following injury or disease, muscle fibers may become inexcitable without other features of denervation, such as spontaneous fibrillation and positive sharp wave potentials (see p. 253). In such instances, when excitability returns, it does so without any electrophysiological signs that the motor nerve fibers have grown back or undergone remyelination. Thus, instead of the muscle evoked responses being temporally dispersed, as happens when there has been regeneration or remyelination, the responses retain normal configurations.

This peculiar behavior has occasionally been seen following crush or stretch injuries to peripheral nerves or nerve roots (McComas, Jorgensen, & Upton, 1974; McComas, 1977) and, more frequently, in patients recovering from overactivity of the thyroid gland (thyrotoxicosis; McComas, Sica, McNabb, Goldberg, & Upton, 1974). To explain these puzzling findings, it has been necessary to postulate the existence of *silent synapses*, that is, of neuromuscular junctions that are

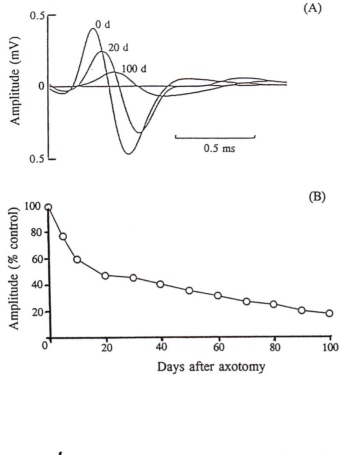

(A)

(B)

Figure 16.1 (A) Recordings of compound action potentials in a cat common peroneal nerve at 0, 20, and 100 days after distal section. (B) Mean sizes of compound action potentials at different times after distal nerve section. Adapted from Gordon, Gillespie, Orozco, and Davis (1991).

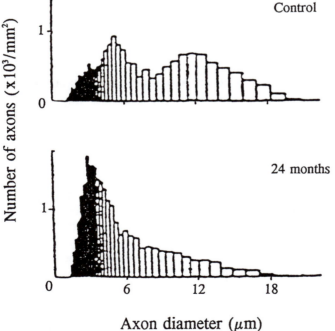

Control

24 months

Figure 16.2 Diameters of myelinated fibers in an intact cat sciatic nerve and in a nerve examined 24 months after distal section. Adapted from Dyck, Nukada, Lais, and Karnes (1984).

largely intact but in which there is insufficient depolarization of the muscle fiber membranes to generate action potentials. Under more natural circumstances, the phenomenon is seen when muscle fibers have been multiply innervated during normal embryonic development (see chapter 6, Box 6.1). In animal experiments, silent synapses have been produced by irradiating motoneurons in the spinal cords of rats

(Fewings, Harris, Johnson, & Bradley, 1977) and by immobilizing cat hindlimbs (Robinson, Enoka, & Stuart, 1991); in the latter situation, stimulating single motor axons elicits no force, or very small responses (see Figure 19.5 in chapter 19; also McComas, De Bruin, & Quartly, 1991).

Changes in Muscle Fibers

If the nerve to a muscle is severed, the muscle will gradually waste over a period of weeks. This process is termed *denervation atrophy*; it demonstrates that the muscle fibers are dependent upon the motoneuron for the maintenance of their normal structure. This sustaining action depends on the contractile activity, which the motoneuron induces through impulse bombardment of the muscle fiber. This is not the whole story, however, for the motoneuron also sends chemical messages to the muscle fiber through the microtubules and neurofilaments that run along the axon (see chapter 8). These influences are referred to as *trophic*; it will be seen in chapter 18 that the trophic influence is mutual, because the motoneuron is itself affected by the muscle fibers to which it is connected. The gross changes which take place in a muscle following denervation have been studied on many occasions (see, for example, Adams, Denny-Brown, & Pearson, 1962; Sunderland & Ray, 1950; Tower, 1939). The changes probably involve every part of the muscle fiber; some of them have been summarized in Figure 16.3.

Denervation Causes Marked Atrophy of All Muscle Fibers

The most obvious change in the muscle following denervation is its reduction in size. This atrophy can be detected at about the 3rd day and is rapid in the ensuing 2 months. At the end of this period, only 20% to 40% of the original muscle mass remains (Sunderland & Ray, 1950); much of this will be connective tissue, since the latter accounts for 10% to 25% of the initial muscle weight. Although further atrophy may occur, it is now a much slower process. According to Stonnington and Engel (1973), the denervation atrophy affects red and white fibers equally. It is of interest that in the hemidiaphragm denervation atrophy is preceded by a 1- to 2-week phase of hypertrophy, in which the cross-sectional areas of the fibers are almost doubled and new myofibrils are formed (Miledi & Slater, 1969). This curious response is due to excessive stretch of the denervated half of the diaphragm by the contractions of the innervated remainder.

Myonuclei Take Up Central Positions in the Fibers

As early as the 2nd day after denervation, the myonuclei start to become rounded instead of narrow and elongated; at the same time, the nucleoli become enlarged and more prominent. Many of the nuclei then move into the centers of the fibers where they line up to form chains. There is dispute as to whether the number of nuclei actually increases; much of the apparent preponderance of the nuclei is due to their preservation within atrophying fibers, but a true increase has also been reported (Bowden & Gutmann, 1944). In late atrophy (e.g. 6 months), all that remains of some fibers are chains or clumps of nuclei surrounded by thin cylinders of cytoplasm.

Necrosis (Death) of Fibers May Also Occur

In addition to the changes of "simple" atrophy, previously described, a variable proportion of the muscle fibers undergoes further degenerative changes after several months. These changes are probably irreversible and may result in the death of the fiber; not uncommonly they are restricted to part of a fiber. The affected fibers swell; the nuclei also enlarge and then begin to fragment. Vacuoles appear in the cytoplasm of the fiber, and elsewhere the cytoplasm stains darkly because of increased basophilia. The cross-striations become less distinct, and the plasmalemma thickens before disintegrating. Mononuclear cells appear at the site of necrosis and subject the accumulating fiber debris to phagocytosis. Although complete destruction of the muscle fiber may eventually occur, Schmalbruch, Al-Amood, and Lewis (1991) have suggested that, in long-term denervation, there are repeated cycles of regeneration and necrosis, with the necrotic phase occupying only 2 days or so. As the muscle fibers atrophy or degenerate, the connective tissue within the muscle becomes increasingly prominent, with large fat cells occupying spaces between the surviving fibers and fiber bundles.

Ultrastructural Studies Show That All Components of the Muscle Fibers Become Smaller

Several excellent studies have been published of the fine structure of the denervated muscle fiber, as examined with the electron microscope (Miledi & Slater, 1969; Pellegrino & Franzini, 1963; Schmalbruch et al., 1991; Stonnington & Engel, 1973). The account by Stonnington and Engel is notable for the measurements of cross-sectional areas of the various fiber organelles (see Figure 16.4). In this last study, the soleus and superficial medial gastrocnemius muscles of rats were

INNERVATED

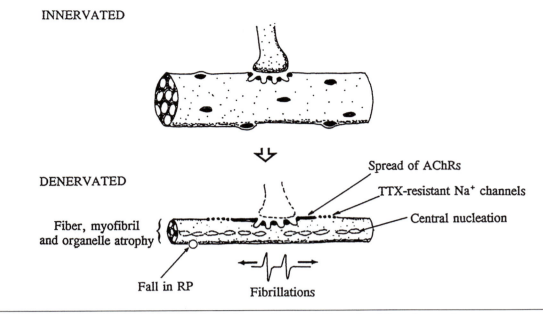

DENERVATED

Fiber, myofibril
and organelle atrophy {

Spread of AChRs

TTX-resistant Na⁺ channels

Central nucleation

Fall in RP

Fibrillations

Figure 16.3 Some of the structural and functional changes in muscle fibers that follow denervation.

studied between 1 and 84 days after section of the sciatic nerve. Such studies show that, during the period of atrophy, the change in fiber area is matched by a fall in the mean area of the myofibrils. The atrophy begins at the periphery of a myofibril, but after the 1st month degeneration also becomes visible in the interior. During the 1st week the mitochondria in both red and white fibers enlarge in the longitudinal axis of the fiber. Subsequently, the organelles shrink and form clusters; some mitochondria undergo frank degeneration and inclusion in autophagic vacuoles. Similarly, the sarcoplasmic reticulum at first enlarges and then diminishes, though to a lesser extent than the fiber itself. Other changes observed with the electron microscope include abnormalities of the Z-disc, irregularities and small papillary projections of the plasmalemma, and focal dilatations of the transverse and sarcoplasmic tubules. Increased numbers of ribosomes can be found between the myofibrils and under the surface membrane.

Enzyme Activities Decrease in the Denervated Fibers

There have been several studies of the effect of denervation on the biochemical and histochemical properties of muscle, and those of Romanul and Hogan (1965) and of Hogan, Dawson, and Romanul (1965) in the rat are particularly thorough. There is a rapid decrease in the high enzyme activities typical of a particular type of fiber; as a result fiber types can no longer be distinguished. In addition, the red muscle fibers (as in soleus) become paler, partly due to a reduction in myoglobin concentration. Gundersen, Leberer, Lømo, Pette, and

Staron (1988) have demonstrated, in rat soleus and extensor digitorum longus muscles, large reductions in the activities of creatine kinase, several glycolytic enzymes (glycogen phosphorylase, lactic dehydrogenase, phosphofructokinase, pyruvate kinase, and glyceraldehyde-3-phosphate dehydrogenase), and two enzymes of the citric acid cycle (citrate synthase and malate dehydrogenase). The only exceptions to these decreases are in the soleus muscles, in which slight increases in Ca^{2+}, Mg^{2+}-ATPase, and parvalbumin occur.

Twitch Contractions Become Slower in Denervated Muscles and the Tetanus-Twitch Ratio Declines

Even though the nerve supply to a muscle may have been interrupted, it is still possible to stimulate the muscle directly. As would be expected, the tensions developed during a single twitch or during a tetanus are reduced, for the muscle will be undergoing denervation atrophy and will have lost some of its myofibrils (see p. 251). In addition the twitch becomes significantly slower, and this is true for a slow-twitch muscle such as soleus as well as for fast-twitch muscles like the flexor hallucis longus (FHL). Lewis (1972) has shown that the slowing occurs relatively abruptly during the 3rd week of denervation; even in chronically denervated animals, however, the twitches of the fast muscles remain considerably faster than those of slow muscles studied in control animals. Lewis also found that in both types of muscle the tetanus-twitch ratio fell and that in the FHL, but not in the soleus, the maximum rate of rise

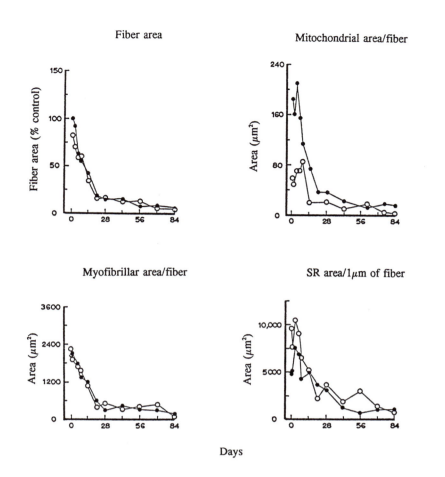

Fiber area

Mitochondrial area/fiber

Myofibrillar area/fiber

SR area/1μm of fiber

Days

Figure 16.4 Quantitative changes in "red" (●, soleus) and "white" (○, superficial medial gastrocnemius) muscle fibers of the rat following denervation (see text). Adapted from Stonnington and Engel (1973).

of tetanic tension was somewhat diminished. Although some of the slowing of twitch might have been due to a reduction in impulse conduction velocity, the most satisfactory way of reconciling the various findings was to attribute them to an increase and prolongation of active state following a single stimulus (see chapter 11). This, in turn, could have been due to such factors as the increased duration of muscle fiber action potential and the greater proportion of sarcoplasmic reticulum relative to the myofibril mass in the atrophied fibers.

The Electrical Properties of the Muscle Plasmalemma Change After Denervation

Following denervation important changes take place at the surface of the fiber in addition to those in the fiber interior. The irregularity of the plasmalemma and the development of small papillary projections have already been described. Their relationship to the altered electrophysiological properties of the sarcolemma described below is uncertain. These changes occur:

1. First, there is a *fall in resting membrane potential* at 2 hr (Albuquerque, Schuh, & Kauffman, 1971). This depolarization begins at the end-plate and spreads outward, eventually amounting to about 20 mV; it is caused

by inhibition of the Na$^+$-K$^+$ pump (Bray, Hawken, Hubbard, Pockett, & Wilson, 1976).

2. The *permeability* of the muscle fiber membrane alters, becoming smaller for both K$^+$ and Cl$^-$ (Klaus, Lüllmann, & Muscholl, 1960; Thesleff, 1963).

3. The *muscle action potential*, which can be initiated by direct stimulation of the muscle fiber, has a lower rate of rise and a longer duration. Unlike the situation in a normal fiber, the action potential can still be elicited in the presence of tetrodotoxin (TTX; see Figure 16.5); this resistance to TTX is first evident at about 36 hr (Harris & Thesleff, 1972). The impulse has a reduced propagation velocity, and the refractory period that follows is prolonged.

4. The *sensitivity of the muscle fiber to ACh*, though remaining highest in the end-plate region, after about 24 hr begins to spread out to involve the remainder of the fiber membrane (Axelsson & Thesleff, 1959). This development results from the formation of new acetylcholine receptors in the previously inert membrane.

5. After an interval of about 1 week, the muscle fiber membrane becomes spontaneously excitable; the impulses are termed *fibrillation potentials* (see Figure 16.6D). These impulses are fully developed action potentials that recur with a frequency of 0.5 to 3 Hz.

(A)

Ringer   Ringer + TTX

(B)

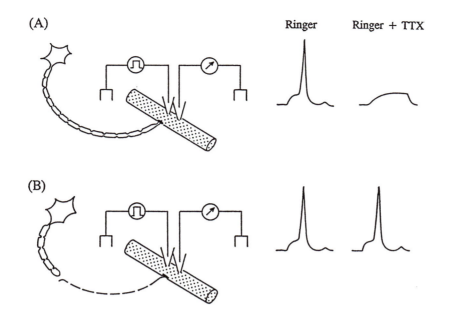

Figure 16.5 The development of tetrodotoxin (TTX) resistance following denervation. (A) A still-innervated muscle fiber has been impaled by stimulating and recording microelectrodes. In a standard Ringer solution, the fiber responds to direct stimulation with an action potential, but after adding TTX, only an electrotonic local potential is seen. (B) Following denervation, an action potential can be elicited by direct stimulation in the presence of TTX.

Belmar and Eyzaguirre (1966) analyzed this activity using an extracellular electrode at different points along the fiber and, from the configuration of the potentials, argued that it must have arisen at the site of the original end-plate. Further evidence supporting this conclusion was that they could alter the frequency of the fibrillations by passing a current through the membrane in this region but not elsewhere. Intracellular recordings have shown that this region of the membrane is electrically unstable, developing spontaneous *oscillations* which may build up and become large enough to initiate action potentials (Purves & Sakmann, 1974; Thesleff & Ward, 1975; Figure 16.6C). A second type of generator activity consists of irregularly occurring sudden depolarizations (*fibrillatory origin potentials* or FOPs). These last potentials are sometimes too small to trigger action potentials (Figure 16.6, A & B) and may possibly correspond to the *positive sharp waves* (Figure 16.6E) that can be recorded with a coaxial EMG electrode from denervated muscle. This second type of activity is also encountered most commonly at the old end-plate zone but, like the oscillatory activity, can sometimes occur elsewhere. Since fibrillation potentials persist after curarization, the underlying membrane depolarizations cannot have resulted from the action of circulating ACh. The spontaneous depolarizations do, however, depend on transient increases in the Na^+ permeability of the membrane, for they are abolished by removing Na^+ from the bathing solution or by applying TTX.

6. The concentration of *cholinesterase* in the end-plate falls by a variable amount that, in the rat sternomastoid muscle, is as much as 50 to 70% of the original value; this change takes place in the first few days following denervation (Guth, Albers, & Brown, 1964).

7. Fast-twitch muscles become sensitive to *caffeine* and respond to this drug with a contracture (Gutmann & Sandow, 1965).

8. The denervated muscle fiber stimulates neighboring intact motor axons to *sprout*. Unlike the situation in an innervated muscle fiber, the denervated muscle fiber membrane becomes receptive to the arrival of an axonal sprout and will permit a new neuromuscular junction to form. Usually the site of the original end-plate is chosen, but if this is prevented by mechanical factors, other regions of the fiber engage in synaptogenesis.

APPLIED PHYSIOLOGY

Of the hundreds of causes of peripheral nerve damage and disease, one of the most fascinating is an autoimmune disorder termed *idiopathic polyradiculoneuritis* or, in honor of those who described it first, the *Guillain-Barré Syndrome* (GBS). This condition, which can occur at any age, may arise either spontaneously or, in approximately half of the cases, following a viral infection some 2 to 4 weeks earlier.

The Guillain-Barré Syndrome, Which May Have a Precipitating Event, Is an Autoimmune Disorder

In some cases the precipitating event may be a surgical operation or an immunizing procedure; many cases occurred after the use of an influenza vaccine prepared from pigs (Schonberger et al., 1979). In addition to the

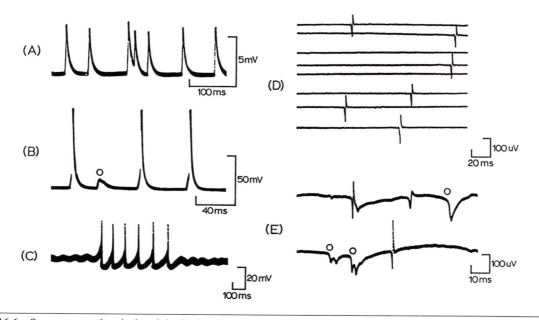

Figure 16.6 Spontaneous electrical activity in denervated muscle fibers. (A, B, & C) show intracellular recordings from denervated rat diaphragm muscle maintained in organ culture. (A) Subthreshold FOPs (see text). In (B), the FOPs are large enough to initiate action potentials except on one occasion (O); note that the amplification is lower than in (A). (C) Spontaneous oscillation of muscle fiber membrane potential, resulting in a burst of action potentials. Adapted from Purves and Sakmann (1974). (D & E) Recordings made with coaxial needle electrodes from two patients with severe denervation. Fibrillation potentials are evident in both records, but in (E) there are also simple and complex positive sharp waves (O).

suggestive history, evidence that the disorder is immunologically based comes from the finding of circulating antibodies to peripheral nerve (Melnick, 1963). Further, sera from patients will cause unfolding and degeneration of the myelin sheaths of axons maintained in tissue culture (Cook, Murray, Whitaker, & Dowling, 1969; Dubois-Dalcq, Buyse, Buyse, & Gorce, 1971) or of sciatic nerve fibers following intraneural injection in rats (Feasby, Hahn, & Gilbert, 1982).

The circulating lymphocytes can also be shown to be myelinotoxic (Arnäson, Winkler, & Hadler, 1969) and will transform on exposure to peripheral nerve antigen (Knowles, Currie, Saunders, Walton, & Field, 1969). Finally, there is the finding of immunoglobulin deposits and of lymphocytes in the nerve trunks of affected patients. The likeliest hypothesis for the pathogenesis of the disorder is that, during the preceding injection, viruses enter the Schwann cells and replicate. The emergent viruses may then contain Schwann cell antigen, presumably myelin basic protein or ganglioside, on their outer surfaces; alternatively, the virus and myelin sheath may possess an identical antigenic complex. At any rate, the exposed antigen promotes the production of antibody and lymphocytes by the patient; these then react with the myelin sheath and cause its degeneration. The immunological disorder is discussed further in relation to

experimental allergic neuritis, which closely resembles GBS in humans (see Box 16.1 and also Rostami, 1993; Willison & Kennedy, 1993).

GBS Patients Develop Ascending Paralysis and Numbness

Patients who are developing GBS neuropathy typically become aware of numbness and weakness of the legs. These symptoms start in the feet and then ascend to involve the remainder of the legs; a day or so later similar symptoms commence in the hands and spread along the arms. In some cases there may be involvement of the cranial nerves and the spinal cord. Occasionally the progression of the disease is so rapid and severe that patients require urgent tracheostomy and artificial ventilation to compensate for the paralysis of the respiratory muscles. Recovery usually starts in 2 to 4 weeks and may eventually be complete, but in some patients it is delayed and some numbness and weakness persist.

One useful diagnostic test is the examination of the cerebrospinal fluid, for in about half of the cases there are significant rises in protein without any accompanying increases in white cells. Nerve conduction studies are easy to do and will demonstrate involvement of peripheral nerves; the reduced impulse conduction velocities indicate that demyelination has taken place. In

BOX 16.1   A GBS-LIKE ILLNESS CAN BE PRODUCED IN ANIMALS

Further insights into the Guillain-Barré type of nerve disorder have come from the production of an animal model. In 1955 Waksman and Adams reported that they had been able to induce an inflammatory disorder of peripheral nerves and nerve roots by an immunological method, and they noted its resemblance to the Guillain-Barré syndrome. Their technique was to make a suspension of rabbit sciatic nerve and to inject small samples, together with Freund's adjuvant (containing heat-killed tubercle bacilli), into the footpads of other rabbits (Figure 16.7). They observed that after 12 days nearly all the injected animals developed the clinical signs of a polyneuropathy. The animals "tended to lie in a splayed position, with all extremities extended and the head resting on the floor. . . . In hopping they were unsteady and erratic. . . . Upon landing they would often stagger or lurch to one side. . . . The musculature of extremities and trunk was weak and slack." Most of the affected animals recovered spontaneously after several weeks of illness.

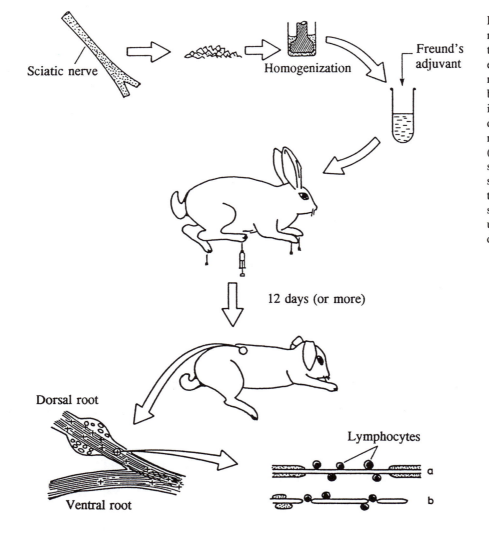

Figure 16.7   Experimental procedure for the induction of experimental allergic neuritis (see text). At bottom left, the relative involvement of the dorsal, ventral, and mixed roots is indicated (+). At bottom right is shown a fiber with segmental demyelination (*a*) and a more severely affected fiber undergoing Wallerian degeneration (*b*).

In the Animal Model, One of the Myelin Basic Proteins Is the Antigen

These important results have been confirmed in many laboratories, and it is now possible to understand in more detail the immunological mechanisms involved in the pathogenesis. In the first place, it appears that the antigenic component of peripheral nerve is one or both of the two basic proteins contained in myelin. The chemical compositions of these proteins have now been determined. The P1 protein has a molecular weight of 18 kD and contains 168 amino acids; it is identical to one of the two basic proteins found in myelin from the central nervous system (Brostoff & Eylar, 1972). The antigenic property of the P1 protein is confined to a small part of the molecule containing only nine amino acids (Westall, Robinson, Caccam, Jackson, & Eylar, 1971). When injected by itself, the P1 protein, or the critical nine-amino acid residue, will induce a demyelinating disorder in both the central and peripheral nervous systems. However, if *intact* peripheral nervous system (PNS) myelin is administered instead, only the peripheral nerves and roots are affected, and the CNS is spared. Wísniewskí, Brostoff, Carter, and Eylar (1974) have explained this curious result by suggesting that the antigenic site on the P1 protein is either hidden within the PNS myelin substructure or is conformationally altered so as to be immunologically inert. They further propose that it is the second basic protein of PNS myelin, the P2 protein, that is responsible for experimental allergic neuritis. This protein has a molecular weight of 11 to 12 kD and contains 101 amino acids; when purified and injected with Freund's adjuvant into an animal, it can be shown to produce an inflammatory polyneuropathy (Brostoff, Burnett, Lambert, & Eylar, 1972).

Both Lymphocytes and Antibodies Participate in Nerve Destruction

There is good evidence that, although antibodies to myelin protein can be demonstrated in experimental allergic neuritis (EAN), the degenerative changes are mostly cell mediated. For example, lymphocytes removed from an affected animal will produce demyelination when injected into another animal (Astrom & Waksman, 1962), and they are also active when applied to myelinated axons in tissue culture (Lampert, 1969). According to Arnason et al. (1969), the lymphocytes cross into the nerve parenchyma through the small veins and, upon exposure to myelin antigen, are transformed into larger cells. These then undergo repeated mitoses before commencing the demyelinating process.

With the electron microscope, the mononuclear cells can be seen to have penetrated the basement membranes of the Schwann cells with processes that eventually surround the myelin sheaths. The Schwann cells are now isolated from their axons but are not themselves attacked by the mononuclear cells. Instead, the mononuclear cells appear to prize apart the tightly bound spiral of the myelin sheath and cause it to undergo a bubbly dissolution (Lampert, 1969). Eventually, the myelin becomes completely stripped from the axis cylinder and is then removed by phagocytes (Figure 16.7, bottom).

In most animals recovery takes place and is achieved by remyelination of the affected axons. In a small proportion of animals, the demyelinating process assumes a chronic course. In such animals the nerve fibers become abnormally thickened so as to form *onion bulbs*; the thickening is caused by excessive numbers of Schwann cells and their processes (Pollard, King, & Thomas, 1975).

(A)

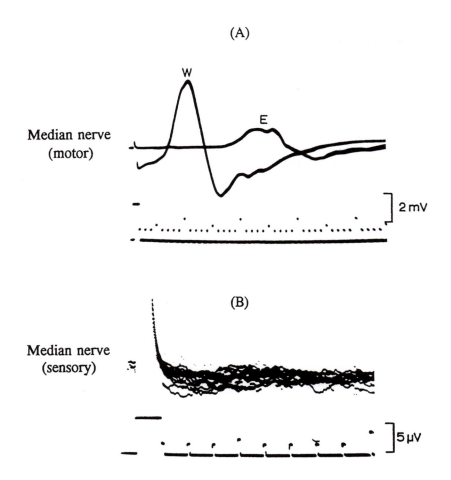

Median nerve
(motor)

W

E

2 mV

(B)

Median nerve
(sensory)

5 µV

Figure 16.8 Electrophysiological findings in a patient with Guillain-Barré syndrome. (A) Recordings made from the thenar muscles with surface electrodes following stimulation of the median nerve at the wrist (response *W*) and at the elbow (response *E*). (B) Absence of detectable sensory nerve action potential when the thumb was stimulated maximally and the recordings were made from the median nerve at the wrist with surface electrodes. For significance of findings, see text.

Figure 16.8 there is evidence not only of slowed impulse conduction, but of complete block in a proportion of the median nerve fibers, presumably because of a demyelinating plaque in the forearm.

As we have seen, paralysis is only one of the consequences when a skeletal muscle is deprived of its in-nervation. Profound as these denervation changes are, they can be reversed, partially or completely, if a new nerve supply can be obtained by the muscle. The next chapter describes the events that take place in the muscle and nerve fibers during reinnervation and the possible stimuli that bring these changes about.

Chapter 17

Recovery of Muscle Innervation

Muscles can recover from denervation in different ways. Either the motor nerve will *regenerate* and grow back to the muscle or, in the case of partial denervation, surviving motor axons will sprout and establish synaptic connections with the denervated muscle fibers, a process termed *collateral reinnervation*. Both of these processes will now be considered.

Nerve Regeneration

If an axon is divided and viewed under the microscope, axoplasm can be seen to bulge out of the end of the central stump; this extruded material probably reflects the continuing slow flow of axoplasm from the cell body (see chapter 8). After a few minutes the extrusion ceases, possibly because some of the myelin lamellae have slipped over each other and sealed the exposed end of the cut axon. The end of the fiber now gradually swells; initially this is the result of degenerative changes, but the later distension is due to the formation of a *growth cone*. This important structure was encountered in chapter 6, in which the outgrowth of axons to their targets in the embryo was discussed. Studies of

nerve regeneration have revealed more of the properties of the growth cone.

The Growth Cone Is Formed at the Cut End of the Axon by Local Mechanisms

The growth cone still forms if the divided axon is separated from the cell body by a second cut. Hence there must be local mechanisms at the tip of the divided axon that control the formation of the cone (Brown & Lunn, 1988). The changes that subsequently take place in the growth cone were described with memorable clarity by Cajal (1928), who also summarized much of the earlier work in this field. In most of his studies, Cajal stained the regenerating fibers with silver. His drawings are not only elegant but display his findings in remarkable detail, especially when it is remembered that all the observations were made with an ordinary light microscope. Within 24 hr of nerve section, the growing axons were seen to have penetrated into the exudate produced by the wound; sometimes the dilated endings remained single, but in other instances they divided into two or more processes. In some nerve fibers additional sprouts formed above the cut ends at nodes of Ranvier.

GAP-43 and Intracellular Ca²⁺ Help to Direct the Growth Cone

Studies with the electron microscope have shown that the growth cones are rich in mitochondria, vesicles, dense bodies, and lamellar figures. The growth cones also contain a novel protein, *GAP-43* (or *B-50*), which is synthesized by the neuronal cell body and then conveyed to the tip of the growing axon. GAP-43 has a molecular weight of about 24 kD, and antibodies labeled with gold show that it is associated with the membrane. Exactly what GAP-43 does in the plasmalemma is not clear, but one suggestion is that it transforms the interactions of receptors with external guidance factors within the cytoplasm of the growth cone (see below). This signal would, in turn, control the laying-down of new cytoskeleton in the cone (Gordon-Weeks, 1989).

Schwann Cells Provide Pathways for Regenerated Axons

Initially the growth cones are free of surrounding Schwann cell cytoplasm, but if they encounter the peripheral nerve stump, they will grow preferentially down the columns formed by the Schwann cells of the stump (*bands of Bungner*). The growing axon tips are guided down the stump by the same guidance factors that play similar roles in embryogenesis. These molecules include *laminin, fibronectin, neural-cell adhesion molecule* (N-CAM), and *N-cadherin*, all of which are found in the basement membranes of the Schwann cells, as well as in the cells themselves.

The Rate of Axon Growth Slows Distally

Axon growth is at first slow but subsequently quickens. Gutmann, Gutmann, Medawar, and Young (1942) calculated an average velocity of 4.3 mm/day during a period between 13 and 25 days following nerve crush. Jacobsen and Guth (1965) measured the extent of regeneration by stimulating the nerve at the site of crush and recording the compound action potential at different positions in a distal direction. They found that the rate of growth increased until it had reached 3.0 mm/day by the 18th day. In humans, Sunderland (1947) was able to make observations on the recovery of nerve function in injured military personnel; he noted that the speed of nerve growth depends not only on the time after injury but also on the site of the lesion, being highest after a proximal wound. For example, in the thigh and upper arm, the average rate is about 3 mm/day, whereas in the hand and foot, values of only 0.5 mm/day are found. Another factor influencing nerve regeneration is previous injury, for recovery is then less complete; in

such cases the Schwann cells divide to form large groups associated with multiple axons of varying diameters (Thomas, 1970).

Growth Cones Are Directed to Old End-Plates

Once the regenerating axons reach the denervated muscle, the growth cones appear to seek out the sites of the old end-plates in order to establish new neuromuscular junctions. The situation does not arise simply because the axons have been guided back by the peripheral nerve stump; even if the axons are forced to take other pathways, they will still run along the denervated muscle fiber until the old end-plate is found (Bennett, McLachlan, & Taylor, 1973). As described later, it is the basement membrane investing the former end-plate that expresses the molecular signals needed to attract the growth cone.

Regenerated Axons Are Thinner Than the Original Ones and Conduct Impulses More Slowly

Once an axon has established successful synaptic connections with the previously denervated muscle, the diameter of the axis cylinder increases (Aitken, Sharman, & Young, 1947). Schröder (1972) found that the myelin sheath also became thicker, though to a lesser extent than the axis cylinder, in studies on the regeneration of sciatic nerves in dogs. At 12 months after suture, the mean diameter of the axis cylinder was 79% of normal, whereas that of the myelin sheath was only 57%. Since the speed of impulse propagation is directly proportional to the diameter of the axis cylinder and is also dependent on the insulating properties of the myelin sheath (see chapter 9), low velocities would be expected in regenerated nerves. Hodes, Larrabee, and German (1948) found values as low as 50% to 60% of normal in human nerves 3 to 4 years after repair by suture. Rather higher, but still reduced, values were noted in a study by Ballantyne and Campbell (1973), which included observations on both sensory and motor fibers. A further change in the appearance of the new axons is that the nodes of Ranvier are situated closer together, though in an irregular manner; this alteration is caused by the proliferation of Schwann cells that takes place during nerve regeneration.

Collateral Reinnervation

There is another way in which the nerve supply to a muscle can be restored, but this mechanism is only applicable in instances in which the denervation has been incomplete. In this reinnervation, the surviving healthy axons form sprouts that grow across to denervated muscle fibers in their vicinity and form new neuromuscular junctions.

Sprouts Can Develop at Several Sites in Spared Axons

The sprouts commonly grow out from the nodes of Ranvier of the healthy axon (*nodal sprouts*), but they can also arise distally, either from the neuromuscular junction itself (*terminal sprouts*) or from the region of motor axon immediately in front of it (*pre-* or *subterminal sprouts*). These three branching systems are all examples of collateral reinnervation (Figure 17.1) The existence of this type of innervation was first postulated by Exner (1885) in order to explain the absence of degeneration in the rabbit cricothyroid muscle following its partial denervation. Much later, interest in collateral reinnervation was revived by the experiments of Weiss and Edds (1945) in the rat, in which partial denervation of leg muscles had been effected by dividing ventral roots entering the lumbosacral plexus. Using a similar preparation, Hoffman (1951) subsequently showed that a nodal sprout might sometimes enter the endoneurial tube vacated by a degenerating axon and follow it down to the original end-plate zone.

Terminal Sprouting Can Enlarge Existing Neuromuscular Junctions

Terminal sprouting has already been mentioned as one of the mechanisms by which an axon can undertake collateral reinnervation of a denervated muscle fiber in its vicinity. The process has a further significance in that it enables a motoneuron to either renew or else enlarge existing synaptic connections with its own colony of muscle fibers. As a result of terminal sprouting, the new fine axon twigs ramify on the surface of the muscle fiber and enlarge the territory of the neuromuscular junction; at the same time, the terminal axonal expansions are often swollen. Many examples of the bizarre and distorted innervations which can result, especially in patients with neuromuscular disorders, were described by Cöers and Woolf (1959), using the methylene blue staining reaction.

Terminal Sprouting Has Been Studied After Poisoning With Botulinum Toxin

Particularly clear examples of terminal sprouting have been observed by Duchen and his colleagues (Duchen,

1970a, 1970b; Duchen, Stolkin, & Tonge, 1972) in muscles poisoned with botulinum or tetanus toxins. The first evidence of sprouting can be seen as early as six days after the injection of botulinum toxin into soleus muscles of mice and well before any degenerative changes are visible in the motor nerve terminals. Figure 17.2 and the accompanying description are reproduced from Duchen's (1970a) paper. Figure 17.2A shows a normal nerve ending with a single preterminal axon (arrow) innervating the end-plate. Terminal sprouting (B) develops after neuromuscular transmission has been blocked by the toxin. The nerve sprouts grow in all directions and branch, while the original end-plate may disappear (C). New nerve terminals are then formed randomly on some of the nerve sprouts (D). It can be seen that the nerve fibers which were at first terminal sprouts in (B) have now become branched preterminal axons in (D).

Having now dealt with the main features of the neural response to muscle denervation, it is appropriate to consider some of the underlying phenomena in more detail.

Muscle Fibers May Normally Suppress Motoneuron Sprouting

So far no attention has been given to two of the most fascinating problems in neurobiology. First, what is the stimulus that causes an axon to sprout, and second, how do the sprouts find their way to the denervated muscle fibers? As will be seen, rather more is known about the second question than the first.

First consider the nature of the stimulus that initiates the sprouting at the central end of a divided nerve fiber. In this situation, a signal from the denervated muscle cannot be involved, since sprouting will still occur even if the nerve is cut at a distance from the affected muscles. For example, if the sciatic nerve is cut in midthigh, sprouting will still take place, although the central nerve stump is surrounded by the still-innervated hamstring muscles and well away from the denervated muscle fibers below the knee. An alternative suggestion is that the nerve is stimulated to sprout by some factor(s) released by the degenerating peripheral stump. Yet this is also unlikely, for sprouting continues at the cut central end of the nerve after the peripheral stump has been removed. Furthermore, Cajal (1928) and others before him had found it impossible to induce sprouting in intact axons by placing a length of degenerating nerve in their vicinity.

A more attractive hypothesis suggests that the sprouting from the central end of a cut axon takes place because the motoneuron is no longer receiving instructions "not to sprout" from the muscle (see Figure 17.3). A carrier for such instructions would be the centripetal

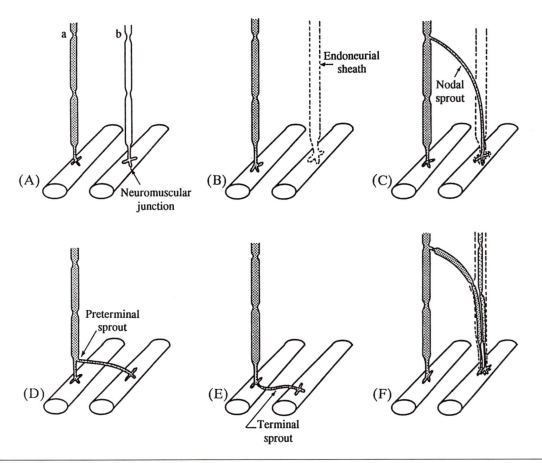

Figure 17.1   Different types of collateral reinnervation. In (A), axon *b* is about to undergo Wallerian degeneration while axon *a* remains healthy. In (B), degeneration of *b* has taken place, leaving the endoneurial sheath intact. In (C), the denervated muscle fiber has been innervated by a nodal sprout from *a*, which has entered and followed the endoneurial sheath. In (D), the axon branch has arisen from *a* prior to the neuromuscular junction, and in (E), the sprout has formed from the terminal axon expansion. In (F), axon *b* has regenerated so that the previously denervated muscle fiber is now supplied by two motoneurons; the respective axons share the same endoneurial sheath.

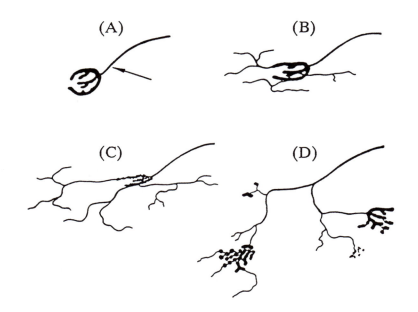

Figure 17.2   Remodeling of muscle fiber innervation following terminal sprouting, as seen in mouse soleus muscle fibers after the injection of botulinum toxin. Reprinted from Duchen (1970).

flow of axoplasm, which is discussed in chapter 8. Furthermore, the surge in production of RNA in the nucleolus of the motoneuron occurs sooner if the axon has been divided close to the cell body. This observation could be interpreted to mean that the synthesis of protein, necessary for axonal sprouting, is normally repressed by a factor traveling in the axon toward the motoneuron soma.

Muscle Inactivity May Stimulate Collateral Reinnervation

The sequence of events responsible for *collateral reinnervation* is equally puzzling. It will be recalled that, in this process, sprouts arise from intact axons, either at the nodes of Ranvier or at the motor nerve terminals. If the third hypothesis is correct, the motoneuron will still be receiving instructions not to sprout from the muscle fibers of its own intact motor unit. These instructions must now be overridden by local signals from the denervated muscle fibers. One such signal is the *inactivity* of the fibers. This role of inactivity can be shown by paralyzing *innervated* muscle fibers with tetrodotoxin (TTX) applied to the nerve fibers, or by one of the other methods described in chapter 19. Regardless of the method employed, axonal sprouts will form in the terminal arborization and will seek new sites for synaptogenesis on the surfaces of the inactive fibers. The terminal sprouting can be prevented if the muscle fibers are stimulated directly (Van Essen & Jansen, 1974).

Factors May Be Released From Degenerating Axons and Denervated Muscle Fibers

A second stimulus for sprouting is likely to be *nerve fiber degeneration*. Indeed, terminal sprouting can occur in muscles in which the motor nerve fibers were

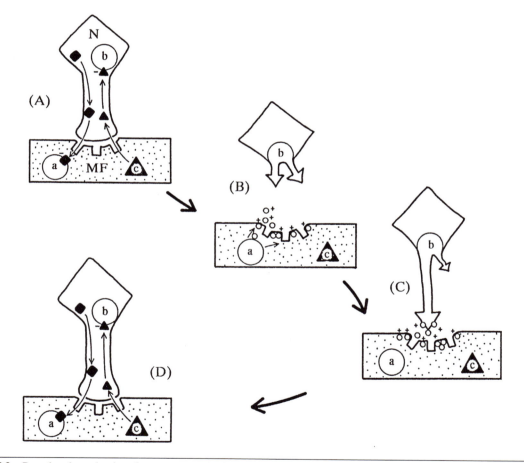

Figure 17.3 Postulated mechanism for axon regeneration and muscle fiber reinnervation following nerve section. (A) In the intact nerve-muscle preparation, the sprouting apparatus (*b*) of the motoneuron (*N*) is inhibited by centripetal delivery of molecular signals from (*c*) in the muscle fiber (*MF*). Similarly, (*a*) in the muscle fiber is prevented from secreting axon-attracting molecules by centrifugal inhibition from the motoneuron. (B) The axon has been sectioned; (*b*) is no longer repressed and starts to produce axon sprouts, and (*a*), also unrestrained, secretes attractants that will guide an axon sprout to the site of the original end-plate (C). (D) Once the motor axon has reestablished a successful neuromuscular connection, the mutual inhibiting mechanism is reimposed.

left intact but the muscle sensory fibers were made to degenerate (Brown, Holland, & Ironton, 1978). It is not known whether the products of nerve degeneration act directly on adjacent neuromuscular junctions or whether other steps are involved (e.g., growth factors might be released by the macrophages digesting the nerve stump).

Yet another possibility is that the denervated muscle fibers release an agent that induces terminal or nodal sprouting. Early, but unsuccessful, attempts to identify such a substance were made by Van Harreveld (1947) and Hoffman (1950). Subsequently, Tweedle and Kabara (1977) reported positive results, probably due to the lipid component of their muscle extracts. More recently Henderson, Huchet, and Changeux (1983) denervated chicken muscles and found that they started to produce a *neurite-promoting factor* within 1 day, when tested on motoneurons in tissue culture. This factor attained a maximal concentration at 3 days and then declined, as the muscle fibers began to degenerate. Other experimental results, consistent with the release of a sprouting factor from denervated muscle, have been obtained by Pockett and Slack (1982) and Slack and Pockett (1982).

The Basement Membrane Attracts the Growth Cone and Directs the Morphological Changes at the New Synapses

The next stage in the reinnervation process is better understood, for there is very good evidence that the basement membrane surrounding the muscle fiber directs the neural growth cone to form a new terminal arborization. It also instructs the muscle fiber to develop synaptic folds. These key roles of the basement membrane were first demonstrated by McMahan and colleagues (Sanes, Marshall, & McMahan, 1978), who, in one type of experiment, denervated the frog cutaneous pectoris muscle and also removed a segment of the muscle; the surrounding regions were prevented from regenerating into the gap by a dose of irradiation. Despite the absence of underlying muscle, the motor nerve fibers regenerated to the sites of the old end-plates and developed terminal arborizations containing ACh and synaptic vesicles (Figure 17.4). Conversely, if the nerve fibers were stopped from growing back, but the muscle fibers were allowed to regenerate, the membranes of the latter underwent the complex infolding characteristic of the secondary synaptic clefts and formed AChRs and acetylcholinesterase (Figure 17.4; McMahan & Slater, 1984). Biochemical analysis of the basement membrane shows that an essential component is the protein *agrin*, which can be produced by both the nerve and the muscle fibers and inserted into the membrane. Other important

components of the basement membrane are *synaptic* (S) *laminin*, *N-CAM*, and *terminal anchorage protein* (TAP_1). In addition, the nerve terminals produce *calcitonin gene-related peptide* (CGRP), which stimulates the synthesis of AChRs in the muscle fiber and promotes their incorporation into the muscle fiber plasmalemma.

Reinnervated Muscle Fibers May Have More Than One Neuromuscular Junction

In normal mammalian extrafusal muscle, each fiber is usually supplied by only one motoneuron. Consider, however, a partially denervated muscle in which the twin phenomena of axon regeneration and collateral reinnervation proceed together. What happens if a regenerating axon returns to a muscle fiber which, through collateral reinnervation, has already formed a synapse with a foreign motoneuron? This situation is likely to arise in those disorders in which the peripheral nerve fiber stump is left in continuity even though the axis cylinder has degenerated. It will occur, for example, after a nerve crush or following motoneuron recovery from a ''dying-back'' type of neuropathy. In humans the observations are few, but in animals several investigations have been directed to this problem. The critical experimental maneuvers and observations may be summarized in the following way.

First of all, provided the original nerve supply to a muscle is left intact, that muscle is unable to accept innervation from a foreign nerve (Elsberg, 1917). Once the original nerve supply is interrupted, either by cutting, crushing, or the application of botulinum toxin, the muscle will readily form new neuromuscular junctions with a foreign nerve presented to it (Fex, Sonesson, Thesleff, & Zelená, 1966; see Figure 17.5). If the original nerve is now allowed to grow back to the muscle, it is able to reestablish synaptic connections with a large proportion of fibers, even though these fibers already have an ectopic innervation from the foreign nerve. In the study by Frank, Jansen, Lømo, and Westgaard (1974), the proportion of doubly innervated fibers in the muscle was as high as 57% after a nerve crush.

Double Innervation May Persist for Long Periods

The next question is whether the double innervation persists, for in embryonic and neonatal muscle fibers of mammals, any accessory innervation is ultimately rejected (see chapter 6). There is also evidence from cross-innervation experiments on the extraocular muscles of fish that foreign innervation ceases over a period

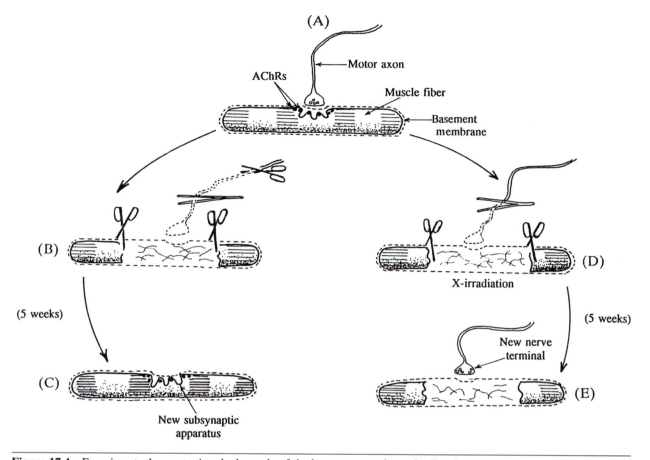

Figure 17.4 Experiments demonstrating the key role of the basement membrane in directing regeneration of the neuromuscular junction. In (B), the muscle fibers are cut on either side of their innervation zones and then allowed to regenerate. Although reinnervation is prevented by cutting and crushing the motor axons, normal subsynaptic apparatus develops in the muscle fibers (C). In (D) regeneration of the muscle fiber is blocked by irradiation, but a new motor nerve terminal is still able to form (E). A normal fiber and axon are shown in (A). Based on experiments of Sanes, Marshall, and McMahan (1978) and McMahan and Slater (1984).

of 2 days after the original nerve supply has been reestablished (Marotte & Mark, 1970). However, the results of single muscle fiber studies on this type of preparation are at variance with this interpretation (Scott, 1977). As far as mammalian muscles are concerned, most of the evidence available is firmly in favor of the persistence of foreign synapses without any suppression by restoration of the original nerve supply (Fex et al., 1966; Tonge, 1974). The longest period of study to date appears to be that of Frank and his colleagues (1974, 1975), who found that foreign innervation of rat soleus muscles by the superficial fibular nerve persisted for 9 months or longer. The induction of these doubly innervated fibers in adult mammalian muscles raises interesting questions. For example, do the two axons belong to motoneurons of the same type (see chapter 12)? If not, does the muscle fiber contain two regions of contrasting biochemical and contractile properties? If, on the contrary, the muscle fiber has identical properties throughout its

length, which of the two motoneurons is the dominant one?

In Some Conditions Supernumerary Innervation May Be Suppressed

Even though the evidence for the maintenance of doubly innervated fibers is convincing, it is possible that in some circumstances suppression of synapses occurs in reinnervated mammalian muscle. The results of Brown and Butler (1974) on reinnervated intrafusal muscle fibers may prove to be a case in point. These authors found that, in muscle spindles of the cat, the pattern of sensory fiber discharges following motor fiber stimulation indicated that the initial reinnervation came exclusively from α-motor axons. Subsequently, the spindle responses changed to a form that suggested that the slower growing γ-axons had now reached the spindle

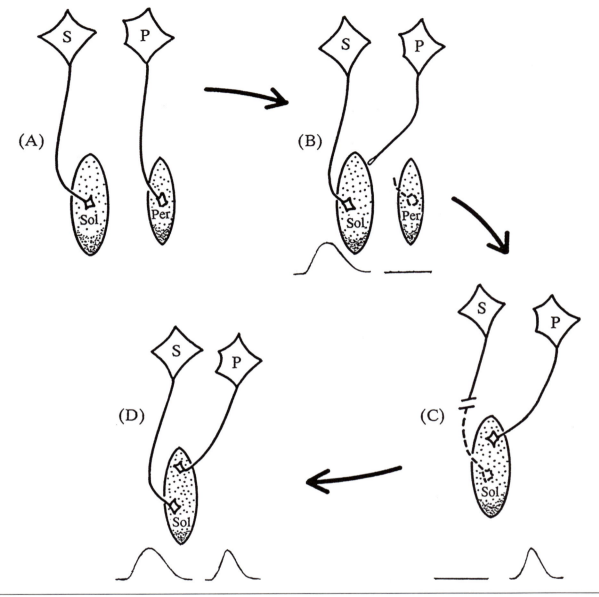

Figure 17.5 Hyperinnervation experiment. (A) shows the normal soleus (*Sol*) and peroneal muscles (*Per*) with their corresponding motoneurons (*S* and *P* respectively). In (B), the peroneal nerve has been cut and the central stump applied to the soleus; no innervation takes place (note absent twitch response below). If the soleus nerve is now divided or crushed, as in (C), the peroneal nerve rapidly forms neuromuscular junctions on the soleus muscle. (D) If the soleus nerve is now allowed to grow back, many muscle fibers will acquire a double innervation. Based on studies of Fex, Sonesson, Thesleff, and Zelená (1966) and others.

and restored the correct innervation of the intrafusal fibers.

Changes in the Muscle Fibers Following Reinnervation

As far as is known, all the effects of denervation on muscle fibers are reversible, provided the new nerve is

introduced within a reasonable period of time. Exactly what constitutes a "reasonable" period is not clear, for the requisite experiments do not appear to have been performed.

Delay Makes Successful Reinnervation Less Likely

One factor, that is time-dependent, is the necrosis that takes place in a small portion of the denervated muscle fibers. In some of the other fibers, the myonuclei may survive with only vestigial collars of cytoplasm, and it is doubtful if these fibers can recover their function. Another reason why reinnervation is more likely to be successful when attempted early rather than late is that fewer motoneurons will have undergone irreversible retrograde degeneration (see chapter 18). Also, with increasing time the peripheral nerve stump retracts, the endoneurial spaces decrease in size, and the Schwann cells become less active (Gutmann, 1971). With these qualifications in mind, there is no doubt that an excellent return of muscle function can result if reinnervation is able to take place within a few months.

Reinnervation After Nerve Crush May Be Complete in 1 Month

McArdle and Albuquerque (1973) produced temporary denervation in the hindleg of the rat by crushing selected nerves as close as possible to the muscles. In such animals there is a return of MEPPs 9 days later, initially with a low discharge frequency. At about this time, the resting membrane potential starts to rise and reaches normal values at 30 days (see Figure 17.6). The ionic permeability of the membrane and the sensitivity of the membrane to ACh assume normal levels by about 30 days.

The New Neuromuscular Junctions Are Initially Inexcitable

In the early stages of reinnervation, the new neuromuscular junctions pass through a phase in which they are unable to excite the muscle fibers, even though adequate supplies of transmitter are available in the axon terminals. This point was first demonstrated by Miledi (1960b), who found that MEPPs are present at a time when no evoked response can be detected in the muscle fiber following nerve stimulation. Originally shown in the frog, the presence of nontransmitting synapses during reinnervation has since been established in the rat (Dennis & Miledi, 1974) and in the fowl (Bennett, Pettigrew, & Taylor, 1973), though not in the mouse (Tonge, 1974). The transmission failure may be caused by a block of impulse conduction in the distal motor axon, since in some instances a second electric shock to the nerve, given 4 ms after the first, is able to elicit a muscle fiber response (Dennis & Miledi, 1974). In the frog, the nontransmitting stage appears to last only a day or two, and in the rat, its duration is shorter still. Of significance for an understanding of trophic mechanisms (chapter 18) is the observation that, even before the synapse has become operational, some of the denervation phenomena are already beginning to disappear from the muscle

fiber. For example, the ACh sensitivity of the membrane starts to resume its normal pattern, and a change can be detected in the contracture responses of frog slow muscle fibers to K$^+$ and ACh (Elul, Miledi, & Stefani, 1968). These findings can only be explained by postulating that some neural factor other than impulse-induced activity in the muscle fiber is exerting a trophic action on the muscle.

Denervated Muscle Fibers Will Accept Innervation From Autonomic Axons

The reversal of denervation characteristics described in the previous section depends only upon the reappearance of a functioning neuromuscular junction and is not affected by the origin of the motor axon. Suppose that, instead of a motor axon, a cholinergic autonomic axon is presented to a denervated skeletal muscle fiber. Would the muscle fiber be able to accept the new axon because, like the original motor nerve, it is cholinergic— or would its autonomic origin in some way preclude it from establishing a functional synapse? In an ingenious experiment, Landmesser (1971, 1972) was able to introduce the (autonomic) preganglionic fibers of the frog vagus nerve to a denervated sartorius muscle transplant. She found that the vagal fibers can make functional synaptic connections and usually do so at the site of the original end-plates. The vagal neurons are able to maintain the normal structure and electrical properties of the sartorius fibers but cannot induce the formation of cholinesterase at the neuromuscular junction. Because of the discrepancy between the cholinesterase and other properties of the reinnervated muscle, more than one nerve trophic factor must be involved (see chapter 18).

Motoneurons Can Convert Muscle Fibers From One Type to Another

For a more detailed analysis of the effects of cross-innervation, it is necessary to return to studies of mammalian skeletal muscles. What happens if muscle fibers capture axons from motoneurons that normally supply a different type of fiber (e.g., type I as opposed to type II)? Initially it was assumed that a reinnervated muscle fiber would regain all its old characteristics (color, enzyme profile, twitch speed, etc.). In an elegant experiment, which is described in chapter 18, Buller, Eccles, and Eccles (1960b) showed that this supposition was incorrect, and that motoneurons were capable of transforming the contractile properties of the muscle fibers. Other pioneering experiments, performed a few years later, have demonstrated that the histochemical staining properties and chemical compositions of the muscle

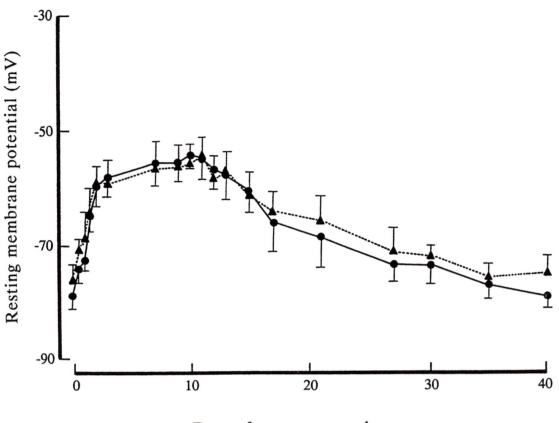

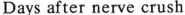

Days after nerve crush

Figure 17.6   Fall of resting membrane potential following temporary denervation produced by nerve crush. After reinnervation has taken place, the resting potential is eventually restored to its original level for both the extensor digitorum longus (●) and soleus (▲) muscles. Adapted from McArdle and Albuquerque (1973).

fibers are also modified by cross-innervation (Drahota & Gutmann, 1963; Dubowitz, 1967; Romanul & Van Der Meulen, 1966). In order to bring about these changes, the motoneuron evidently regulates the expression of genes in the myonuclei which code for those proteins (e.g., myosin heavy chains) characteristic of the different fiber types. The nature of this regulation is discussed in chapter 18.

Motor Unit Properties Following Reinnervation

It will be recalled that in normal muscle the fibers of the different motor units overlap (see chapter 12). Further, because several types of muscle fiber can be distinguished histochemically, the normal appearance of a stained cross section of muscle is a mosaic (Figure 17.7A). Once reinnervation begins in a partially denervated muscle, many neighboring fibers may come to be innervated by the same axon. Since the motoneuron

controls the biochemical composition of the muscle fibers, all the fibers in the same unit will exhibit similar histochemical staining. Thus, instead of the normal mosaic appearance of the muscle, large groups of fibers will be found with similar staining characteristics. This *fiber-type grouping*, when seen in a muscle biopsy, is one of the most useful signs of previous denervation (Figure 17.7B).

The Sizes of New Units Are Proportional to the Numbers of Denervated Fibers Available

A leg muscle such as the rat tibialis anterior muscle can be partially denervated by sectioning one of the ventral roots. Following reinnervation by fibers of the remaining root, it is found that, on average, the new units are four times larger (Kugelberg, Edström, & Abbruzzese, 1970). Since the enlarged motor unit boundaries correspond to those of the muscle fiber fascicles, it seems probable that connective tissue septae within the muscle belly impede the growth of the axonal sprouts. More evidence of the power of collateral reinnervation

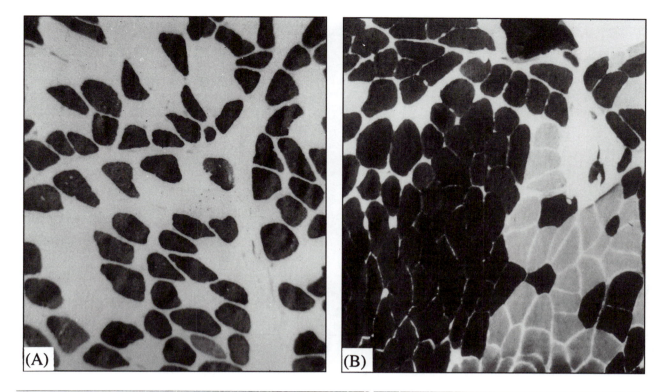

Figure 17.7 (A) Normal mosaic pattern of muscle fiber types in a human quadriceps muscle, stained for myosin ATPase activity at pH 10.0. The type II fibers are dark, but the intervening type I fibers are unstained. (150x). (Courtesy of Dr. John Maguire.) (B) Fiber-type grouping in a patient with spinal muscular atrophy; the staining method was the same as in (A). (150x).

can be found in patients with motoneuron disease and other chronic denervating disorders, by taking advantage of the fact that the mean sizes of the motor unit potentials are proportional to the numbers of component muscle fibers. In such patients a hyperbolic relationship between the sizes and numbers of surviving motor units is observed; that is, the sizes of the new units are always proportional to the number of denervated fibers available for adoption (see Figure 17.8; McComas, Sica, Campbell, & Upton, 1971).

The Enlarged Motor Units Develop More Tension

When the maximal twitch tensions of partially denervated muscles are measured, it is found that the values remain within the normal range until fewer than 20% of the normal mean population of units are left (see Figure 17.9). In the most severely denervated muscles, the surviving units generate, on average, seven times the normal tension (Gordon, Yang, Ayer, Stein, & Tyreman, 1993; McComas, Sica, et al., 1971). Among these enlarged motor units may be found others that are unusually small and may be those that, having previously sprouted, are now beginning to fail (Dengler, Konstanzer, Hesse, Schubert, & Wolf, 1989).

Abnormally large motor unit tensions can also be demonstrated in animal muscles following reinnervation, with values up to 16 to 19 times normal being reached (Luff, Hatcher, & Torkko, 1988). In such muscles, as in those of patients with chronic denervation, there may also be some very small units (Bagust & Lewis, 1974). Whether a partially denervated muscle is as efficient in continuous work, as opposed to single twitches, is less clear, since Milner-Brown, Stein, and Lee (1974a) found no evidence of increased motor unit forces when examining patients with peripheral nerve injuries. One factor that may have contributed to Milner-Brown et al.'s negative findings is that in some partially denervated muscles there may be impaired neuromuscular function. This abnormality may result from the inability of a parent motoneuron to maintain its increased population of neuromuscular junctions satisfactorily. Also, impulse conduction in the slender new preterminal axons also may be unusually hazardous.

After Recovery From Nerve Division, the Normal Recruitment Pattern Is Initially Lost but May Be Restored

In reinnervated or partially denervated muscles, the order of motor unit recruitment during voluntary contractions is of special interest. In a normal muscle,

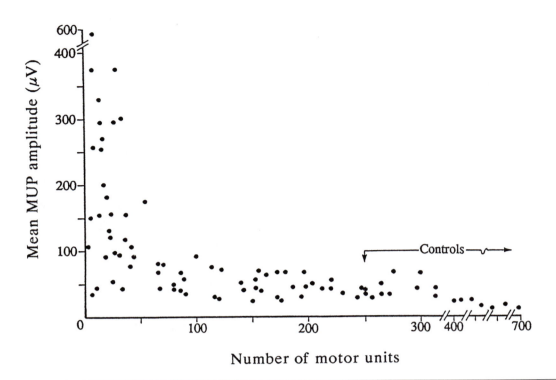

Figure 17.8   Sizes of motor units in partially denervated hypothenar muscles of ALS patients. The sizes have been expressed in terms of the mean potential amplitudes of the surviving motor units, since the greater the number of fibers in a motor unit, the larger will be the potential of the motor unit. Note the hyperbolic relationship between the numbers and sizes of units. Reprinted from Dantes and McComas (1991).

progressively larger motor units are called into activity with increasing effort (*size principle*, see chapter 13), but this pattern may be lost in muscles reinnervated after total nerve section (Milner-Brown, Stein, & Lee, 1974b). In the disease ALS, however, in which the muscles undergo progressive denervation and *collateral reinnervation* takes place, the orderly recruitment pattern is maintained. The explanation is that, in nerve regeneration following axon division, the occurrence and extent of reinnervation will depend largely on such simple mechanical factors as the accessibility of the Schwann cells in the peripheral stump to the newly formed axon sprouts. For example, in patients in whom the ulnar nerve has been accidentally divided at the wrist and then rejoined, nerve fibers that had previously supplied the first dorsal interosseous muscle may now go to the adductor pollicis or to the lumbricals and other interossei (Thomas, Stein, Gordon, Lee, & Elleker, 1987). However, the situation is not simple because, in the case of median nerve regeneration following section at the wrist, the normal relationship between motor unit size and recruitment threshold is restored—despite many axons having been misdirected from other muscles.

In animal muscles, too, there is evidence that, under favorable circumstances, restoration of motor unit properties may take place after random reinnervation. For example, 2 months after nerve section and resuture, cat triceps surae motor units have mixed populations of muscle fibers. By 9 months, however, the different types of motor unit can be easily recognized (Gordon & Stein, 1982). Thus, the largest motoneurons (as gauged by their axon potentials) develop the largest forces and have the fastest twitches, as would be expected of type FF motor units (Figure 17.10). Such studies even suggest that muscle fibers may be redistributed among the motor units—like the intrafusal muscle fibers after reinnervation (Brown & Butler, 1974).

APPLIED PHYSIOLOGY

In this section we will briefly consider the relevance of collateral reinnervation to neuromuscular disorders and its implications for the diagnosis and prognosis of patients.

Collateral Reinnervation Is Prevalent in Neuromuscular Disorders

The importance of collateral reinnervation lies in the fact that it is not only a very powerful compensatory

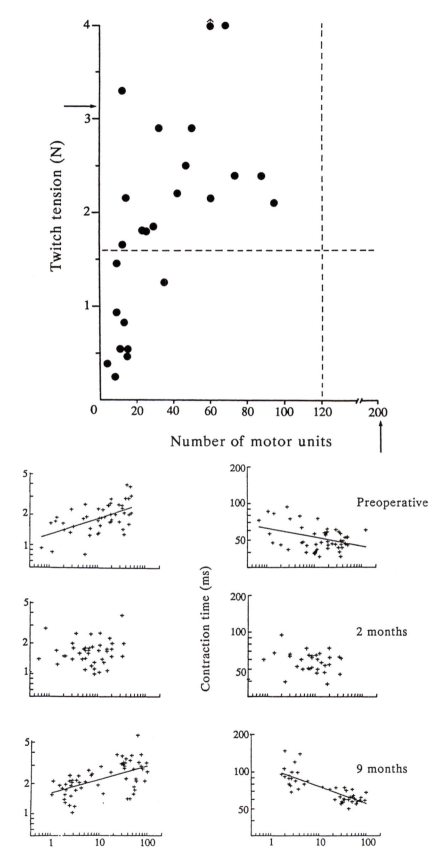

Figure 17.9 Maximum isometric twitch tensions of 25 extensor hallucis brevis muscles with varying degrees of denervation. The horizontal and vertical dashed lines indicate the lower limits of the respective normal ranges, and the arrows identify the control mean values. Note that muscles with severe denervation may still develop normal force. Reprinted from McComas, Sica, Campbell, and Upton (1971).

Figure 17.10 Recovery of normal motor unit properties in cat triceps surae muscles after nerve section and suture. The axon potential (left) is an indirect measure of axon diameter and hence of the size of the motoneuron cell body. Note that the normal correlations among axon potential, contraction time, and twitch tension, which are seen in the preoperative (*control*) muscles, are missing at 2 months of recovery but have reappeared at 9 months. Adapted from Gordon and Stein (1982).

mechanism but also an extremely common phenomenon after nerve injuries and in neuromuscular diseases. Evidence of its presence is found in almost every case of partial denervation, irrespective of etiology; thus, it is to be seen in patients with ALS (see chapter 22), spinal muscular atrophy (see chapter 6), and old poliomyelitis, as well as in cases of trauma or disease of peripheral nerves. It must not be imagined that axonal regeneration and collateral reinnervation are mutually exclusive recovery processes. On the contrary, they occur simultaneously in any reversible neuromuscular disorder in which muscle denervation has been incomplete. Thus, while damaged axons are growing back toward the muscle, intact healthy axons are already sprouting and undertaking collateral reinnervation. This situation is one that will occur, for example, after nerve damage or in the Guillain-Barré syndrome (see chapter 16).

In a chronic disorder, such as ALS or spinal muscular atrophy, collateral reinnervation may be so effective that it may mask the presence of the disease until it is far advanced. Not only is there no wasting of the muscles, but strength may be preserved until fewer than 20% of the axons remain (see p. 269).

Reinnervated Motor Units Are Especially Susceptible to the Effects of Aging

An interesting question is whether reinnervated motor units are as viable as normal ones. The answer appears to be "no," at least in those instances in which the motor units are much larger than normal. This conclusion has been reached by Pachter and Eberstein (1992), who have examined rat plantaris muscles 12 months after severe partial denervation. At this time many signs of recent denervation are present; these include degenerating end-plates, muscle fibers without terminal arborization, angulated muscle fibers, and grouped muscle fiber atrophy. In those fibers that still retain their innervation, the numbers of terminal nerve branches are reduced at the end-plates. These important findings could well explain why, for example, human patients may develop weakness and. wasting of skeletal muscles 15 to 20 years after recovery from poliomyelitis (Cashman et al., 1987).

Chapter 18

Neurotrophism

In this chapter we will try to understand how a motoneuron and the colony of muscle fibers that it innervates exert a sustaining action on each other. It will be seen that, in the case of the muscle fibers, the effects of the motoneuron are mediated in part by impulse-induced stretching of fibers. The remainder of the effects, and those achieved by the muscle on the motoneurons, are brought about by chemical messengers.

Motoneuron Effects on Muscle

When the nerve supply to a muscle is interrupted by section, crush, or disease, a number of important changes take place in the muscle fibers (see chapter 16). The fibers become thinner and develop smaller tetanic tensions, while their surface membranes become more widely sensitive to ACh and start discharging action potentials spontaneously (*fibrillations*). These denervation phenomena can be viewed in a different way, however, for they indicate that, under normal circumstances, the motoneurons are preventing such changes from taking place. That is, the motoneurons, in addition to exciting muscle fibers and causing them to contract,

have a sustaining, or *trophic*, action that regulates many of the structural, biochemical, and physiological features of the fibers.

Motoneurons Exert a Sustaining Influence on Muscle Fibers

Further evidence for the existence of neurotrophism comes from cross-innervation experiments of the kind pioneered by Buller, Eccles, and Eccles (1960b). These authors had intended to examine the possible role of the muscle in determining the shape of the motoneuron action potential. Their experiments involved selecting two muscles in the cat hindlimb with contrasting contractile properties, cutting their respective nerves, and then cross-connecting the nerves (Figure 18.1). Thus, the flexor digitorum longus (FDL) was connected to the proximal stump of the nerve that had previously supplied the soleus (SOL) muscle, while the latter muscle was connected to the spinal cord through the nerve to FDL. After several months had elapsed, to allow the nerve fibers to regenerate, the authors stimulated the reconstructed nerves and found, to their surprise, that the FDL, which normally has a fast twitch, now had a slow twitch. Conversely the SOL, which is the slowest contracting muscle in the hindlimb, now had a much faster twitch (Figure 18.1). Clearly, the contractile properties of the muscle fibers had been transformed by the new nerve supply. This classic cross-innervation experiment has been repeated many times, most often in the cat and rat, and the same results have always been obtained. Subsequent histochemical analyses of cross-innervated muscles, first reported by Dubowitz (1967) and by Romanul and Van Der Meulen (1967), also indicated that nerves determine the phenotypic characteristics of postnatal muscle fibers.

The Sustaining Influence Is Largely Brought About by Impulse Patterns

How is the trophic action of the motoneurons exerted? An obvious clue comes from the results of disuse. If a muscle is no longer used fully, either as the intended result of an experiment or, in human subjects, because of an injury to a limb, the muscles may undergo striking atrophy and become weaker (see chapter 19). On the basis of such observations, it is safe to conclude that, in some way, the nerve impulse patterns must normally be responsible for maintaining muscle bulk and strength. In a second type of experiment Lømo and Rosenthal (1972) showed that impulse activity could prevent many of the features of denervation from taking place. These authors cut the nerve to the soleus and extensor digitorum longus (EDL) muscles of the rat and then stimulated the muscle fibers directly for 5 days. They found that they were able to reduce the muscle atrophy and

loss of contractile force that would otherwise have occurred. In later experiments, Hennig and Lømo (1987) showed that direct electrical stimulation of denervated muscle improved the tetanic tension in the soleus 37-fold and that of the fast-twitch EDL 8-fold. In addition, direct electrical stimulation prevented the spread of ACh sensitivity along the muscle fibers—another consequence of denervation (Lømo & Rosenthal, 1972; see also chapter 16).

A third type of observation, also pointing to the importance of impulse activity, comes from experiments that are the logical extension of the cross-innervation studies already described and involve changing the nerve impulse patterns received by the muscle fibers. The design of these experiments is influenced by the fact that, when an animal is standing quietly, soleus motoneurons tend to discharge steadily at low frequencies, thereby keeping the soleus muscle under sufficient tension to prevent the ankle from flexing under the weight of the animal. The motoneuron discharge to a fast-twitch muscle is quite different and consists of high-frequency bursts of impulses as the animal moves forward (Salmons & Vrbová, 1969).

In the first such investigation, Salmons and Vrbová fastened stimulating electrodes around the nerve to the tibialis anterior (TA) and EDL muscles in the rabbit and cat, and attached the leads to a stimulator implanted beneath the skin of the abdomen (Figure 18.2). The stimulator was programmed to deliver an intermittent stream of shocks at 10 Hz. At the end of 6 weeks, Salmons and Vrbová found that the twitches of the TA and EDL had become considerably slower. Confirmation of the importance of the impulse pattern was ob-

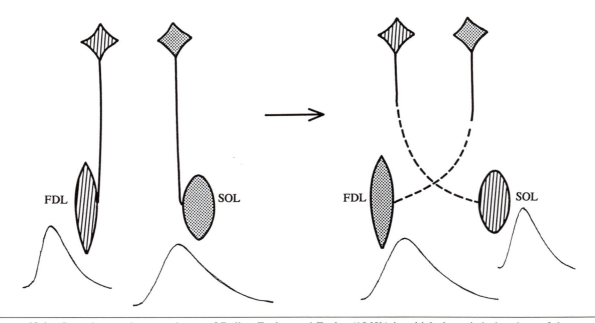

Figure 18.1   Cross-innervation experiment of Buller, Eccles, and Eccles (1960b) in which the twitch durations of the cat soleus (*SOL*) and flexor digitorum longus (*FDL*) muscles were largely reversed. See text.

tained by Al-Amood, Buller, and Pope (1973), who attached a stimulator to the spinous process of a cat vertebra and used it to excite ventral root fibers supplying the FDL muscle. The shocks were delivered at a frequency of 10 Hz for 8 weeks, and, once again, the stimulated muscle was found to have become slower.

Impulse Patterns Determine the Speed of Contraction and Fatigability

In the chronic stimulation experiments considered in the previous section, was it the frequency of the impulses or simply the extra number of impulses that was responsible for transforming a muscle from fast-twitch to slow-twitch? This issue has been thoroughly explored by Lømo and colleagues using different frequencies and periods of stimulation applied to denervated rat hindlimb muscles. The experiments of Gorza, Gundersen, Lømo, Schiaffino, and Westgaard (1988) included the use of bursts of high-frequency stimuli (60 shocks at 100 Hz every 60 s) and of short or long epochs of low frequency stimuli (10 Hz). They concluded that

- *low* frequency (10-15 Hz) stimulation induces *slow*-twitch characteristics and

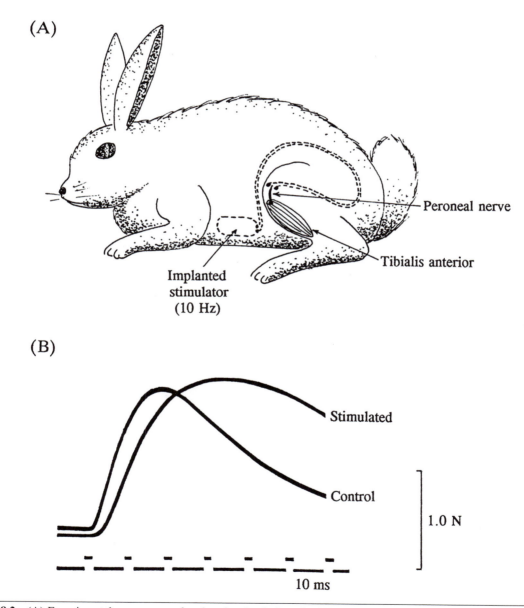

Figure 18.2 (A) Experimental arrangement for chronic stimulation of rabbit tibialis anterior muscle. (B) Examples of twitches from a muscle that had been stimulated at 10 Hz for 6 weeks, and from the control, unstimulated muscle in the opposite leg; note the much slower twitch in the stimulated muscle. Adapted from Salmons & Vrbová (1969).

- *high* frequency (100 Hz) stimulation induces *fast-twitch* characteristics.

In a separate study, Westgaard & Lømo (1988) demonstrated that

- the *total number* of impulses determines the *fatigability* of the fibers.

Thus, the greater the number of stimuli delivered during the day, regardless of frequency, the more resistant the fibers become to fatigue (Figure 18.3).

It is important to note, however, that the ability of stimulus frequency to convert muscle fibers from one type to another is not always as clear-cut as in the case of the rat soleus; both the species and the previous phenotypic expression of the muscle will influence the responsiveness of the fibers to changes in excitation pattern (for discussion, see Gorza et al., 1988).

Impulse Patterns Also Determine Histochemical and Immunocytochemical Features of the Muscle Fibers

The speed of contraction and the susceptibility to fatigue are not the only properties of the muscle fiber that are controlled by impulse patterns. For example, Pette et al. (1975) have shown that the histochemical staining properties are also affected, in that chronic stimulation of rabbit TA muscle converts the fibers from type II to type I. Similarly, high frequency stimulation will change

type I fibers to type IIC (2X; Gundersen, Leberer, Lømo, Pette, & Staron, 1988). The alterations in staining and in the speed of contraction are due to the substitution of a different isoform of the myosin heavy chain (MHC). Thus, in fibers acquiring the fast isoform, the cycling of the myosin cross-bridges with the actin filaments becomes quicker, because the heavy chains can split ATP faster (see chapter 11). In the experiments of Lømo and colleagues (Gorza et al., 1988), immunocytochemical techniques demonstrated that the fast MHC isoform appears in slow-twitch fibers by the 7th day of appropriate stimulation and increases thereafter, even though the fibers still retain small amounts of slow isoforms. Similar results have been obtained by other workers studying transformations between slow- and fast-twitch muscles (see chapter 19). Not only the isoforms of the MHCs, but also those of the light chains and troponins are altered by patterned stimulation. In addition, the fibers that acquire slow-twitch characteristics develop greater mitochondrial volumes and increased capillary densities. All of these changes make the slow-twitch fibers better equipped to carry out prolonged work under aerobic conditions.

The Trophic Effects of Impulse Patterns Are Mediated by Stretch of the Muscle Fiber

The appearance of new isoforms in a muscle fiber, following patterned stimulation, indicates that protein synthesis may have been modified at the level of gene

(A)   (B)   (C)

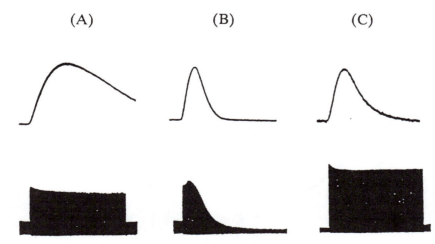

Figure 18.3 Contractile properties of three surgically denervated soleus muscles following 2 months of direct stimulation. The muscles received either many pulses per day at 15 Hz (muscle A), few pulses per day at 100 Hz (muscle B), or many pulses per day at 100 Hz (muscle C). The top row of recordings shows the respective twitches, and the lower row the responses to 2 min of intermittent tetanic stimulation at 77 Hz. The sweeps are slower and the force amplification is less in the lower row of records than in the upper row. Note that each muscle differs from the other two; for example, although B and C both have fast twitches, only B fatigues rapidly. Adapted from Westgaard & Lømo (1988).

transcription. Until recently it was difficult to imagine what sort of coupling might be involved between the electrical and genetic events. It now seems probable that the link is stretching of the muscle fiber membrane (or of the extracellular matrix) and that, through the release of soluble factors, several second messenger systems are activated. In addition, Ca^{2+}, released into the cytosol from the sarcoplasmic reticulum, may also act as a second messenger. The second messengers then activate immediate-early genes in the myonuclei and the latter, in turn, permit transcription of the genes coding for the new isoforms of the various contractile proteins to take place. This subject is discussed more fully in chapter 19.

Some Trophic Effects Are Not Dependent on Impulse Patterns

It would be surprising if all the trophic effects of nerve and muscle could be accounted for solely on the basis of impulse patterns, since some motor units have such high thresholds that they would rarely be excited during everyday life. This situation applies particularly to muscles of the human upper arm, for many individuals may pass an entire day without having to lift the arm above shoulder level, let alone make a maximal contraction. Even Lømo, the strongest protagonist of the trophic influence of impulse activity, has drawn attention to the small amounts of activity exhibited by some motor units, and it is possible that there may have been other units, with still higher thresholds, that would not have been detected in his experiments (Hennig & Lømo, 1985).

There is a further reason why impulse patterns are unlikely to be the sole explanation for *all* the trophic effects of motoneurons on muscle fibers. In experiments on animals, for example, the consequences of disuse on various muscle properties are usually less than those of surgical denervation (see chapters 16 and 19). In addition to these general observations, four other specific findings point to the presence of trophic influences not mediated by impulses:

• *Block of axoplasmic flow*. Using an implanted cuff technique, it is possible to block axoplasmic flow with colchicine without interrupting impulse propagation (Albuquerque, Schuh, & Kauffman, 1971; Cangiano, 1973; Hofmann & Thesleff, 1972). The deprived muscle fibers will then acquire features of denervation—a fall in resting membrane potential, a spread of ACh sensitivity, and the development of TTX-resistant action potentials.

• *Effect of nerve stump length*. The onset of denervation features can be delayed if a nerve is cut proximally so as to leave an appreciable stump attached to the muscle (Figure 18.4). This interesting phenomenon

was first shown for fibrillation activity (Luco & Eyzaguirre, 1955) and subsequently for the resting membrane potential, spread of AChRs, and development of TTX-resistant Na^+ channels (Albuquerque et al., 1971; Harris & Thesleff, 1972). The importance of this observation is that it cannot be explained by an absence of impulse activity, because the latter will occur as soon as the nerve is cut and will be independent of the site of section. An explanation can be provided in terms of axoplasmic flow, however, by suggesting that, following nerve section, the muscle fiber is able to use up the trophic material in the distal nerve stump. A longer stump will contain a greater amount of material, and the onset of denervation phenomena will be correspondingly delayed.

• *Partial denervation of single fibers*. The muscle fibers of the frog sartorius muscle receive a double innervation from the obturator nerve. If impulse patterns were the sole trophic mechanism, division of one nerve branch should not affect the activity of the muscle appreciably, because the fibers would still be excited through the remaining nerve branch. In such fibers, however, there is a definite increase in ACh sensitivity around the denervated end-plates (Miledi, 1960a), suggesting that some factor other than muscle fiber activity is normally controlling the distribution of AChRs.

• *Fibrillations and positive sharp waves*. When the nerve supplying a muscle is severed, there is an interval of several days, and then the denervated muscle fibers begin to discharge spontaneously. As recorded with a needle electrode, this spontaneous activity appears as either fibrillations or positive sharp waves (see chapter 16). However, if a human muscle is subjected to disuse rather than denervation, the spontaneous activity does not develop (an example of this absence is given in Box 19.1 in chapter 19).

The Nonimpulse Trophic Effects Are Brought About by Neurotrophins

What is the nature of the trophic mechanism that is not associated with impulse patterns? The general consensus is that it consists of different types of trophic molecules (*neurotrophins*), which are synthesized in the motoneuron soma and transported along the axon to its terminal branches. In some instances, the transport velocity can be established by observing the effects of cutting the nerve at different levels. For example, the fast fraction of axoplasmic transport appears to contain those factors responsible for maintaining the motor nerve terminals and for preventing fibrillations and the synthesis of TTX-resistant Na^+ channels in the muscle fibers. Until recently, one of the arguments against the existence of neurotrophic molecules was that none had

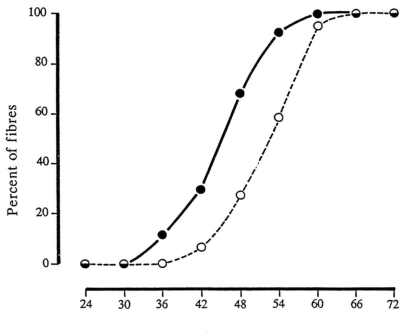

Figure 18.4 The development of resistance to tetrodotoxin (TTX) in mammalian muscle fibers that have been denervated by nerve section either close to the muscle (●) or else well away from the muscle (○). The ordinate shows the percentage of TTX-resistant fibers. Adapted from Harris and Thesleff (1972).

been isolated and characterized. To be accepted as a trophic substance, Lømo and Gundersen (1988) insist that the following criteria be satisfied, namely, that the putative agent must

- be released from motor nerve endings,
- not be supplied by blood or nonneuronal cells, and
- through its absence, result in an abnormality of skeletal muscle fibers that persists despite electrical stimulation.

While the first two criteria are reasonable, the third requires qualification, for the electrical stimulation should only be sufficient to match the naturally occurring excitation of the innervated fibers. If some motor units hardly ever fire, the results of bombarding their denervated fibers with stimuli may be misleading.

Application of the second criterion would exclude *sciatin*. Isolated from sciatic nerve, sciatin was found to promote maturation and survival of muscle cells *in vitro*, and to induce the synthesis of AChRs and acetylcholinesterase (AChE; Markelonis & Oh, 1979). Subsequent studies showed that sciatin was identical to *transferrin*, a plasma protein that regulates iron transport into cells. Certain other substances, however, appear as stronger candidates for neurotrophic mediators. Two of these are *agrin* and *ARIA*, both peptides that have been found to regulate the aggregation of AChRs and AChE at the neuromuscular junction (Fischbach et al., 1989; Godfrey, Nitkin, Wallace, Rubin, & McMahan, 1984). Another candidate is *calcitonin gene-related peptide*

(CGRP), which also acts at the neuromuscular junction and controls the numbers of AChRs (Fontaine, Klarsfeld, Hökfelt, & Changeux, 1986). Thesleff, Molgó, and Tågerud (1990) have speculated that *ACh* may have a facilitatory role, by assisting in the uptake of neuropeptides by the muscle fiber. With more sensitive collection and purification procedures, it is possible, if not probable, that other neurotrophic proteins and peptides will be identified; possibly mRNA is itself a trophic molecule.

In assessing the relative importance of neurotrophic molecules and impulse activity, it is important to bear in mind the possibility that the two may be related, at least at some junctions; thus Musick and Hubbard (1972) found that the release of substances from the nerve endings was enhanced by electrical stimulation. How do the trophic agents enter the muscle fiber? Libelius and Tågerud (1984) have reported that, following denervation, muscle fibers show increased endocytotic activity in the vicinity of the end-plate, and it is possible that this is the means by which access to the fiber is obtained. If so, the myonuclei in that region might be those most influenced by the neurotrophic molecules.

How do the neurotrophins work? As previously noted, some neurotrophins might gain access to the muscle fiber interior through endocytosis. Alternatively, others might combine with receptors on the surface of the muscle fiber membrane. Regardless of the means of access, it is probable that the neurotrophins use second messenger systems to activate immediate-early genes in the myonuclei, thereby controlling the expression of muscle protein genes (see Figure 18.5).

Muscle Effects on Motoneurons

Can muscle influence nerve? In view of the extensive use of feedback systems by the body, the expected answer would be "yes." Indeed, it would be unnecessarily wasteful if, following a peripheral nerve injury, a motoneuron continued to manufacture trophic material for a colony of muscle fibers to which it was no longer connected. Nor could a motoneuron gear its metabolic activity to a higher level, in order to support an enlarged muscle fiber colony, in the absence of a message from the periphery informing the cell that collateral reinnervation was taking place (see chapter 17). The very existence of microtubules within the axon suggests that there might be a centripetal, as well as a centrifugal, flow of information between the motoneuron and the muscle fiber. Thus, if only centrifugal signals were required, the cell could achieve this simply by bulk movement of axoplasm, though this would be a rather slow process.

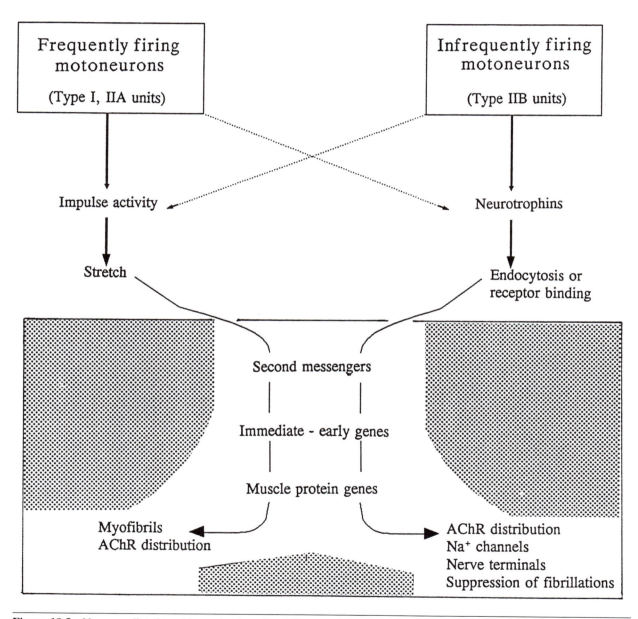

Figure 18.5 Nerve-mediated trophic mechanisms for different muscle fiber types. Type I and IIA fibers are heavily dependent on impulse activity, and the resulting stretch is especially important for the synthesis of myofibrillar proteins. In contrast, type IIB fibers probably rely more on neurotrophins, and these are more effective in suppressing fibrillations, maintaining the motor nerve terminal, and controlling Na$^+$ channel synthesis. Both stretch and neurotrophins regulate AChR distribution.

Axonal Damage Alters the Appearance of the Motoneuron Cell Body

The possible existence of a two-way traffic in trophic effects between the cord and periphery has exercised a number of investigators, among whom one of the earliest was Nissl (1892). Nissl avulsed facial nerves in rabbits and studied the changes visible in neurons of the facial nucleus under the light microscope (Figure 18.6). He noticed especially striking alterations in the granular material situated in the region of the axon hillock. This material, since called Nissl substance, is now known to consist of RNA. Following axotomy, the granules become smaller and are dispersed to the periphery of the motoneuron soma (*chromatolysis*). At the same time, the neuron becomes swollen and rounded in outline. The nucleus is usually pushed to the edge of the soma opposite the axon hillock; within the nucleus the nucleoli enlarge. These changes are complete at 7 to 10 days and are sufficiently prominent for the affected neurons to be readily distinguished from the other cells.

Subsequent workers have confirmed Nissl's observations and have described additional features. For example, the glial cells in the vicinity of the affected motoneurons swell (see Watson, 1972), and lysosomes may appear in the neuronal cytoplasm. In addition, there is retraction of the dendrites and loss of the synaptic connections that they normally make with the axon terminals (boutons) of other neurons; these changes probably underlie the longer latencies and increased temporal dispersion of reflex discharges.

After a while, however, the motoneuron dendrites extend and even develop growth cones, as if they were searching for afferent nerve fibers with which to make synaptic connections (Rose & Tourond, 1993). Microelectrode recordings from axotomized motoneurons have revealed little change in resting potential, although the overshoot of the action potential becomes larger and there is a considerable reduction in impulse conduction velocity along the axon stump (Eccles, Libet, & Young, 1958; Kuno, Miyata, & Muñoz-Martinez, 1974a). Normally the motoneurons supplying the fast- and slow-twitch muscles can be differentiated on the basis of the hyperpolarization that follows the action potential; after axotomy this difference becomes less marked, but it is restored as soon as reinnervation takes place (Kuno, Miyata, & Muñoz-Martinez, 1974b; Titmus & Faber, 1990).

RNA Increases in the Motoneuron Cell Body After Axotomy

In a series of elegant experiments on the hypoglossal nerve and tongue muscles of the rat, Watson analyzed the nature and course of some of the biochemical events

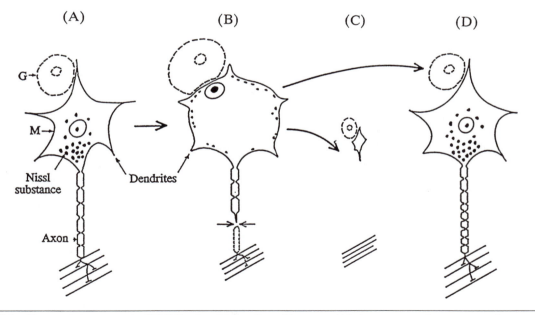

Figure 18.6 Changes in motoneuron (*M*) following interruption of its axon. (A) shows a normal motoneuron innervating several muscle fibers. (B) represents the situation about 7 days after axotomy (at arrows); note the swollen motoneuron soma with its displaced nucleus and enlarged nucleolus. The Nissl substance is dispersed, and the dendrites have retracted. By this time the distal axon stump is degenerating, and the muscle fibers have started to atrophy. The glial cell (*G*) close to the motoneuron is also enlarged. (D) If reinnervation of muscle is successful, normal neuronal architecture is restored, though the axon now has shorter internodal segments. (C) Failure to innervate is associated with progressive atrophy of the motoneuron and, in some instances, leads to its eventual disappearance.

taking place within the motoneuron following axotomy. He showed that the first detectable change within the motoneuron is an increase in ribosomal RNA within the nucleolus (Watson, 1968b). It is probable that this RNA is required for the synthesis of proteins as part of the forthcoming repair process in the axon. The latency of this change in RNA depends on the site of nerve injury, being least following a lesion close to the hypoglossal nucleus. In another type of experiment, Watson (1969) injected botulinum toxin, a substance that is known to interfere with the release of ACh (and probably other trophic factors) at the end-plate. Since a rise in nucleolar RNA could be detected in these animals also, the phenomenon cannot have been due to axonal injury but rather to a loss of functional contact between the nerve ending and muscle fiber. Thus, it seems probable, but has not yet been proved, that the toxin prevents not only the centrifugal, but also the centripetal, transfer of trophic material at the neuromuscular junction.

Gutmann (1971) has pointed out that the vigorous metabolic response of the motoneuron to disconnection from its target organ hardly deserves the term "retrograde degeneration" and that Nissl's own description, "primäre Zellreizung" (primary cell excitation), is more appropriate. Subsequent studies (e.g., Hoffman & Lasek, 1980) have shown that the increased messenger RNA codes for proteins, such as actin and tubulin, that are associated with slow axonal transport, as well as for GAP-43 (B 50) and other novel proteins found in the growth cones of the regenerating axons (for review, see Skene, 1989). In contrast, levels of messenger RNA for neurofilament proteins and for transmitter-synthesizing enzymes decrease after nerve section (Grafstein & McQuarrie, 1978).

In an extension of his earlier work, Watson (1970) was able to show that, if a motoneuron is allowed to reinnervate muscle fibers through axonal sprouting, a second rise in its metabolic activity takes place. During this phase, the dendrites return to their original lengths and acquire fresh synaptic connections. Successful reinnervation enables the motoneuron to resume a healthy appearance in other respects—the cell size becomes normal, the nucleus regains its central position, and the Nissl substance re-forms at the axon hillock; at the same time, the regenerating axon enlarges its diameter (Sanders & Young, 1946). In contrast, if reinnervation of muscle is prevented, many motoneurons will disintegrate or else gradually atrophy, particularly in young animals.

Signals Can Be Sent to the Motoneuron Cell Body From the Muscle and Motor Axon by Axoplasmic Transport

Experimental data has accumulated to show that several different types of substance can be conveyed from muscle to nerve by axoplasmic transport (see chapter 8).

For example, Watson (1968a) demonstrated that the injection of [^{3}H]-lysine into muscle was followed after a short interval by its appearance and proximally directed movement in motor nerves. Glatt and Honegger (1973) coupled the fluorescent dye, Evans Blue, with albumin, and within 12 hr of injection of this marker into the rat triceps muscle, motoneurons in the cervical cord could be seen to fluoresce; the length of the neural pathway was 30 mm.

In the study of Kristensson and Olsson (1971), horseradish peroxidase (HRP) was injected into the gastrocnemius muscles of mice. It is known that this electron-dense material is taken up by pinocytosis at the neuromuscular junction and can then be seen with the electron microscope to be contained in coated vesicles within the axon terminal (Zacks & Saito, 1969). Kristensson and Olsson found that HRP, after being transported centripetally in the axons, was distributed in small cytoplasmic granules around the nucleus within 24 hr of injection.

In conclusion, the demonstration that an axon can take up exogenous proteins and transport them to the neuron has some important implications. It provides a possible mechanism for the trophic influence of muscle on nerve, in the absence of which chromatolysis occurs. The process also sheds light on the way in which certain toxins and viruses may spread from the periphery to the central nervous system.

Neurotrophins Are Produced by Muscle Fibers for Motoneurons

Recognizing that muscle was in some way necessary for the upkeep of motoneurons, several attempts have been made to find the factors (*neurotrophins*) involved. One successful approach followed the purification and molecular characterization of a trophic factor found in pig brain. This factor, *brain-derived neurotrophic factor* (BDNF), was known to save sensory neurons in embryonic rats from naturally occurring cell death and was shown by Leibrock et al. (1989) to be a polypeptide with 119 amino acids (cf. Thoenen, 1991). Other groups then looked for neurotrophins with structures similar to BDNF and found two, termed *NT-3* and *NT-4* (also known as NT-5). All three neurotrophins strongly resemble a factor that had been discovered much earlier, was known to be necessary for the growth and maintenance of sympathetic and sensory neurons, and is produced in large amounts by the salivary glands; this is *nerve growth factor* (NGF; Levi-Montalcini, 1987).

Unlike NGF, however, BDNF and neurotrophins 3 and 4/5 are produced by muscle fiber mRNAs and can rescue motoneurons from naturally occurring cell death in the embryo and from the cell death that follows axotomy (Henderson et al., 1993; Yan, Elliot, & Snider, 1992). By analogy with NGF (Meakin & Shooter, 1992), the muscle neurotrophins are thought to combine with receptors on the nerve terminals, and then to cluster together. Next, the clusters are taken into the axoplasm within membrane-bound vesicles. While in the nerve terminals, the neurotrophin-receptor complexes may exert local effects, such as the potentiation of synaptic transmission in developing neuromuscular junctions of the embryo (Lohof, Ip, & Poo, 1993). Otherwise, the neurotrophin-receptor complexes are transported along microtubules to the motoneuron cell bodies to act on the genetic machinery in the nucleus.

The receptors for the neurotrophins on the motor nerve terminals have been identified recently and consist of high- and low-affinity types (Meakin & Shooter, 1992). The *high-affinity receptors* appear to be the ones necessary for the biological actions of the neurotrophins and are *tyrosine kinases* (Trks), produced by *Trk proto-oncogenes*. BDNF and NT-4/5 are bound by TrkB, while NT-3 combines with TrkC; the binding results in phosphorylation of the tyrosine kinase (Figure 18.7).

Although the situation has not been studied, it is safe to assume that the same kinds of neurotrophin receptor occur on the dendrites of the motoneuron, where they receive trophic inputs from other nerve cells. Loss of these signals may have a disastrous effect on the motoneuron (see Box 18.1).

Other Polypeptides Can Sustain Motoneurons

Muscle fibers also produce other polypeptides which, although unrelated to neurotrophins in their molecular structure, can nevertheless promote the survival of motoneurons, at least *in vitro* (Figure 18.7). These polypeptides are *fibroblast growth factor-5*, *leukemia-inhibiting factor*, *insulin-like growth factor*, and *growth promoting activity* (Mudge, 1993). Yet another polypeptide, *ciliary neurotrophic factor* (CNTF), is known to have a powerful sustaining effect on motoneurons but is produced by Schwann cells rather than muscle fibers. CNTF has been observed to rescue facial motoneurons in rats following axotomy (Sendtner, Kreutzberg, & Thoenen, 1990), and to increase the survival of motoneurons during embryonic cell death (Oppenheim, Prevette, Qin-Wei, Collins, & MacDonald, 1991); it also prolongs the life expectancies of mice with hereditary motoneuron degeneration (Sendtner et al., 1992). Another neurotrophic polypeptide, with a molecular weight of 22 kD, is *choline acetyltransferase factor* (CDF). CDF increases the level of choline acetyltransferase activity in cultured spinal cord neurons and can protect embryonic

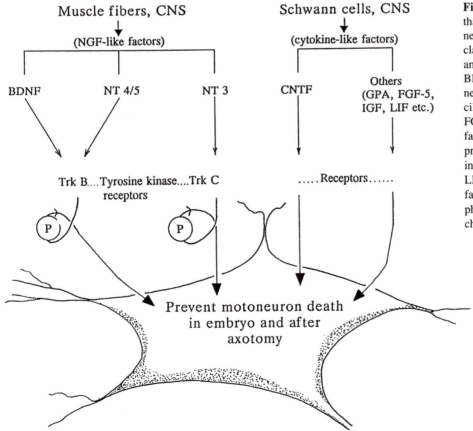

Figure 18.7 Neurotrophins that act on motoneurons. The neurotrophins fall into two classes, those resembling NGF and those resembling cytokines. BDNF = brain-derived neurotrophic factor; CNTF = ciliary neurotrophic factor; FGF-5 = fibroblast growth factor-5; GPA = growth promoting activity; IGF = insulin-like growth factor; LIF = leukemia inhibiting factor; NT = neurotrophin; P = phosphate. Not shown is CDF, choline acetyltransferase factor.

motoneurons from naturally occurring cell death (McManaman, Oppenheim, Prevette, & Marchetti, 1990). Figure 18.8 shows the greater survival of motoneurons in the spinal cord of the chick embryo following treatment with CDF.

In view of the functional and trophic interactions between glia and neurons, it is not surprising that a neurotrophin should have been identified in glia. This factor, *glial-cell-line-derived neurotrophic factor* (GDNF), is as effective as BDNF in preventing motoneurons from undergoing cell death in the embryo or after axotomy (Henderson et al., 1994; Yan, Matheson, & Lopez, 1995).

APPLIED PHYSIOLOGY

The trophic effects of motoneurons and muscle fibers on each other, considered in this chapter, are dramatic examples of a widespread phenomenon in the nervous system. It seems very likely that all neurons in the brain and spinal cord exert trophic influences on each other, and that similar influences exist between neurons and glial cells. It is because of the loss of trophic effects that a lesion in one tract of nerve fibers may cause secondary neuronal degeneration "downstream." For example, damage to the optic nerve produces degeneration not only of the optic nerve fibers, as expected, but also of the neurons in the lateral geniculate nucleus, with which the optic nerve fibers have synaptic connections (Cook, Walker, & Barr, 1951). In Box 18.1 we review an experiment of nature that appears to demonstrate the trophic effects of neurons in the cerebral cortex on motoneurons in the spinal cord.

Now that we have studied denervation and neurotrophism, it is logical to examine the effects of disuse on muscle. This is the subject of chapter 19.

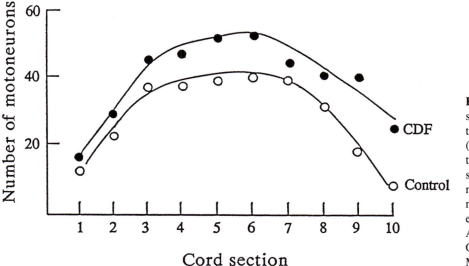

Figure 18.8 Motoneuron survival in chick embryos treated with CDF or saline (*control*). The lumbar region of the cord was divided into 10 sections, and the number of motoneurons in the lateral motor column was counted in each; section 1 is most rostral. Adapted from McManaman, Oppenheim, Prevette, and Marchetti (1990).

BOX 18.1   MOTONEURONS BECOME DYSFUNCTIONAL AFTER STROKES

Are motoneurons trophically dependent not only on muscle, but also on other neurons from which they receive synaptic connections? One way of exploring this possibility would be to examine patients who have had cerebral hemorrhages or thromboses resulting in weakness of the muscles on the opposite side of the body (*hemiplegia*). Vascular lesions of this kind are known to destroy axons descending from the motor cortex to motoneurons of the brainstem and spinal cord. Patients with hemiplegia have not only weakness but also atrophy of muscles on the affected side. Although it is reasonable to attribute the wasting to disuse, examination of such muscles with a needle recording electrode will often reveal spontaneous muscle fiber discharges, or *fibrillations* (see chapter 16; Goldkamp, 1967). Since this finding normally indicates the presence of muscle fiber denervation, it raises the possibility that some motoneurons have undergone dysfunctional or degenerative changes, secondary to loss of trophic input from the motor cortex. The definitive test would be to determine the number of functioning motoneurons supplying one of the muscles on the side of the hemiplegia, and to relate it to the value for the corresponding muscle on the normal side. Such estimates of motoneuron numbers can be made by comparing the potentials or twitch tensions of single motor units with the response of the entire muscle (see Box 12.1 in chapter 12). When this type of analysis is performed on hemiplegic patients, it is found that, for the first 2 months after the stroke, there is no significant difference between the numbers of motor units (and functional motoneurons) on the two sides of the body (McComas, Sica, Upton, & Aguilera, 1973). Soon after that time, however, the number of motor units on the hemiplegic side drops by half, and remains reduced (Figure 18.9).

It is not known why some motoneurons should be more resistant to loss of descending input from the motor cortex than others. What is clear, however, is that there are regions of the brain that can compensate for the disruption of connections from the motor cortex. Thus, if motor unit estimates are performed on patients with severe spinal cord injuries, in whom *all* descending motor pathways have been divided, the reduction in functioning motor units is very much greater (Brandstater & Dinsdale, 1976). These findings are of more than theoretical interest, because they suggest that one therapeutic approach to the problem of stroke would be to prevent the secondary loss of functioning motoneurons.

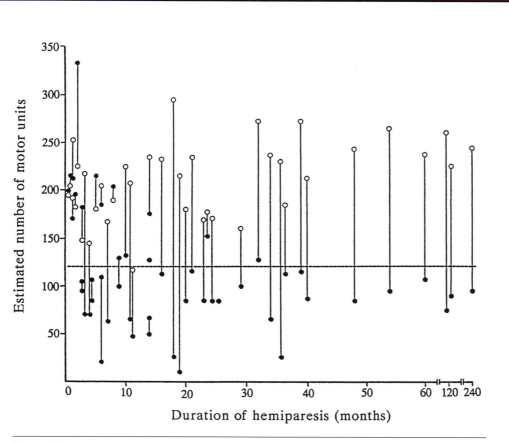

Figure 18.9   Numbers of functioning motor units in the extensor digitorum brevis muscles of hemiparetic patients. Filled and open circles indicate values for hemiparetic and normal sides respectively, and the horizontal line indicates lower limit of normal range.

Reprinted from McComas, Sica, Upton, and Aguilera (1973).

Chapter 19

Disuse

No tissue in the body is more responsive to changes in usage than skeletal muscle. The ability of exercise programs to increase strength or endurance is described in chapter 20; the present chapter deals with the effects of disuse. Inevitably, studies in animals have provided more detailed information than those in humans and have allowed the pathophysiological mechanisms underlying some of the disuse effects to be analyzed more completely.

The Effects of Disuse on Human and Animal Muscles and on Their Motor Innervation

In view of the voluminous literature and the diversity of findings, some generalizations should be made at the outset:

- The effects of disuse are more marked in animals than in humans. The effects of disuse are the opposite of those induced by training.

- The consequences of disuse depend, to some extent, on the experimental model employed; the absence of load bearing is the most important factor.
- In animals, slow-twitch (type I) fibers are more susceptible to disuse than fast-twitch (type II) fibers.
- Not only do disused fibers become smaller, but they may exhibit other morphological changes.
- The motor innervation may also be affected by disuse.
- The effects of disuse are reversible.

Studies in Human Subjects

It is not difficult to find examples of the effects of disuse on human muscle because most people, at one time or another, will have had a relevant experience. As will become evident, however, few of the human studies have been sufficiently comprehensive to include both morphometric data and measurements of muscle contractions.

Bed Rest and Space Travel Affect Muscles Similarly

Anyone who has been obliged to rest in bed because of illness or injury will have noticed a feeling of weakness on standing or walking; further, some leg muscles, the quadriceps in particular, show obvious wasting after as little as 3 to 4 days. A recent and more exotic source of data has been the observations made on astronauts returning from periods of space travel. It should be noted that the amount of EMG activity and joint movement generated by astronauts in performing their mission tasks inside the space shuttle may be considerable. It is clearly the lack of load bearing, due to the low gravitational forces in space, that is the provocative factor in producing loss of muscle mass and weakness. Unfortunately, accessible scientific reports on the effects of space travel on muscle structure and function in human subjects have yet to appear (see Buchanan & Convertino, 1989). Recently, an attempt to simulate the effects of reduced gravity has been made by subjecting healthy individuals to 30 days of bed rest with the head tilted downwards by 6° (Dudley, Duvoisin, Convertino, & Buchanan, 1989; Hikida, Gollnick, Dudley, Convertino, & Buchanan, 1989). With this exception, the only systematic observations on the effects of disuse in humans have been made on subjects whose limbs were immobilized as part of the treatment for injuries, or as an experimental strategy.

Muscle Wasting Is Initially Rapid and Depends on Muscle Length and Previous Usage

Muscle activity can be estimated, rather crudely, by measuring the girth or volume of a limb and correcting for the thickness of the subcutaneous fat. More accurate determinations can be made by visualizing the muscles with computerized axial tomography (CAT) or magnetic resonance imaging (MRI); for convenience, however, and also because of its absence of irradiation, ultrasonography is the preferred method. The results of these various techniques show that limb immobilization produces rapid muscle wasting, first detectable by 3 days (Lindboe & Platou, 1984) and then gradually slowing.

It should be noted that the amount of atrophy will depend on the usage of the muscles prior to immobilization and will therefore always be greater in antigravity muscles than in their antagonists; hence the quadriceps will show more wasting than the hamstrings. Second, because of sarcomere resorption (see p. 70), the length of the muscle at the time of fixation is critical; for example, wasting will be accentuated in the quadriceps because the leg is usually immobilized with the knee

extended, a position that allows the muscle to shorten and therefore encourages loss of sarcomeres. A final point is that estimates of muscle belly shrinkage may underestimate the fiber atrophy if there is an increase in the amount of connective tissue (Jokl & Konstadt, 1983; Tomanek & Lund, 1974).

Both Fast-Twitch and Slow-Twitch Fibers May Atrophy

Using specimens of muscle, most often obtained by needle biopsy, attempts have been made to determine whether there is greater atrophy of slow-twitch (type I) or fast-twitch (type II) fibers following limb immobilization. The results so obtained are conflicting, and probably depend on the choice of muscle, its degree of previous usage, and the amount of isometric muscle contraction that takes place within the cast. As previously noted, atrophy will be greatest in antigravity muscles fixed in shortened positions; if only a small amount of isometric contraction is allowed, this will better maintain the fast-twitch fibers, which are normally used intermittently, than the more constantly employed slow-twitch fibers. In many studies, however, there is significant wasting of both fiber types. As an example of work in this field, Sargeant, Davies, Edwards, Maunder, and Young (1977) found substantial atrophy in both type I and type II fibers in the quadriceps muscles of patients whose legs had been immobilized for fractures; the cross-sectional areas of the type I and type II fibers were reduced by 46% and 37% respectively. Like other estimates of this kind, these values will be rather high because the fibers in the contralateral legs, which are used for comparison, will exhibit some degree of hypertrophy due to the extra weight bearing imposed on that limb. The investigation by MacDougall, Elder, Sale, Moroz, and Sutton (1980) was free from this limitation, in that it was carried out on an arm muscle, the triceps brachii; also, unlike most other studies, it involved healthy subjects rather than patient volunteers. These authors found that, in the long head of the triceps, 5 to 6 weeks of immobilization reduced the cross-sectional areas of type I and type II fibers by 25% and 30% respectively.

An example of preferential type I fiber involvement is a study of soleus muscles in patients with ankles immobilized for 6 weeks after rupture of the Achilles tendon (Häggmark & Eriksson, 1979). Further, if human soleus muscles are examined within 2 weeks of such an injury, some type I fibers can be seen to have developed central cores (Dr. J. Maguire, personal communication, 1994), in keeping with similar observations on the immobilized soleus muscles of small mammals (Karpati, Carpenter, & Eisen, 1972). The observation of White and Davies (1984), that the twitch of the human triceps surae complex is briefer in an immobilized leg than in its normal counterpart, would also be consistent with preferential atrophy of type I fibers.

Does Fiber Type Conversion Occur in Immobilized Human Muscles?

There are now several reports of a decrease in the percentage of human type I fibers after immobilization. For example, this change has been noted in vastus lateralis muscles 4 to 6 weeks after knee immobilization for acute ligamentous injuries (Halkjaer-Kristensen & Ingemann-Hansen, 1985; Ingemann-Hansen & Halkjaer-Kristensen, 1983; Young, Hughes, Round, & Edwards, 1982). Similarly, the proportion of type IIB fibers was increased in soleus muscles of legs immobilized for similar periods of time (Häggmark & Eriksson, 1979). However, the large fiber type sampling error inherent in the needle biopsy technique must be taken into account (Elder, Bradbury, & Roberts, 1982). If these results are valid, however, they corroborate results of experiments with animals (see p. 290).

Muscle Weakness May Exceed Atrophy Because of Poor Motor Unit Recruitment

There have been few studies of stimulated force after a period of limb immobilization, but in general, the decrease in tension (see Figure 19.1) is commensurate with the degree of muscle wasting or rather greater than that expected. White and Davies (1984) attribute the latter finding to a reduction in the specific tension of the atrophied fibers (i.e., force generated for a given cross-sectional area). In their study of the adductor pollicis muscle, Duchateau and Hainaut (1987) noted that, although tetanic force was reduced by 33% after 6 weeks of thumb immobilization, the twitch force remained normal.

Of greater interest is the finding of a more severe reduction in voluntary strength than in stimulated force. An example of this discrepancy is the study of arm fixation in healthy subjects by Sale, McComas, MacDougall, and Upton (1982), in which the median-innervated thenar muscles exhibited a normal twitch but a striking loss of voluntary force; significantly, these muscles did not appear wasted, despite having been in a cast for 5 weeks. Very similar results were found by Duchateau and Hainaut (1987), also in the thumb muscles. The explanation for this rather surprising finding is that, after a period of disuse, the subject is unable to recruit the motor unit population fully—an example of the nervous system "forgetting" a motor task. Thus, Fuglsang-Frederiksen and Scheel (1978) observed that the density of the EMG interference pattern was reduced in comparison with that

Twitch responses - ankle plantarflexors

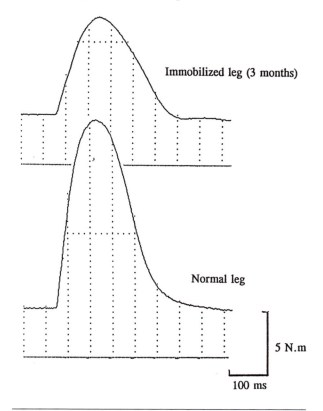

Figure 19.1 Reduced twitch force after immobilization of human plantarflexor muscles. The subject had sustained comminuted fractures of the right tibia and fibula; following surgery the leg was placed in a below-knee cast for 3 months. At the end of this time, the twitch force in the treated leg (top trace) was only half that of the opposite limb (lower trace); the similar contraction and half-relaxation times suggest that the considerable muscle wasting was due to atrophy of both fast-twitch and slow-twitch fibers. Reprinted from McComas (1994).

in the normal leg, when maximal voluntary contraction of the quadriceps muscle was examined with a coaxial recording electrode after a period of disuse (see also Duchateau and Hainaut, 1987). Other evidence of impaired motoneuron activation is the ability to elicit interpolated twitches (see chapter 13) in subjects who, for one reason or another, have reduced physical activity; clinical examples of this phenomenon are patients left moderately disabled after polio (Allen et al., 1994). The functional nature of the problem is illustrated by the fact that motor unit recruitment improves after repeated attempts at maximal contraction.

Disused Human Muscle Is More Fatigable

In their investigation of muscle changes after thumb immobilization, Duchateau and Hainaut (1987) observed that tetanic force declined more rapidly in the treated hand than in the contralateral control (see Figure 19.2). This very clear difference between the two sides is in contrast to the mixed results obtained in animal experiments (see p. 294).

The various findings in disused human muscles are summarized in Table 19.1

Studies in Animals

The experimental methods devised to examine the consequences of disuse in mammalian muscles are remarkable for variety and ingenuity. Some of the techniques have required diligent and skillful nursing of the animals on a daily basis, often for several months. The methods, summarized in Figure 19.3, fall into three categories.

Table 19.1   Summary of Disuse Findings in Human Muscles

Property	Involvement
Muscle atrophy	Greatest for leg antigravity muscles, least for hand muscles
Fiber type atrophy	Both type I and type II fibers affected
Twitch force	May remain normal
Tetanic force	Reduced
Voluntary force	Very reduced
Fatigability	Increased

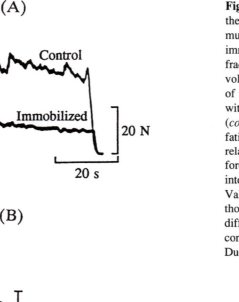

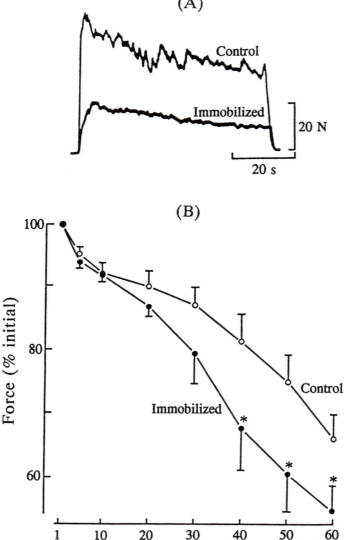

(A)

(B)

Time (s)

Figure 19.2   Disuse effects in the human adductor pollicis muscle, following 6 weeks of immobilization (for forearm fractures). (A) Reduced voluntary torque during 1 min of maximum effort, compared with the untreated hand (*control*). (B) Increased fatigability in two subjects, relative to the control hand. The forces were elicited by intermittent 30 Hz tetani. Values are means ± SEs, and those marked with asterisks differed significantly from controls. Adapted from Duchateau and Hainaut (1987).

1. *Removal of load bearing*

   - Animals in space
   - Hindlimb suspension

2. *Decrease in EMG activity (and in muscle contraction)*

   - General anesthesia
   - Spinal cord isolation
   - Local anesthesia
   - Neurapraxia
   - Botulinum toxin application
   - Curarization
   - Bungarotoxin application
   - Tenotomy

3. *Limb immobilization*

   - Joint pinning
   - Casting

Many of the foregoing strategies were intended to clarify the nature of neurotrophism, that is, whether contractile activity or the delivery of trophic factors by the motoneuron is responsible for maintaining the biochemical and physiological properties of the muscle

fibers (see chapter 18). Consequently, observations usually were made of twitch and tetanic contractions and of muscle fiber histochemistry, rather than of muscle morphology; a notable exception was the pioneering study of Tower (1937), using the isolated cord preparation. A second comment is that the mechanisms involved in the disuse model were often mixed; for example, in hindlimb suspension, there is not only an absence of weight bearing but also a reduction in EMG activity and muscle contraction. A final point is that virtually all the observations have been made on hindlimb muscles of the cat and rat, and there must be some hesitation in applying the findings to other species and to non-weight-bearing muscles, such as those of the human arm. The following commentary is given with these general considerations in mind.

Removal of Load Bearing Affects Slow-Twitch (Type I) Fibers Most

In both the Russian COSMOS flights and those of the NASA program, rats have been subjected to space travel. These studies have shown that the slow-twitch

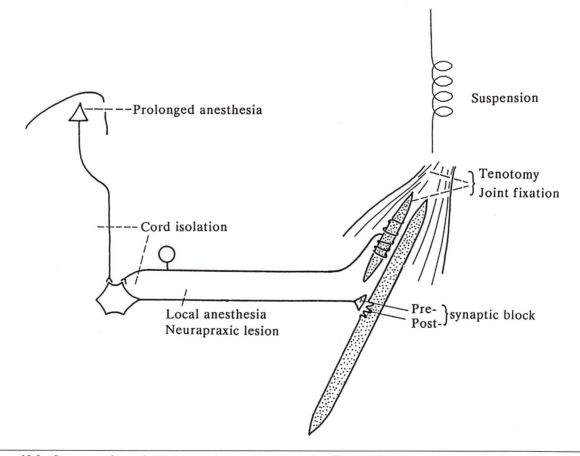

Figure 19.3 Summary of experimental strategies employed to study effects of disuse on animal muscles (see text).

muscle fibers, especially those in soleus, are those most susceptible to the effects of weightlessness; enzyme assays suggest that there is a tendency for type I fibers to become IIA fibers (Martin, Edgerton, & Grindeland, 1988).

Coincident with the recognition of the muscular consequences of space travel has been the development of an animal model of weightlessness, in which the rear of the animal is suspended, compelling it to propel itself around the cage by the forelimbs (Musacchia, Deavers, Meininger, & Davis, 1980; Figure 19.4A). It is evident from chronic EMG recordings with implanted electrodes that the loss of weight bearing is associated with considerable reduction in impulse activity in the hindlimb muscles (e.g., Blewett & Elder, 1993); some of this residual activity produces periodic extension of the hindlimbs, but often no movement can be observed. The results of the various studies employing this model have been unanimous in identifying the type I (slow-twitch) fibers as having the greatest atrophy and loss of tetanic force. If young animals are treated in this way, the incidence of type II fibers is reduced in the soleus, and the muscle has a briefer than normal twitch (e.g., Elder & McComas, 1987; Figure 19.4B).

Quiescence of Motoneurons Produces Atrophy and Other Degenerative Changes in Muscle Fibers

A decrease in EMG activity can be achieved in a noninvasive manner by keeping an animal under *prolonged general anesthesia*. Davis and Montgomery (1977) found that, after 3 weeks, there was little change in the twitch speeds of cat hindlimb muscles. A demanding surgical technique is *spinal cord isolation*, conceived by Sarah Tower (1937). To deprive motoneurons of their normal synaptic bombardment, she divided the spinal cord above and below the segments of interest and also sectioned the dorsal root fibers. Tower observed marked decreases in the diameters of the paralyzed muscle fibers. The fibers failed to stain well, and there were reductions in the numbers and sizes of the myonuclei; the motor end-plate regions of the fibers were well preserved. Associated with the muscle fiber atrophy was considerable proliferation of connective tissue.

Use of the isolated cord preparation was subsequently made by Johns and Thesleff (1961), who found only a small spread of ACh sensitivity in the muscle fibers. Atrophy of muscle fibers was noted by Klinkerfuss and Haugh (1970) and by Karpati and Engel (1968); the last authors also observed features in the muscles that would normally be described as "myopathic." These features included necrosis of muscle fibers, proliferation of endomysial connective tissue, and the presence of muscles

with obvious distortions of their myofibrillar architecture (*ringed* and *snake-coil* fibers).

In recent years, Eldridge (1984) has studied muscle disuse, using cat spinal cords isolated for as long as 3 years. In such animals, the neuromuscular junctions exhibit spread of AChRs and sprouting of the motor nerve terminals (Eldridge, 1984). In addition, the twitches of slow muscles become faster, due to the substitution of fast for slow isoforms of myosin light chains, tropomyosin, and troponin (Steinbach, Schubert, & Eldridge, 1980).

When kittens, rather than adult animals, are subjected to cord isolation, there is a failure of the normal differentiation of slow-twitch muscles; instead the muscles become fast contracting (Buller, Eccles, & Eccles, 1960a) and contain abnormally high proportions of type II fibers (Karpati & Engel, 1968).

Impulses Can Be Chronically Blocked in Nerve Fibers by the Prolonged Infusion of Local Anesthetic or Tetrodotoxin

Robert and Oester (1970) introduced a technique for implanting a plastic cuff containing *local anesthetic* around a muscle nerve. The local anesthetic is gradually released from the cuff and is able to block impulse conduction from the motoneurons for a week or more. In rats treated in this way, there is a spread of ACh sensitivity in the muscle fibers (Lømo & Rosenthal, 1972); further, the paralyzed fibers will readily accept a fresh innervation (Jansen, Lømo, Nicolaysen, & Westgaard, 1973). A possible drawback of such experiments is that the local anesthetic may interfere with axoplasmic transport (Bisby, 1975; Byers, Fink, Kennedy, Middaugh, & Hendrickson, 1973), but this complication can be avoided if *tetrodotoxin* (TTX; see chapter 9) is given instead by intraneural injection. TTX is also found to increase the distribution of AChRs (Pestronk, Drachman, & Griffin, 1976) and to induce terminal sprouting at neuromuscular junctions (Brown & Ironton, 1977).

Pressure Block of Impulse Conduction Causes More Severe Effects in Rodents Than in Larger Mammals

Either through accident or by experimental design, nerve fibers may be compressed to an extent which leaves their axis cylinders in continuity, while temporarily depriving their membranes of excitability and hence of the ability to conduct impulses (*neurapraxia*). Weir Mitchell, a surgeon in the American Civil War, described transient palsies in some of his patients after bullet wounds: "This condition of local shock is very curious. A man is shot in the thigh, the ball passes near

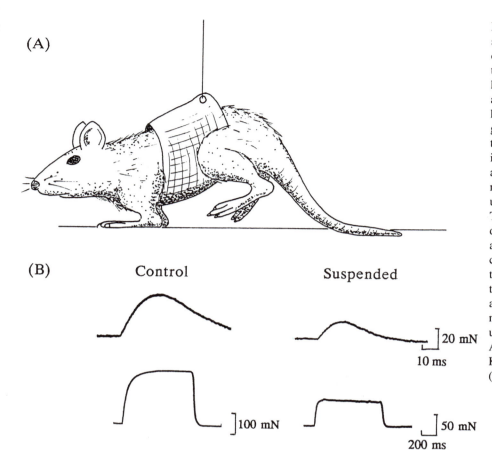

(A)

(B)   Control   Suspended

]20 mN

10 ms

]100 mN

]50 mN

200 ms

Figure 19.4 (A) Hindlimb suspension model of disuse. A cloth harness is fitted around the body of the animal (rat or hamster) and is suspended from a mobile arm, so that the hindpaws are kept clear of the ground. The animal is still able to move around the cage using its forepaws and to feed, drink, and sleep; the body weight is rather lower than that of untreated animals, however. (B) Twitch and tetanic forces developed by soleus muscles in a suspended hamster and in a control of the same age. Note the smaller forces and faster twitches in the suspended animal; the fast-twitch plantaris muscle (not shown) was unaffected by suspension. Adapted from Corley, Kowalchuk, and McComas (1984).

the sciatic nerve, and instantly the limb is paralysed; within a few minutes, or at the close of a day or a week, the volitional control in part returns'' (Mitchell, Morehouse, & Keen, 1864). Denny-Brown and Brenner (1944) produced neurapraxic lesions in cats by compressing the sciatic nerve with a bag containing mercury at known pressures. In muscles that had been paralyzed for as long as 3 weeks, they observed that there was an absence of atrophy and of fibrillation activity. Therefore, they reasoned "... anatomic continuity of nerve, not receipt of impulses, prevents atrophy and fibrillation of muscle." A clinical example of neurapraxia is shown in Box 19.1. Extensive use of nerve compression, as an experimental tool, was subsequently made by Gilliatt and colleagues. They were able to demonstrate that impulse block results from the sliding of a node of Ranvier under the membrane of the adjacent segment of nerve (*intussusception*; Ochoa, Danta, Fowler, & Gilliatt, 1971). In a later study, also in the baboon, Gilliatt and colleagues showed that, following a neurapraxic lesion, the spread of ACh sensitivity is considerably less than that after surgical denervation (Gilliatt, Westgaard, & Williams, 1978). The neurapraxic model has also been employed, by Kowalchuk and McComas (1987), to investigate the effects of disuse on rat hindlimb muscles. After only 1 week of impulse block, there

is marked wasting of slow- and fast-twitch muscles, with significant prolongation of the twitch in the fast-twitch muscle; in this instance, the changes are as severe as those reported in rat muscles after surgical denervation.

Certain Neurotoxins Block Neuromuscular Transmission, Producing Muscle Fiber Atrophy and Nerve Terminal Sprouting

Botulinum toxin is produced by soil bacteria and is responsible for an especially dangerous type of food poisoning; it interferes with the emptying of ACh from synaptic vesicles in the motor nerve terminal (see chapter 10). Both fast-twitch and slow-twitch muscles undergo considerable atrophy after treatment with botulinum toxin. The fast muscles, in particular, show a number of degenerative features (Thesleff, Molgó, & Tågerud, 1990). In addition, the resting membrane potentials of the fibers fall, and the twitch durations increase. At the surfaces of the fibers, botulinum toxin induces a spread of ACh sensitivity beyond the endplate region (Thesleff, 1960); the fibers also synthesize Na^+ channels that are abnormally resistant to TTX. Outside the neuromuscular junction, acetylcholinesterase disappears from the fiber surface, but the synaptic moiety, in contrast, is largely retained. The situation can be summarized by stating that the effects induced by botulinum toxin are qualitatively similar to those following surgical denervation of the muscle fiber, but they are less pronounced.

Another toxin employed in the study of disuse is β-*bungarotoxin*, which is one of the toxins in the venom of the banded krait, *Bungarus multicinty*, a poisonous snake indigenous to parts of southern Asia. Like botulinum toxin, β-bungarotoxin prevents ACh release and causes similar changes to take place in the muscle fiber membrane (Hofmann & Thesleff, 1972). It is possible that presynaptic blocking agents such as botulinum toxin and β-bungarotoxin, in addition to depriving the muscle fiber of ACh and impulse activity, also prevent the release of neurotrophins. This potential complication is avoided if postsynaptic blocking agents, such as *d-tubocurarine*, *succinylcholine*, and α-*bungarotoxin*, are used instead. Berg and Hall (1975) have used these substances to paralyze rats, maintaining the animals for 3 days with artificial respiration. At the end of this time, the muscle fibers show such denervation features as a fall in resting potential, spread of ACh sensitivity, and resistance to TTX.

Striking Changes Follow Tenotomy and Joint Fixation, but Are Difficult to Interpret

The purpose of *tenotomy* is to allow the muscle to shorten passively and thereby diminish any excitatory input to the motoneurons from the muscle spindles; the motoneurons should therefore become quiescent. Following tenotomy in the rabbit, there is pronounced atrophy, particularly of type I fibers, and this is associated with a speeding-up of the twitch (Vrbová, 1963). However, the experimental situation is not ideal because the motoneurons may be subjected to increased, rather than reduced, sensory bombardment (Hník, 1972). Again, by depriving the muscle fibers of passive stretch, atrophic changes may be anticipated anyway (Gutmann, Schiaffino, & Hanzlíková, 1971).

Experimental immobilization, achieved by *bone pinning*, has been applied to animals by Fischbach and Robbins (1969). A very considerable reduction in background impulse activity is produced in muscles, and marked speeding of the twitch can be seen. At the single motor unit level, it has been possible to confirm the susceptibility of S (type I) units to immobilization, but it appears that the FR (IIA) units are even more affected, as judged by reductions in their twitch and tetanic tensions. Although smaller changes in force generation are observed in FF (IIB) units, both these and S (I) units exhibit a speeding-up of their twitches (Mayer et al., 1981; see also St.-Pierre & Gardiner, 1985).

Perplexing results from disuse experiments have been reported by Robinson, Enoka, and Stuart (1991), who immobilized the hindlimbs of cats for 3 weeks by unilateral *casting*. While tetanic force decreased in slow-twitch motor units of the tibialis posterior, as might have been anticipated from previous studies (e.g., St.-Pierre & Gardiner, 1985), the twitch and tetanic tensions actually *increased* in fast-twitch units. However, this apparent anomaly could have arisen, in part, from collateral reinnervation of functionally denervated fibers (see p. 260).

Disuse Has Produced Variable Effects on Muscle Fatigability

In keeping with the discordant findings in single motor units (see the preceding sections), animal models of disuse have produced variable effects on muscle fatigability. Increased susceptibility to fatigue has been reported after hindlimb suspension, by Fell, Gladden, Steffen, and Masacchia (1985), although decreased fatigability was observed by others (Haida et al., 1989; Robinson et al., 1991). No significant changes were noted by Mayer et al. (1981) or by Witzmann, Kim, and Fitts (1983). However, in those studies in which increased fatigability was seen, the findings are consistent with similar observations in humans (Duchateau & Hainaut, 1987) and with reports of decreased oxidative capacity in animal muscles after disuse (Booth, 1977; Rifenberick, Gamble, and Max, 1973).

Disuse Also Modifies the Neuromuscular Junction

After as few as 5 days of immobilization, motor axon terminals begin to sprout and to become distorted in their longitudinal axes (Fahim & Robbins, 1986). With the electron microscope, other signs of degeneration can be found in the neuromuscular junctions of type I and type II fibers; these include exposure of the synaptic folds and disruption of nerve terminals (Pachter & Eberstein, 1984). The presence of several small axons over the same primary cleft suggests that some nerve terminals may also be regenerating.

These morphological changes could well be related to a surprising finding by Robinson et al. (1991). In immobilized cat tibialis posterior muscles, they discovered that 12% of motor units developed no measurable force when their motor axons were stimulated (see Figure 19.5). The concept of nontransmitting neuromuscular junctions ("silent synapses") is one that has been developed for human muscles in a variety of clinical conditions (see chapter 16).

The Lengths of the Immobilized Muscle Fibers Will Affect the Number of Sarcomeres

Tabary, Tabary, Tardieu, Tardieu, and Goldspink (1972) showed that the fibers in the cat soleus muscle could respond to immobilization of the hindlimb in a plaster cast by altering the number of sarcomeres. If the muscles were fixed in a shortened position, sarcomeres were removed, while lengthening the muscles caused sarcomeres to be added. Williams and Goldspink (1978) subsequently demonstrated that this remarkable plasticity was taking place at the ends of the fibers. The functional advantages of the adaptation are evident, if the artificiality of immobilization is ignored; since the modified fibers can once again operate at normal sarcomere lengths, the actin and myosin filaments will be able to overlap optimally and to generate maximal tensions.

It is important, in the rehabilitation of orthopedic patients, to appreciate that much of the stiffness of a limb, newly removed from a cast, is due to shortening of the muscle fibers, as well as to changes in the joint capsules and ligaments. As the shortened muscle is stretched, damage to the myofilaments will inevitably occur; however, the end result of stretching is myofibrillar protein synthesis and the eventual restoration of the missing sarcomeres.

Some Disuse Effects Are Brought About by Changes in Gene Expression

Since new proteins may be produced in the muscle fibers as a result of disuse, it is evident that gene expression must be modified at the transcriptional or translational levels. In the case of the myofibrillar proteins, immobilization in a shortened position causes the fast (IIB) myosin heavy chain gene to be expressed in the slow soleus muscle of the rat (Loughna, Izumo, Goldspink, & Nadal-Ginard, 1990). Since the same gene is repressed if the soleus is fixed in a lengthened position, it is clear that it is the absence of passive stretch which enables this gene to be expressed in this type of situation.

Other Biochemical Changes Occur in Disuse

Most of the biochemical studies of disuse have been conducted on animal, rather than human, muscles (for review, see St.-Pierre & Gardiner, 1987). As would be anticipated from the presence of muscle atrophy, there is a loss of myofibrillar proteins from the fibers, but, in addition, the concentration of these proteins is reduced in the residual muscle tissue. This last finding may explain why, in some studies, the twitch and tetanic forces developed by disused muscles are diminished in

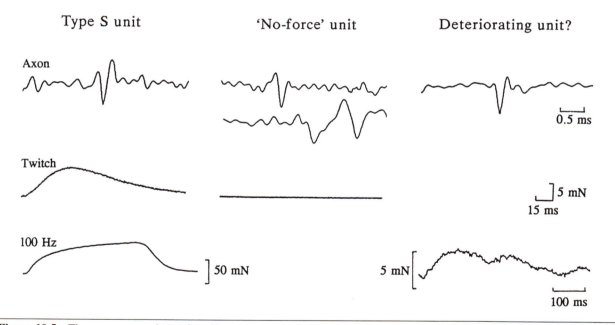

Figure 19.5 The appearance of "no-force" motor units in the cat tibialis posterior muscle following hindlimb immobilization. The results of stimulating three different ventral root fibers are shown. In each case the upper trace displays the averaged axon potential, and the middle and lower traces, if present, show the isometric twitch and the 100-Hz tetanic tension respectively. The type S (type I) unit gave normal responses. The "no-force" unit generated an axon potential but no twitch force. The additional trace at the top (middle row) shows an averaged intramuscular recording of axon (or EMG?) activity. The unit on the right developed very little tetanic force, possibly because it was becoming inactive. From "Immobilization-induced changes in motor unit force and fatigability in the cat," by G.A. Robinson, R.M. Enoka, and D.G. Stuart, 1991, *Muscle and Nerve*, **14**, pp. 566, 571. Copyright 1991 by John Wiley & Sons, Inc. Reprinted with permission.

relation to the cross-sectional areas of the muscles (i.e., reduced "specific tension"). In contrast to the myofibrillar proteins, there are increases in connective tissue proteins (Herbison, Jaweed, & Ditunno, 1978; Jokl & Konstadt, 1983). The concentrations of glycolytic and oxidative enzymes may also be lessened by disuse, but the length at which the muscle was immobilized is critical; thus, decreases in concentration are more likely to be found if the muscle is fixed in a shortened position.

Atrophy Is the Main Consequence of Disuse in Animal Muscle Fibers

It is time to summarize the effects of disuse on animal muscles, bearing in mind that there will be some variation depending on the methodological model employed as well as on the choices of muscle and species (Figure 19.6).

As far as the *muscle fibers* are concerned, the most striking consequence of disuse is atrophy, especially in fibers of the slow-twitch (type I) variety. In addition, a minority of fibers undergo necrosis, and there is an increase in the endomysial and perimysial connective tissue. The muscles develop smaller twitch and tetanic tensions, even beyond those expected on the basis of fiber atrophy. There is also a tendency for slow-twitch fibers to be transformed into fast-twitch fibers, with attendant changes in the isoforms of the myofibrillar proteins.

At the surfaces of the disused fibers, there is a spread of AChRs beyond the neuromuscular junction, and the resting membrane potential is diminished. The motor

nerve terminals are abnormal in showing signs of degeneration in some places but also evidence of sprouting in others.

Finally, there is a loss of motor drive after a period of disuse, such that the motor units cannot be recruited fully.

APPLIED PHYSIOLOGY

In this section we will make some brief comments about the possible contributions of disuse to patient disability. Following this, Box 19.1 presents an example of muscle disuse in a young patient with prolonged impulse block after an unusual mishap.

In Patients With Disabling Illnesses, Important Secondary Effects Can Be Caused by Disuse

Patients often complain of weakness and increased fatigability, and it is essential to recognize that much of the problem may be due to disuse, rather than to the primary illness. Disuse will produce weakness through atrophy of muscle fibers and loss of myofibrillar protein. Of equal importance, however, is the inability of the motor centers in the brain to recruit motoneurons fully, due to disuse of the descending motor pathways (see p. 289). Finally, disuse will increase the fatigability of those motor units that are still functional, as can be seen in Figure 19.7.

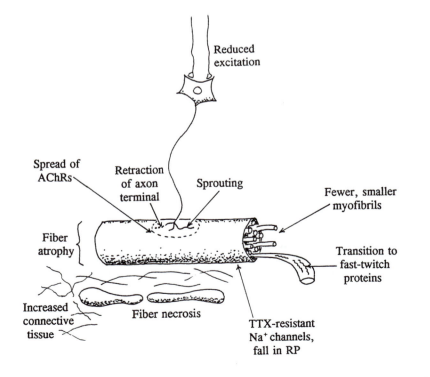

Figure 19.6 Summary of changes in muscle and nerve fibers that may follow disuse; RP = resting potential. See text.

The opposite of disuse is, of course, overuse, and this is the subject of the next chapter, which is devoted to the effects of training. We will see that, as with disuse, there have been some ingenious experiments in animals and some interesting, if incomplete, observations in humans.

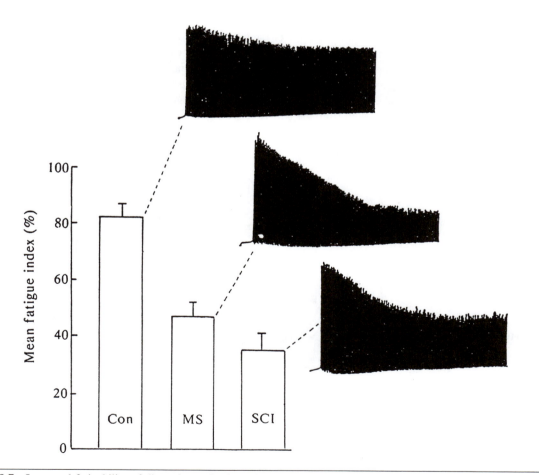

Figure 19.7 Increased fatigability of disused muscles in patients with injury or disease of the spinal cord. The histograms show the respective mean fatigue indices (+ SEMs) for patients with multiple sclerosis (*MS*) or spinal cord injury (*SCI*), and for control subjects (*Con*). Each histogram is linked to a typical recording of the forces developed by intermittent stimulation at 30 Hz for 3 min. The fatigue index is the amplitude of the last response as a percentage of that of the first response. From "Muscle fatigue in some neurological disorders," by R.A.J. Lenman, F.M. Tulley, G. Vrbová, M.R. Dimitrijevic, and J.A. Towle, 1989, *Muscle and Nerve*, **12**, p. 940. Copyright 1989 by John Wiley & Sons, Inc. Adapted with permission.

BOX 19.1   A CLINCIAL EXAMPLE OF NEURAPRAXIA

This example of neurapraxia occurred in an 11-year-old schoolgirl who, 2 months before the EMG examination, had caught her left arm between the rollers of a clothes wringer; the arm had been drawn in to the midforearm level. Immediately after the accident, she had noticed numbness of the whole hand and muscle weakness; the latter affected all the intrinsic muscles of the left hand together with the extensors of the fingers and wrist. Both sensation and muscle strength had been improving up to the time of the EMG study.

Figure 19.8 shows the responses evoked in the thenar and hypothenar muscles following supramaximal stimulation of the median and ulnar nerves. It can be seen that the muscle potentials were much larger when the nerves were excited at the wrist rather than at the elbow. The compact forms and normal configurations of the diminished potentials suggest that the discrepancy had not arisen from dispersion of the muscle responses due to slowed nerve impulse conduction. Instead the findings must have resulted from neurapraxic lesions in the forearm, which had caused local inexcitability of median and ulnar nerve fibers. In relation to an understanding of neurotrophic phenomena (see chapter 18), it is significant that, in spite of the interruption of the centrifugal flow of impulses, the paralyzed muscle fibers showed no other evidence of denervation. Thus, the muscles did not appear wasted, nor could fibrillation potentials and sharp-wave activity be detected during exploration of the muscle with a needle electrode (Figure 19.8, bottom). It also was unlikely that any collateral reinnervation had occurred, for the amplitudes of the potentials evoked from the still-functioning motor units were not enlarged.

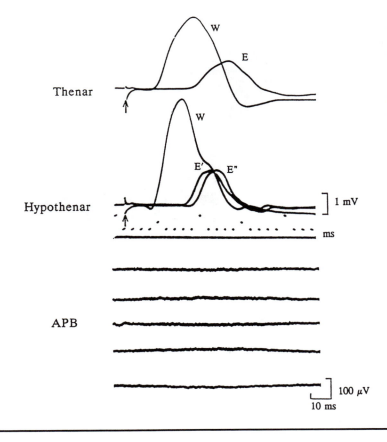

Figure 19.8 Responses evoked from the hand muscles of a patient with neurapraxia, using a stimulating and recording arrangement similar to that shown in Figure 9.10. The thenar and hypothenar responses are much larger when the median and ulnar nerves are stimulated at the wrist (*W*) than at the elbow (*E*) or just below and above the elbow (*E'*, *E"*). The differences in size indicate the presence of nonconducting nerve lesions in the forearm. The lower part of the figure shows the absence of fibrillation activity when a needle electrode is inserted into the abductor pollicis brevis (APB) muscle.

Chapter 20

Muscle Training

A number of important biochemical changes take place within the muscle fiber during contractile activity, but the normal milieu within the fiber is usually restored by a variety of enzymatic processes within the next minutes or hours. However, if the contractions are unusually strong or prolonged, and if they are repeated, then a combination of structural and biochemical adaptations take place that ensure that the muscle is better suited to the type of work demanded of it. Thus, if force or power is required, the muscles become stronger; if, on the contrary, the muscle activity is of long duration, then the muscles become less readily fatigued. Although the changes are most obvious in the muscle fibers, the motoneurons are also affected in terms of their patterns of recruitment and impulse discharges. When the neuromuscular adaptations are intended, the bouts of exercise can be designed so as to constitute effective training programs. The purpose of this chapter is to consider the nature of the adaptive changes and the methods by which they can be induced and investigated. As in so many other situations, the human and animal studies have different advantages and disadvantages, and so tend to complement each other.

Human Strength Training

Thomas DeLorme (1945), a captain in the U.S. Army during the Second World War, had many patients in his care with severe wasting and weakness of the quadriceps muscles caused by disuse following knee injuries. De-Lorme reasoned that it was important to involve as much of the muscle fiber population as possible in the exercise program. Therefore, rather than subject his patients to many repetitions of low-intensity contractions, as had been the normal practice, he made them lift a heavy load instead. The load was one that could only just be raised 10 times in a sequence of contractions; as the muscles became stronger, the load was increased accordingly. DeLorme found that a simple way of exercising was to attach the weights to a metal boot (see Figure 20.1, top).

Small Numbers of Intensive Contractions Are the Most Effective Means of Improving Muscle Bulk and Strength

The progressive resistance training program has since become accepted as the most effective means of improving muscle bulk and strength and, as such, has

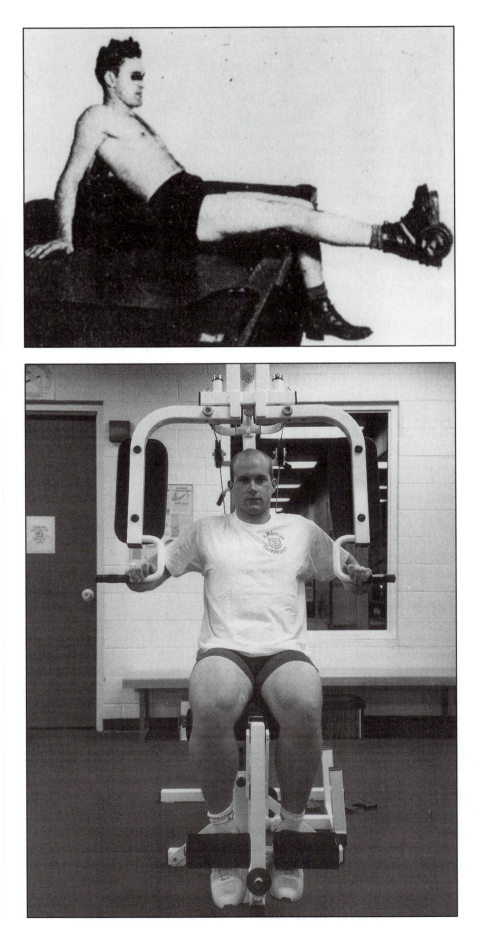

Figure 20.1 Different methods for improving muscle bulk and strength. (Top) Metal boot with detachable weights devised by DeLorme (1945, p. 646). (Bottom) Modern strength training, as in fitness clubs.

been adopted widely by body builders and various types of athlete. An important point is that, whether exercising with weights or on one of the newer strength-training machines (see Figure 20.1, bottom), the muscles not only contract concentrically, as the load is lifted, but also eccentrically, as the load is returned to its original position. Since the muscle fibers are liable to be damaged by eccentric contractions (Fridén, Sjöström, & Ekblom, 1983), it is possible that cellular repair mechanisms promote the gain in muscle bulk; on the other hand, eccentric exercise has not been found superior to concentric in increasing muscle strength (Jones & Rutherford, 1987). So effective is the progressive resistance training program that no more than 10 repetitions a day, using loads from 60 to 90% of maximum, will increase strength by 0.5 to 1.0% per day over a period of several weeks (Jones, Rutherford, & Parker, 1989). An example is given in Figure 20.2, taken from a study in which healthy subjects trained by lifting near-maximal loads on a leg-extension machine. It can be seen that over a period of 12 weeks the mean maximum strength almost tripled (Rutherford & Jones, 1986).

Maximal Muscle Enlargement Requires Long-Term Training

Although a progressive resistance training program can produce quite striking muscle enlargement in dedicated athletes (Figure 20.3, left), scientific studies have resulted in only modest muscle changes. In these studies, the cross-sectional areas of the muscles were reliably outlined by modern diagnostic imaging machines such as ultrasound, computed transaxial tomography, and magnetic resonance. In a number of training studies lasting 3 to 5 months, the increases in muscle cross-sectional areas increased by only 9 to 23% (cf. Frontera, Meredith, O'Reilly, Knuttgen, & Evans, 1988; Ikai & Fukunaga, 1970; MacDougall, Ward, Sale, & Sutton, 1977).

In contrast, MacDougall, Sale, Alway, and Sutton (1984) found that the cross-sectional areas of the biceps brachii muscles were, on average, 76% greater in body builders than in untrained controls. The discrepancy between these results is probably due to the much longer training times undertaken by the body builders in comparison with the experimental subjects, since the former group will have exercised for a period of years rather than months. However, the situation is complicated by the possibility that anabolic steroids may have been used by the body builders and "power" athletes.

Type II (Fast-Twitch) Fibers Show the Greatest Hypertrophy

By performing needle biopsies of muscles before and after training, and then staining the cut sections histochemically, it is possible to determine to what extent the different types of fiber hypertrophy. The results of four such studies are given in Table 20.1 (McComas, 1994). In three of these, the type II (fast-twitch) fibers show much larger increases in cross-sectional area than the type I (slow-twitch) fibers. The greater areas of the fibers, in turn, are due to increased numbers and sizes of myofibrils; it is not known whether some of these are synthesized *de novo* or whether all are formed by the splitting of thickened myofibrils (Goldspink, 1965). Associated with the new myofibrils are increases in the numbers of mitochondria and in the amounts of T-tubular and sarcoplasmic reticular membranes.

Transformation of Fibers From Type IIB to IIA May Occur During Strength Training

There is evidence that the incidence of slow-twitch (type I) fibers is significantly reduced in sprinters, but not in other types of power athletes, such as throwers, weight lifters and high jumpers (Saltin, Henriksson, Nygaard, & Anderson, 1977). In this study, it was not known whether the reduction in type I fibers in sprinters was due to training or to genetic endowment.

In a subsequent investigation in which sprinting was performed on a cycle ergometer, it was found that training could cause a reduction in type I fibers (Jansson, Esbjörnsson, Holm, & Jacobs, 1990), though not of the same magnitude as that observed in the earlier study. However, the simple identification of fibers as type I or type II may mask certain induced changes in the pattern of myosin heavy chain (MHC) isoforms. For example, although the proportions of type I and type II fibers are normal in body builders, there may be an almost total absence of fibers with the MHC IIB isoform but a much higher than normal incidence of fibers with the IIA isoform (Klitgaard, Zhou, & Richter, 1990). Similarly, Staron, Malicky, and colleagues (1990) have found that the incidence of type IIA fibers can increase at the expense of IIB fibers; they studied a group of 24 women who underwent strength training over a 20-week period.

Muscle Fiber Hyperplasia Can Occur in Birds but Probably Not in Human Muscles

In the past, there has been argument as to whether the muscle enlargement is entirely due to fiber hypertrophy or whether there is an associated increase in the

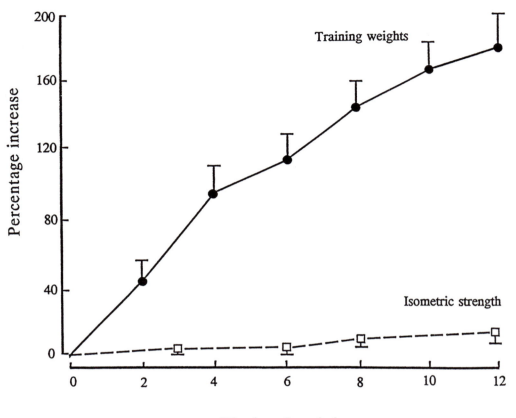

Figure 20.2 Marked improvement in the ability of the knee extensors to lift weights, following the onset of a training program (upper curve). The lower curve shows the relatively small increase in isometric strength, measured with the knee at a right angle, in the same subjects. Adapted from Rutherford and Jones (1986).

number of muscle fibers (*hyperplasia*). The latter situation might be expected to arise in one of two ways—by splitting of the muscle fibers or by satellite cell proliferation following muscle fiber damage. In relation to the first possibility, it is certainly true that hypertrophied fibers may split into two or more daughter fibers. In this process, the myonuclei first move into the center of the fibers and there direct the laying down of membranes (Hall-Craggs, 1970). It is unlikely, however, that the daughter fibers become fully separated, for serial sections show that the splits do not usually run the full length of the parent fiber (Isaacs, Bradley, & Henderson, 1973). Further evidence against hyperplasia is that the cross-sectional areas of the muscle fibers, measured in biopsy specimens taken before and after training, increase in the same proportion as the cross-sectional area of the whole muscle (MacDougall et al., 1984). Although MacDougall and colleagues found considerable variation in fiber numbers (172,000-419,000) of the biceps brachii among

their subjects, this variation was equally prominent in control subjects and body builders, and the mean numbers of fibers were not significantly different in the two groups. Therefore the number of fibers in a muscle appears to be determined genetically and not to be increased by strength training.

The situation in human subjects is in contrast to that in birds, in which chronic load bearing induces marked fiber hyperplasia (see p. 307). Perhaps the difference between the species is related to the fact that some avian muscles are largely composed of *slow tonic* fibers. These are fibers which give graded electrical and mechanical responses following nerve stimulation, and over which the motor nerve terminals are extensively distributed rather than grouped together in a single end-plate (Ginsborg, 1960). While slow tonic fibers are a prominent feature of avian and amphibian muscles, in mammals they are only found in external ocular and inner ear muscles (Hess, 1970) and in muscle spindles.

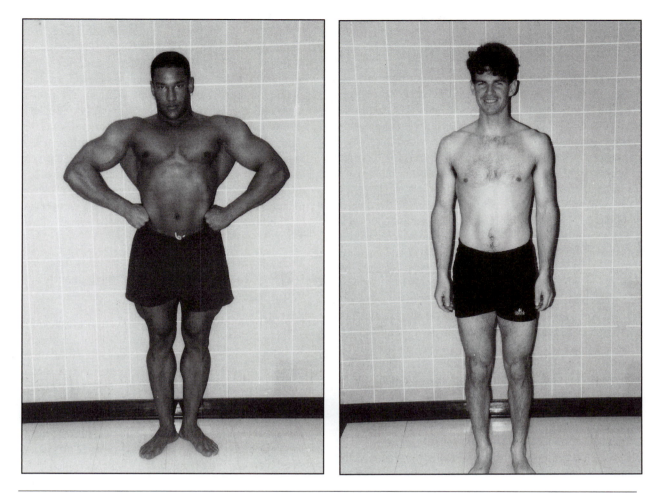

Figure 20.3 Contrasting body shapes of a body builder (left) and a long-distance runner (right).

Table 20.1 Effects of Strength Training on Cross-Sectional Areas of Muscle Fibers Belonging to Main Histochemical Types

Authors	Number of male (m) & female (f) subjects	Muscle	Duration of training (weeks)	Changes in mean fiber area (%)		
				I	IIA	IIB
Hather, Tesch, Buchanan, & Dudley (1991)	8 m	Vastus lateralis	19	+14	----+32----	
Houston, Froese, Valeriote, Green, & Ranney (1983)	6 m	Vastus lateralis	10	+3	+21	+18
MacDougall, Elder, Sale, Moroz, & Sutton (1980)	7 m	Triceps brachii	22-26	+15	----+17----	
Staron, Malicky, et al., (1990)	24 f	Vastus lateralis	20	+15	+45	+57[a]

[a]Includes type IIAB fibers.

Note. Reprinted from McComas (1994).

Some of the Increase in Muscle Strength Is Brought About by Neural Mechanisms

Two puzzling but related findings from progressive resistance training are that, during the early stages, the increase in strength not only exceeds the change in muscle girth but is largely restricted to the training task (Enoka, 1988; Sale, 1988). Thus, when the trained subjects with the results portrayed in Figure 20.2 carried out maximal isometric contractions of their quadriceps muscles, their mean strength was found to have increased only 15%. More remarkable still, the power output of the same muscles, measured with a cycle ergometer, was unchanged. These discrepancies in performance after training illustrate a general rule, namely: The greatest changes are found in the training exercise itself (*specificity of the training response*).

The possible reasons for this specificity have been considered in some detail by Jones et al. (1989); they include differences in optimum muscle length between the training exercise and other tasks, and differences in the relative contributions of synergistic muscles. The main factor, however, appears to be a *neural adaptation* that may be expressed in various ways—for example, by higher initial motoneuron discharge frequencies and by more persistent firing of high-threshold motor units (Grimby, Hannerz, & Hedman, 1981). Also, antagonist muscles, which may initially have been coactivated in the performance of the task, become less responsive as the training performance proceeds (Carolan & Cafarelli, 1992), presumably through inhibition of the α-motoneurons (see Figure 20.4).

Human Endurance Training

It is an everyday experience that frequent repetition of a motor task improves endurance, and this is reflected in the long training distances covered by competitive runners, cyclists, and swimmers. Nor are the beneficial effects restricted to young adults and to those in good health; impressive gains in endurance can be achieved by the elderly (for review, see Vandervoort, Hayes, & Bélanger, 1986) and by patients with previous myocardial infarctions (Todd, Wosornu, Stewart, & Wild, 1992). There have been few physiological studies of human muscles before and after an endurance training program, but Figure 20.5 shows one example.

Endurance Training Produces Muscles That Have Greater Resistance to Fatigue and May Also Be Thinner Than Before

Hainaut and Duchateau (1989) asked their subjects to perform 200 submaximal adductions of the thumb every day over a 3-month period. At the end of this time, tetanic stimulation of the adductor pollicis showed that force was maintained better than before when tested over a 1-min sampling time. In more proximal muscles of the limbs, endurance training results in muscles that are not only more effective during sustained activity, but also, in the case of long-distance runners, more slender (Figure 20.3, right). The explanation for the smaller girth is that the myofibrils, and the fibers themselves, are reduced in cross-sectional area. It is likely that this adaptation allows better diffusion of metabolites and nutrients between the contractile filaments and the cytoplasm, and between the cytoplasm and the interstitial fluid.

Decreased Fatigability Is Partly Due to Increased Numbers of Mitochondria and Capillaries

Ingjer (1979) trained 7 young women in cross-country running for 24 weeks. In muscle biopsies, he observed that the numbers of *capillaries* had increased significantly around fibers of all histochemical types, while the type I fibers had gained the most *mitochondria*. During exercise, the more plentiful mitochondria would be able to keep the fibers better supplied with ATP, using aerobic metabolism. Similarly, the more extensive capillary bed would be expected to improve the delivery of oxygen and circulating energy sources (glucose, free fatty acids) to the fibers, while the products of muscle activity, especially H^+, K^+, and lactate, would be removed more efficiently. Surprisingly, it appears that the increased size of the capillary bed is associated with a correspondingly greater muscle blood flow only during *severe* exercise (Hudlicka, 1990). During less extreme exercise, however, it is conceivable that oxygen extraction is enhanced by the larger contact area between the capillary blood and the interstitial space, even though the blood flow rate in individual capillaries may be lower than that in untrained muscles.

Endurance Training Is Associated With Increased Percentages of Type I and Type IIA Muscle Fibers, and of Intermediate Fibers

In four studies of endurance athletes, the incidences of type I muscle fibers were increased, and those of type IIB fibers decreased, in comparison with controls (see Table 20.2). These results leave open the question of whether differences in fiber composition are due to transformation from one fiber type to another, or whether athletes are able to excel in endurance events because they are genetically endowed with a preponderance of type I and type IIA fibers. To resolve this

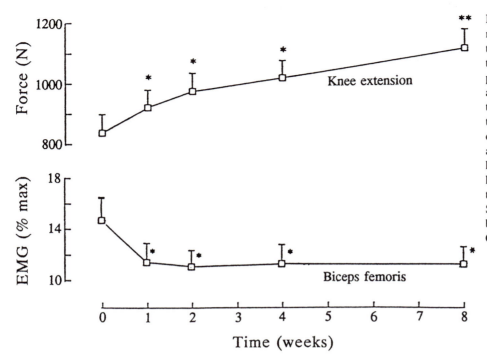

Figure 20.4 One of the neural mechanisms involved in strength training. Twenty male subjects trained their knee extensors to perform isometric contractions, and the improvement is seen in the top curve. During the training period, there was a decrease in the associated activity in the biceps femoris, a hamstring muscle that opposes knee extension, as reflected in the EMG (bottom curve). Significant changes are shown by asterisks. Adapted from Carolan and Cafarelli (1992).

issue, longitudinal studies have been undertaken in which previously untrained subjects engaged in various types of prolonged physical activity—for example, cross-country running, skiing, or cycling. Some of these results are shown in Table 20.3. It can be seen that, in contrast to the rather variable and generally unimpressive results of human strength training regimens (previously discussed), endurance training produces well-defined, consistent changes. Like electrical stimulation in animals, the results of human endurance training are to make some of the type II fibers acquire the physiological, biochemical, and structural features of type I fibers; other fast-twitch fibers alter their myosin ATPase activity and so convert from type IIB to type IIA.

By employing antibody staining and gel electrophoresis, it has been possible to study in more detail the changes in MHC isoforms that are responsible for the altered histochemical staining. As an example, Schantz and Dhoot (1987) studied the triceps brachii muscles of 6 subjects who skied and pulled sledges for 800 km in mountainous territory over a 36-day period. At the end of this training period, they found that a significant proportion of fibers contained both fast and slow isoforms of the MHCs, and hence could be regarded as intermediate fibers. In parallel with the changes in the MHCs was the appearance of slow isoforms of troponins I, T, and C. Similar results have been found by Klitgaard, Zhou, et al. (1990) and by Baumann, Jäggi, Soland, Howald, and Schaub (1987); in the latter study, in which

subjects trained on a cycle ergometer, there was also a transformation of MHCs from the fast to the slow variety. The general conclusion from the different studies is that, if fibers transform as a result of endurance training, they do so in the following sequence: IIB → IIA → IIC → I.

In Infancy, Weight Bearing May Promote Conversion of Type II Fibers to Type I

An indirect method of studying the fiber-type composition of a muscle is the measure of time occupied by a twitch contraction, since type I fibers have slower twitches than type II fibers. The technique has the merit of being noninvasive and of being applicable to the same muscle on a later occasion; its disadvantage is that the twitch duration depends as much on the intracellular Ca^{2+} kinetics as on the rate of myosin ATPase reaction at the cross-bridges. In one study, in which the adductor pollicis muscle was stimulated supramaximally for 3 hr a day for 6 weeks, the twitch half-relaxation time did not alter (Rutherford & Jones, 1988). In infants, however, a distinct prolongation of the twitch can be observed in the ankle plantarflexor muscles between 6 and 12 months of age (see Figure 5.14 in chapter 5). This change comes at a time when the child is beginning to stand and to support the body weight through the antigravity muscles, including the plantarflexors. In contrast, the ankle dorsiflexors,

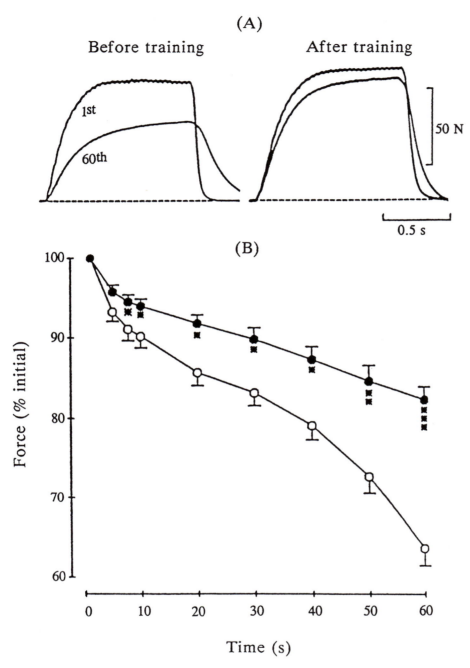

(A)

Before training   After training

1st

60th

50 N

0.5 s

(B)

Force (% initial)

100

90

80

70

60

0   10   20   30   40   50   60

Time (s)

Figure 20.5 Responses of human thumb adductor muscles to endurance training. (A) shows forces generated during 1st and 60th 30 Hz tetani. (B) compares mean (± SE) losses in force during successive 30 Hz tetani in 8 subjects before (○) and after (●) training. It can be seen that the trained muscles have become less fatigable. Means differed significantly at $p < 0.05$ (■), <0.01 (■■), and <0.001 (■■■). From "Muscle fatigue, effects of training and disuse," by K. Hainaut and J. Duchateau, 1989, *Muscle and Nerve*, **12**, p. 664. Copyright 1989 by John Wiley & Sons, Inc. Reprinted with permission.

which are not antigravity in their actions, show no change in twitch duration (Elder & Kakulas, 1994).

Training Studies in Animals

As noted earlier, animal studies have advantages over human investigations in that they can be more comprehensive; in addition, they lend themselves to large numbers of observations with good controls. Four strategies have been employed to alter muscle fiber properties in animals (see Timson, 1990, for review):

- Exercise
- Weight lifting
- Internal loading (tenotomy, ablation)
- Chronic stimulation

The ingenuity of an experimenter in devising an animal model of exercise has often been exceeded by that of the rat (the usual choice) in minimizing effort. A rat on a treadmill may prefer to be carried against the backstop and to allow the belt to slip underneath its body, while a rat placed in a water tank may elect not to swim but to rest on the bottom and come up for air as required!

Treadmill running, however, was used successfully by Gollnick, Timson, Moore, & Riedy (1981) to produce

Table 20.2   Alterations in Fiber-Type Composition (%) in the Vastus Lateralis Muscles of Endurance-Trained Athletes

Type of training	(n)	I	(P<)	IIA	(P<)	IIB	(P<)	Authors
				%Fiber types				
Controls	69	54		32		13		Jansson & Kaijser (1977)
Orienteers	8	68	0.01	24	n.s.	3.3	0.01	
Controls	6	51		41		7.1		Howald (1982)
Long distance runners	9	78	0.05	19	0.05	2.5	0.05	
Controls	4	38		31		26		Fridén, Sjöström, & Ekblom (1984)
Cross-country runners	6	52	n.s.	35	n.s.	12	0.01	
Sedentary controls	4	51		33		13		Baumann et al. (1987)
Professional cyclists	13	80	0.001	17	0.001	0.6	0.001	

n.s. = not statistically significant.

Note. Adapted from Baumann, Jäggi, Soland, Howald, and Schaub (1987).

weight increases in the rat plantaris and soleus muscles of 44% and 21% respectively.

Striking Enlargements Can Be Induced in Avian Muscles by Weights

Weight-lifting models have required rats to maintain their positions in inclined tubes despite having weights attached to their tails (Exner, Staudte, & Pette, 1973), to climb vertical poles with progressively larger weights strapped to their backs (Gordon, 1967), to lift weighted food baskets (Goldspink, 1964), and to stand upright with weights attached to their waists or to their necks (Ho et al., 1980; Klitgaard, 1988). Some of these experiments were unsuccessful in increasing muscle weight, and in others the changes, though significant statistically, were modest (8-21%). In birds, however, weight training can be very effective, as when the weight is attached to a wing (Figure 20.6A). In one such study, the wet weight of the chicken anterior latissimus dorsi (ALD) muscle was found to increase by 180% within 5 weeks (Sola, Christensen, & Martin, 1973). Such increases are due to the addition of new sarcomeres at the ends of existing muscle fibers as well as to thickening of the fibers; in addition, the number of fibers may increase by 50%, following proliferation of the satellite cells (*hyperplasia*; Alway, Winchester, Davis, & Gonyea, 1989; Figure 20.6B). There is also a change in the MHC isoforms (Kennedy,

Kamel, Tambone, Vrbová, & Zak, 1986). The slow tonic ALD muscle exhibits two types of fibers, one specific for the SM-1 heavy myosin isoform and the other for the SM-2 form. Following a 4-week weight-lifting regimen, all the fibers reacted positively for the SM-2 isoform, while the SM-1 form was eliminated in the hypertrophied ALD muscle. Thus, the genetic expression of one particular heavy myosin isoform can be suppressed following overload.

Muscle Enlargement Results From Removing or Tenotomizing Synergistic Muscles

The term *internal loading* describes a surgical manipulation that results in a muscle having to carry an increased load whether supporting body weight, moving a limb, or contracting against an antagonist. Two methods have been devised. *Tenotomy* of synergistic muscles can produce an appreciable weight gain in the remaining agonist muscle during the next 7 days. However, the enlargement is due to edema rather than to the formation of myofibrillar proteins (Armstrong, Marum, Tullison, & Saubert, 1979). An additional disadvantage is that the tenotomized muscle tends to reattach itself to the residual muscle bellies through the growth of connective tissue. A better model is *ablation* of the synergistic muscles. For example, the medial and lateral gastrocnemii can be removed, so that plantarflexion of the foot is necessarily undertaken largely by the soleus and

Table 20.3   Alterations in Fiber-Type Composition (%) in Longitudinal Studies of Endurance-Trained Athletes

Type of training	(n)	I	(P<)	% Fiber types IIA	(P<)	IIB	(P<)	Authors
8-wk high-intensity endurance	12	b 41 a 43	n.s.	37 42	0.05	19 14	0.05	Andersen & Henriksson (1977)
8-wk high-intensity endurance	4	b 49 a 48.8	n.s.	38.3 43.3	n.s.	12.8 8	0.05	Baumann, Jäggi, Soland, Howald, & Schaub (1987)
24-wk cross-country running	7	b 58 a 57	n.s.	26 32	0.005	9.2 3.4	0.005	Ingjer (1979)
6-wk high-intensity endurance	10	b 50 a 56	0.05	37 34	n.s.	12.4 9.6	0.05	Howald, Hoppeler, Claassen, Mathieu, & Straub (1985)
15-wk high-intensity continuous/ interval work	24	b 41 a 47	0.01	42 42	n.s.	17 11	0.01	Simoneau et al. (1985)
5-wk skiing with 80-kg sledge (500 mi)	7	b 29 a 28	n.s.	48 42	n.s.	21 14	n.s.	Schantz & Henriksson (1983)

n = number of subjects.

b = before training.

a = after training.

n.s. = not significantly different.

Note. Adapted from Baumann, Jäggi, Soland, Howald, and Schaub (1987).

plantaris. With this approach, increases in muscle mass are usually in the 30% to 50% range and are sustained; they correspond to the changes in muscle fiber cross-sectional area (Timson, Bowlin, Dudenhoeffer, & George, 1985). The increase in tetanic tension in the hypertrophied muscles is associated with prolongation of the isometric twitch, and with the conversion of a proportion of fast-twitch fibers to slow-twitch (Noble, Dabrowski, & Ianuzzo, 1983). In contrast, the activities of glycolytic and oxidative enzymes are mostly unaltered.

Chronic Stimulation Induces All the Properties of Slow-Twitch Muscles

A muscle can be chronically stimulated through electrodes placed around the motor nerve. A pioneering study was that of Salmons and Vrbová (1969), who attempted to change the twitch durations of the rabbit tibialis anterior by stimulating muscles continuously at 10 Hz. In these experiments, the stimulator was powered by a mercury cell and was embedded in epoxy resin; it was implanted in the abdomen and connected to stimulating electrodes around the peroneal nerve. After several weeks, the twitch duration was significantly increased (see Figure 18.2 in chapter 18).

The same type of experiment was repeated by Al-Amood, Buller, and Pope (1973), who used a miniature stimulator mounted on a vertebral spinous process to excite one of the ventral roots of the cat; these authors also found that slowing of contraction could be made to occur in a previously fast muscle. In keeping with the altered contractile properties of chronically stimulated muscles has been the demonstration, by Sreter, Gergely, Salmons, and Romanul (1973), of a change in the structure of myosin. In the slowed tibialis anterior muscle of the rabbit, these workers were able to show by gel electrophoresis that the myosin had acquired the light chain pattern normally characteristic of slow muscle.

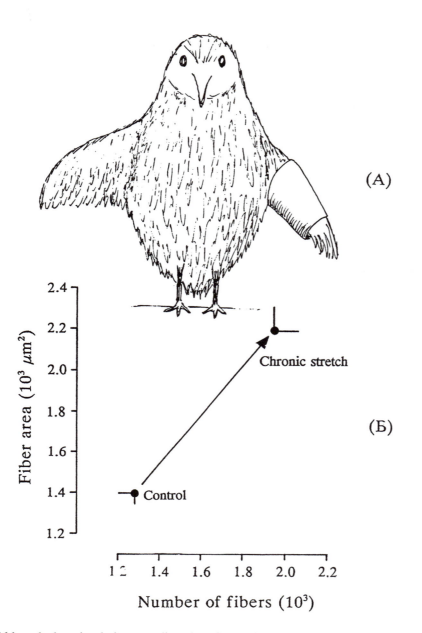

Figure 20.6 (A) Japanese quail with tubular weight attached to one wing. (B) Increased numbers and cross-sectional areas of fibers in stretched ALD muscles, compared with controls; the mean values have been given with their SEs. Data of Alway, Winchester, Davis, and Gonyea (1989).

Although the stimulation paradigm has the obvious drawback that the motor units are activated without regard to their normal recruitment pattern, it has the advantage that the temporal pattern and amount of superimposed impulse activity are known precisely. From the experiments of Lømo and colleagues (Gorza, Gundersen, Lømo, Schiaffino, & Westgaard, 1988; Westgaard & Lømo, 1988), it is evident that the slowing of the twitch is dependent on the low frequency of the applied stimuli, while the resistance to fatigue is a consequence of the total number of stimuli applied (see Figure 18.3 in chapter 18).

The Changes in the Muscles Occur in an Orderly Sequence

The chronic stimulation model has made it possible to determine the times at which the changes in the muscle, characteristic of endurance training, occur. The sequence is as follows for rat muscle stimulated at 10 Hz (Lieber, 1988):

Time	Change
3 hours	Swelling of the sarcoplasmic reticulum
4 days	Increased numbers and sizes of mitochondria, leading to a rise in oxidative enzyme activity Capillary formation, leading to a rise in muscle blood flow
14 days	Increased width of Z-line Decreased Ca^{2+}-ATPase activity in the SR
28 days	Appearance of slow-twitch isoforms of myosin heavy and light chains and of troponin Decrease in muscle bulk, associated with reduced muscle fiber cross-sectional area

At 28 days, a chronically stimulated muscle, if previously fast twitch, will have completed its conversion to slow twitch. The conversion is evident both in the prolongation of the isometric twitch and in the greater incidence of type I fibers, as shown by myosin ATPase staining.

The Volume and Enzymatic Activity of the Mitochondria Are Greatly Increased

The changes in the mitochondria resulting from chronic stimulation are impressive; for example, in rabbit tibialis anterior muscles stimulated at 10 Hz during alternate hours, there is a sevenfold rise in mitochondrial volume after 28 days (Reichmann, Hoppeler, Mathieu-Costello, Von Bergen, & Pette, 1985). Despite this large increase, the ultrastructure of the organelles is well preserved, and the rise in citric acid enzyme activity is proportional to the new volume. If the stimulation is continued beyond 5 to 6 weeks, however, the mitochondrial volume declines, presumably because the muscle fibers, having been converted to type I, are now more efficient in terms of energy utilization.

Adaptive Changes in DNA and RNA Processing

Regardless of whether the training program has been carried out in human subjects or experimental animals, the resulting structural changes in the muscle fibers must involve increases or decreases in protein formation. In turn, the protein effects must be a consequence of alterations in the processing of DNA or RNA in the fibers. With the recent application of molecular biology techniques to this area of study, it is now possible to understand some of the DNA and RNA events.

Alterations in Both Gene Transcription and in mRNA Translation May Occur as a Result of Training

In those situations in which new types of protein are produced in the muscle fibers (i.e., myosin heavy and light chains and troponin), it is clear that the genes for the respective isoforms must have been *transcribed*, with the appearance of the corresponding messenger RNAs (mRNAs) in the sarcoplasm. In some situations, however, synthesis of a type of protein, already present in the muscle fiber, may change too rapidly to be explained by an alteration in the amount of mRNA. For example, protein synthesis in the gastrocnemius muscle can be completely halted during the first 10 min of

contractile activity (Bylund-Fellenius et al., 1984), a time that is much shorter than the half-life of mRNAs. Conversely, increases in the rates of synthesis of contractile and mitochondrial proteins may occur too soon to be due to the transcription of new mRNAs. In such situations it would appear that the *translation* of mRNA is modified. In many instances, however, it is probable that protein synthesis is regulated at both transcriptional and translational levels, as in the case of the increase in citric synthase activity that follows 12-hr/day stimulation of rat fast-twitch muscle. Figure 20.7 shows that in the first 6 days there is a significant rise in enzyme activity without any change in mRNA level; between 7 and 10 days, however, the mRNA increases six- to sevenfold (Seedorf, Leberer, Kirschbaum, & Pette, 1986).

Muscle Stretch Is an Important Stimulus for Altering DNA and mRNA Processing

Thomsen and Luco (1944) found that if the ankles of cats were fixed in the dorsiflexed position, the stretched soleus muscles enlarged during the next 7 days, before regressing. Similarly, if one half of the diaphragm is denervated, it will undergo hypertrophy due to the stretching produced by the contracting half. More recently, Goldspink and colleagues (1991) showed that lengthening the rabbit tibialis anterior, by fixation of the ankle in a cast, could increase the muscle wet weight by 20% and the RNA content fourfold in as short a time as 4 days. If the lengthened muscle is stimulated, both muscle mass and RNA content increase still further. Associated with these changes is the appearance of slow MHCs, due to expression of the corresponding gene; at the same time, the gene for the fast MHCs appears to be repressed. To account for the rapid growth of the tibialis anterior, at least 30,000 MHC molecules must be synthesized by each myonucleus per minute (Goldspink, 1985). The converse experiment, in which the rabbit soleus muscle is fixed in a shortened position, causes expression of the IIB MHC as early as 2 days (Goldspink et al., 1992).

Soluble Factors, Released by Stretching the Muscle Fiber or the Extracellular Matrix, Activate Second Messenger Systems

Experiments performed on cultured fibers, derived from skeletal or heart muscle, have identified some of the cellular mechanisms that are triggered by stretching. In one system, cells are grown in a culture medium coated onto a silicone sheet; the sheet can then be stretched by known amounts for specified times. It turns out that stretch is a potent stimulus, for a 10% stretch, applied

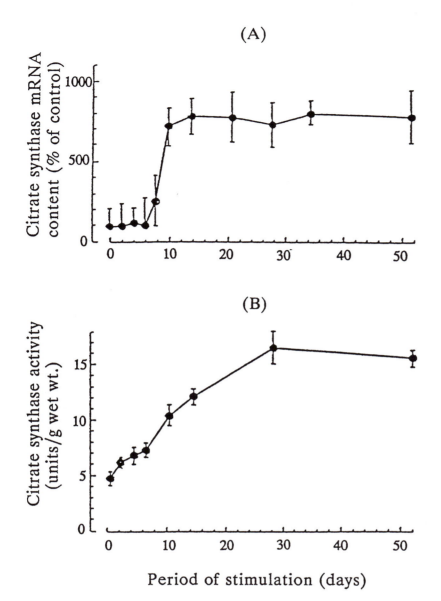

(A)

(B)

Period of stimulation (days)

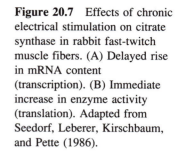

Figure 20.7 Effects of chronic electrical stimulation on citrate synthase in rabbit fast-twitch muscle fibers. (A) Delayed rise in mRNA content (transcription). (B) Immediate increase in enzyme activity (translation). Adapted from Seedorf, Leberer, Kirschbaum, and Pette (1986).

for only 1 min, is able to affect gene transcription 30 min later (Sadoshima & Izumo, 1993). It seems probable that stretch operates by releasing soluble factors from the muscle fibers or the extracellular matrix, and that some of these factors are prostaglandins (Vandenburgh, Hatfalundy, Karlisch, & Shansky, 1991).

Second Messengers Activate Immediate-Early Genes, Which Then Modify the Genes for Muscle Proteins

The soluble factors released by stretch act on the nucleus either directly, as in the case of the prostaglandins, or through other second messenger systems, as shown in Figure 20.8 (see also Sadoshima & Izumo, 1993). Ca^{2+} will also be available as a second messenger when, as in the intact organism, the stretch of the muscle is the outcome of impulse activity and is associated with a 20- to 100-fold rise in intracellular Ca^{2+} concentration (Allen, Westerblad, Lee, & Lännergren, 1992). The first genes to be affected are immediate-early genes, such as *c-fos* (Sadoshima & Izumo, 1993). These genes, in turn, control the transcription of other genes in the nucleus, including those that code for all the special proteins needed for the transformation of the muscle fibers (e.g., myosin heavy and light chains, troponins and other Ca^{2+}-binding proteins, and the various muscle enzymes). Master regulatory genes (see chapter 5) may also be involved, to ensure that the genes responsible for the different types of fast or slow protein are expressed together. Finally, the increase in capillaries, which accompanies endurance training, may be the result of greater production of fibroblast growth factor (Kraus & Williams, 1990). Figure 20.9 summarizes some of the pathways that are likely to be involved in bringing about the adaptive changes in skeletal muscle fibers.

Motoneuron Metabolism and Endurance Exercise

The changes in the motoneurons that take place during training programs are not limited to impulse firing patterns (see p. 304), for there are also changes in the histochemistry of the cells and in their fast axonal transport. In histochemical studies, Gerchman, Edgerton, and Carrow (1975) exercised rats by swimming and demonstrated that acid phosphatase activity was diminished

in motoneurons, while the reaction for glucose-6-phosphate dehydrogenase was intensified. In a similar type of study, this time using treadmill running over a period of 8 weeks, there was a pronounced increase in the oxidative enzyme activity of rat soleus motoneurons (Suzuki et al., 1991). Just as an increase in oxidative enzyme activity enables muscle fibers to function more effectively during prolonged contractile efforts by improving the supply of ATP, so would there be a similar benefit to motoneurons in the same circumstances. Treadmill running has also been found effective in increasing both the velocity of fast axonal transport and the amount of material carried down the axon (Jasmin, Lavoie, & Gardiner, 1988). It is possible that the extra protein is needed to repair any damage that may have

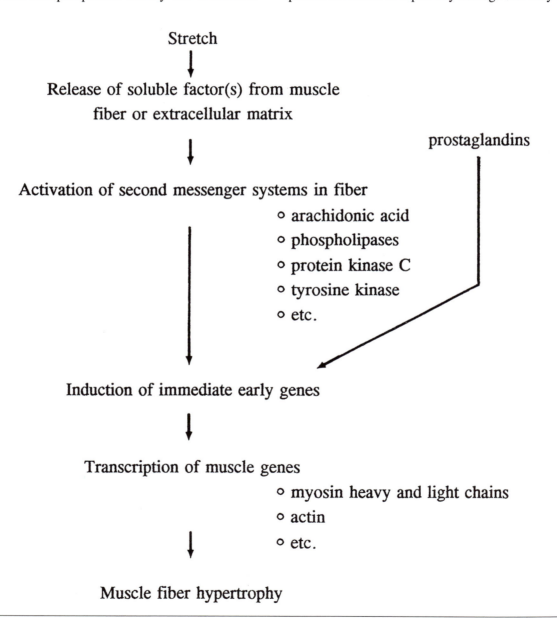

Figure 20.8   Sequence of intracellular events that results in muscle fiber hypertrophy following stretch.

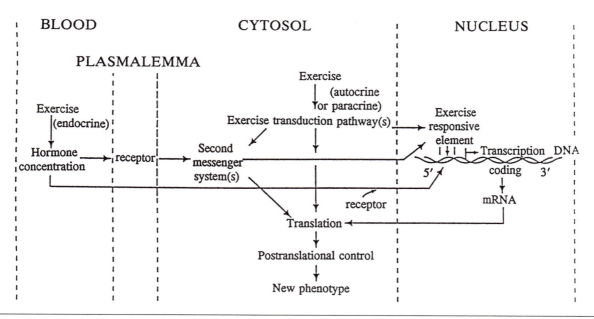

BLOOD CYTOSOL NUCLEUS

PLASMALEMMA

Figure 20.9 Simplified scheme of pathways involved in response of muscle fibers to training. Adapted from Booth (1988).

taken place in the motor nerve terminals during contractile activity.

APPLIED PHYSIOLOGY

Exercise Programs Can Benefit Patients With Neuromuscular Diseases

In this section we will briefly consider whether or not exercise training programs are effective in patients with chronic degenerative neuromuscular disorders, such as old polio, muscular dystrophy, and spinal muscular atrophy. On the one hand, it might be supposed that an appropriate training program would enlarge those muscle fibers that were still relatively healthy, and enable them to develop more force. On the other hand, as will be shown in the next chapter, exercise can damage healthy muscle if it is excessive in intensity or duration. Were this to happen in muscles that were already weak through disease, the consequences could be catastrophic in terms of the activities of daily living. Tasks which could previously have been accomplished with some difficulty might now become impossible. With these considerations in mind, there is an obvious need to carry out studies on the effects of exercise programs in patients. Though this might appear a simple task, it is

usually difficult to collect a group of patients having the same diagnosis, with moderate rather than severe weakness, and with the willingness and opportunity to travel to the gym for supervised daily training sessions.

In one small study involving patients with various diagnoses, significant improvements in strength were observed (McCartney, Moroz, Garner, & McComas, 1988). More striking, though, was the ability of training to decrease fatigability. This last effect is especially important, for an improvement in stamina may be more useful in many daily tasks than an increase in strength. Another relevant study, this time restricted to persons with old polio, also found increased muscle performance following an exercise program (Feldman & Soskolne, 1987). It cannot be stressed sufficiently, however, that in any exercise program patients must start with small loads and proceed slowly, for fear of overtaxing the already damaged muscles. Finally, as pointed out in chapter 19, an exercise program will be able to reverse the effects of disuse in moderately disabled patients, not only on the muscle fibers but on their neural activation also.

In the next chapter, we will examine the muscle damage that can be brought about by the mechanical stresses of intensive exercise, even in healthy individuals. We will also look at the way the fibers are able to repair themselves.

Chapter 21

Injury and Repair

There are many ways in which muscle fibers can be damaged (Table 21.1). *External causes* include crushing and laceration injuries to the body, and extremes of heat and cold. *Internal causes* include muscle tears and tendon ruptures following sudden forceful contractions. Muscle tears may be associated with considerable bleeding into the muscle belly, due to breaching of the walls of the intramuscular blood vessels. Unaccustomed exercise, especially that involving lengthening (eccentric) contractions, may also damage muscles (see the following sections). Finally, degeneration or necrosis (death) of muscle fibers is a feature of a number of *diseases*, particularly those due to inflammation of the muscles (polymyositis, dermatomyositis) or to inherited defects (e.g., Duchenne muscular dystrophy, malignant hyperthermia). Especially alarming is the muscle necrosis that may accompany local infections with virulent strains of streptococcus A. So rapidly may the necrosis spread that, in the case of an infected limb, amputation may be required as a life-saving measure. As will be shown, the degenerative and necrotic processes that come into play are similar, regardless of the nature of the provocative event. Further, in almost all cases there is an attempt at regeneration, such that entirely new fibers, or segments of fibers, can be formed. Especially in young

fibers subjected to mechanical injury, the regeneration is rapid and successful.

We will commence our study of muscle injury by examining the changes that may develop after strenuous exercise in untrained, but otherwise healthy, individuals.

Muscle Contraction Damage

Repeated eccentric contractions can cause surprisingly severe morphological changes in the muscle fibers. From chapter 11 it will be recalled that eccentric contractions are those that take place while the muscle is being lengthened. Naturally occurring examples are contractions in the biceps brachii as a heavy load is being slowly lowered, and the activity in the gluteus maximus that checks the forward swing of the leg during running and walking. In terms of the sliding filament mechanism of muscle contraction, the myosin crossbridges make repeated connections with the actin filaments throughout the duration of the active state in the muscle fiber. However, the actin filaments, instead of being propelled toward the center of the myosin filament, are pulled in the opposite direction by the external

Table 21.1   Causes of Muscle Fiber Necrosis

Alcohol intake
Chemicals and drugs
  • Bupivicaine
  • Rifampicin
  • Quinacrine
  • Calvacin
  • Others
Diseases
  • Polymyositis, dermatomyositis
  • Muscular dystrophy
  • Biochemical disorders (myophosphorylase, carnitine
    deficiencies)
Exercise (especially eccentric)
Irradiation
Ischemia
Mechanical injuries (stretching, crushing, cutting)
Thermal injuries (heating, freezing)
Biological toxins (venoms, streptococcus A toxin)

forces on the muscle. The external forces may be a large load, or the contraction of an antagonist muscle. With reference to the pattern of striations along the muscle fiber (see Figures 1.1 in chapter 1 and 11.2 in chapter 11), the H-region in the M-band will widen as the sarcomere lengthens. At any instant, some sarcomeres will be stretched more than others.

Eccentric Contractions Can Produce Fiber Necrosis

The changes in muscles that result from repeated eccentric contraction have been studied in human subjects by Fridén, Sjöström, and Ekblom (1983) and by Newham, McPhail, Mills, and Edwards (1983). The changes are barely evident within the first 24 hr of exercise and are then restricted to *streaming of the Z-lines*. By 10 to 15 days, however, a significant proportion of the muscle fibers have undergone necrosis, and there are prominent infiltrations of mononuclear *inflammatory cells* in and around the degenerating fibers (Jones, Newham, Round, & Tolfree, 1986). After an additional 2 to 3 weeks, much of the muscle damage has been repaired by regeneration of fiber segments; nevertheless, many fibers still have central nuclei and vary considerably in diameter.

The tendency of eccentric contractions to cause severe inflammatory and degenerative changes in muscles has been confirmed in the animal study of McCully and Faulkner (1985). The extensor digitorum longus (EDL) muscles of anesthetized mice were repeatedly tetanized while being stretched by a servomotor. Three days later, 37% of fibers had degenerated, and the maximum tetanic tension was only 22% of the control value (Figure

21.1); at this time inflammatory cells were prominent, but one day later regenerative processes were already under way (Figure 21.2; see also Faulkner, Jones, & Round, 1989).

The Fiber Damage Is Due to Ca^{2+} and Free Radicals in the Cytoplasm

As previously noted, during eccentric contractions, the sarcomeres, especially in the central regions of the muscle fibers, are being overstretched. Jones, Jackson, McPhail, and Edwards (1984) believe that the entry of excessive amounts of Ca^{2+} into the muscle fiber from the interstitial fluid is a critical step in producing damage. Jones et al. reached this conclusion on the basis of experiments in mouse muscles that had been either stimulated to exhaustion or else treated with various metabolic inhibitors (cyanide, dinitrophenol, or iodoacetic acid). They found that the muscle fiber damage which would normally occur under these conditions could be prevented by removing Ca^{2+} from the bathing fluid. In muscles that had performed eccentric contractions, it would be reasonable to suppose that Ca^{2+} gained access to the overstretched regions of muscle fibers through minute tears in the plasmalemma. Inside the cytosol, Ca^{2+} ions might activate a *protease* and a *phospholipase*, which could digest structural proteins and lipid membranes respectively. Jackson, Jones, and Edwards (1984) consider that, of the two calcium-activated enzymes, phospholipase is the most likely culprit, and they suggest that additional damage ensues from the liberation of fatty acids and oxidation of the latter to *free radicals* (see Jones & Round, 1990). An experimental finding consistent with free radical involvement is that intraperitoneal injection of *superoxide dismutase*, an enzyme which scavenges free radicals, significantly reduces the loss of tension in mouse EDL muscles subjected to eccentric contractions (Zebra, Komorowski, & Faulkner, 1990).

Exercise and Fiber Damage Release Muscle Enzymes Into the Circulation

Heavy muscular exercise of any kind can cause significant rises in the plasma levels of muscle enzymes; of these enzymes, the most sensitive indicator is *creatine kinase*. Increases in enzyme titer of two to five times can be observed 24 hr after exercise, but the values have usually returned to normal after another day or two. Following prolonged eccentric contractions, there is another, much larger, rise in plasma enzymes (up to 100 times normal), which reaches a peak at about the 5th day. It is probable that the large amounts of enzyme

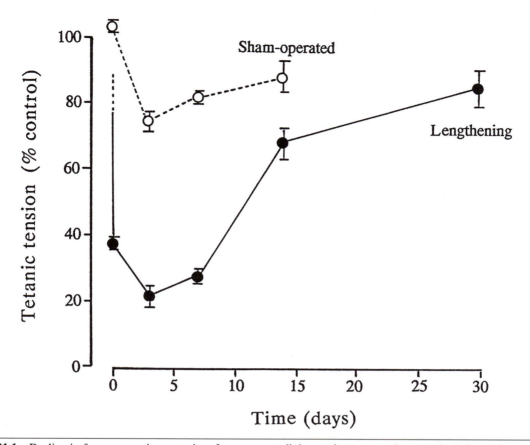

Figure 21.1   Decline in force-generating capacity of rat extensor digitorum longus muscles at various times after repeated tetani, delivered with the muscle lengthened. The decline was significantly greater and longer lasting than those following shortening and isometric contractions (not shown). Results for sham-operated animals are also given; all values are means ± SEs. Adapted from McCully and Faulkner (1985).

are being released from regions of muscle fibers undergoing necrosis (see p. 318). The rises in muscle enzyme are prevented if subjects are trained in the eccentric exercise, and there is also much less fiber damage (Jones & Round, 1990); the basis for this increased resistance is not known.

Eccentric Exercise Induces Delayed Muscle Discomfort

Especially in untrained subjects, eccentric contractions may cause appreciable muscle discomfort or even pain. The discomfort is not usually evident during the period of the exercise, but appears on the following day, and can last for several more days. In view of the latent period, the discomfort is referred to as *delayed onset muscle soreness* (DOMS). While the muscles may be somewhat swollen and also tender on palpation, the discomfort is felt most when the muscles are engaged in activities that involve further lengthening contractions. When the soreness has passed, further bouts of eccentric exercise produce comparatively minor discomfort, in

keeping with the reduction in fiber damage observed with the microscope (discussed earlier).

Muscle Injury From External Causes

The previous section explored the muscle changes that result from eccentric exercise in human subjects and experimental animals. We must now consider damage from external sources—usually mechanical injury of one kind or another, or the application of a necrotizing chemical such as the local anesthetic *bupivicaine*. The cellular events that take place in the muscle fibers are qualitatively similar to those occurring in eccentric exercise, but they usually result in much more necrosis. The nature of the cellular mechanisms, which permit the muscle fibers to repair themselves, will also be considered.

One of the techniques for studying the effects of muscle injury was pioneered by Studitsky (1952) and

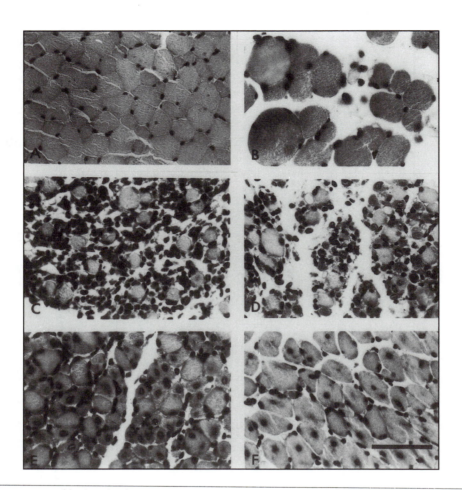

Figure 21.2   Damage caused by lengthening (eccentric) contractions in mouse extensor digitorum longus muscles. The sections are from (A) a control muscle, and (B) at 1 day, (C) 3 days, (D) 4 days, (E) 7 days, and (F) 14 days after the stimulated contractions. In (B), the fibers are already swollen and rounded, and in (C) and (D), many are undergoing necrosis and have been invaded by inflammatory cells. In (E), regeneration is well under way, and in (F), the regenerated fibers can be recognized by their central nuclei. Photographs courtesy of Dr. John Faulkner and Dr. Richard Hinkle; see also McCully and Faulkner (1985). Calibration bar = 100 µm.

subsequently exploited by others (e.g., Carlson, 1970); it is to remove a muscle from an animal, mince it into small cubes, and insert the pieces into the former bed (see Figure 21.3). Alternatively, the muscle can be excised and replaced intact. In both instances, the transplanted muscles are *autografts*. In contrast, a *homograft* is a muscle transferred to the normal anatomical position, but in another animal.

Less radical procedures than grafting are to divide or pinch a muscle; if a slender muscle with longitudinally running fibers is chosen, the microscopic analysis of subsequent events is simplified. All these methods, and some of those listed in Table 21.1, have the advantage that the full repertoire of possible cellular events can be examined, including those involving the immune system. In contrast, *in vitro* studies of cultured muscle fibers enable more precise observations to be made by interference or phase-contrast microscopy; in addition,

the fates of individual fibers can be followed by time-lapse photography.

The Influx of Ca²⁺ and Complement Promotes Fiber Necrosis

Following a sharply localized pinch or incision, a *necrotic zone* develops at the site of damage and extends for a few millimeters on either side (Figure 21.4a). The torn plasmalemma allows Ca^{2+} ions to enter the fiber cytoplasm from the interstitial fluid; as described in a previous section, Ca^{2+}, once admitted to the fiber, activates proteases that break down proteins in the myofibrils, mitochondria, and cytoskeleton. Circulating *complement* components also gain access to the fiber through gaps in the plasmalemma, are deposited, and assist in the destruction of the myofibrils, organelles, and tubular systems (Arahata & Engel, 1985).

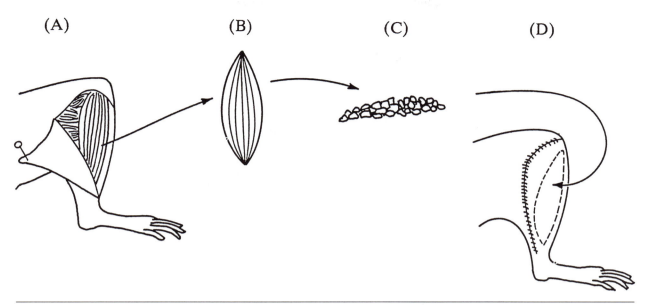

Figure 21.3 The transplantation of minced skeletal (tibialis anterior) muscle using the techniques of Studitsky (1952) and Carlson (1970). The muscle is (A) exposed, (B) removed, (C) cut into small pieces, and (D) packed into the tissue bed left by the same or a different muscle.

Some of the products, formed by the actions of the Ca^{2+}-proteases and complement fixation, become molecular signals that attract mononuclear cells from the capillary circulation. A number of these cells, the *macrophages*, invade the fiber, continue the digestion of its contents, and then remove the debris (Figure 21.4a). *T-lymphocytes* may also arrive at the site of injury, presumably in response to the liberation of cytokines. Figure 1.15, in the section on Duchenne muscular dystrophy in chapter 1, illustrates in schematic form these cellular events.

On either side of the central necrotic zone, the uninjured myofibrils retract and, with the organelles and tubular systems, form a stump that becomes covered with newly formed membrane. The membrane evidently has a protective role, in that it prevents the entry of Ca^{2+} and complement.

Distal to the Necrotic Region, the Fiber Is Functionally Denervated

Once the plasmalemma is torn and the underlying region of the fiber begins to degenerate, the distal part of the fiber loses communication with the motor end-plate. Neither electrical impulses nor trophic factors are able to span the damaged segment, and hence the distal sarcomeres are functionally denervated (McComas & Mrożek, 1967). Although few electrophysiological studies have been made of this situation, the isolated region would be expected to exhibit all the features of denervation, including a fall in resting membrane potential and

the synthesis of AChRs. Further, the release of a diffusible molecular signal appears capable of causing neighboring motor nerve fibers to sprout and to form new neuromuscular junctions on the denervated segments (see chapter 17).

Satellite Cells Are Responsible for Forming New Fiber Segments

At the same time that the central region is undergoing necrosis, this and neighboring parts of the fiber begin the repair process. The first important step is the activation of *satellite cells*. These cells, which can be recognized by electron microscopy, consist of nuclei with very little cytoplasm. Like the myonuclei, they lie at the periphery of the muscle fiber, but they are surrounded by their own plasmalemma and are separated from the fiber by basement membrane (see Figure 1.12 in chapter 1). Normally there are few of these cells bordering a single muscle fiber; in human muscle, for example, they constitute 4% to 11% of the muscle nuclei (e.g., Wakayama, 1976). In response to an unknown signal from the damaged region of a fiber, the previously dormant satellite cells first proliferate and then migrate into the necrotic area before differentiating into *myoblasts*. As in the embryological development of muscle, the myoblasts fuse to form *myotubes* (Figure 21.4b). With the electron microscope, the myoblasts can be recognized not only by their thin profiles and central nuclei, but also by the abundance of ribosomes. The control of cell proliferation, differentiation, and fusion, and the accompanying synthesis

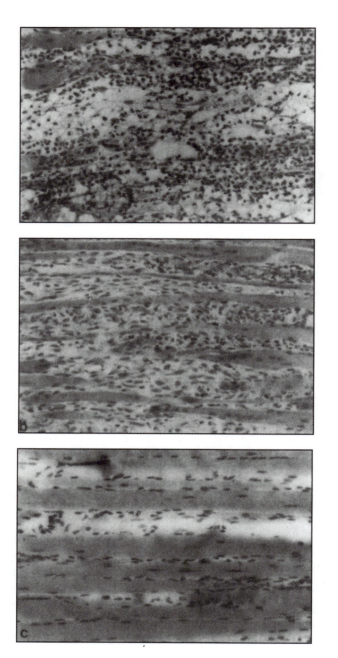

Figure 21.4 Histological changes in rat semitendinosus muscle fibers after crush injury. (a) At 2 days, the damaged fiber segments have undergone necrosis, with digestion and removal by invading macrophages. (b) At 5 days, several newly formed slender myotubes, with central nuclei, can be seen in the damaged region. (c) At 10 days, the myotubes have transformed into muscle fibers, many of which have already linked up with the fiber stumps on either side. Reprinted from Stuart, McComas, Goldspink, and Elder (1981). (200x).

of muscle-specific proteins, is under genetic control. Although molecular biology studies of muscle regeneration do not appear to have been made, it is probable that the interaction of myogenic genes and growth factors are similar to those responsible for muscle development in the embryo (see chapter 5).

The Myotubes Link the Separated Muscle Fiber Stumps Together

As the ends of the myotubes extend, they reach the intact stumps of the damaged fibers, and union occurs, with dissolution of the respective plasmalemmae; the fibers are once more in continuity (see Figure 21.4c). The time-course of the bridging process can be estimated by a simple electrophysiological technique, in which the muscle is stimulated on one side of the lesion and action potentials are recorded on the other side, using microelectrodes to impale individual fibers (Stuart, McComas, Goldspink, & Elder, 1981). In rat semitendinosus muscles, crushed with watchmaker forceps, reinnervation can be detected at 5 days and is complete by 30 days (see Figure 21.5). The architecture of the muscle is best restored if the endomysial sheaths surrounding the damaged fibers have been preserved, since the endomysium can act as a scaffold for the containment and alignment of the myotubes. If the endomysium is ruptured, however, satellite cells can escape into the interstitial space and form new fibers.

APPLIED PHYSIOLOGY

In this section, we will see how muscle grafting has been taken out of the animal research laboratory into the patient operating theater, and used for the correction of physical deformity.

Muscle Grafting Can Be Used in Patients With Facial Palsy

The technique of muscle grafting has, in recent years, become more widely used in plastic surgery. However, there might be little benefit to the patient if the fibers in the graft, having been deprived of their blood supply, degenerated and died with no certainty of successful regeneration afterwards. For this reason, microvascular anastomoses are performed under the operating microscope, the artery and vein in the cut pedicle (stalk) to the muscle being delicately sutured to corresponding vessels in the tissue bed prepared for the graft. Perhaps nowhere is the challenge of grafting greater than in the face, for success or failure is all too visible. Figure 21.6 shows the design of a grafting procedure for restoring mobility to one side of the face following irreparable damage to the facial nerve on that side. Such damage

may be the result of inflammation of the nerve (Bell's palsy) or the inevitable consequence of removing a large tumor from the adjacent eighth cranial nerve within the cranial cavity.

In the restorative operation, a block of gracilis muscle is taken from the inner thigh, complete with its pedicle, and is grafted onto a tissue bed underneath the cheek. Either at the same time, or in a later operation, a branch of the "healthy" facial nerve on the normal side of the face is divided and joined to a length of sural nerve taken from the calf. The sural nerve graft is led through the soft subcutaneous tissues between the nose and upper lip, and sutured to the nerve stump in the gracilis muscle graft (Figure 21.6; see also Harii, Ohmori, & Torii, 1976). The completely severed axons in the sural nerve graft will die, but facial nerve axons from the normal side of the face will eventually regenerate through the graft and establish synaptic connections with muscle fibers in the transplanted gracilis. Henceforth, the two sides of the face will contract together under the influence of the single facial nerve. The results of the operation can be very satisfactory to the patient, for both in repose and in smiling evidence of paralysis may be lost.

Grafted Human Muscles May Not Fully Transform

From our understanding of the trophic influences exerted by motoneurons on muscle fibers (see chapter 18),

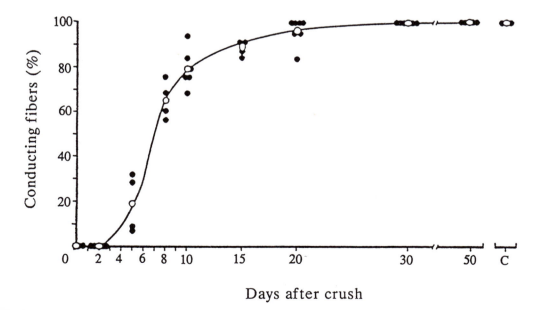

Figure 21.5 Incidence of muscular fibers with restored functional continuity, following crush injury. The assessment was determined by recording action potentials in single fibers after stimulating the muscle on the other side of the crush. C = control (intact muscle). Open circles denote mean values. Reprinted from Stuart, McComas, Goldspink, and Elder (1981).

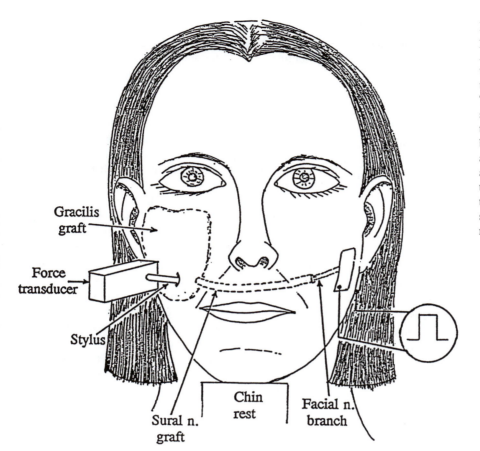

Figure 21.6 Double grafting procedure for facial nerve palsy. A piece of gracilis muscle is transplanted from the thigh to the paralyzed side of the face. The graft receives its nerve supply from the opposite side of the face, through a facial nerve branch that regenerates through a sural nerve graft. The figure also shows the stylus of a force transducer indenting the skin, so that the contraction of the stimulated muscle can be recorded.

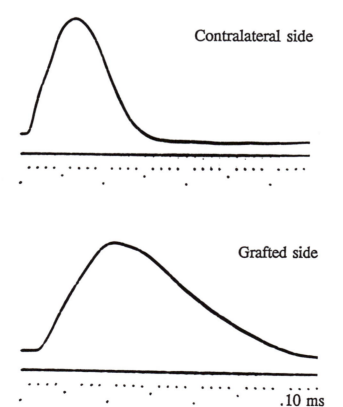

Figure 21.7 Comparison of muscle contractions on the two sides of the face in a patient who had undergone the nerve and muscle grafting procedure shown in Figure 21.6. On stimulating the facial nerve, the twitch was much slower on the side of the graft.

we would expect the slower twitch muscle fibers of the gracilis to transform into the fast-twitch fibers characteristic of healthy facial muscle. Curiously, this may not happen, as in the case of the patient whose results are illustrated in Figure 21.7. It can be seen that when the healthy facial nerve is stimulated, the twitch remains much slower in the grafted gracilis muscle than in the normal facial muscle on the opposite side. Possibly, some human muscles are less ''plastic,'' in response to altered innervation, than the muscles that have been studied extensively in animals.

There is one more topic for consideration before our survey of skeletal muscle is complete. This is the appearance and the physiological properties of muscle in old age. As we shall see, there are very marked changes both in the muscle fibers and in their nerve supply. Even so, muscle performance can still be enhanced through training.

Chapter 22

Aging

Not only are men and women living significantly longer than before, but many are able to pursue full and active lives in their later years. Nevertheless, important changes in the motor system accompany aging; muscles become thinner, strength decreases, and movements become slower and less precise. The study of world records indicates that peak performance is reached between the midteens (for female gymnasts and swimmers) and the late 20s (for marathon runners) and is maintained for only 5 years or so. This decline in performance is evident in Figure 22.1, which shows the best speeds achieved for the 200 m and marathon distances by male runners of different ages. The various features of the aging motor system will now be considered.

Changes in Muscle Strength and Structure

Comparisons of maximal isometric voluntary contractions (MVCs) in young and elderly subjects have been made for a number of muscle groups, as shown in Table 22.1. In this type of study, it is important not only to examine elderly subjects in good health but to establish whether or not the voluntary contractions are indeed maximal. One way of resolving the latter uncertainty is to employ the twitch interpolation technique (see chapter 13). For muscles of the ankle it appears that, with encouragement from an observer, the elderly can recruit motor units as effectively as younger subjects (Vandervoort & McComas, 1986). It is therefore likely that the substantial reductions in strength, shown for the various muscle groups in Table 22.1, are the result of changes in the muscle fibers and their nerve supply, rather than in the motor commands generated in the central nervous system. The loss of strength affects men and women equally and involves all the muscle groups tested. The decline is, as expected, greatest in the oldest subjects, in whom some muscles can only develop half their former force.

In contrast to the numerous studies of isometric contractions, little attention has been given to the forces developed during isokinetic movements in the elderly. It would appear, however, that strength is compromised to a greater extent at high velocities (e.g., 180°/s) than at low ones, or in comparison with isometric conditions (Larsson, Grimby, & Karlsson, 1979). A diminution in both force

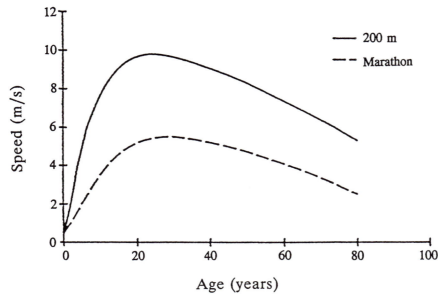

Figure 22.1 Maximal running speeds for short and long distances at different ages. Exponential curves were fitted to world record times for men's 200 m and marathon (42 km). Reprinted from Moore (1975).

and velocity of shortening would reduce power to an even greater extent (see Figure 11.18 in chapter 11).

Muscle Twitches Become Smaller and Slower

There is argument as to when loss of strength commences, but agreement that only modest declines can be detected prior to 60 years. The results of Vandervoort and McComas (1986) suggest that ankle dorsiflexor and plantarflexor muscles lose approximately 1.3% of their former strength each year beyond the age of 52 years (Figure 22.2), and it is likely that similar conclusions would apply to other muscle groups. Another approach to investigating muscle strength in the elderly is to measure the force evoked by electrical stimulation. Since tetanic stimulation is painful, measurements have usually been made of twitch tension instead. The twitch results also show a decline in force with aging, though it is proportionately less than the reduction in maximal voluntary strength (e.g., Vandervoort & McComas, 1986). The twitch studies are of interest for another reason, for as muscles age, there is a progressive prolongation of the contraction and half-relaxation times. This feature has been demonstrated not only in small muscles of the hand and foot (Botelho, Candler, & Guiti, 1954; Campbell, McComas, & Petito, 1973; Newton & Yemm, 1986) but in larger ones, such as the dorsiflexors and plantarflexors of the ankle (Davies & White, 1983; Keh-Evans et al., 1992; Vandervoort & McComas, 1986) and can be seen in Figure 22.3.

The cellular mechanisms responsible for the prolongation may involve the sarcoplasmic reticulum, since its volume and Ca^{2+}-pumping capacity are reduced in aged rodent muscle (Larsson & Salviati, 1989). Although motoneuron firing rates decrease in elderly human subjects (Borg, 1981; Soderberg, Minor, & Nelson, 1991), the prolongation of the contractile response

still enables fused contractions to occur at the lower excitation frequencies.

Muscle Mass Is Also Reduced in the Elderly, Though to a Lesser Extent Than Strength

Muscle mass can be estimated in life from cross-sectional images made by ultrasound, computerized transaxial tomography, or magnetic resonance. In the investigation of Klitgaard, Mantoni et al. (1990), however, cross sections were cut in muscles of elderly men who had died from accidents. All such studies provide clear evidence of reduced muscle bulk in elderly subjects, as set out in Table 22.2.

In some muscles, the reductions in fiber volume may be considerably greater than the percentage declines in cross-sectional area given in Table 22.2, due to the replacement of contractile elements by fat and connective tissue (Inokuchi, Ishikawa, Iwamoto, & Kimura, 1975; Overend, Cunningham, Paterson, & Lefcoe, 1992). This substitution for contractile tissue explains why the force developed per unit cross-sectional area may be significantly lower in the elderly, in comparison with younger subjects (Vandervoort & McComas, 1986).

Substantial Losses of Muscle Fibers Occur

It would be a daunting prospect to count all the muscle fibers in one of the larger human muscles. Instead Lexell, Henriksson-Larsén, Winblad, and Sjöström (1983) examined every 48th mm^2 of a cross section and estimated that the vastus lateralis muscles of elderly men (70-73 years) contain, on average, 25% fewer fibers than the corresponding muscles of young men (19-37 years). Still older subjects, in their 80s, have approximately half the number

Table 22.1 Summary of Studies on Changes in Isometric Muscle Strength With Aging

Studies (categorized by muscle action)	Sex	Age composition of elderly subject group[a]	n	% Decline in strength of elderly vs. young adults[b]
Ankle plantarflexion				
Davies & White (1983)	M	x̄ = 70 ± 1.3	13	43
Vandervoort & McComas (1986)	M	80-100	13	45
Vandervoort & McComas (1986)	F	80-100	8	55
Knee extension				
Larsson, Grimby, & Karlsson (1979)	M	60-69	16	25
Clarkson, Kroll, & Melchionda (1981)	M	55-73	15	39
Young, Stokes, & Crowe (1984)	F	71-81	25	35
Young, Stokes, & Crowe (1985)	M	70-79	12	39
Murray, Gardner, Mollinger, & Sepic (1980)	M	70-86	24	45
Murray, Duthie, Gambert, Sepic, & Mollinger (1985)	F	70-86	24	37
Handgrip				
Fisher & Birren (1947)	M	53-68	20	17
Shephard (1969)	M	60-69	10	18
Mathiowetz et al. (1985)	M	75-94	25	53
Mathiowetz et al. (1985)	F	75-94	26	59
Kallman, Plato, & Tobin (1990)	M	60-69	158	12
Kallman et al. (1990)	M	80-89	42	66
Elbow flexion				
McDonagh, White, & Davies (1984)	M	x̄ = 71 ± 3.7	11	20
Elbow extension				
Davies, Thomas, & White (1986)	M	x̄ = 70 ± 2.8	20	39
Davies et al. (1986)	F	x̄ = 69 ± 2.9	11	28

[a]Age range in years is presented when available, otherwise the mean is age in years ± SD.

[b]Compiled from published data, using the mean strength values of young and old subjects.

Note. Adapted from Vandervoort, Hayes, and Bélanger (1986).

of fibers of young subjects (Lexell, Taylor, & Sjöström, 1988; see also Grimby & Saltin, 1983).

The study by Inokuchi et al. (1975) differs from the preceding ones in having been carried out on the rectus abdominis, a trunk muscle, and in the use of an automated scanning microscope to count the fibers. This investigation, performed on muscles from victims of accidental death, revealed that the loss of fibers is considerably greater in the rectus than in the quadriceps. The loss commences in, or before, the 3rd decade in both men and women (Figure 22.4). Despite increases in fat cells and connective tissue, the cross-sectional

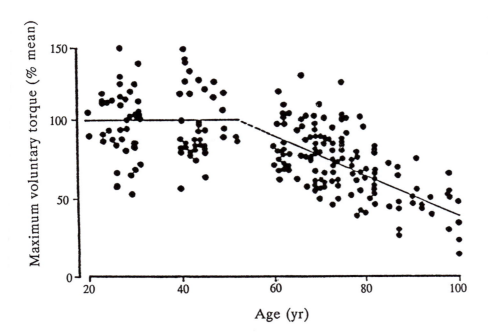

Figure 22.2 Maximal voluntary torques of ankle dorsiflexor and plantarflexor muscles in male and female subjects of different ages. The results have been expressed as percentages of respective mean values for the youngest age group (20-30 years). The horizontal line for subjects aged 52 years or less signifies the lack of an age effect before this time. Reprinted from Vandervoort and McComas (1986).

areas of the muscles decrease by 40% in men and 32% in women between the 3rd and 9th decades. As Table 22.3 shows, evidence of muscle fiber loss with increasing age has been found in most animal studies.

Type II Fibers Undergo the Greatest Atrophy

The cross-sectional areas of human muscle fibers have been measured in the vastus lateralis, biceps brachii, and tibialis anterior. Tissue has been obtained either from needle biopsies during life, or from transverse sections of whole muscles obtained postmortem. The picture that emerges from studies of this kind is that muscle fibers maintain their sizes into the 7th decade; beyond that time there is progressive shrinkage of the type II fibers, with lesser changes taking place in the type I fibers (for example Klitgaard, Mantoni, et al.,

1990; Lexell, 1993). A small proportion of fibers undergo hypertrophy, however, and may then start to split longitudinally.

Despite the Presence of Fiber-Type Grouping, Normal Fiber-Type Proportions Are Maintained

In elderly subjects there is sometimes the appearance of an unusually large incidence of type I fibers in muscle cross sections. Two quantitative investigations, in the vastus lateralis and tibialis anterior, have supported this impression (Larsson, 1983; Jakobsson, Borg, Edström, & Grimby, 1988). The majority of studies, however, have *not* shown any significant alteration in fiber-type proportions with age (Grimby, Danneskiold-Samsøe, Hvid, & Saltin, 1982; Lexell et al., 1988).

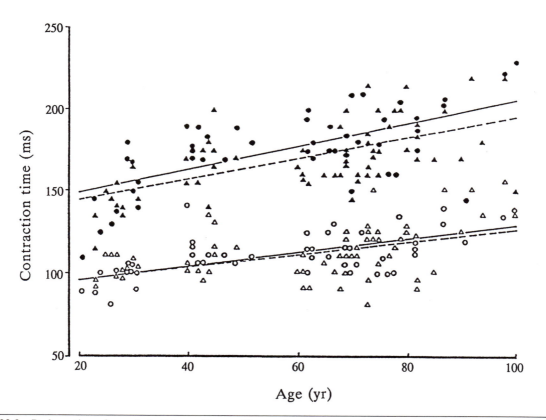

Figure 22.3 Prolongation of muscle twitches with age. The results are expressed as contraction times (see Figure 11.15 in chapter 11) of ankle dorsiflexor and plantarflexor muscles. ▲, ● = male and female plantarflexor values respectively; Δ, ○ = male and female dorsiflexor values respectively. Reprinted from Vandervoort and McComas (1986).

In contrast to these negative findings, muscle histochemistry in the elderly often shows a loss of the normal mosaic pattern. Instead, fibers of similar types tend to be grouped together, a finding usually indicative of chronic denervation (see chapter 17). Using monoclonal antibodies to myosin heavy chains (MHCs), Klitgaard, Zhou, Schiaffino, et al. (1990) have been able to demonstrate, in vastus lateralis muscles from elderly men, unusually high incidences of fibers containing more than one MHC isoform; for example, type IIA and IIB isoforms may be found together, or type I isoforms with those of type IIA. This association of isoforms would be expected if fibers were changing from one histochemical type to another, as part of the process responsible for fiber-type grouping.

Aged Muscles Show Other Degenerative Features

Some of the histopathological features of muscles in the elderly have been mentioned already; these include muscle fiber atrophy (particularly of type II fibers), fiber-type grouping, and the replacement of fibers by fat and connective tissue. To these must be added the following changes in a proportion of fibers (Rubinstein,

1960; Serratrice, Roux, & Aquaron, 1968; Tomlinson, Walton, & Rebeiz, 1969):

- Hyaline degeneration, vacuoles at ends of fibers, lipofucshin pigmentation, and loss of myofibrils so as to form spirals (Ringbinden) or central masses surrounded by cytoplasm
- Necrosis, with infiltrations of macrophage cells in and around the degenerating fibers
- Groups of very small fibers with dense (pyknotic) nuclei, also angulated fibers
- Central nucleation of fibers
- Hypertrophy and splitting of muscle fibers

With the electron microscope, other changes have been observed (Orlander, Kiessling, Larsson, Karlsson, & Aniansson, 1978; Shafiq, Lewis, Dimino, & Schutta, 1978; Tomonaga, 1977):

- Increased endomysial connective tissue
- Disorganization of sarcomere spacing, with streaming of the Z-bands
- Reductions in the sizes of mitochondria
- Accumulation of tubular material from the sarcoplasmic reticulum and T-tubules

Table 22.2   Human Studies Demonstrating Loss of Muscle Mass With Age[a]

Study	Sex	Age	n	% Decline
Triceps surae				
Vandervoort & McComas (1986)	M	82-100	5	23
	F		5	23
Quadriceps femoris				
Klitgaard, Mantoni, et al. (1990)	M	68 ± 0.5	7	24
Young, Stokes, & Crowe (1984)	F	71-81	25	33
Young, Stokes, & Crowe (1985)	M	70-79	12	25
Vastus lateralis				
Lexell, Henriksson-Larsén, Winblad, & Sjöström (1983)	M	70-73	6	18
Biceps brachii, brachialis				
Klitgaard, Zhou, Schiaffino, et al. (1990)	M	68 ± 0.5	7	20

[a]Values are for cross-sectional area, as determined by imaging.

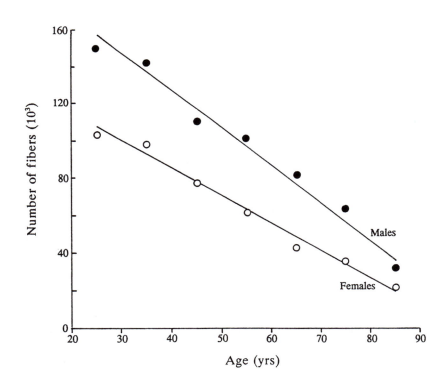

Figure 22.4 Loss of muscle fibers with age. The total numbers of fibers were counted at a single level of the rectus abdominis muscle. The mean values are shown for males (●) and parous females (○), for the midpoint of each decade, rather than for mean age. The two straight lines were fitted by eye. In comparison with these results, the values for nulliparous females (not shown) are better maintained until the 8th decade. Data of Inokuchi, Ishikawa, Iwamoto, & Kimura (1975).

Table 22.3   Animal Studies Demonstrating Loss of Muscle Fibers With Age

Study	Species	Sex	Age (mo)	n	% Decline
Soleus					
Alnaqeeb & Goldspink (1987)	Rat	M	24	5	13
Gutmann & Hanzlíková (1965)	Rat	—	24	—	25
Caccia, Harris, & Johnson (1979)	Rat	F	30	4	27
Rowe (1969)	Mouse	M	25	—	6
	Mouse	F	25	—	21
Extensor digitorum longus					
Alnaqeeb & Goldspink (1987)	Rat	M	24	8	21
Caccia et al. (1979)	Rat	F	30	4	15
Rowe (1969)	Mouse	M	25	—	12
	Mouse	F	25	—	16
Sternomastoid					
Rowe (1969)	Mouse	M	25	—	5
	Mouse	F	25	—	7
Hooper (1981)	Mouse	M	18	12	7
Tibialis anterior					
Rowe (1969)	Mouse	M	25	—	2
	Mouse	F	25	—	(+4)
Hooper (1981)	Mouse	M	18	12	19
Biceps brachii					
Rowe (1969)	Mouse	M	25	—	13
	Mouse	F	25	—	(+4)
Hooper (1981)	Mouse	M	18	12	15
Peroneus digiti quiniti					
Tuffery (1971)	Cat	—	216	1	50
	Cat	—	228	1	63

Degenerative Changes Occur Also in Aged Rodents

The same kinds of abnormality have been reported in the muscles of aged rodents that have been examined with the light and electron microscopes (Alnaqeeb & Goldspink, 1987; Gutmann & Hanzlíková, 1966). In addition, Cardasis (1983), studying electron micrographs of the rat soleus, has described accumulations of enlarged mitochondria, with indistinct cristae, underneath the plasmalemmae of the muscle fibers; in these areas of the fibers, there are often deposits of glycogen also. Lysosomes are frequently seen at the poles of the myonuclei, and polysomes are to be found in close association with the myofilaments.

Motoneuron Losses in Aging

A question which now arises is whether the changes described are due to intrinsic degeneration of the muscle fibers or whether some result from denervation. Histological study of muscle itself by conventional methods is of limited value, for many of the features previously described can occur as part of the so-called secondary "myopathic" complications of denervation (see Drachman, Murphy, Nigam, & Hills, 1967), as well as in a primary muscle disorder. Although it is accepted that the presence of clusters of atrophied muscle fibers (Tomlinson et al., 1969), fiber-type grouping (Jennekens, Tomlinson, & Walton, 1971), or angulated fibers are suggestive of denervation, some of the relevant studies are still subject to criticism over the source of material. Thus, unselected postmortem specimens of muscle are not acceptable since almost any terminal illness may be expected to affect the neuromuscular system—whether as a result of disuse (through rest in bed), malnutrition, chronic infection, or ischemia. Similarly, the neuromuscular system may be involved as a remote effect of cancer, or as a consequence of metabolic disturbances such as renal failure and diabetes. Even elderly subjects killed in accidents may have been harboring serious chronic illness beforehand.

Losses of Motoneurons and Ventral Root Axons Can Be Demonstrated in the Elderly

Although the source of the material may still be problematic, a more direct and satisfactory approach to the investigation of muscle innervation during aging is to count the numbers of motoneurons in the spinal cord or the numbers of large axons in the ventral nerve roots. The former method is not without difficulty, for care must be taken not to count the same motoneuron twice in successive cross sections or to include cells that are interneurons rather than γ-motoneurons or α-motoneurons. The method has been used successfully by Tomlinson and Irving (1977), who have found a reduction in the population of lumbosacral motoneurons that begins in the 7th decade; by the 9th decade, the loss is approximately 29% (see also Kawamura, Okazaki, O'Brien, & Dyck, 1977).

One can more easily count ventral root fibers than motoneurons, remembering that those with large diameters are the axons of α-motoneurons and that smaller

fibers belong to the γ-motoneurons. Unfortunately, criticism can be made of the choice of material used by Gardner (1940); nevertheless, his data strongly suggest that about a quarter of the motor axons are lost in old age (see Table 22.4). In a more recent study, directed to the eighth cervical nerve root, Mittal and Logmani (1987) have also found a reduction in the number of myelinated nerve fibers, together with diminished nerve fiber diameters.

Motor Unit Counting Reveals Losses of Functioning Motoneurons With Aging

Motoneuron populations can also be estimated by the electrophysiological technique described in Box 12.1 in chapter 12, in which the size of the maximal evoked muscle action potential is compared with that of an average motor unit potential. The advantage of this approach is that, unlike the anatomical ones previously described, it provides information as to the numbers of *functioning* motoneurons and excludes any axons that might still be present in an aged spinal cord but no longer capable of exciting muscle fibers. The first application of this technique was to the extensor digitorum brevis muscle (EDB) by Campbell and McComas (1970) and Campbell et al. (1973), who were careful to study only those subjects considered to be in good physical condition for their age (see Figure 22.5). As would be expected of such an approximate technique, the results for any one age exhibit considerable scatter. Nevertheless, when the results for different ages are compared, a striking feature emerges. Until the age of 60, the number of functioning motor units shows little

Table 22.4 Numbers of Axons in Eighth and Ninth Thoracic Ventral Roots in Postmortem Specimens From 60 Subjects of Varying Ages

	Mean no. of axons	
Age (yr)	(T8)	(T9)
10-19	5,698	6,081
20-29	6,205	6,204
30-39	5,816	5,648
40-49	5,495	5,506
50-59	5,361	5,293
60-69	4,805	4,845
70-79	4,420	4,628
80-89	4,923	5,086
Max loss (%)	28.8	25.4

Results for each decade have been averaged.

Note. Adapted from "Decrease in Human Neurones With Age," by E. Gardner, 1940, *Anatomical Record*, **77**, p. 532.

change; beyond the age of 60, there is a progressive fall in the number of functioning units, and by 70, the population of units is reduced to less than half its original size. In some of the oldest subjects, only a very small number of motor units remained, and Figure 22.6 shows deterioration of the sole surviving EDB unit in a man of 92.

The same electrophysiological method has also been applied to the median-innervated thenar muscles and the hypothenar group (Brown, 1972; Sica, McComas, Upton, & Longmire, 1974), as well as to larger muscles such as the soleus (Sica, Sanz, & Colombi, 1976; Vandervoort & McComas, 1986) and the biceps-brachialis (Doherty, Stashuk, & Brown, in press). In these muscles also, there are losses of units after the age of 60, although they are less obvious in the biceps (Dr. Victoria Galea, personal communication, 1995) and hypothenar group than in the other muscles.

Aging Affects Motor Units Selectively

The finding of a reduced number of motor units in old age is not in itself sufficient evidence for denervation. For example, the reduction could arise by a gradual paring away of muscle fibers within motor units, as part of a myopathic process, until some units cease to exist;

a "destitute" motoneuron would be left. Were this the case, one would expect the surviving motor units, because of their reduced fiber populations, to generate smaller potentials than normal. In fact, the mean amplitudes of the motor unit potentials are significantly larger in the elderly than in controls (Campbell et al., 1973). Considered together, these observations suggest that, although some motoneurons (or axons) cease to function, others successfully adopt denervated muscle fibers by sending out new axonal branches. In old age, the capacity of the surviving motoneurons to reinnervate is limited, however, for it is found that the motor unit potentials are not as large as those in younger subjects with similar amounts of denervation.

Animal Studies Also Demonstrate Motoneuron Depletion

In animals, there is also evidence of a loss of motoneurons during aging. For example, in rats aged between 300 and 800 days, a mean loss of 10% of ventral root axons was demonstrated by Duncan (1934). In still older rats, aged from 900 to 1,100 days, Bari and Andrew (1964, cited in Andrew, 1971) found that there was an 18% to 38% loss of ventral horn neurons. This last study showed that degenerative changes may be present

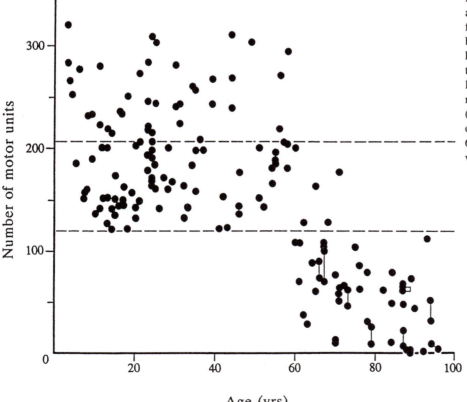

Figure 22.5 Loss of functioning motor units with age. The results were obtained from the extensor digitorum brevis (EDB) muscles of 207 healthy subjects aged 7 months to 97 years. The upper and lower horizontal lines show respectively the mean value (210 units) and the lower limit of the range for subjects below 60 years (120 units). Linked values are bilateral.

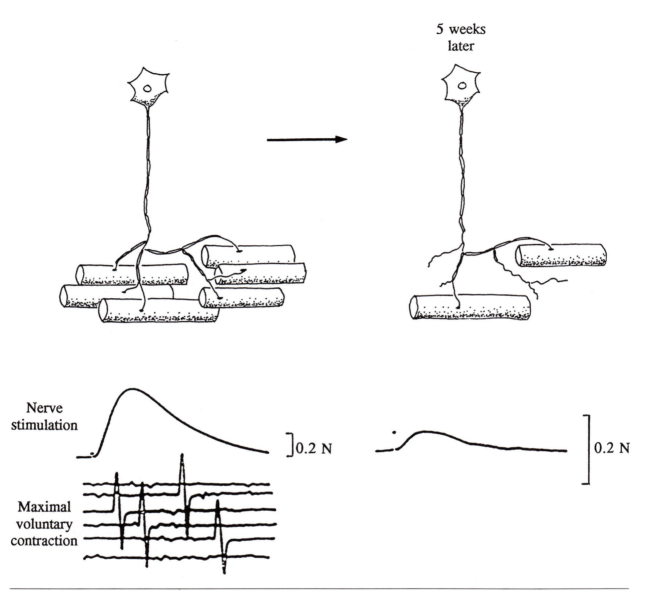

Figure 22.6 Severe denervation in a 92-year-old man. During maximal voluntary contractions, only a single motor unit is seen to discharge. When stimulated through its axon, the same unit developed a large twitch, reflecting substantial collateral reinnervation. Five weeks later, the twitch force had fallen to only 5% of its previous value, suggesting the imminent death of the motor unit. Adapted from Campbell, McComas, & Petito (1973).

in a high proportion of the surviving neurons. Cell bodies are smaller and accumulate pigment, the Nissl substance is reduced in amount, and the nuclei are pale and vacuolated.

Neurons can also be counted in squash preparations of single segments of mouse spinal cord. The total number of large ventral horn neurons (presumably mainly motoneurons) is unchanged up to 50 weeks of age but falls by 15% to 20% during the next 60 weeks (Wright & Spink, 1959). Finally, motoneurons can be labeled by retrograde transport of horseradish peroxidase. Such studies in aged rats also demonstrate reduced numbers

of motoneurons; in addition, it appears that the most susceptible motoneurons are those innervating type IIB muscle fibers (Isihara & Araki, 1988; Isihara, Naitoh, & Katsuta, 1987). The findings of Gutmann and Hanzlíková (1966) stand in contrast, for these authors found normal numbers of motor axons in the soleus nerves of aged rats. However, it is possible that some axons were no longer functional, in view of the reduced electrophysiological estimates of motor units for this muscle (Caccia, Harris, & Johnson, 1979) and for the rat plantaris (Pettigrew & Gardiner, 1987) and medial gastrocnemius (Einsiedel & Luff, 1992).

Motor Units Are Larger in Elderly Subjects

Human motor unit potentials are enlarged through collateral reinnervation in old age, regardless of whether the potentials are evoked by electrical stimulation and recorded with a surface electrode (Campbell et al., 1973) or recruited during voluntary contraction and picked up with needle electrodes having small or large lead-off areas (Howard, McGill, & Dorfman, 1988; Stålberg & Fawcett, 1982; Stålberg et al., 1989). Although the evidence is indirect, the greater sizes of the motor unit potentials strongly suggest that the motor units are themselves enlarged through collateral reinnervation.

Few studies have been made of motor unit twitches in aging. In one such investigation, carried out on the EDB muscle, Campbell et al. (1973) found that the twitch tensions were often increased; similar observations have been made by Galganski, Fuglevand, and Enoka (1993) in the first dorsal interosseous muscle of the hand and by Doherty, Vandervoort, Taylor, and Brown (1993) in the thenar muscle group. Another finding in the elderly is that the contraction times of the motor units are prolonged (Newton, Yemm, & McDonagh, 1988), presumably due to delay in the pumping of Ca^{2+} back into the sarcoplasmic reticulum.

In Aged Animals Also, Motor Unit Twitches Are Larger and Prolonged

Surprisingly few physiological studies of single motor units have been made in aged animals, but the information gained has been detailed and has enabled correlations to be made between the contractile and histochemical properties of the units (Larsson, Ansved, Edström, Gorza, & Schiaffino, 1991; Pettigrew & Gardiner, 1987). As in the human investigations, prolongation of the twitch is found, and the twitch and tetanic tensions of the units are increased. That the motor units contain more fibers than in young animals can be shown by the glycogen depletion method; Larsson et al. (1991) found the type IIB units to have increased their territories and their muscle fiber complements by 40% and 31% respectively. These authors also detected an increased proportion of aged muscle fibers containing the novel myosin heavy chain, IIX, and some units containing fibers of more than one histochemical type, suggesting that the fibers were undergoing transitions. Although motor units varied in their susceptibility to fatigue, there were no significant differences between the results in young and old animals.

The Cause of Neuronal Death in Aging Remains Unknown

What is it that causes motoneurons to die in old age? One possibility is that some factor outside the motoneuron may be responsible. For example, there might be ischemia of the ventral horn, or perhaps the body might gradually accumulate a neurotoxin; alternatively, a deficiency might develop of some humoral or neurotrophic factor necessary to sustain motoneurons. A different type of explanation is that the neurons are genetically programmed to die at a certain age, which varies from one species to another. Indirect evidence for such a mechanism is that tissue culture techniques have succeeded in isolating strains of central nervous system neurons that are capable of indefinite survival in conventional media.

Other Neurons Resemble Motoneurons by Degenerating in Later Life

Do the changes in motoneurons reflect a more widespread loss of neural function with advancing age? On the basis of frequently quoted studies on cerebellar Purkinje cells, cerebral cortex, olfactory bulb, and optic nerve, it is generally believed that there is a steady loss of neurons throughout adult life. Wright and Spink (1959) have argued that, in the mouse cord, the loss of motoneurons is delayed until relatively late in life. The ingeniously simple experiments on whole brains of mice by Johnson and Erner (1972) shed new light on this controversy. After fixation in formalin, the brains were mashed in water and then further broken up into a fine suspension by ultrasonic vibration. Using a standard dilution, the suspension was stained with thionine, and samples were examined in a standard hemocytometer chamber. The somata of the neurons were well preserved and could be identified and counted; the total brain neuron content was determined by a multiplication factor appropriate for the dilution of the suspension. Johnson and Erner found that a marked loss of neurons did take place during aging but that this only occurred relatively late (see Figure 22.7). Allowing for differences in the lifespans of the species, these results are similar to those of Campbell and associates (1973) for human motor units, discussed earlier.

Changes in Axons and Neuromuscular Junctions

The loss of axons in the ventral roots of elderly subjects has already been commented on. In this section, we shall see that the surviving axons may show morphological changes and that the neuromuscular junctions, too, are different in old age. The possibility of diminished function, raised by these modifications, will be explored.

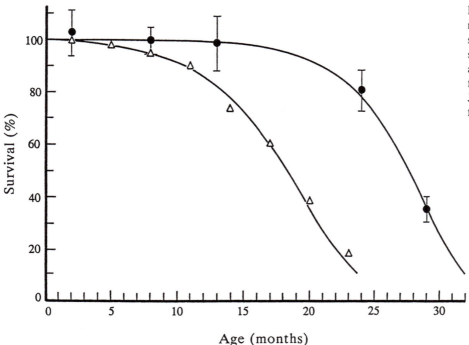

Figure 22.7 Percentages of mice (Δ) and their neurons (●) surviving at different ages. The smooth curves have been drawn according to a "Gompertz" function as explained by Johnson and Erner. Adapted from Johnson and Erner (1972).

Nerve Impulses Are Conducted More Slowly in Elderly Subjects

It is easy to measure the maximal impulse conduction velocities in human subjects (see chapter 9), and a consistent finding in the elderly is that the values decrease (Falco, Hennessey, Braddom, & Goldberg, 1992; Norris, Shock, & Wagman, 1953). Histological examination of nerve fibers, including teased specimens, reveal certain features that contribute to the lower velocities; these features include segmental demyelination, remyelination, decreased internodal length, and drop-out of the fibers with largest diameters (Arnold & Harriman, 1970; Dyck, 1975).

Neuromuscular Junctions Become More Complex With Age

In human muscles, the intramuscular motor axons can be examined by the intravital methylene blue method. The dye is applied to the surface of the muscle over the innervation zone, and the specimen is excised 5 min afterward; the nerve fibers are stained blue (Cöers & Woolf, 1959). This technique has revealed that the neuromuscular junctions are often complex, due to the presence of additional nerve twigs from the original motor axons (Harriman, Taverner, & Woolf, 1970). In the preterminal axons, spherical swellings are evident, and there is an increased incidence of axon sprouting, suggesting that denervation of muscle fibers may be occurring with subsequent reinnervation by surviving motor axons.

In cat muscle fibers, teased and impregnated with silver, there is also evidence of complex end-plates in older animals (Barker & Ip, 1966). As many as five axon sprouts may grow from the nodes of Ranvier to supply the same end-plate (Tuffery, 1971; see Figure 22.8).

As in human muscles, some cat axons have dilatations. The presence of swollen, retracted axon terminals suggests that remodeling of the junctions continues throughout life. It is probable that, in aging, the axon sprouts are needed to maintain the safety margin for synaptic transmission. In addition, by opening up new areas of contact for the transfer of trophic material, the new branches may boost the declining influence of the motoneuron on the muscle fiber, and vice versa.

In Aged Rodents, Nerve Endings May Withdraw in Their Synaptic Gutters

Studies in small, short-lived mammals, such as the mouse and rat, have confirmed that aging neuromuscular junctions have abnormal structures. However, the additional axonal branches derived from nodal sprouts do not appear to be present so that, although the end-plate area is enlarged, the axonal arborization is less dense than in young animals (Figure 22.9). Ultraterminal sprouts *are* found, though (Rosenheimer, 1990). With the electron microscope, it can be seen that some of the synaptic gutters are empty, indicating that the motor nerve terminals have withdrawn (Cardasis, 1983; Wernig et al., 1984). The remaining nerve endings tend

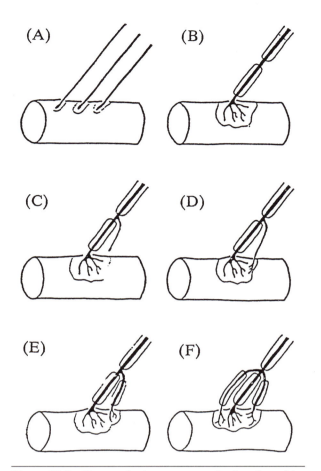

Figure 22.8 Changes in the neuromuscular junction with age, as studied in the cat. At birth (A), muscle fibers are multiply innervated, but within the next few weeks (B), the surplus motor axons are lost. In adulthood, a sprout may form at a node of Ranvier (C), reach the end-plate (D), and increase the complexity of the end-plate (E). The new sprout, and others that appear later, become myelinated (F). Reprinted from Tuffery (1971).

to be smaller than in young animals and may contain fewer synaptic vesicles, though more neurofilaments and microtubules (Banker, Kelly, & Robbins, 1983). More than one ending may lie within the same synaptic gutter. At some junctions, Schwann cells are partially or completely interposed between the motor nerve terminals and the synaptic folds, reducing the possibility of effective synaptic transmission. On the postsynaptic side of the junction, the most obvious changes during aging are the widening of the secondary clefts, with the inclusion of collagen, and the shallowness of the primary clefts. Surprisingly, although the amount of synaptic folding is decreased, there is no reduction in the numbers of ACh receptors (Banker et al., 1983). The various changes in the neuromuscular junction and motor axon with aging are summarized in Figure 22.10 and in Table 22.5.

Although Altered in Their Morphology, Aged Neuromuscular Junctions Appear to Function Well

How well do the neuromuscular junctions function in old age? Gutmann, Hanzlíková, and Vyskocil (1971), in a microelectrode examination of neuromuscular junction in old rats, reported that the frequency of MEPPs was reduced, but other authors have not found this to be true of all muscles (Banker et al., 1983; Smith, 1984). Despite the relative paucity of synaptic vesicles in the nerve terminal (see the preceding section), the numbers that are released with each impulse may be greater than normal (Banker et al., 1983; Smith, 1984). It is likely that the presence of more than one (small) axon terminal over the same synaptic cleft is responsible for this over-compensation. Another factor may be an increased influx of Ca^{2+} into the terminal during the impulse (Alshuaib & Fahim, 1990). In human muscles, single fiber EMG recordings have shown that neuromuscular jitter (see chapter 10) increases with age (Stålberg & Trontelj, 1979), although muscle responses to repetitive nerve stimulation are well maintained.

Nerve Regeneration Is Slower in Aged Animals

Drahota and Gutmann (1961) crushed rat sciatic nerves and observed that axonal regeneration is considerably delayed and much less vigorous in older animals than in younger ones. The most complete study is that of Pestronk, Drachman, and Griffin (1980), who examined terminal sprouting in rat soleus muscles after pharmacological denervation had been produced by the application of botulinum toxin. In contrast to the striking increases in motor end-plate length and in terminal arborization that occurred in younger animals, there was no significant sprouting response in the oldest (28-month-old) rats. Axon regeneration after a sciatic nerve crush was also studied, using axonal transport as a marker of the outgrowth from the site of the lesion. It was found that some axons in older rats grow back as quickly as those in young animals, but the average rate of regeneration is considerably lower (Figure 22.11). Although these results showed the impaired response of aged motoneurons to injury, they do not distinguish between an intrinsic failure of the nerve cells to generate new axonal branches and a lack of stimulatory cues (e.g., growth factors) supplied by the denervated tissues and Schwann cells.

APPLIED PHYSIOLOGY

In this section, there are two topics for consideration. The question as to whether exercise programs are effective in the elderly is an important one. In a society that

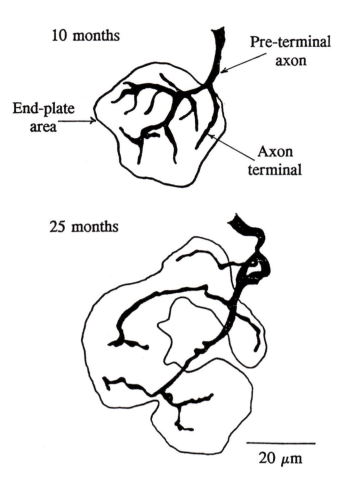

10 months

Pre-terminal axon

End-plate area

Axon terminal

25 months

20 μm

Figure 22.9 Changes in motor nerve endings with age, as studied in the rat. By 25 months, an old age for a rat, the terminals are longer but have lost some of their side branches. The total area occupied by the end-plate is larger. Adapted from Rosenheimer (1990).

encourages comfort and passive entertainment, it is all too easy for the elderly to adopt sedentary lifestyles and to be at risk for certain types of injury and illness. We will also learn about a deadly disease in older people in which there is a rapid loss of motoneurons.

Elderly Subjects Benefit From Training Programs

Given the various degenerative changes in the muscles and motoneurons of elderly people, can improvements in strength and endurance come from training programs? This question has been addressed by several investigators who have applied heavy resistance and isometric training techniques to a variety of large and small limb muscles. Very substantial improvements in strength were reported in most of these studies, as Table 22.6 shows; it is likely that the changes resulted from a combination of neural adaptation and muscle hypertrophy. According to Grimby (1988), the type II fibers show the most prominent increases in cross-sectional area following strength training in the elderly. Although none of the studies appears to have studied changes in endurance in a systematic manner, it is an everyday observation that some elderly athletes are able to run

considerable distances, and even post-myocardial infarction patients have been trained to run marathons (Todd, Wosornu, Stewart, & Wild, 1992).

In Amyotrophic Lateral Sclerosis, Motoneurons Die Prematurely

Amyotrophic lateral sclerosis, or ALS, is a fatal disease in which the motoneurons appear to age unusually quickly. The age-adjusted incidence increases in the 6th and 7th decades, when motor unit populations normally start to decline (see Figure 22.5) and continues to rise thereafter. Although age is undoubtedly a factor in causation, in some patients other influences are at play. Thus, a small proportion of cases are familial, and in others there are biochemical or immunological abnormalities. Occasionally, there is a striking association with a severe electric shock or with mechanical trauma.

Among the well known who have died from this condition was Lou Gehrig, who played at first base for the New York Yankees baseball team between 1925 and 1939, and after whom the disease is named. Others have included Ezzard Charles, the former heavyweight boxing champion of the world, and David Niven, the British film actor. Figure 22.12 shows how rapidly the

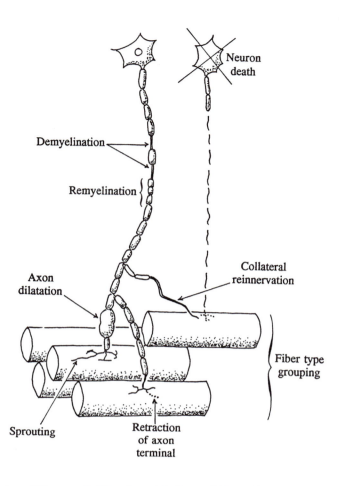

Figure 22.10 Summary of aging changes that may occur in the motor innervation of mammalian muscle fibers.

Table 22.5   Electron Microscopic Morphometric Data of Nerve Terminals in Young and Old Mouse Extensor Digitorum Longus Muscles

Parameter investigated	Young	Old	Percentage change in old versus young
Number of junctional regions (number of animals)	28 (5)	27 (5)	
Junctional length of nerve terminal (μm)	4.6 ± 3.0	3.9 ± 3.0	15↓
Nerve terminal area (μm^2)	6.3 ± 3.2	2.9 ± 2.9	55↓
Number of vesicles per terminal	415.1 ± 180.8	157.6 ± 117.0	62↓
Number of vesicles per μm^2 terminal area free of mitochondria	111.2 ± 42.6	78.5 ± 33.0	29↓
Number of SER[a] profiles per terminal	3.6 ± 2.0	5.9 ± 5.2	65↑
Number of cisternae per terminal	3.6 ± 2.1	4.4 ± 5.6	22↑
Number of coated vesicles per terminal	5.0 ± 3.0	7.5 ± 4.6	49↑
Number of mitochondria per terminal	16.6 ± 8.5	3.9 ± 5.1	77↓
Mitochondrial area (μm^2)	2.2 ± 1.5	0.4 ± 0.5	84↓
Mitochondrial area as a percentage of the nerve terminal area	33.3 ± 12.0	11.0 ± 12.0	67↓
Synaptic cleft width (nm)	55.7 ± 16.6	62.0 ± 24.1	11↑

[a]SER = smooth endoplasmic reticulum.

All data are expressed as mean ± SD. All percentage changes are significant at the $p = {<}0.05$ level.

Note. Reprinted from Fahim and Robbins (1982).

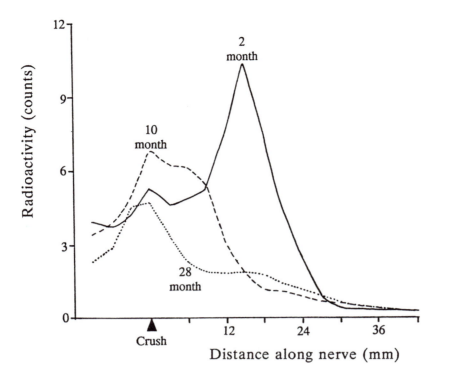

Figure 22.11 Delayed regeneration in motor axons of old animals. Rat sciatic nerves were crushed 9 days previously, and the spinal cords were injected with radioactive amino acid (^{35}S-methionine), so that the radioactivity could be incorporated into the regenerating axons. The majority of axons, as reflected in the peak of radioactivity, have grown farthest in the 2-month-old animals and least in the 28-month-old rats. See also Figure 8.2 in chapter 8. Adapted from Pestronk, Drachman, and Griffin (1980).

Table 22.6 Summary of Strength-Training Studies in the Elderly

Study reference	Sex	Age (yr)	n	Type of control	Muscle action	Type of training	Type of test	Training sessions per wk	No. of wks	Increase in strength (%)
Perkins & Kaiser (1961)	M	62-86	5	None	Plantarflexion	Isometric (n = 10)	Same	3	6	50
	F		15		Knee extension					57
					Hip extension					29
					Plantarflexion	Concentric weight lifting (n = 10)	Same	3	6	63
					Knee extension					41
					Hip extension					64
Chapman, de Vries, & Swezey (1972)	M	63-88	20	Contralateral limb	Index finger flexion	Concentric weight lifting	Isometric	3	6	33
Aniansson & Gustaffson (1981)	M	69-74	12	Matched cohort	Knee extension	Calisthenics	Isokinetic 30°/s-180°/s	3	12	14-22
							Isometric			9
Aniansson, Ljungberg, Rundgren, & Wattequis (1984)	F	63-84	15	Matched cohort	Knee extension	Calisthenics, elastic bands	Isokinetic 30°/s-180°/s	3	26	7-11
							Isometric			13
Kaufman (1985)	F	65-73	10	None	Little finger abduction	Isometric	Isometric	3	6	72

Note. Reprinted from Vandervoort, Hayes, and Bélanger (1986).

motoneurons degenerate; once a motoneuron pool is affected, approximately half the cells will cease to function within 6 months. The disease may present in a variety of ways—for example, clumsiness of the hands (as in Lou Gehrig), weakness of the legs, difficulty in swallowing, or even a change in voice. As the disease progresses, the muscles become increasingly weak and wasted, and motor units may begin to twitch spontaneously (*fasciculation*). The mean life expectancy is only 3 years from the time of diagnosis.

Much research is now being devoted to establishing the etiology of this condition, and a recent success has been the identification of a gene on chromosome 21 as one cause of the familial cases (Siddique et al., 1991). This gene codes for an enzyme, cytosolic *superoxide dismutase*, that disposes of free oxygen radicals in the cell body. If the radicals are allowed to linger because of the enzyme deficiency, it is easy to imagine that they would damage the metabolic and genetic machinery of

the cell, and cause its premature death (McNamara & Fridovich, 1993). Attractive though this explanation is, current opinion holds that it may not be the full story and that the enzyme abnormality may exert its effect in some other way.

This exploration of aging changes brings to an end the study of skeletal muscle and its nerve supply. If parts of the study have appeared complex, it is nonetheless inevitable that yet more detail will emerge about many of the physiological processes and their ultrastructural correlates. At the same time, there can be satisfaction that so much is known about the many facets of muscle and nerve fibers, and that this knowledge has extended to the identification of precise abnormalities responsible for an increasing number of neuromuscular disorders.

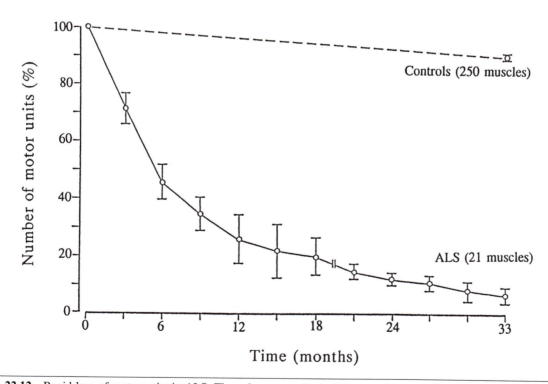

Figure 22.12 Rapid loss of motor units in ALS. The values were obtained from various muscles using the electrophysiological method described in chapter 12. From "The extent and time course of motoneuron involvement in amyotrophic lateral sclerosis," by M. Dantes and A.J. McComas, 1991, *Muscle and Nerve*, **14**, p. 419. Copyright 1991 by John Wiley & Sons, Inc. Reprinted with permission.

References

Adams, R.D., Denny-Brown, D., & Pearson, C. (1962). *Diseases in muscle. A study in pathology* (2nd ed.). New York: Harper and Row. [16]

Adrian, E.D., & Bronk, D.W. (1929). The discharge of impulses in motor nerve fibres. II. The frequency of discharge in reflex and voluntary contractions. *Journal of Physiology*, **67**, 119-151. [13]

Adrian, R.H., & Bryant, S.H. (1974). On the repetitive discharge in myotonic muscle fibres. *Journal of Physiology*, **240**, 505-515. [7]

Adrian, R.H., Costantin, L.L., & Peachey, L.D. (1969). Radial spread of contraction in frog muscle fibres. *Journal of Physiology*, **204**, 231-257. [11, 15]

Aguayo, A.J., Attiwell, M., Trecarten, J., Perkins, S., & Bray, G.M. (1977). Abnormal myelination in transplanted Trembler mouse Schwann cells. *Nature*, **265**, 73-75. [2]

Aguayo, A.J., Perkins, S., Bray, G., & Duncan, I. (1978). Transplantation of nerves from patients with Charcot-Marie-Tooth (CMT) disease into immune-suppressed mice. *Journal of Neuropathology and Experimental Neurology*, **37**, 582. [2]

Aitken, J.T., Sharman, M., & Young, J.Z. (1947). Maturation of regenerating nerve fibres with various peripheral connexions. *Journal of Anatomy*, **81**, 1-22. [17]

Al-Amood, W.S., Buller, A.J., & Pope, R. (1973). Long-term stimulation of cat fast-twitch skeletal muscle. *Nature*, **244**, 225-227. [18, 20]

Albani, A., Lowrie, M.B., & Vrbová, G. (1988). Reorganization of motor units in reinnervated muscles of the rat. *Journal of the Neurological Sciences*, **88**, 195-206. [6]

Albe-Fessard, D., Liebeskind, J., & Lamarre, Y. (1965). Projection au niveau du cortex somato-moteur du singe d'afférences provenant des recepteurs musculaires. *Comptes Rendus de l'Academie des Sciences*, **261**, 3891-3894. [4]

Albers, R.W. (1967). Biochemical aspects of active transport. *Annual Review of Biochemistry*, **36**, 727-756. [7]

Alberts, B., Bray, D., Lewis, J., Raff, M., Roberts, K., & Watson, J.D. (1983; 1989). *Molecular biology of the cell* (1st & 2nd eds.). New York: Garland Publishing. [1, 14]

Albuquerque, E.X., Schuh, F.T., & Kauffman, F.C. (1971). Early membrane depolarization of the fast mammalian muscle after denervation. *Pflügers Archiv*, **328**, 36-50. [16, 18]

Allen, D.G., Lee, J.A., & Westerblad, H. (1989). Intracellular calcium and tension during fatigue in isolated single muscle fibres from *Xenopus laevis*. *Journal of Physiology*, **415**, 433-458. [11, 15]

Allen, D.G., Westerblad, H., Lee, J.A., & Lännergren, J. (1992). Role of excitation-contraction coupling in muscle fatigue. *Sports Medicine*, **13**, 116-126. [20]

Allen, F., & Warner, A. (1991). Gap junction communication during neuromuscular junction formation. *Neuron*, **6**, 101-111. [6]

Allen, G.M., Gandevia, S.C., Neering, I.R., Hickie, I., Jones, R., & Middleton, J. (1994). Muscle performance, voluntary activation and perceived effort in normal subjects and patients with prior poliomyelitis. *Brain*, **117**, 661-670. [19]

Allt, G., & Cavanagh, J.B. (1969). Ultrastructural changes in the regions of the node of Ranvier in the rat caused by diphtheria toxin. *Brain*, **92**, 459-468. [9]

Alnaes, E., & Rahamimoff, R. (1975). On the role of mitochondria in transmitter release from motor nerve terminals. *Journal of Physiology*, **248**, 285-306. [3]

Alnaqeeb, M.A., & Goldspink, G. (1987). Changes in fibre type, number and diameter in developing and ageing skeletal muscle. *Journal of Anatomy*, **153**, 31-45. [22]

Alshuaib, W.B., & Fahim, M.A. (1990). Aging increases calcium influx at motor nerve terminal. *International Journal of Developmental Neuroscience*, **8**, 655-666. [22]

Alway, S.E., Winchester, P.K., Davis, M.E., & Gonyea, W.J. (1989). Regionalized adaptations and muscle fiber proliferation in stretch-induced enlargement. *Journal of Applied Physiology*, **66**, 771-781. [20]

Amos, L.A., Linck, R.W., & Klug, A. (1976). Molecular structure of flagellar microtubules. In R. Goldman, T. Pollard, & J. Rosenbaum (Eds.), *Cell motility. Book C. Microtubules and related proteins* (pp. 847-868). New York: Cold Spring Harbor Laboratory. [2]

Andersen, P., & Henricksson, J. (1977). Training induced changes in the subgroups of human type II skeletal muscle fibres. *Acta Physiologica Scandinavica*, **99**, 123-125. [20]

Anderson, D.C., King, S.C., & Parsons, S.M. (1982). Proton gradient linkage to active uptake of [^{3}H] acetylcholine by *Torpedo* electric organ synaptic vesicles. *Biochemistry*, **21**, 3037-3043. [10]

Anderson, H.J., Churchill-Davidson, H.C., & Richardson, A.J. (1953). Bronchial neoplasm with myasthenia: Prolonged apnoea after administration of succinylcholine. *Lancet*, **ii**, 1291-1293. [10]

Andreassen, S., & Arendt-Nielsen, L. (1987). Muscle fibre conduction velocity in motor units of the human anterior tibial muscle: A new size principle parameter. *Journal of Physiology*, **391**, 561-571. [9, 12]

Andrew, W. (1971). *The anatomy of ageing in man and animal*, p. 231. New York: Grune and Stratton. [22]

Aniansson, A., & Gustaffson, E. (1981). Physical training in elderly men with specific reference to quadriceps muscle strength and morphology. *Clinical Physiology*, **1**, 87-98. [22]

Aniansson, A., Ljungberg, P., Rundgren, A., & Wattequist, H. (1984). Effect of a training programme for pensioners on condition and muscular strength. *Archives of Geron-tology and Geriatrics*, **3**, 229-241. [22]

Arahata, K., & Engel, A.G. (1985). Endomysial killer/natural killer (K/NK) cells, T cells and macrophages (MF) in polymyositis (PM), inclusion body myositis (IBM), and Duchenne dystrophy (DD). *Neurology*, **35**(Suppl. 1), 205. [21]

Armstrong, C.M., & Bezanilla, F. (1973). Currents related to movement of the gating particles of the sodium channels. *Nature*, **242**, 459-461. [7]

Armstrong, C.M., Bezanilla, F., & Horowicz, P. (1972). Twitches in the presence of ethylene glycol bis(β-aminoethylene)-N,N^1-tetraacetic acid. *Biochimica et Biophysica Acta*, **267**, 605-608. [11]

Armstrong, C.M., Bezanilla, F., & Rojas, E. (1973). Destruction of sodium conductance inactivation in squid axons perfused with pronase. *Journal of General Physiology*, **62**, 375-391. [7]

Armstrong, R.B., Marum, P., Tullison, P., & Saubert, C.W., IV. (1979). IV. Acute hypertrophic response of skeletal muscle to removal of synergists. *Journal of Applied Physiology*, **46**, 835-842. [20]

Arnáson, B.G.W., Winkler, G.F., & Hadler, N.M. (1969). Cell-mediated demyelination of peripheral nerve in tissue culture. *Laboratory Investigation*, **21**, 1-10. [16]

Arnold, N., & Harriman, D.G.F. (1970). The incidence of abnormality in control human peripheral nerves studied by single axon dissection. *Journal of Neurology, Neurosurgery and Psychiatry*, **33**, 55-61. [22]

Ashley, C.C., & Ridgway, E.B. (1968). Simultaneous recording of membrane potential, calcium transient and tension in single muscle fibres. *Nature*, **219**, 1168-1169. [11]

Ashley, C.C., & Ridgway, E.B. (1970). On the relationships between membrane potential, calcium transient and tension in single barnacle muscle fibres. *Journal of Physiology*, **209**, 105-130. [11]

Åstrand, P.-O. (1967). Diet and athletic performance. *Federation Proceedings: Experimental Biology*, **26**, 1772-1777. [14]

Aström, K.E., & Waksman, B.H. (1962). The passive transfer of experimental allergic encephalomyelitis and neuritis with living lymphoid cells. *Journal of Pathology and Bacteriology*, **83**, 89-106. [16]

Auld, V.J., Goldin, A.L., Kraft, D.S., Catterall, W.A., Lester, H.A., Davidson, N., & Dunn, B.J. (1990). A neutral amino acid change in segment 11 S4 dramatically alters the gating properties of the voltage-dependent sodium channel. *Proceedings of the National Academy of Sciences* (U.S.), **87**, 323-327. [7]

Auld, V.J., Goldin, A.L., Kraft, D.S., Marshall, J., Dunn, J.M., Catterall, W.A., Lester, H.A., Davidson, N., & Dunn, R.J. (1988). A rat brain Na$^+$ channel α subunit with novel gating properties. *Neuron*, **1**, 449-461. [7]

Axelsson, J., & Thesleff, S. (1959). A study of supersensitivity in denervated mammalian skeletal muscle. *Journal of Physiology*, **147**, 178-193. [16]

Bagust, J. (1974). Relationships between motor nerve conduction velocities and motor unit contraction characteristics in a slow-twitch muscle of the cat. *Journal of Physiology*, **238**, 269-278. [13]

Bagust, J., & Lewis, D.M. (1974). Isometric contractions of motor units in self-reinnervated fast and slow-twitch muscles of the cat. *Journal of Physiology*, **237**, 91-102. [17]

Baichwal, R.R., Bigbee, J.W., & DeVríes, G.H. (1988). Macrophage-mediated myelin-related mitogenic factor for cultured Schwann cells. *Proceedings of the National Academy of Sciences* (U.S.), **85**, 1701-1705. [16]

Baker, L.P., Chen, O., & Peng, H.B. (1992). Induction of acetylcholine receptor clustering by native polystyrene beads. *Journal of Cell Biology*, **102**, 543-555. [3]

Baker, P.F., Hodgkin, A.L., & Shaw, T.I. (1962a). The effects of changes in internal ionic concentrations on the electrical properties of perfused giant axons. *Journal of Physiology*, **164**, 355-374. [9]

Baker, P.F., Hodgkin, A.L., & Shaw, T.I. (1962b). Replacement of the axoplasm of giant nerve fibres with artificial solutions. *Journal of Physiology*, **164**, 330-354. [9]

Balice-Gordon, R.J., & Lichtman, J.W. (1994). Long-term synapse loss induced by focal blockade of postsynaptic receptors. *Nature*, **372**, 519-524. [6]

Ballantyne, J.P., & Campbell, M.J. (1973). Electrophysiological study after surgical repair of sectioned human peripheral nerves. *Journal of Neurology, Neurosurgery and Psychiatry*, **36**, 797-805. [17]

Bangsbo, J., Graham, T.E., Kiens, B., & Saltin, B. (1992). Elevated muscle glycogen and anaerobic energy production during exhaustive exercise in man. *Journal of Physiology*, **451**, 205-227. [14]

Banister, E.W., & Cameron, B.J.C. (1990). Exercise-induced hyperammonemia: Peripheral and central effects. *International Journal of Sports Medicine*, **11**(Suppl. 2), S129-S142. [14]

Banker, B.Q. (1986). The congenital myopathies. In A.G. Engel & B.Q. Banker (Eds.), *Myology. Basic and clinical* (pp. 1527-1581). New York: McGraw-Hill. [5]

Banker, B.Q., Kelly, S.S., & Robbins, N. (1983). Neuromuscular transmission and correlative morphology in young and old mice. *Journal of Physiology*, **339**, 355-375. [22]

Bär, A., & Pette, D. (1988). Three fast myosin heavy chains in adult rat skeletal muscle. *FEBS Letters*, **235**, 153-155. [12]

Barchi, R.L. (1988). Probing the molecular structure of the voltage-dependent sodium channel. *Annual Review of Neuroscience*, **11**, 455-495. [7]

Barcroft, H., & Dornhorst, A.C. (1949). The blood flow through the human calf during rhythmic exercise. *Journal of Physiology*, **109**, 402-411. [13]

Barcroft, H., & Millen, J.L.E. (1939). The blood flow through muscle during sustained contraction. *Journal of Physiology*, **97**, 17-31. [13, 15]

Barker, D. (1974). The morphology of muscle receptors. In C.C. Hunt (Ed.), *Handbook of sensory physiology. Muscle receptors* (Vol. 3, pp. 1-190). Berlin: Springer-Verlag. [4]

Barker, D., & Ip, M.C. (1966). Sprouting and degeneration of mammalian motor axons in normal and deafferentated skeletal muscle. *Proceedings of the Royal Society of London, Series B*, **163**, 538-554. [22]

Barker, D., & Saito, M. (1981). Autonomic innervation of receptors and muscle fibres in cat skeletal muscle. *Proceedings of the Royal Society of London, Series B*, **212**, 317-332. [15]

Barnard, E.A., Dolly, J.O., Porter, C.W., & Albuquerque, E.X. (1975). The acetylcholine receptor and the ionic conductance modulation system of skeletal muscle. *Experimental Neurology*, **48**, 1-28. [3]

Basmajian, J.V. (1963). Control and training of individual motor units. *Science*, **141**, 440-441. [13]

Bastian, J., & Nakajima, S. (1974). Action potential in the transverse tubules and its role in the activation of skeletal muscle. *Journal of General Physiology*, **63**, 257-278. [11]

Bateson, D.S., & Perry, D.J. (1983). Motor units in a fast-twitch muscle of normal and dystrophic mice. *Journal of Physiology*, **345**, 515-523. [12]

Baumann, H., Jäggi, M., Soland, F., Howald, H., & Schaub, M.C. (1987). Exercise training induces transitions of myosin isoform subunits within histochemically typed human muscle fibres. *Pflügers Archiv*, **409**, 349-360. [20]

Beam, K.G., Caldwell, J.H., & Campbell, D.T. (1985). Na channels in skeletal muscle concentrated near the neuromuscular junction. *Nature*, **313**, 588-590. [7]

Beam, K.G., Knudson, C.M., & Powell, J.A. (1986). A lethal mutation in mice eliminates the slow calcium current in skeletal muscle cells. *Nature*, **320**, 168-170. [11]

Beeson, D., & Barnard, E. (1990). Acetylcholine receptors at the neuromuscular junction. In A. Vincent & D. Wray (Eds.), *Neuromuscular transmission: Basic and applied aspects* (pp. 157-181). Manchester: Manchester University Press. [3]

Bélanger, A.Y., & McComas, A.J. (1981). Extent of motor unit activation during effort. *Journal of Applied Physiology*, **51**, 1131-1135. [13, 15]

Bélanger, A.Y., & McComas, A.J. (1983). Contractile properties of muscles in myotonic dystrophy. *Journal of Neurology, Neurosurgery and Psychiatry*, **46**, 625-631. [15]

Bellemare, F., & Bigland-Ritchie, B. (1984). Assessment of human diaphragm strength and activation using phrenic nerve stimulation. *Respiratory Physiology*, **58**, 263-277. [13]

Bellemare, F., Bigland-Ritchie, B., & Woods, J.J. (1986). Contractile properties of the human diaphragm *in vivo*. *Journal of Applied Physiology*, **61**, 1153-1161. [12]

Bellemare, F., Woods, J.J., Johansson, R., & Bigland-Ritchie, B. (1983). Motor-unit discharge rates in maximal voluntary contractions of three human muscles. *Journal of Neurophysiology*, **50**, 1380-1392. [12, 13]

Belmar, J., & Eyzaguirre, C. (1966). Pacemaker site of fibrillation potentials in denervated mammalian muscle. *Journal of Neurophysiology*, **29**, 425-441. [16]

Benezra, R., Davis, R.L., Lassar, A., Tapscott, S., Thayer, M., Lockshon, D., & Weintraub, H. (1990). Id: A negative regulator of helix-loop-helix DNA binding proteins. Control of terminal myogenic differentiation. *Annals of the New York Academy of Sciences*, **599**, 1-11. [5]

Bennett, M.K., Calakos, N., & Scheller, R.H. (1992). Syntaxin: A synaptic protein implicated in docking of synaptic vesicles at presynaptic active zones. *Science*, **257**, 255-259. [10]

Bennett, M.K., & Scheller, R.H. (1993). The molecular machinery for secretion is conserved from yeast to neurons. *Proceedings of the National Academy of Sciences* (U.S.), **90**, 2559-2563. [10]

Bennett, M.R., McLachlan, E., & Taylor, R.S. (1973). The formation of synapses in reinnervated mammalian striated muscle. *Journal of Physiology*, **233**, 481-500. [17]

Bennett, M.R., & Pettigrew, A.G. (1974). The formation of synapses in striated muscle during development. *Journal of Physiology*, **241**, 515-545. [6]

Bennett, M.R., Pettigrew, A.G., & Taylor, R.S. (1973). The formation of synapses in reinnervated and cross-reinnervated adult avian muscle. *Journal of Physiology*, **230**, 331-357. [17]

Benoit, P., & Changeux, J.-P. (1978). Consequences of blocking the nerve with a local anaesthetic on the evolution of multi-innervation at the regenerating neuromuscular junction of the rat. *Brain Research*, **149**, 89-96. [6]

Berg, D.K., & Hall, Z.W. (1975). Increased extra-junctional acetylcholine sensitivity produced by chronic post-synaptic neuromuscular blockage. *Journal of Physiology*, **244**, 659-676. [19]

Bergström, J. (1962). Muscle electrolytes in man. *Scandinavian Journal of Clinical Laboratory Investigation*, **68**(Suppl.), 1-110. [13]

Bergström, J., Hermansen, L., Hultman, E., & Saltin, B. (1967). Diet, muscle glycogen and physical performance. *Acta Physiologica Scandinavica*, **71**, 140-150. [14]

Berliner, E., Young, E.C., Anderson, K., Mahtani, H.K., & Gelles, J. (1995). Failure of a single-headed kinesin to track parallel to microtubule protofilaments. *Nature*, **373**, 718-721. [8]

Bessou, P., & Pagès, B. (1972). Intracellular potentials from intrafusal muscle fibres evoked by stimulation of static and dynamic fusimotor axons in the cat. *Journal of Physiology*, **227**, 709-727. [4]

Betz, W.J., Caldwell, J.H., & Ribchester, R.R. (1979). The size of motor units during post-natal development of rat lumbrical muscle. *Journal of Physiology*, **297**, 463-478. [12]

Bevan, S., Chiu, S.Y., Gray, P.T., & Ritchie, J.M. (1985). The presence of voltage-gated sodium, potassium and chloride channels in rat cultured astrocytes. *Proceedings of the Royal Society of London, Series B*, **225**, 299-313. [2]

Bigland, B., & Lippold, O.C.J. (1954). Motor unit activity in the voluntary contraction of human muscle. *Journal of Physiology*, **125**, 322-335. [13]

Bigland-Ritchie, B., Johansson, R., Lippold, O.C.J., Smith, S., & Woods, J.J. (1983). Changes in motoneurone firing rates during sustained maximal voluntary contractions. *Journal of Physiology*, **340**, 335-346. [13]

Bigland-Ritchie, B., Kukulka, C.G., Lippold, O.C.J., & Woods, J.J. (1982). The absence of neuromuscular transmission failure in sustained maximal voluntary contractions. *Journal of Physiology*, **330**, 265-278. [15]

Bigland-Ritchie, B.R., Dawson, N.J., Johansson, R.S., & Lippold, O.C.J. (1986). Reflex origin for the slowing of motoneurone firing rates in fatigue of human voluntary contractions. *Journal of Physiology*, **376**, 451-459. [4, 15]

Bigland-Ritchie, B.R., Johansson, R., Lippold, O.C.J., & Woods, J.J. (1983). Speed and EMG changes during fatigue of sustained maximal voluntary contractions. *Journal of Neurophysiology*, **50**, 314-324. [15]

Billeter, R., Heizmann, C.W., Howald, H., & Jenny, E. (1981). Analysis of myosin light and heavy chain types in single human skeletal muscle fibres. *European Journal of Biochemistry*, **116**, 389-395. [12]

Bisby, M.A. (1975). Inhibition of axonal transport in nerves chronically treated with local anesthetics. *Experimental Neurology*, **47**, 481-489. [19]

Bischoff, R., & Holtzer, H. (1969). Mitosis and the process of differentiation of myogenic cells *in vitro*. *Journal of Cell Biology*, **4**, 188-200. [5]

Bischoff, R., & Lowe, M. (1974). Cell surface components and the interaction of myogenic cells. In A.T. Milhorat (Ed.), *Exploratory concepts in muscular dystrophy II* (pp. 17-29). Amsterdam: Excerpta Medica. [5]

Black, J.A., Kocsis, J.D., & Waxman, S.G. (1990). Ion channel organization of the myelinated fiber. *Trends in Neurosciences*, **13**, 48-54. [7, 9]

Blasi, J., Chapman, E.R., Link, E., Binz, T., Yamasaki, S., DeCamilli, P., Südhof, T.C., Niemann, H., & Jahn, R. (1993). Botulinum neurotoxin A selectively cleaves the synaptic protein SNAP-25. *Nature*, **365**, 160-163. [10]

Blewett, C., & Elder, G. (1993). Quantitative EMG analysis in soleus and plantaris during hindlimb suspension and recovery. *Journal of Applied Physiology*, **74**, 2057-2066. [19]

Blinks, J.R., Rüdel, R., & Taylor, S.R. (1978). Calcium transients in isolated amphibian skeletal muscle fibres: Detection with aequorin. *Journal of Physiology*, **277**, 291-323. [15]

Blumcke, S., & Niedorf, H.R. (1965). Elektronenoptische untersuchungen an Wachstumsendkolben regenerierender peripherer nervenfasern. *Virchow's Archiv für Pathologische Anatomie und Physiologie*, **340**, 93-104. [3]

Bodian, D. (1964). An electromicroscopic study of the monkey spinal cord. I. Fine structure of normal motor column. II. Effects of retrograde chromatolysis. III. Cytologic effects of mild and virulent poliovirus infection. *Bulletin of Johns Hopkins Hospital*, **114**, 13-119. [2]

Bodine-Fowler, S., Garfinkel, A., Roy, R.R., & Edgerton, V.R. (1990). Spatial distribution of muscle fibers within the territory of a motor unit. *Muscle & Nerve*, **13**, 1133-1145. [12]

Bongiovanni, L.G., & Hagbarth, K.-E. (1990). Tonic vibration reflexes elicited during fatigue from maximal voluntary contractions in man. *Journal of Physiology*, **423**, 1-14. [15]

Booth, F.W. (1977). Time course of muscular atrophy during immobilization of hindlimbs of rats. *Journal of Applied Physiology*, **43**, 656-661. [19]

Booth, F.W. (1988). Perspectives on molecular and cellular exercise physiology. *Journal of Applied Physiology*, **65**, 1461-1471. [20]

Borg, J. (1981). Properties of single motor units in the extensor digitorum brevis in elderly humans. *Muscle & Nerve*, **4**, 429-434. [22]

Borg, T.K., & Caulfield, J.B. (1980). Morphology of connective tissue in skeletal muscle. *Tissue & Cell*, **12**, 197-207. [1]

Bostock, H., & Sears, T.A. (1978). The internodal axon membrane: Electrical excitability and continuous conduction in segmental demyelination. *Journal of Physiology*, **280**, 273-301. [9]

Botelho, S.Y., Candler, L., & Guiti, N. (1954). Passive and active tension-length diagrams of intact skeletal muscle in normal women of different ages. *Journal of Applied Physiology*, **7**, 93-98. [22]

Bowden, R.E.M., & Gutmann, E. (1944). Denervation and re-innervation of human voluntary muscle. *Brain*, **67**, 273-313. [16]

Boyd, I.A. (1966). The behaviour of isolated mammalian muscle spindles with intact innervation. *Journal of Physiology*, **186**, 109-110P. [4]

Boyd, I.A., & Davey, M.R. (1968). *Composition of peripheral nerves*. Edinburgh: Livingstone. [12]

Brady, S.T. (1985). A novel brain ATPase with properties expected for the fast axonal transport motor. *Nature*, **317**, 73-75. [8]

Brandstater, M.E., & Dinsdale, S.M. (1976). Electrophysiological studies in assessment of spinal cord lesions. *Archives of Physical Medicine and Rehabilitation*, **57**, 70-74. [18]

Brandstater, M.E., & Lambert, E.H. (1969). A histochemical study of the spatial arrangement of muscle fibres in single motor units within rat tibialis anterior muscle. *Bulletin of the American Association of Electromyography and Electrodiagnosis*, **82**, 15-16. [12]

Brandstater, M.E., & Lambert, E.H. (1973). Motor unit anatomy. Type and spatial arrangement of muscle fibres. In J.E. Desmedt (Ed.), *New developments in electromyography and clinical neurophysiology* (Vol. 1, pp. 14-22). Basel: Karger. [12]

Braun, T., Rudnicki, M.A., Arnold, H.-H., & Jaenisch, R. (1992). Targeted inactivation of the muscle regulatory gene myf-5 results in abnormal rib development and perinatal death. *Cell*, **71**, 369-382. [5]

Bray, J.J., Hawken, M.J., Hubbard, J.I., Pockett, S., & Wilson, L. (1976). The membrane potential of rat diaphragm muscle fibres and the effect of denervation. *Journal of Physiology*, **255**, 651-667. [16]

Brimijoin, S. (1975). Stop-flow: A new technique for measuring axonal transport, and its application to the transport of dopamine-β-hydroxylase. *Journal of Neurobiology*, **6**, 379-394. [8]

Broman, H., DeLuca, C.J., & Mambrito, B. (1985). Motor unit recruitment and firing rates interaction in the control of human muscles. *Brain Research*, **337**, 311-319. [13]

Brooke, M.H., & Engel, W.K. (1969). The histographic analysis of human muscle biopsies with regard to fibre types. 1. Adult male and female. *Neurology*, **19**, 221-233. [5]

Brooke, M.H., & Kaiser, K.K. (1970). Muscle fibre types: How many and what kind? *Archives of Neurology*, **23**, 369-379. [12, 13]

Brooke, M.H., & Kaiser, K.K. (1974). The use and abuse of muscle histochemistry. *Annals of the New York Academy of Sciences*, **228**, 121-144. [12]

Brooks, J.E. (1969). Hyperkalaemic periodic paralysis. Intracellular electromyographic studies. *Archives of Neurology*, **20**, 13-18. [7]

Brooks, V.B., & Thies, R.E. (1962). Reduction of quantum content during neuromuscular transmission. *Journal of Physiology*, **162**, 298-310. [15]

Brostoff, S., Burnett, P., Lambert, P., & Eylar, E.H. (1972). Isolation and characterization of a protein from sciatic nerve myelin responsible for experimental allergic neuritis. *Nature New Biology*, **235**, 210-212. [16]

Brostoff, S.W., & Eylar, E.H. (1972). The proposed amino acid sequence of the P1 protein of sciatic nerve myelin. *Archives of Biochemistry and Biophysics*, **153**, 590-598. [16]

Brown, G.L., & Harvey, A.M. (1939). Congenital myotonia in the goat. *Brain*, **62**, 341-363. [7]

Brown, M.C., & Butler, R.G. (1974). Evidence for innervation of muscle spindle intrafusal fibres by branches of α-motoneurones following nerve injury. *Journal of Physiology*, **238**, 41-43P. [17]

Brown, M.C., Holland, R.L., & Ironton, R. (1978). Degenerating nerve products affect innervated muscle fibres. *Nature*, **275**, 652-654. [17]

Brown, M.C., & Ironton, R. (1977). Motor neurone sprouting induced by prolonged tetrodotoxin block of nerve action potentials. *Nature*, **265**, 459-461. [19]

Brown, M.C., Jansen, J.K.S., & Van Essen, D. (1976). Polyneuronal innervation of skeletal muscle in new-born rats and its elimination during maturation. *Journal of Physiology*, **261**, 387-422. [6]

Brown, M.C., & Lunn, E.R. (1988). Mechanism of interaction between motoneurons and muscles. In *Plasticity of the Neuromuscular System, CIBA Symposium*, **138**, 78-96. Chichester: Wiley. [17]

Brown, W.F. (1972). A method for estimating the number of motor units in thenar muscles and the changes in motor unit counting with ageing. *Journal of Neurology, Neurosurgery and Psychiatry*, **35**, 845-852. [22]

Browne, K., Lee, J., & Ring, P.A. (1954). The sensation of passive movement at the metatarsophalangeal joint of the great toe in man. *Journal of Physiology*, **126**, 448-458. [4]

Bryant, S.H. (1969). Cable properties of external intercostal muscle fibres from myotonic and normal goats. *Journal of Physiology*, **204**, 539-550. [7]

Bryant, S.H., & Morales-Aguilera, A. (1971). Chloride conductance in normal and myotonic muscle fibres and the action of monocarboxylic aromatic acids. *Journal of Physiology*, **219**, 367-383. [7, 9]

Buchanan, P., & Convertino, V.A. (1989). A study of the effects of prolonged simulated microgravity on the musculature of the lower extremities in man: An introduction. *Aviation, Space, and Environmental Medicine*, **60**, 649-652. [19]

Buchthal, F., & Clemmesen, S. (1941). On the differentiation of muscle atrophy by electromyography. *Acta Psychiatrica Neurologica*, **16**, 143-181. [13]

Buchthal, F., Guld, C., & Rosenfalck, P. (1955). Propagation velocity in electrically activated muscle fibres in man. *Acta Physiologica Scandinavica*, **34**, 75-89. [9]

Buchthal, F., Guld, C., & Rosenfalck, P. (1957). Multi-electrode study of the territory of a motor unit. *Acta Physiologica Scandinavica*, **39**, 83-104. [12]

Buchthal, F., Rosenfalck, A., & Behse, F. (1975). Sensory potentials of normal and diseased nerves. In P.J. Dyck, P.K. Thomas, & E.H. Lambert (Eds.), *Peripheral neuropathy* (Vol. 1, pp. 442-464). Philadelphia: W.B. Saunders. [2]

Buchthal, F., & Rosenfalck, P. (1958). Rate of impulse conduction in denervated human muscle. *Electroencephalography and Clinical Neurophysiology*, **10**, 521-526. [9]

Buchthal, F., Rosenfalck, P., & Erminio, F. (1960). Motor unit territory and fiber density in myopathies. *Neurology*, **10**, 398-408. [9]

Buchthal, F., & Schmalbruch, H. (1970). Contraction times and fibre types in intact human muscle. *Acta Physiologica Scandinavica*, **79**, 435-452. [12]

Buller, A.J., Dornhorst, A.C., Edwards, R., Kerr, D., & Whelan, R.F. (1959). Fast and slow muscles in mammals. *Nature*, **183**, 1516-1517. [12]

Buller, A.J., Eccles, J.C., & Eccles, R.M. (1960a). Differentiation of fast and slow muscles in the cat hindlimb. *Journal of Physiology*, **150**, 399-416. [5, 18, 19]

Buller, A.J., Eccles, J.C., & Eccles, R.M. (1960b). Interaction between motoneurones and muscles in respect of the characteristic speeds of their responses. *Journal of Physiology*, **150**, 417-439. [17, 18]

Burke, R.E. (1967). Motor unit types of cat triceps surae muscle. *Journal of Physiology*, **193**, 141-160. [12, 13]

Burke, R.E. (1975). A comment on the existence of motor unit "types." In D.B. Tower (Ed.), *The nervous system. The basic neurosciences* (Vol. 1, pp. 611-619). New York: Raven Press. [12]

Burke, R.E. (1986). Physiology of motor units. In A.G. Engel & B.Q. Banker (Eds.), *Myology. Basic and clinical* (Vol. 1, pp. 419-443). New York: McGraw-Hill. [13]

Burke, R.E., Dunn, R.P., Fleshman, J.W., Glenn, L.L., Lev-Tov, A., O'Donovan, M.J., & Pinter, M.J. (1982). An HRP study of the relation between cell size and motor unit type in cat ankle extensor motoneurons. *Journal of Comparative Neurology*, **209**, 17-28. [12]

Burke, R.E., Levine, D.N., Salcman, M., & Tsairis, P. (1974). Motor units in cat soleus muscle: Physiological, histochemical and morphological characteristics. *Journal of Physiology*, **238**, 503-514. [12]

Burke, R.E., Levine, D.N., Tsairis, P., & Zajac, F.E., III. (1973). Physiological types and histochemical profiles in motor units of the cat gastrocnemius. *Journal of Physiology*, **234**, 723-748. [12, 15]

Burke, R.E., Levine, P.N., & Zajac, F.E., III. (1971). Mammalian motor units: Physiological-histochemical correlation in three types in cat gastrocnemius. *Science*, **174**, 709-712. [12, 13]

Burke, R.E., & Tsairis, P. (1973). Anatomy and innervation ratios in motor units of cat gastrocnemius. *Journal of Physiology*, **234**, 749-765. [12]

Butler, J., Cauwenbergs, P., & Cosmos, E. (1986). Fate of brachial muscles of the chick embryo innervated by inappropriate nerves: Structural, functional and histochemical analyses. *Journal of Embryology and Experimental Morphology*, **95**, 147-168. [6]

Butler, J., Cosmos, E., & Brierley, J. (1982). Differentiation of muscle fiber types in aneurogenic brachial muscles of the chick embryo. *Journal of Experimental Zoology*, **224**, 65-80. [5]

Butler, J., Cosmos, E., & Cauwenbergs, P. (1988). Positional signals: Evidence for a possible role in muscle fibre-type patterning of the embryonic avian limb. *Development*, **102**, 763-772. [5]

Butler-Browne, G.S., Eriksson, P.-O., Laurent, C., & Thornell, L.-E. (1988). Adult human masseter muscle fibers express myosin isozymes characteristic of development. *Muscle & Nerve*, **11**, 610-620. [12]

Byers, M.R., Fink, B.R., Kennedy, R.D., Middaugh, M.E., & Hendrickson, A.E. (1973). Effects of lidocaine on axonal morphology, microtubules, and rapid transport in rabbit vagus nerve *in vitro*. *Journal of Neurobiology*, **4**, 125-143. [19]

Bylund-Fellenius, A.-C., Ojamaa, K.M., Flaim, K.E., Li, J.B., Wassner, S.J., & Jefferson, L.S. (1984). Protein synthesis versus energy state in contracting muscles of perfused rat hindlimb. *American Journal of Physiology*, **246**, E297-E305. [20]

Caccia, M.R., Harris, J.B., & Johnson, M.A. (1979). Morphology and physiology of skeletal muscle in aging rodents. *Muscle & Nerve*, **2**, 202-212. [22]

Cady, E.B., Jones, D.A., Lynn, J., & Newham, D.J. (1989). Changes in force and intracellular metabolites during fatigue of human skeletal muscle. *Journal of Physiology*, **418**, 311-325. [15]

Cajal, S.R. (1909). *Histologie du systeme nerveux de l' homme et des vertebrés* (Vol. 1, pp. 485-489). Paris: Maloine. [4]

Cajal, S.R. (1928). *Degeneration and regeneration of the nervous system* (Vol. 1, R.M. May, Trans.). London: Oxford University Press. [17]

Calancie, B., & Bawa, P. (1985). Voluntary and reflexive recruitment of flexor carpi radialis motor units in humans. *Journal of Neurophysiology*, **53**, 1194-1200. [12, 13]

Campbell, L.R., Dayton, D.H., & Sohal, G.S. (1986). Neural tube defects: A review of human and animal studies on the etiology of neural tube defects. *Teratology*, **34**, 171-187. [6]

Campbell, M.J., & McComas, A.J. (1970). The effects of ageing on muscle function. In *5th Symposium on Current Research on Muscular Dystrophy and Related Disease* (Abstract No. 6). London: Muscular Dystrophy Group of Great Britain. [22]

Campbell, M.J., McComas, A.J., & Petito, F. (1973). Physiological changes in ageing muscles. *Journal of Neurology, Neurosurgery and Psychiatry*, **36**, 174-182. [22]

Cangiano, A. (1973). Acetylcholine supersensitivity: The role of neurotrophic factors. *Brain Research*, **58**, 255-259. [18]

Carafoli, E., & Penniston, J.T. (1985). The calcium signal. *Scientific American*, **253**(5), 70-78. [7]

Cardasis, C.A. (1983). Ultrastructural evidence of continued reorganization at the aging (11-26 months) rat soleus neuromuscular junction. *Anatomical Record*, **207**, 399-415. [22]

Carlson, B.M. (1970). Regeneration of the rat gastrocnemius muscle from sibling and non-sibling muscles fragments. *American Journal of Anatomy*, **128**, 21-31. [21]

Carolan, B., & Cafarelli, E. (1992). Adaptations in coactivation after isometric resistance training. *Journal of Applied Physiology*, **73**, 911-917. [20]

Carrington, J.L., & Fallon, J.F. (1988). Initial limb budding is independent of apical ectodermal ridge activity; Evidence from a limbless mutant. *Development*, **104**, 361-367. [5]

Casey, E.B., Jellife, A.M., Le Quesne, P.M., & Millett, Y.L. (1973). Vincristine neuropathy: Clinical and electrophysiological observations. *Brain Research*, **96**, 69-86. [8]

Cashman, N.R., Maselli, R., Wollmann, R.L., Roos, R., Simon, R., & Antel, J.P. (1987). Late denervation in patients with antecedent paralytic poliomyelitis. *New England Journal of Medicine*, **317**, 7-12. [17]

Castle, N.A., Haylett, D.G., & Jenkinson, D.H. (1989). Toxins in the characterization of potassium channels. *Trends in Neurosciences*, **12**, 59-65. [7]

Catterall, W.A. (1988). Structure and function of voltage-sensitive ion channels. *Science*, **242**. 50-61. [7]

Chandler, W.K., Rakowski, R.F., & Schneider, M.F. (1976). Effects of glycerol treatment and maintained depolarization on charge movement in skeletal muscle. *Journal of Physiology*, **254**, 285-316. [11]

Changeux, J.-P., & Danchin, A. (1976). Selective stabilisation of developing synapses as a mechanism for the specification of neuronal networks. *Nature*, **264**, 705-712. [6]

Chapman, E.A., de Vries, H.A., & Swezey, R. (1972). Joint stiffness: Effects of exercise on young and old men. *Journal of Gerontology*, **27**, 218-221. [22]

Chapman, R.A., & Tunstall, J. (1987). The calcium paradox of the heart. *Progress in Biophysics and Molecular Biology*, **50**, 67-96. [15]

Charlton, M.P., Silverman, H., & Atwood, H.L. (1981). Intracellular potassium activities in muscles of normal and dystrophic mice: An *in vivo* electrometric study. *Experimental Neurology*, **71**, 203-219. [9]

Charlton, M.P., Smith, S.J., & Zucker, R.S. (1982). Role of presynaptic calcium ions and channels in synaptic facilitation and depression at the squid giant synapses. *Journal of Physiology*, **323**, 173-193. [10]

Chevallier, A., Kieny, M., & Mauger, A. (1977). Limb-somite relationships: Origin of the limb musculature. *Journal of Embryology and Experimental Morphology*, **41**, 245-258. [5]

Chong-Hiyo, T.P., Pruitt, J.H., & Bennett, D. (1989). A mouse model for neural tube defects: The curtailed (T^c) mutation produces spina bifida occulta in T^c/+ animals and spina bifida with meningomyelocele in T^c/t. *Teratology*, **39**, 303-312. [6]

Chow, I., & Cohen, M.W. (1983). Developmental changes in the distribution of acetylcholine receptors in the myotomes of *Xenopus laevis*. *Journal of Physiology*, **339**, 553-571. [6]

Christensen, E.H., & Hansen, O. (1939). Arbeitsfahigkeit und ernahrung. *Skandinavische Archiv fur Physiologie*, **81**, 160-171. [14]

Chu, L.W. (1954). A cytological study of anterior horn cells isolated from human spinal cord. *Journal of Comparative Neurology*, **100**, 381-413. [2]

Clark, A.W., Mauro, A., Longenecker, H.E., & Hurlbut, W.P. (1970). Effects of black widow spider venom on the frog neuromuscular junction. Effects on the fine structure of the frog neuromuscular junction. *Nature*, **225**, 703-705. [3]

Clarke, D.M., Loo, T.W., Inesi, G., & MacLennan, D.H. (1989). Location of high affinity Ca^{2+}-binding sites within the predicted transmembrane domain of the sarcoplasmic reticulum Ca^{2+}-ATPase. *Nature*, **339**, 476-478. [7]

Clarkson, P.M., Kroll, W., & Melchionda, A.M. (1981). Age, isometric strength, rate of tension development and fiber type composition. *Journal of Gerontology*, **36**, 648-653. [22]

Clausen, T. (1986). Regulation of active Na^+-K^+ transport in skeletal muscle. *Physiological Reviews*, **66**, 542-580. [7]

Clausen, T. (1990). Significance of Na^+-K^+ pump regulation in skeletal muscle. *News in Physiological Sciences*, **5**, 148-151. [15]

Clausen, T., & Everts, M.E. (1989). Regulation of the Na,K-pump in skeletal muscle. *Kidney International*, **35**, 1-13. [7]

Close, R. (1967). Properties of motor units in fast and slow skeletal muscles of the rat. *Journal of Physiology*, **193**, 45-55. [12]

Cöers, C., & Woolf, A.L. (1959). *The innervation of muscle: A biopsy study*. Oxford: Blackwell Scientific Publications. [17, 22]

Cole, K.S., & Curtis, H.J. (1939). Electric impedance of the squid giant axon during activity. *Journal of General Physiology*, **22**, 649-670. [9]

Cole, K.S., & Moore, J.W. (1960). Ionic current measurements in the squid giant axon membrane. *Journal of General Physiology*, **44**, 123-167. [7]

Colling-Saltin, A. (1978). Enzyme histochemistry on skeletal muscle of the human foetus. *Journal of the Neurological Sciences*, **39**, 169-185. [5]

Colquhoun, D., Rang, H.P., & Ritchie, J.M. (1974). The binding of tetrodotoxin and α-bungarotoxin to normal and denervated mammalian muscle. *Journal of Physiology*, **240**, 199-226. [9]

Condon, K., Silberstein, L., Blau, H.M., & Thompson, W.J. (1990). Differentiation of fiber types in aneural musculature of the prenatal rat hindlimb. *Developmental Biology*, **138**, 275-295. [5]

Cook, J.A. (1777). *A voyage towards the South Pole and around the world* (Vol. ii, pp. 112-113). London: Straham and Cadell. [7]

Cook, S.D., Murray, M.N., Whitaker, J.N., & Dowling, P. (1969). Myelinotoxic antibody in the Guillain-Barré syndrome. *Neurology* (Minn.), **19**, 284. [16]

Cook, W.H., Walker, J.H., & Barr, M.L. (1951). A cytological study of transneuronal atrophy in the cat and the rabbit. *Journal of Comparative Neurology*, **94**, 267-291. [18]

Cooke, R., & Pate, E. (1985). The effects of ADP and phosphate on the contraction of muscle fibres. *Biophysical Journal*, **48**, 789-798. [15]

Cooper, R.G., Stokes, M.J., & Edwards, R.H.T. (1989). Myofibrillar activation failure in McArdle's disease. *Journal of the Neurological Sciences*, **93**, 1-10. [14]

Coote, J.H., Hilton, S.M., & Perez-Gonzalez, J.M. (1971). The nature of the pressor response to muscular exercise. *Journal of Physiology*, **215**, 789-804. [4]

Corley, K., Kowalchuk, N., & McComas, A.J. (1984). Contrasting effects of suspension on hindlimb muscles in the hamster. *Experimental Neurology*, **85**, 30-40. [19]

Costantin, L.L. (1970). The role of sodium current in the radial spread of contraction in frog muscle fibers. *Journal of General Physiology*, **55**, 703-715. [11]

Courbin, P., Koenig, J., Ressouches, A., Beam, K.G., & Powell, J.A. (1989). Rescue of excitation-contraction coupling in dysgenic muscle by addition of fibroblasts in vitro. *Neuron*, **2**, 1341-1350. [11]

Creese, R., Hashish, S.E.E., & Scholes, N.W. (1958). Potassium movements in contracting diaphragm muscle. *Journal of Physiology*, **143**, 307-324. [15]

Creese, R., Head, S.D., & Jenkinson, D.F. (1987). The role of the sodium pump during prolonged end-plate currents in guinea-pig diaphragm. *Journal of Physiology*, **384**, 377-403. [9]

Crick, F.H.C. (1970). Diffusion in embryogenesis. *Nature*, **225**, 420-422. [5]

Crowe, A., & Matthews, P.B.C. (1964). The effects of stimulation of static and dynamic fusiform fibres on the response to stretching of the primary endings of muscle receptors. *Journal of Physiology*, **174**, 109-131. [4]

Cummins, K.L., Dorfman, L.J., & Perkel, D.H. (1979). Nerve fiber conduction-velocity distributions. II. Estimation based on two compound action potentials. *Electroencephalography and Clinical Neurophysiology*, **46**, 647-658. [9]

Curtis, H.J., & Cole, K.S. (1938). Transverse electric impedance of the squid giant axon. *Journal of General Physiology*, **21**, 757-765. [9]

Dan, Y., & Poo, M.-M. (1992). Hebbian depression of isolated neuromuscular synapses in vitro. *Science*, **256**, 1570-1573. [6]

Dantes, M., & McComas, A.J. (1991). The extent and time course of motoneuron involvement in amyotrophic lateral sclerosis. *Muscle & Nerve*, **14**, 416-421. [17, 22]

Dasgupta, A., & Simpson, J.A. (1962). Relation between firing frequency of motor units and muscle tension in the human. *Electromyography*, **2**, 117-128. [13]

David, J.D., See, W.M., & Higginbotham, C.-A. (1981). Fusion of chick embryo skeletal myoblasts: Role of calcium influx preceding membrane union. *Developmental Biology*, **82**, 297-307. [5]

Davies, C.T.M., Thomas, D.O., & White, M.J. (1986). Mechanical properties of young and elderly human muscle. *Acta Medical Scandinavica*, **Suppl. 711**, 219-226. [22]

Davies, C.T.M., & White, M.J. (1983). Contractile properties of elderly human triceps surae. *Gerontology*, **29**, 19-25. [22]

Davis, C.J.F., & Montgomery, A. (1977). The effect of prolonged inactivity upon the contraction characteristics of fast and slow mammalian twitch muscle. *Journal of Physiology*, **270**, 581-594. [19]

Dawes, G.S. (1941). The vaso-dilator action of potassium. *Journal of Physiology*, **99**, 224-238. [13]

Dawson, M.J., Gadian, D.G., & Wilkie, D.R. (1978). Muscular fatigue investigated by phosphorus nuclear magnetic resonance. *Nature*, **274**, 861-866. [15]

Dean, R.B. (1941). Theories of electrolyte equilibrium in muscle. *Biological Symposium*, **3**, 331-339. [7]

DeBarsy, T., & Hers, H.-G. (1990). Normal metabolism and disorders of carbohydrate metabolism. *Baillière's Clinical Endocrinology and Metabolism*, **4**, 499-522. [14]

DeLorme, T.L. (1945). Restoration of muscle power by heavy resistance exercises. *Journal of Bone and Joint Surgery*, **27**, 645-667. [20]

DeLuca, C.J., LeFever, R.S., McCue, M.P., & Xenakis, A.P. (1982). Behaviour of human motor units in different muscles during linearly varying contractions. *Journal of Physiology*, **329**, 113-128. [13]

Denborough, M.A., & Lovell, R.R.H. (1960). Anaesthetic deaths in a family. *Lancet*, **ii**, 45. [11]

Dengler, R., Konstanzer, A., Hesse, S., Schubert, M., & Wolf, W. (1989). Collateral nerve sprouting and twitch forces of single motor units in conditions with partial denervation in man. *Neuroscience Letters*, **97**, 118-122. [17]

Dengler, R., Stein, R.B., & Thomas, C.K. (1988). Axonal conduction velocity and force of single human motor units. *Muscle & Nerve*, **11**, 136-145. [12]

Dennis, H.J., Ziskind-Conhaim, L., & Harris, A.J. (1981). Development of neuromuscular functions in rat embryos. *Developmental Biology*, **81**, 266-279. [6]

Dennis, M.J., & Miledi, R. (1974). Non-transmitting neuromuscular junctions during an early stage of end-plate reinnervation. *Journal of Physiology*, **239**, 553-570. [17]

Denny-Brown, D. (1929). The histological features of striped muscle in relation to its functional activity. *Proceedings of the Royal Society of London, Series B*, **104**, 371-411. [12]

Denny-Brown, D., & Brenner, C. (1944). Paralysis of nerve induced by direct pressure and by tourniquet. *Archives of Neurology and Psychiatry* (Chicago), **51**, 1-26. [19]

Desmedt, J.E., & Hainaut, K. (1977). Inhibition of the intracellular release of calcium by dantrolene in barnacle giant muscle fibres. *Journal of Physiology*, **265**, 565-585. [11]

Detwiler, S.R. (1920). On the hyperplasia of nerve centers resulting from excessive peripheral loading. *Proceedings of the National Academy of Sciences* (U.S.), **6**, 96-101. [6]

Devreotes, P.N., & Fambrough, D.M. (1975). Acetylcholine receptor turnover in membranes of developing muscle fibers. *Journal of Cell Biology*, **65**, 335-358. [3]

de Vries, J.I.P. (1987). *Development of specific movement patterns in the human fetus*. Groningen: Drukkerij Van Denderen B V. [5]

de Vries, J.I.P., Visser, G.H.A., & Prechtl, H.F.R. (1982). The emergence of fetal behaviour. I. Qualitative aspects. *Early Human Development*, **7**, 301-322. [5]

DiMauro, S., & DiMauro, P.M.M. (1973). Muscle carnitine palmityltransferase deficiency and myoglobinuria. *Science*, **182**, 929. [14]

Dodge, F.A., & Rahamimoff, R. (1967). Co-operative action of calcium ions in transmitter release at the neuromuscular junction. *Journal of Physiology*, **193**, 419-432. [10]

Doherty, T.J., Komori, T., Stashuk, D.W., Kassam, A., & Brown, W.F. (1994). Physiological properties of single thenar motor units in the F-response of younger and older adults. *Muscle & Nerve*, **17**, 860-872. [9]

Doherty, T.J., Stashuk, D.W., & Brown, W.F. (in press). Methods to estimate the numbers of motor units in human muscles. In Y.I. Kim & N. Thakor (Eds.), *Neural engineering*. New York: Springer-Verlag. [22]

Doherty, T.J., Vandervoort, A.A., Taylor, A.W., & Brown, W.F. (1993). Effects of motor unit losses on strength in older men and women. *Journal of Applied Physiology*, **74**, 868-874. [22]

Donaldson, S.K.B., & Hermansen, L. (1978). Differential, direct effects of H^+ on Ca^{2+}-activated force of skinned fibres from the soleus, cardiac and adductor magnus muscles of rabbits. *Pflügers Archiv*, **376**, 55-65. [15]

Donoghue, M.J., Morris-Valero, R., Johnson, Y.R., Merlie, J.P., & Sanes, J.R. (1992). Mammalian muscle cells bear a cell-autonomous, heritable memory of their rostrocaudal position. *Cell*, **69**, 67-77. [5]

Drachman, D.B., Murphy, S.R., Nigam, M.P., & Hills, J.R. (1967). Myopathic changes in chronically denervated muscles. *Archives of Neurology*, **16**, 14-24. [22]

Draeger, A., Weeds, A.G., & Fitzsimons, R.B. (1987). Primary, secondary and tertiary myotubes in developing skeletal muscle: A new approach to the analysis of human myogenesis. *Journal of the Neurological Sciences*, **81**, 19-43. [5, 12]

Drahota, Z., & Gutmann, E. (1961). The influence of age on the course of reinnervation of muscle. *Gerontologica*, **5**, 88-109. [22]

Drahota, Z., & Gutmann, E. (1963). Long-term regulatory influence of the nervous system on some metabolic differences in muscles of different function. *Physiologica Bohemoslovaca*, **12**, 339-348. [17]

Droz, B., & Leblond, C.P. (1963). Axonal migration of proteins in the central nervous system and peripheral nerves as shown by radioautography. *Journal of Comparative Neurology*, **121**, 325-346. [8]

Droz, B., Rambourg, A., & Koenig, H.L. (1975). The smooth endoplasmic reticulum: Structure and role in the renewal of axonal membrane and synaptic vesicles by fast axonal transport. *Brain Research*, **93**, 1-13. [8]

Dubois-Dalcq, M., Buyse, M., Buyse, G., & Gorce, F. (1971). The action of Guillain-Barré syndrome serum on myelin. A tissue culture and electron microscopic analysis. *Journal of the Neurological Sciences*, **13**, 67-83. [16]

Dubowitz, V. (1967). Cross-innervated mammalian skeletal muscle: Histochemical, physiological and biochemical observations. *Journal of Physiology*, **193**, 481-496. [17, 18]

Dubowitz, V. (1985). *Muscle biopsy. A practical approach*. London: Baillière Tindall. [12]

Duchateau, J., & Hainaut, K. (1987). Electrical and mechanical changes in immobilized human muscle. *Journal of Applied Physiology*, **62**, 2168-2173. [19]

Duchen, L.W. (1970a). Changes in motor innervation and cholinesterase localization induced by botulinum toxin in skeletal muscle of the mouse: Differences between fast and slow muscles. *Journal of Neurology, Neurosurgery and Psychiatry*, **33**, 40-54. [17]

Duchen, L.W. (1970b). The effect in the mouse of nerve crush and regeneration on the innervation of skeletal muscles paralyzed by *Clostridium* botulinum toxin. *Journal of Pathology*, **102**, 9-14. [17]

Duchen, L.W., & Stefani, E. (1971). Electrophysiological studies of neuromuscular transmission in hereditary "motor end-plate disease" of the mouse. *Journal of Physiology*, **212**, 535-548. [15]

Duchen, L.W., Stolkin, C., & Tonge, D.A. (1972). Light and electronmicroscopic changes in slow and fast skeletal muscle fibres and their motor end-plates in the mouse after the local injection of tetanus toxin. *Journal of Physiology*, **222**, 136-137P. [17]

Duchenne, G.B.A. (1861, 1872). *De l'électrisation localisée et de son application à la physiologie à la pathologie et à thérapeutique*. (2nd and 3rd eds.) Paris: Baillière et fils. [1]

Dudley, G.A., Duvoisin, M.R., Convertino, V.A., & Buchanan, P. (1989). Alterations of the *in vivo* torque-velocity relationship of human skeletal muscle following 30 days exposure to simulated microgravity. *Aviation, Space, and Environmental Medicine*, **60**, 659-663. [19]

Dulhunty, A.F., & Franzini-Armstrong, C. (1975). The relative contributions of the folds and caveolae to the surface membrane of frog skeletal fibres at different sarcomere lengths. *Journal of Physiology*, **250**, 513-539. [1]

Duncan, D. (1934). A determination of the number of nerve fibres in the eight thoracic and the largest lumbar ventral roots of the albino rat. *Journal of Comparative Neurology*, **59**, 47-60. [22]

Dyck, P.J. (1975). Pathological alterations of the peripheral nervous system of man. In P.J. Dyck, P.K. Thomas, & E.H. Lambert (Eds.), *Peripheral neuropathy* (pp. 296-336). Philadelphia: W.B. Saunders. [22]

Dyck, P.J., Lais, A.C., & Low, P.A. (1978). Nerve zenografts to assess cellular expression of the abnormality of myelination in inherited neuropathy and Friedreich ataxia. *Neurology*, **28**, 261-265. [2]

Dyck, P.J., Nukada, H., Lais, A.C., & Karnes, J.L. (1984). Permanent axotomy: A model of chronic neuronal degeneration preceded by axonal atrophy, myelin remodelling and degeneration. In P.J. Dyck, P.K. Thomas, E.H. Lambert, & R. Bunge (Eds.), *Peripheral neuropathy* (pp. 666-690). Philadelphia: W.B. Saunders. [16]

Eaton, L.M., & Lambert, E.H. (1957). Electromyography and electric stimulation of nerves in diseases of motor unit. Observations on the myasthenic syndrome associated with malignant tumors. *Journal of the American Medical Association*, **163**, 1117-1124. [10]

Eberstein, A., & Sandow, A. (1963). Fatigue mechanisms in muscle fibres. In E. Gutmann & P. Hník (Eds.), *The effects of use and disuse on neuromuscular functions* (pp. 515-526). Prague: Publishing House of the Czechoslovak Academy of Sciences. [15]

Eccles, J.C., Eccles, R.M., & Lundberg, A. (1958). Action potentials of alpha motoneurones supplying fast and slow muscles. *Journal of Physiology*, **142**, 275-291. [12]

Eccles, J.C., Libet, B., & Young, R.R. (1958). The behaviour of chromatolysed motoneurones studied by intracellular recording. *Journal of Physiology*, **143**, 11-40. [18]

Eccles, J.C., & Liley, A.W. (1959). Factors controlling the liberation of acetylcholine at the neuromuscular junction. *American Journal of Physical Medicine*, **38**, 96-103. [10]

Eccles, J.C., & Sherrington, C.S. (1930). Numbers and contraction values of individual motor units examined in some muscles of the limb. *Proceedings of the Royal Society of London, Series B*, **106**, 326-357. [12]

Edman, K.A., & Lou, F. (1992). Myofibrillar fatigue versus failure of activation during repetitive stimulation of frog muscle fibres. *Journal of Physiology*, **457**, 655-673. [15]

Edström, L., & Kugelberg, E. (1968). Histochemical composition, distribution of fibres and fatigability of single motor units. Anterior tibial muscle of the rat. *Journal of Neurology, Neurosurgery and Psychiatry*, **31**, 424-433. [12]

Edwards, R.H.T. (1981). Human muscle function and fatigue. In R. Porter & J. Whelan (Eds.), *Human muscle fatigue: Physiological mechanisms* (pp. 1-18). London: Pitman Medical. [15]

Edwards, R.H.T., Hill, D.K., & Jones, D.A. (1975). Heat production and chemical changes during isometric contractions of the human quadriceps muscle. *Journal of Physiology*, **251**, 303-315. [13]

Edwards, R.H.T., Hill, D.K., Jones, D.A., & Merton, P.A. (1977). Fatigue of long duration in human muscle after exercise. *Journal of Physiology*, **272**, 769-778. [15]

Einsiedel, L.J., & Luff, A.R. (1992). Alterations in the contractile properties of motor units within the aging rat medial gastrocnemius. *Journal of Neurological Sciences*, **112**, 170-177. [22]

Eisenberg, B.R. (1974). Quantitative ultrastructural analysis of adult mammalian skeletal muscle fibers. In A.T. Milhorat (Ed.), *Exploratory concepts in muscular dystrophy II* (pp. 258-269). Amsterdam: Excerpta Medica. [12]

Eisenberg, B.R., & Kuda, A.M. (1976). Discrimination between fiber populations in mammalian skeletal muscle by using ultrastructural parameters. *Journal of Ultrastructure Research*, **54**, 76-88. [12]

Eisenberg, R.S., & Gage, P.W. (1969). Ionic conductances of the surface and transverse tubular membranes of frog sartorius fibers. *Journal of General Physiology*, **53**, 279-297. [9]

Ekstedt, J. (1964). Human single muscle fibre action potentials. *Acta Physiologica Scandinavica*, **61**(Suppl. 226), 1-96. [12]

Ekstedt, J., & Stålberg, E. (1967). Myasthenia gravis. Diagnostic aspects by a new electrophysiological method. *Opuscula Medica*, **12**, 73-76. [10]

Elder, G.C., Dean, D., McComas, A.J., Paes, B., & De Sa, D. (1983). Infantile centronuclear myopathy. Evidence suggesting incomplete innervation. *Journal of Neurological Sciences*, **60**, 79-88. [5]

Elder, G.C.B., Bardbury, K., & Roberts, R. (1982). Variability of fiber type distributions within human muscles. *Journal of Applied Physiology*, **53**, 1473-1480. [12, 19]

Elder, G.C.B., & Kakulas, B. (1993). Histochemical and contractile property changes during human muscle development. *Muscle & Nerve*, **16**, 1246-1253. [5, 20]

Elder, G.C.B., & McComas, A.J. (1987). Development of rat muscle during short- and long-term hindlimb suspension. *Journal of Applied Physiology*, **62**, 1917-1923. [19]

Eldridge, L. (1984). Lumbosacral spinal isolation in cat: Surgical preparation and health maintenance. *Experimental Neurology*, **83**, 318-327. [19]

Elmqvist, D., Hofmann, W.W., Kugelberg, J., & Quastel, D.M.J. (1964). An electrophysiological investigation of neuromuscular transmission in myasthenia gravis. *Journal of Physiology*, **174**, 417-434. [10]

Elsberg, C.A. (1917). Experiments on motor nerve regeneration and the direct neurotization of paralysed muscles by their own and foreign nerves. *Science*, **45**, 318-320. [17]

Elul, R., Miledi, R., & Stefani, E. (1968). Neurotrophic control of contracture in slow muscle fibres. *Nature*, **217**, 1274-1275. [17]

Emerson, C.J., Jr. (1993). Skeletal myogenesis: Genetics and embryology to the fore. *Current Opinion in Genetics and Development*, **3**, 265-274. [5]

Endo, M. (1966). Entry of fluorescent dyes into the sarcotubular system of the frog muscle. *Journal of Physiology*, **185**, 224-238. [11]

Engel, A.G. (1987). Molecular biology of end-plate diseases. In M.M. Salpeter (Ed.), *The vertebrate neuromuscular junction* (pp. 361-424). New York: Alan R. Liss. [10]

Engel, A.G., & Angelini, C. (1973). Carnitine deficiency of skeletal muscle with associated lipid storage myopathy: A new syndrome. *Science*, **173**, 899. [14]

Engel, A.G., Tsujihata, M., Lindstrom, J.M., & Lennon, V.A. (1976). The motor end-plate in myasthenia gravis and in experimental auto-immune myasthenia. A quantitative ultrastructural study. *Annals of the New York Academy of Sciences*, **274** 60-79. [10]

Engel, A.G., Walls, T.J., Nagel, A., & Uchitel, O. (1990). Newly recognized congenital myasthenic syndromes: I. Congenital paucity of synaptic vesicles and reduced quantal release. II. High-conductance fast-channel syndrome. III. Abnormal acetylcholine receptor (AChR) interaction with acetylcholine. IV. AChR deficiency and short channel-open time. *Progress in Brain Research*, **84**, 125-137. [10]

Engel, W.K. (1962). The essentiality of histo- and cytochemical studies of skeletal muscle in the investigation of neuromuscular disease. *Neurology*, **12**, 778-784. [12]

England, P.J. (1986). Intracellular calcium receptor mechanisms. *British Medical Bulletin*, **42**, 375-383. [7]

Enoka, R.M. (1988). Muscle strength and its development. New perspectives. *Sports Medicine*, **6**, 146-168. [20]

Enoka, R.M. (1994). *Neuromechanical basis of kinesiology*. Champaign, IL: Human Kinetics. [1]

Entwistle, A., Zalin, R.J., Bevan, S., & Warner, A.E. (1988). The control of chick myoblast fusion by ion channels operated by prostaglandins and acetylcholine. *Journal of Cell Biology*, **106**, 1693-1702. [5]

Entwistle, A., Zalin, R.J., Warner, A.E., & Bevan, S. (1988). A role for acetylcholine receptors in the fusion of chick myoblasts. *Journal of Cell Biology*, **106**, 1703-1712. [5]

Erlanger, J., & Schloepfle, G.M. (1946). A study of nerve degeneration and regeneration. *American Journal of Physiology*, **147**, 550-581. [16]

Everts, M.F., Retterstøl, K., & Clausen, T. (1988). Effects of adrenaline on excitation-induced stimulation of the sodium-potassium pump in rat skeletal muscle. *Acta Physiologica Scandinavica*, **134**, 189-198. [7]

Exner, G.U., Staudte, H.W., & Pette, D. (1973). Isometric training of rats—Effects upon fast and slow muscle and modifications by an anabolic hormone (nandrolone decanoate). I. Male rats. *Pflügers Archiv*, **345**, 15-22. [20]

Exner, S. (1885). Notiz zu der Fage von der Faserverthelung mehreren nerven in einem Muskeln. *Pflüger's Archiv Fuer Die Gesamte Physiologie Des Menschen und der Tiere*, **36**, 572-576. [17]

Fabiato, A., & Fabiato, F. (1975). Contractions induced by a calcium-triggered release of calcium from the sarcoplasmic reticulum of single skinned cardiac cells. *Journal of Physiology*, **249**, 469-495. [7]

Fahim, M.A., Holley, J.A., & Robbins, N. (1983). Scanning and light microscopic study of age changes at a neuromuscular junction in the mouse. *Journal of Neurocytology*, **12**, 13-25. [3]

Fahim, M.A., & Robbins, N. (1982). Ultrastructural studies of young and old mouse neuromuscular junctions. *Journal of Neurocytology*, **11**, 641-656. [22]

Fahim, M.A., & Robbins, N. (1986). Remodelling of the neuromuscular junction after total disuse. *Brain Research*, **383**, 353-356. [19]

Falco, F.J., Hennessey, W.J., Braddom, R.L., & Goldberg, G. (1992). Standardized nerve conduction studies in the upper limb of the healthy elderly. *American Journal of Physical Medicine and Rehabilitation*, **71**, 263-271. [22]

Fambrough, D., Hartzell, H.C., Rash, J.E., & Ritchie, A.K. (1974). Receptor properties of developing muscle. *Annals of the New York Academy of Sciences*, **228**, 47-61. [1, 5]

Fambrough, D.M. (1979). Control of acetylcholine receptors in skeletal muscle. *Physiological Reviews*, **59**, 165-227. [3]

Fambrough, D.M., Wolitzky, B.A., Tamkun, M.M., & Takeyasu, K. (1987). Regulation of the sodium pump in excitable cells. *Kidney International*, **32**(Suppl. 23), S-97–S-112. [7, 14, 15]

Farmer, T.W., Buchthal, F., & Rosenfalck, P. (1960). Refractory period of human muscle after the passage of a propagated action potential. *Electroencephalography and Clinical Neurophysiology*, **12**, 455-466. [9]

Fatt, P., & Katz, B. (1951). An analysis of the end-plate potential recorded with an intracellular electrode. *Journal of Physiology*, **115**, 320-370. [10]

Fatt, P., & Katz, B. (1952). Spontaneous subthreshold activity at motor nerve endings. *Journal of Physiology*, **117**, 109-128. [3]

Faulkner, J., Jones, D.A., & Round, J.M. (1989). Injury to skeletal muscles of mice by forced lengthening during contractions. *Quarterly Journal of Experimental Physiology*, **74**, 661-670. [21]

Faulkner, J.A., Maxwell, L.C., Ruff, G.L., & White, T.P. (1979). The diaphragm as a muscle. Contractile properties. *American Review of Respiratory Disease*, 119(Suppl. 2), 89-92. [12]

Feasby, T.E., Hahn, A.F., & Gilbert, J.J. (1982). Passive transfer studies in Guillain-Barré polyneuropathy. *Neurology*, 32, 1159-1167. [16]

Feinstein, B., Lindegård, B., Nyman, E., & Wohlfart, G. (1955). Morphologic studies of motor units in normal human muscles. *Acta Anatomica*, 23, 127-142. [1, 12]

Feldman, R.M., & Soskolne, C.L. (1987). The use of non-fatiguing strengthening exercises in post-polio syndrome. *Birth Defects: Original Article Series*, 23, 335-341. [20]

Fell, R.D., Gladden, L.B., Steffen, J.M., & Musacchia, X.J. (1985). Fatigue and contraction of slow and fast muscles in hypokinetic/hypodynamic rats. *Journal of Applied Physiology*, 58, 65-69. [19]

Fenn, W.O. (1936). Electrolytes in muscle. *Cold Spring Harbor Symposium on Quantitative Biology*, IV, 252-259. [15]

Fenton, J., Garner, S., & McComas, A.J. (1991). Abnormal M-wave responses during exercise in myotonic muscular dystrophy: A Na$^+$-K$^+$ pump defect? *Muscle & Nerve*, 14, 79-84. [7]

Fertuck, H.C., & Salpeter, M.M. (1976). Quantification of junctional and extrajunctional acetylcholine receptors by electron microscope autoradiography after ^{125}I-α-bungarotoxin binding at mouse neuromuscular junctions. *Journal of Cell Biology*, 69, 144-158. [10]

Fewings, J.D., Harris, J.B., Johnson, M.A., & Bradley, W.G. (1977). Progressive denervation of skeletal muscle induced by spinal irradiation in rats. *Brain*, 100, 157-183. [16]

Fex, S., Sonesson, B., Thesleff, S., & Zelená, J. (1966). Nerve implants in botulinum poisoned mammalian muscles. *Journal of Physiology*, 184, 872-882. [17]

Fill, M., Coronado, R., Mickelson, J.R., Vilven, J., Ma, J., Jacobson, B.A., & Louis, C.F. (1990). Abnormal ryanodine receptor channels in malignant hyperthermia. *Biophysical Journal*, 57, 471-475. [11]

Finer, J.T., Simmons, R.M., & Spudich, J.A. (1994). Single myosin molecule mechanics: Piconewton forces and nanometre steps. *Nature*, 368, 113-119. [11]

Fischbach, G., & Robbins, N. (1969). Changes in contractile properties of disused soleus muscles. *Journal of Physiology*, 201, 305-320. [19]

Fischbach, G.D. (1972). Synapse formation between dissociated nerve and muscle cells in low density cell cultures. *Developmental Biology*, 28, 407-429. [6]

Fischbach, G.D., Harris, D.A., Falls, D.L., Dubinsky, J.M., English, K.L., & Johnson, F.A. (1989). The accumulation of acetylcholine receptors at developing chick nerve-muscle synapses. In L.C. Sellin, R. Libelius, & S. Thesleff (Eds.), *Neuromuscular junction* (pp. 515-532). Amsterdam: Elsevier Science Publishers. [18]

Fischbach, G.D., Nameroff, M., & Nelson, P.G. (1971). Electrical properties of chick skeletal muscle fibres developing in cell culture. *Journal of Cellular Physiology*, 78, 289-300. [5]

Fisher, M.B., & Birren, J.E. (1947). Age and strength. *Journal of Applied Physiology*, 31, 490-497. [22]

Flucher, B.E., & Daniels, M.P. (1989). Distribution of Na$^+$ channels and ankyrin in neuromuscular junctions is complementary to that of acetylcholine receptors and the 43 kd protein. *Neuron*, 3, 163-175. [10]

Fong, C.N., Atwood, H.L., & Charlton, M.P. (1986). Intracellular sodium-activity at rest and after tetanic stimulation in muscles of normal and dystrophic (dy^{2J}/dy^{2J}) C57BL/6J mice. *Experimental Neurology*, 93, 359-368. [9]

Fontaine, B., Khurana, T.S., Hoffman, E.P., Bruns, G.A.P., Hains, J.L., Trofatter, J.A., Hanson, M.P., Rich, J., McFarlane, H., McKenna Yasek, D., Romano, D., Gusella, J.F., & Brown, R.H. (1990). Hyperkalemic periodic paralysis and the adult muscle sodium channel α-subunit gene. *Science*, 250, 1000-1002, [7]

Fontaine, B., Klarsfeld, A., Hökfelt, T., & Changeux, J.-P. (1986). Calcitonin gene-related peptide, a peptide present in spinal cord motoneurones, increases the number of acetylcholine receptors in primary cultures of chick embryo myotubes. *Neuroscience Letters*, 71, 59-65. [18]

Fosset, M., Jaimovich, E., Delpont, E., & Lazdunski, M. (1983). Nitrendipine receptors in skeletal muscle. Properties and preferential localization in transverse tubules. *Journal of Biological Chemistry*, 258, 6086-6092. [11]

Francke, U., Ochs, H.D., de Martinville, B., Giacalone, J., Lindgren, V., Distèche, C., Pagon, R.A., Hofker, M.H., Van Ommen, G.-J.B., Pearson, P.L., & Wedgwood, R.J. (1985). Minor Xp 21 chromosome deletion in a male associated with expression of Duchenne muscular dystrophy, chronic granulomatous disease, retinitis pigmentosa, and McLeod syndrome. *American Journal of Human Genetics*, 37, 250-267. [1]

Frank, E., Jansen, J.K., Lømo, T., & Westgaard, R. (1974). Maintained function of foreign synapses on hyperinnervated skeletal muscle fibres of the rat. *Nature*, 247, 375-376. [17]

Frank, E., Jansen, J.K.S., Lømo, T., & Westgaard, R.H. (1975). The interaction between foreign and original motor nerves innervating the soleus muscle of rats. *Journal of Physiology*, 247, 725-743. [17]

Frank, G.B. (1958). Inward movement of calcium as a link between electrical and mechanical events in contraction. *Nature*, 182, 1800-1801. [11]

Franzini-Armstrong, C. (1970). Studies of the triad. I. Structure of the junction in frog twitch fibers. *Journal of Cell Biology*, 47, 488-499. [11]

Franzini-Armstrong, C. (1980). Structure of sarcoplasmic reticulum. *Federation Proceedings*, 39, 2403-2409. [11]

Franzini-Armstrong, C., & Porter, K. (1964). Sarcolemmal invagination constituting the T-system on fish muscle fibers. *Journal of Cell Biology*, 22, 675-696. [1]

Freimann, R. (1954). Untersuchungen über Zahl und Anordnung der Muskelspindeln in den Kaumuskeln des Menschen. *Anatomischer Anzeiger*, 100, 258-264. [4]

Freund, H.-J. (1983). Motor unit and muscle activity in voluntary motor control. *Physiological Reviews*, 63, 387-436. [13]

Freund, H.-J., Buedingen, H.-J., & Dietz, V. (1975). Activity of single motor units from forearm muscles during voluntary isometric contractions. *Journal of Neurophysiology*, **38**, 933-946. [13]

Fridén, J., Sjoström, M., & Ekblom, B. (1983). Myofibrillar damage following intense eccentric exercise in man. *International Journal of Sports Medicine*, **4**, 170-176. [20, 21]

Fridén, J., Sjoström, M., & Ekblom, B. (1984). Muscle fibre type characteristics in endurance trained and untrained individuals. *European Journal of Applied Physiology*, **52**, 266-271. [20, 21]

Frontera, W.R., Meredith, C.N., O'Reilly, K.P., Knuttgen, H.G., & Evans, W.J. (1988). Strength conditioning in older men: Skeletal muscle hypertrophy and improved function. *Journal of Applied Physiology*, **64**, 1038-1044. [20]

Fu, Y-.H., Pizzuti, A., Fenwick, R.G., Jr., King, J., Rajnarayan, S., Dunne, P.W., Dubel, J., Nasser, G.A., Ashizawa, T., DeJong, P., Wieringa, B., Korneluk, R., Perryman, M.P., Epstein, H.F., & Caskey, C.T. (1992). An unstable triplet repeat in a gene related to myotonic muscular dystrophy. *Science*, **255**, 1256-1258. [7]

Fuglsang-Frederiksen, A., & Scheel, U. (1978). Transient decrease in number of motor units after immobilization in man. *Journal of Neurology, Neurosurgery and Psychiatry*, **41**, 924-929. [19]

Fujii, J., Otsu, K., Zorzato, F., DeLeon, S., Khanna, V.K., Weiler, J.E., O'Brien, P.J., & MacLennan, D.H. (1991). Identification of a mutation in porcine ryanodine receptor associated with malignant hyperthermia. *Science*, **253**, 448-451. [11]

Fukunaga, H., Engel, A.G., Osame, M., & Lambert, E.H. (1982). Paucity and disorganization of presynaptic membrane active zones in the Lambert-Eaton myasthenic syndrome. *Muscle & Nerve*, **5**, 686-697. [10]

Fulks, R.M., Li, J.B., & Goldberg, A.L. (1975). Effects of insulin, glucose and amino acids on protein turnover in rat diaphragm. *Journal of Biological Chemistry*, **250**, 290-298. [5]

Fürst, D.O., Osborn, M., & Weber, K. (1989). Myogenesis in the mouse embryo: Differential onset of expression of the myogenic proteins and the involvement of titin in myofibril assembly. *Journal of Cell Biology*, **109**, 517-527. [5]

Gage, P.W., & Eisenberg, R.S. (1969a). Action potentials, after potentials, and excitation-contraction coupling in frog sartorius fibers without transverse tubules. *Journal of General Physiology*, **53**, 298-310. [11]

Gage, P.W., & Eisenberg, R.S. (1969b). Capacitance of the surface and transverse tubular membrane of frog sartorius muscle fibers. *Journal of General Physiology*, **53**, 265-278. [9]

Galavazi, G., & Szirmai, J.A. (1971). Cytomorphometry of skeletal muscle: The influence of age and testosterone on the rat m. levator ani. *Zeitschrift fur Zellforschung und Mikrokopische Anatomie*, **121**, 507-530. [5]

Galea, V., De Bruin, H., Cavasin, R., & McComas, A.J. (1991). The numbers and relative sizes of motor units estimated by computer. *Muscle & Nerve*, **14**, 1123-1130. [12]

Galea, V., & McComas, A.J. (1991). Effects of ischemia on M-wave potentiation in human biceps brachii muscles. *Journal of Physiology*, **438**, 212P. [7]

Galea, V., McFadden, L., Cupido, C., & McComas, A.J. (1993). Delayed depression of human muscle excitability following fatigue. *Canadian Journal of Neurological Sciences*, **20**, 359. [15]

Galganski, M., Fuglevand, A.J., & Enoka, R.M. (1993). Reduced control of motor output in a human hand muscle of elderly subjects during submaximal contractions. *Journal of Neurophysiology*, **69**, 2108-2115. [22]

Gandevia, S.C., & McCloskey, D.I. (1976). Joint sense, muscle sense, and their combination as position sense, measured at the distal interphalangeal joint of the middle finger. *Journal of Physiology*, **260**, 387-407. [4]

Gandevia, S.C., McCloskey, D.I., & Burke, D. (1992). Kinaesthetic signals and muscle contraction. *Trends in Neurosciences*, **15**, 62-65. [4]

Gans, C., & Gaunt, A.S. (1991). Muscle architecture in relation to function. *Journal of Biomechanics*, **24**(Suppl. 1), 53-65. [1]

Gardner, E. (1940). Decrease in human neurones with age. *Anatomical Record*, **77**, 529-536. [22]

Garland, S.J., Garner, S.H., & McComas, A.J. (1988a). Reduced voluntary electromyographic activity after fatiguing stimulation of human muscle. *Journal of Physiology*, **401**, 547-556. [15]

Garland, S.J., Garner, S.H., & McComas, A.J. (1988b). Relationship between numbers and frequencies of stimuli in human muscle fatigue. *Journal of Applied Physiology*, **65**, 89-93. [15]

Garland, S.J., & McComas, A.J. (1990). Reflex inhibition of human soleus muscle during fatigue. *Journal of Physiology*, **429**, 17-27. [15]

Garner, S.H., Hicks, A.L., & McComas, A.J. (1989). Prolongation of twitch potentiating mechanism throughout human muscle fatigue and recovery. *Experimental Neurology*, **103**, 277-281. [15]

Garnett, R., & Stephens, J.A. (1980). The reflex responses of single motor units in human first dorsal interosseous muscle following cutaneous afferent stimulation. *Journal of Physiology*, **303**, 351-364. [13]

Garnett, R.A.F., O'Donovan, M.J., Stephens, J.A., & Taylor, A. (1979). Motor unit organization of human medial gastrocnemius. *Journal of Physiology*, **287**, 33-43. [12, 13]

Gatev, V., Stamatova, L., & Angelova, B. (1977). Contraction time in skeletal muscles of normal children. *Electromyography and Clinical Neurophysiology*, **17**, 441-451. [5]

Gauthier, G.F. (1986). Skeletal muscle fiber types. In A.G. Engel & B.Q. Banker (Eds.), *Myology. Basic and clinical* (pp. 255-283). New York: McGraw-Hill. [12]

Gelfan, S., & Carter, S. (1967). Muscle sense in man. *Experimental Neurology*, **18**, 469-473. [4]

Gerchman, L.-R.B., Edgerton, V.R., & Carrow, R.E. (1975). Effects of physical training on the histochemistry and morphology of ventral motor neurons. *Experimental Neurology*, **49**, 790-801. [20]

Geren, B.B. (1954). The formation from the Schwann cell surface of myelin in the peripheral nerves of chick embryos. *Experimental Cell Research, 7,* 558-562. [2]

Gielen, C.C.A.M., & Denier van der Gon, J.J. (1990). The activation of motor units in coordinated arm movements in humans. *News in Physiological Sciences, 5,* 159-163. [13]

Gillespie, C.A., Simpson, D.R., & Edgerton, V.R. (1974). Motor unit recruitment as reflected by muscle fibre glycogen loss in a prosimian (bushbaby) after running and jumping. *Journal of Neurology, Neurosurgery and Psychiatry, 37,* 817-824. [13]

Gillespie, M.J., Gordon, T., & Murphy, P.R. (1986). Reinnervation of the lateral gastrocnemius and soleus muscles in the rat by their common nerve. *Journal of Physiology, 372,* 485-500. [12]

Gillespie, M.J., & Stein, R.M. (1983). The relationship between axon diameter, myelin thickness and conduction velocity during atrophy in mammalian peripheral nerves. *Brain Research, 259,* 41-56. [16]

Gilliatt, R.W., & Hjorth, R.J. (1972). Nerve conduction during Wallerian degeneration in the baboon. *Journal of Neurology, Neurosurgery and Psychiatry, 35,* 335-341. [16]

Gilliatt, R.W., & Taylor, J.C. (1959). Electrical changes following section of the facial nerve. *Proceedings of the Royal Society of Medicine, 52,* 1080-1083. [16]

Gilliatt, R.W., Westgaard, R.H., & Williams, I.R. (1978). Extrajunctional acetylcholine sensitivity of inactive muscle fibres in the baboon during prolonged nerve pressure block. *Journal of Physiology, 280,* 499-514. [19]

Ginsborg, B.L. (1960). Some properties of avian skeletal muscle fibres with multiple neuromuscular junctions. *Journal of Physiology, 193,* 581-598. [20]

Gitlin, G., & Singer, M. (1974). Myelin movements in mature mammalian peripheral nerve fibres. *Journal of Morphology, 143,* 167-186. [2]

Glatt, H.R., & Honegger, C.G. (1973). Retrograde axonal transport for cartography of neurons. *Experientia, 29,* 1515-1517. [8, 18]

Glover, J.C., Petursdottir, G., & Jansen, J.K.S. (1986). Fluorescent dextran-amines used as axonal tracers in the nervous system of the chick embryo. *Journal of Neuroscience Methods, 18,* 243-254. [12]

Godfrey, E.W., Nitkin, R.M., Wallace, B.G., Rubin, L.L., & McMahan, U.J. (1984). Components of torpedo electric organ and muscle that cause aggregation of acetylcholine receptors on cultured muscle cells. *Journal of Cell Biology, 99,* 615-627. [18]

Goldberg, A.L. (1968). Role of insulin in work-induced growth of skeletal muscle. *Endocrinology, 83,* 1071-1073. [5]

Goldberg, A.L., & Goodman, H.M. (1969). Relationship between growth hormone and muscular work in determining muscle size. *Journal of Physiology, 200,* 655-666. [5]

Goldberg, L.J., & Derfler, B. (1977). Relationship among recruitment order, spike amplitude, and twitch tension of single motor units in human masseter muscle. *Journal of Neurophysiology, 40,* 879-890. [13]

Goldkamp, O. (1967). Electromyography and nerve conduction studies in 116 patients with hemiplegia. *Archives of Physical Medicine, 48,* 59-63. [18]

Goldman, D.E. (1943). Potential, impedance and rectification in membranes. *Journal of General Physiology, 27,* 37-60. [9]

Goldspink, G. (1964). The combined effects of exercise and reduced food intake on skeletal muscle fibres. *Journal of Cellular and Comparative Physiology, 63,* 209-216. [20]

Goldspink, G. (1965). Cytological basis of decrease in muscle strength during starvation. *American Journal of Physiology, 209,* 100-114. [5, 20]

Goldspink, G. (1970). The proliferation of myofibres during muscle fibre growth. *Journal of Cell Science, 6,* 593-603. [5]

Goldspink, G. (1985). Malleability of the motor system: A comparative approach. *Journal of Experimental Biology, 115,* 375-391. [20]

Goldspink, G., Scutt, A., Loughna, T., Wells, D.J., Jaenicke, T., & Gerlach, G.F. (1992). Gene expression in skeletal muscle in response to stretch and force generation. *American Journal of Physiology, 262,* R356-R363. [20]

Goldspink, G., Scutt, A., Martindale, J., Jaenicke, T., Turay, L., & Gerlach, G.-F. (1991). Stretch and force generation induce rapid hypertrophy and myosin isoform gene switching in adult skeletal muscle. *Biochemical Society Transactions, 19,* 368-373. [20]

Golgi, C. (1903). Sui nervi dei tendini dell'uomo e di altri vertebrati e diun nuovo organo nervoso terminale musculo-tendineo. Reprinted in his *Opera omnia* (Vol. 1, pp. 171-198). Milan: Ultrico Hoepli. (Original work published in 1880.) [4]

Gollnick, P.D., Karlsson, J., Piehl, K., & Saltin, B. (1974). Selective glycogen depletion in skeletal muscle fibres of man following sustained contraction. *Journal of Physiology, 241,* 59-67. [13]

Gollnick, P.D., Piehl, K., & Saltin, B. (1974). Selective glycogen depletion in human muscle fibres after exercise of varying intensity and at varying pedalling rates. *Journal of Physiology, 241,* 45-57. [13]

Gollnick, P.D., Sjödin, B., Karlsson, J., Jansson, E., & Saltin, B. (1974). Human soleus muscle: A comparison of fiber composition and enzyme activities with other leg muscles. *Pflügers Archiv, 348,* 247-255. [12]

Gollnick, P.D., Timson, B.F., Moore, R.L., & Riedy, M. (1981). Muscular enlargement and number of fibers in skeletal muscles of rats. *Journal of Applied Physiology, 50,* 936-943. [20]

González-Serratos, H. (1971). Inward spread of activation in vertebrate muscle fibres. *Journal of Physiology, 212,* 777-799. [11]

González-Serratos, H., Somlyo, A., McClellan, G., Shuman, H., Borrero, L.M., & Somlyo, A.P. (1978). Composition of vacuoles and sarcoplasmic reticulum in fatigued muscle: Electron probe analysis. *Proceedings of the National Academy of Sciences* (U.S.), *75,* 1329-1333. [15]

Good, P.J., Richter, K., & David, I.B. (1990). Studies on embryonic induction: Establishing molecular markers for neural development. In A.P. Mahowald (Ed.), *Genetics of pattern formation and growth control* (pp. 125-135). New York: Wiley-Liss. [6]

Goodwin, G.M., McCloskey, D., & Matthews, P.B.C. (1972). Proprioceptive illusions induced by muscle vibration:

Contribution by muscle spindles to perception. *Science*, **175**, 1382-1384. [4]

Gordon, A.M., Huxley, A.F., & Julian, F.J. (1966a). Tension development in highly stretched vertebrate muscle fibres. *Journal of Physiology*, **184**, 143-169. [11]

Gordon, A.M., Huxley, A.F., & Julian, F.J. (1966b). The variation in isometric tension with sarcomere length in vertebrate muscle fibres. *Journal of Physiology*, **184**, 170-192. [11]

Gordon, E.E. (1967). Anatomical and biochemical adaptations of muscles to different exercises. *Journal of the American Medical Association*, **201**, 755-758. [20]

Gordon, T., Gillespie, J., Orozco, R., & Davis, L. (1991). Axotomy-induced changes in rabbit hindlimb nerves and the effects of chronic electrical stimulation. *Journal of Neuroscience*, **11**, 2157-2169. [16]

Gordon, T., & Stein, R.B. (1982). Reorganization of motor-unit properties in reinnervated muscles of the cat. *Journal of Neurophysiology*, **48**, 1175-1190. [17]

Gordon, T., Yang, J.F., Ayer, K., Stein, R.B., & Tyreman, N. (1993). Recovery potential of muscle after partial denervation: A comparison between rats and humans. *Brain Research Bulletin*, **30**, 477-482. [17]

Gordon-Weeks, P.R. (1989). GAP-43—What does it do in the growth cone? *Trends in Neurosciences*, **12**, 363-365. [17]

Gorza, L., Gundersen, K., Lømo, T., Schiaffino, S., & Westgaard, R.H. (1988). Slow-to-fast transformation of denervated soleus muscles by chronic high-frequency stimulation in the rat. *Journal of Physiology*, **402**, 627-649. [18, 20]

Grabowski, W., Lobsiger, E.A., & Lüttgau, H.C. (1972). The effect of repetitive stimulation at low frequencies upon the electrical and mechanical activity of single muscle fibres. *Pflügers Archiv*, **334**, 222-239. [15]

Grafstein, B., & Forman, D.S. (1980). Intracellular transport in neurons. *Physiological Reviews*, **60**, 1167-1283. [8]

Grafstein, B., & McQuarrie, I.G. (1978). Role of the nerve cell body in axonal regeneration. In C.W. Cotman (Ed.), *Neuronal plasticity* (pp. 155-195). New York: Raven. [18]

Green, H.J., Daub, B., Houston, M.E., Thomson, J.A., Fraser, I., & Ranney, D. (1981). Human vastus lateralis and gastrocnemius muscles. A comparative histochemical analysis. *Journal of the Neurological Sciences*, **52**, 200-201. [12]

Green, L.S., Donoso, J.A., Heller-Bettinger, I.E., & Samson, F.E. (1977). Axonal transport disturbances in vincristine-induced peripheral neuropathy. *Annals of Neurology*, **1**, 255-262. [8]

Grieshammer, U., Sassoon, D., & Rosenthal, N. (1992). A transgene target marks early rostrocaudal specification of myogenic lineages. *Cell*, **69**, 79-93. [5]

Grimby, G. (1988). Physical activity and effects of muscle training in the elderly. *Annals of Clinical Research*, **20**, 62-66. [22]

Grimby, G., Danneskiold-Samsøe, B., Hvid, K., & Saltin, B. (1982). Morphology and enzymatic capacity in arm and leg muscles in 78-81 year old men and women. *Acta Physiologica Scandinavica*, **115**, 125-134. [22]

Grimby, G., & Saltin, B. (1983). The aging muscle. *Clinical Physiology*, **3**, 209-218. [22]

Grimby, L. (1984). Firing properties of single human motor units during locomotion. *Journal of Physiology*. **346**, 195-202. [13]

Grimby, L., & Hannerz, J. (1968). Recruitment order of motor units in voluntary contraction: Changes induced by proprioceptive afferent activity. *Journal of Neurology, Neurosurgery and Psychiatry*, **31**, 565-573. [13]

Grimby, L., Hannerz, J., & Hedman, B. (1981). The fatigue and voluntary discharge properties of single motor units in man. *Journal of Physiology*, **316**, 545-554. [15, 20]

Gundersen, K., Leberer, E., Lømo, T., Pette, D., & Staron, R.S. (1988). Fibre types, calcium-sequestering proteins and metabolic enzymes in denervated and chronically stimulated muscles of the rat. *Journal of Physiology*, **398**, 177-189. [16, 18]

Gurdon, J.B., Harger, P., Mitchell, A., & Lemaire, P. (1994). Activin signalling and response to a morphogen gradient. *Nature*, **371**, 487-492. [5]

Gurdon, J.B., Mohun, T.J., Sharpe, C.R., & Taylor, M.V. (1989). Embryonic induction and muscle gene activation. *Trends in Genetics*, **5**, 51-56. [5]

Guth, L., Albers, R.W., & Brown, W.C. (1964). Quantitative changes in cholinesterase activity of denervated fibers and sole plates. *Experimental Neurology*, **10**, 236-250. [16]

Gutmann, E. (1971). Histology of degeneration and regeneration. In S. Licht (Ed.), *Electrodiagnosis and electromyography* (pp. 113-133). Baltimore: Waverly Press. [17, 18]

Gutmann, E., Gutmann, L., Medawar, P.B., & Young, J.Z. (1942). The rate of regeneration of a nerve. *Journal of Experimental Biology*, **19**, 14-44. [17]

Gutmann, E., & Hanzlíková, V. (1965). Age changes of motor end-plates in muscle fibres of the rat. *Gerontologica*, **11**, 12-24. [22]

Gutmann, E., & Hanzlíková, V. (1966). Motor unit in old age. *Nature*, **209**, 921-922. [22]

Gutmann, E., Hanzlíková, V., & Vyskocil, F. (1971). Age changes in cross striated muscle of the rat. *Journal of Physiology*, **219**, 331-343. [22]

Gutmann, E., & Holubár, J. (1950). The degeneration of peripheral nerve fibres. *Journal of Neurology, Neurosurgery and Psychiatry*, **13**, 89-105. [16]

Gutmann, E., & Holubár, J. (1952). Degenerace terminálních orgánu v príene pruhovaném a hladkém svalstvu. *Physiologica Bohemoslovenica*, **1**, 168-175. [16]

Gutmann, E., & Sandow, A. (1965). Caffeine induced contracture and potentiation of contraction in normal and denervated muscle. *Life Sciences*, **4**, 1149-1156. [16]

Gutmann, E., Schiaffino, S., & Hanzlíková, V. (1971). Mechanism of compensatory hypertrophy in skeletal muscle of the rat. *Experimental Neurology*, **31**, 451-464. [19]

Haggar, R.A., & Barr, M.L. (1950). Quantitative data on the size of synaptic end-bulbs in the cat's spinal cord. *Journal of Comparative Neurology*, **93**, 17-35. [2]

Häggmark, T., & Eriksson, E. (1979). Hypotrophy of the soleus muscle in man after Achilles tendon rupture. *American Journal of Sports Medicine*, **7**, 121-126. [19]

Haida, N., Fowler, W.M., Jr., Abresh, R.T., Larson, D.B., Sharman, R.B., Taylor, R.G., & Entrikin, R.K. (1989). Effect of hind-limb suspension on young and adult skeletal muscle. I. Normal mice. *Experimental Neurology*, **103**, 68-76. [19]

Hainaut, K., & Duchateau, J. (1989). Muscle fatigue, effects of training and disuse. *Muscle & Nerve*, **12**, 660-669. [20]

Halkjaer-Kristensen, J., & Ingemann-Hansen, T. (1985). Wasting of the human quadriceps muscle after knee ligament injuries. II. Muscle fibre morphology. *Scandinavian Journal of Rehabilitation Medicine*. Suppl. **13**, 12-20. [19]

Hall, Z.W., & Sanes, J.R. (1993). Synaptic structure and development: The neuromuscular junction. *Cell*, **72**, 99-121. [3, 6]

Hall-Craggs, E.C.B. (1970). The longitudinal division of fibres in overloaded rat skeletal muscle. *Journal of Anatomy*, **107**, 459-470. [20]

Hamburger, V. (1947). *A manual of experimental embryology*. Chicago: University of Chicago Press. [6]

Hamburger, V. (1958). Regression versus peripheral control of differentiation in motor hypoplasia. *American Journal of Anatomy*, **102**, 365-409. [6]

Hansen Bay, C.M., & Strichartz, G.R. (1980). Saxitoxin binding to sodium channels of rat skeletal muscles. *Journal of Physiology*, **300**, 89-103. [7]

Harii, K., Ohmori, K., & Torii, S. (1976). Free gracilis muscle transplantation, with microneurovascular anastomoses for the treatment of facial paralysis. A preliminary report. *Plastic & Reconstructive Surgery*, **57**, 133-143. [21]

Harriman, D.G.F., Taverner, D., & Woolf, A.L. (1970). Ekbom's syndrome and burning paraesthesiae. A biopsy study by vital staining and electron microscopy of the intramuscular innervation with a note on age changes in motor nerve endings in distal muscles. *Brain*, **93**, 393-406. [22]

Harris, A.J., & Miledi, R. (1971). The effect of type D botulinum toxin on frog neuromuscular junctions. *Journal of Physiology*, **217**, 497-515. [10]

Harris, J.B., & Luff, A.R. (1970). The resting membrane potentials of fast and slow skeletal muscle fibres in the developing mouse. *Comparative Biochemistry and Physiology*, **33**, 923-931. [5]

Harris, J.B., & Thesleff, S. (1972). Nerve stump length and membrane changes in denervated skeletal muscle. *Nature New Biology*, **236**, 60-61. [7, 16, 18]

Harris, J.B., & Wilson, P. (1971). Mechanical properties of dystrophic mouse muscle. *Journal of Neurology, Neurosurgery and Psychiatry*, **34**, 512-520. [12]

Hartshorne, R.P., Keller, B.U., Talvenheimo, J.A., Catterall, W.A., & Montal, M. (1985). Functional reconstitution of the purified brain sodium channel in planar lipid bilayers. *Proceedings of the National Academy of Sciences* (U.S.), **82**, 240-244. [7]

Hasty, P., Bradley, A., Morris, J.H., Edmondson, D.G., Venuti, J.M., Olson, E.N., & Klein, W.H. (1993). Muscle deficiency and neonatal death in mice with a targeted mutation in the *myogenin* gene. *Nature*, **364**, 501-506. [5]

Hather, B.M., Tesch, P.A., Buchanan, P., & Dudley, G.A. (1991). Influence of eccentric actions on skeletal muscle adaptations to resistance training. *Acta Physiologica Scandinavica*, **143**, 177-185. [20]

Hayward, L., Wesselmann, U., & Rymer, W.Z. (1991). Effects of muscle fatigue on mechanically sensitive afferents of slow conduction velocity in the cat triceps surae. *Journal of Neurophysiology*, **65**, 360-370. [15]

Heilbrunn, L.V., & Wiercinski, F.J. (1947). The action of various cations on muscle protoplasm. *Journal of Cellular and Comparative Physiology*, **29**, 15-32. [11]

Henderson, C.E., Camu, W., Mettling, C., Gouin, A., Poulsen, K., Karihaloo, M., Rullamas, J., Evans, T., McMahon, S.B., Armanini, M.P., Berkemeier, L., Phillips, H.S., & Rosenthal, A. (1993). Neurotrophins promote motor neuron survival and are present in embryonic limb bud. *Nature*, **363**, 266-270. [18]

Henderson, C.E., Huchet, M., & Changeux, J.-P. (1983). Denervation increases a neurite-promoting activity in extracts of skeletal muscle. *Nature*, **302**, 609-611. [17]

Henderson, C.E., Phillips, H.S., Pollock, R.A., Davies, A.M., Lemeulle, C., Armanini, M., Simpson, L.C., Moffet, B., Vandlen, R.A., Koliatsos, V.E., & Rosenthal, A. (1994). GDNF: A potent survival factor for motoneurons present in peripheral nerve and muscle. *Science*, **266**, 1062-1064. [18]

Henneman, E., Shahani, B.T., & Young, R.R. (1976). Voluntary control of human motor units. In M. Shahani (Ed.), *The motor system: Neurophysiology and muscle mechanisms* (pp. 73-78). Amsterdam: Elsevier. [13]

Henneman, E., Somjen, G., & Carpenter, D.O. (1965). Functional significance of cell size in spinal motoneurones. *Journal of Neurophysiology*, **28**, 560-580. [13]

Hennig, R., & Lømo, T. (1985). Firing patterns of motor units in normal rats. *Nature*, **314**, 164-166. [18]

Hennig, R., & Lømo, T. (1987). Effects of chronic stimulation on the size and speed of long-term denervated and innervated rat fast and slow skeletal muscles. *Acta Physiologica Scandinavica*, **130**, 115-131. [18]

Herbison, G.J., Jaweed, M.M., & Ditunno, J.F. (1978). Muscle fiber atrophy after cast immobilization in the rat. *Archives of Physical Medicine and Rehabilitation*, **59**, 301-305. [19]

Hess, A. (1970). Vertebrate slow muscle fibers. *Physiological Reviews*, **50**, 40-62. [20]

Heuser, J.E., & Reese, T.S. (1973). Evidence for recycling of synaptic vesicle membrane during transmitter release at the frog neuromuscular junction. *Journal of Cell Biology*, **57**, 315-344. [3, 10]

Heuser, J.E., & Reese, T.S. (1981). Structural changes after transmitter release at the frog neuromuscular junction. *Journal of Cell Biology*, **88**, 564-580. [3]

Heuser, J.E., Reese, T.S., Dennis, M.J., Jan, Y., Jan, L., & Evans, L. (1979). Synaptic vesicle exocytosis captured by quick freezing and correlated with quantal transmitter release. *Journal of Cell Biology*, **81**, 275-300. [10]

Heuser, J.E., Reese, T.S., & Landis, D.M.D. (1974). Functional changes in frog neuromuscular junctions studied with freeze-fracture. *Journal of Neurocytology*, **3**, 109-131. [10]

Hicks, A., & McComas, A.J. (1989). Increased sodium pump activity following repetitive stimulation of rat soleus muscles. *Journal of Physiology*, **414**, 337-349. [7, 9, 15]

Hikida, R.S., Gollnick, P.D., Dudley, G.A., Convertino, V.A., & Buchanan, P. (1989). Structural and metabolic characteristics of human skeletal muscle following 30 days of simulated microgravity. *Aviation, Space, and Environmental Medicine*, **60**, 664-670. [19]

Hille, B. (1992). *Ionic channels of excitable membranes* (2nd ed.). Sunderland, MA: Sinauer Associates. [7]

Hiraoki, T., & Vogel, H.J. (1987). Structure and function of calcium-binding proteins. *Journal of Cardiovascular Pharmacology*, **10**(Suppl. 1), S14-S31. [7]

Hník, P. (1972). Changes in function of muscle afferents during muscle atrophy (de-efferentation or tenotomy). In *Symposium on Structure and Function of Normal and Diseased Muscle and Peripheral Nerve* (Programme abstract no. 20). Warsaw: Polish Academy of Science. [19]

Ho, K.W., Roy, R.R., Tweedle, C.D., Heusner, W.W., Van Huss, W.D., & Carrow, R.E. (1980). Skeletal muscle fiber splitting with weight-lifting exercise in rats. *American Journal of Anatomy*, **157**, 433-440. [20]

Hodes, R., Larrabee, M.G., & German, W. (1948). The human electromyogram in response to nerve stimulation and the conduction velocity of motor axons. Study on normal and on injured peripheral nerves. *Archives of Neurology and Psychiatry*, **60**, 340-365. [17]

Hodgkin, A.L. (1977). Chance and design in electrophysiology: An informal account of certain experiments on nerve carried out between 1934 and 1952. In A.L. Hodgkin, A.F. Huxley, W. Feldberg, W.A.H. Ruston, R.A. Gregory, & R.A. McCance (Eds.), *The pursuit of nature. Informal essays on the history of physiology* (pp. 1-22). Cambridge: Cambridge University Press. [9]

Hodgkin, A.L., & Horowicz, P. (1959). The influence of potassium and chloride ions on the membrane potential of single muscle fibres. *Journal of Physiology*, **148**, 127-160. [9, 15]

Hodgkin, A.L., & Huxley, A.F. (1952a). Currents carried by sodium and potassium ions through the membrane of the giant axon of *Loligo*. *Journal of Physiology*, **116**, 449-472. [9]

Hodgkin, A.L., & Huxley, A.F. (1952b). A quantitative description of membrane current and its application to conduction and excitation in nerve. *Journal of Physiology*, **117**, 500-544. [7, 9]

Hodgkin, A.L., & Katz, B. (1949). The effect of sodium ions on the electrical activity of the giant axon of the squid. *Journal of Physiology*, **108**, 37-77. [9]

Hoffman, E.P., & Kunkel, L.M. (1989). Dystrophin abnormalities in Duchenne/Becker muscular dystrophy. *Neuron*, **2**, 1019-1029. [1]

Hoffman, H. (1950). Local reinnervation in partially denervated muscle: A histophysiological study. *Australian Journal of Experimental Biology*, **28**, 383-397. [17]

Hoffman, H. (1951). Fate of interrupted nerve fibres regenerating into partially denervated muscles. *Australian Journal of Experimental Biology and Medical Science*, **29**, 211-219. [17]

Hoffman, P.N., & Lasek, R.J. (1980). Axonal transport of the cytoskeleton in regenerating motor neurons: Constancy and change. *Brain Research*, **202**, 317-333. [18]

Hoffmann, P. (1918). Ueber die Beeinflussang der Sehnenreflexe durch die willkürliche Contraction. *Medizinische Klinik*, **XIV**, 203. [4]

Hofmann, F., Flockerzi, V., Nastainczyk, W., Ruth, P., & Schneider, T. (1990). The molecular structure and regulation of muscular calcium channels. *Current Topics in Cellular Regulation*, **31**, 223-239. [7]

Hofmann, W.W., Kundin, J.E., & Farrell, D.F. (1967). The pseudomyasthenic syndrome of Eaton and Lambert: An electrophysiological study. *Electroencephalography and Clinical Neurophysiology*, **23**, 214-224. [10]

Hofmann, W.W., & Thesleff, S. (1972). Studies on the trophic influence of nerve on skeletal muscle. *European Journal of Pharmacology*, **20**, 256-260. [18, 19]

Hogan, E.L., Dawson, D.M., & Romanul, F.C.A. (1965). Enzymatic changes in denervated muscle. II. Biochemical studies. *Archives of Neurology*, **13**, 274-282. [16]

Homsher, E. (1987). Muscle enthalpy production and its relationship to actomyosin ATPase. *Annual Review of Physiology*, **49**, 673-690. [14]

Hooper, A.C.B. (1981). Length, diameter and number of ageing skeletal muscle fibres. *Gerontology*, **27**, 121-126. [22]

Hopf, H.C. (1962). Untersuchungen uber die unterschiede in der leitgeschwindigkeit motorischer nervenfasern beim menschen. *Deutsches Zeitschrift fur Nervenheikunde*, **183**, 579-588. [9]

Hopf, H.C. (1963). Electromyographic study in so-called mononeuritis. *Archives of Neurology*, **9**, 307-312. [9]

Horisberger, J.-D., Lemas, V., Kraehenbühl, J.-P., & Rossier, B.C. (1991). Structure-function relationship of Na,K-ATPase. *Annual Review of Physiology*, **53**, 565-584. [7]

Houk, J., & Henneman, E. (1967). Responses of Golgi tendon organs to active contractions of the soleus muscle of the cat. *Journal of Neurophysiology*, **30**, 466-481. [4]

Houston, M.E., Froese, E.A., Valeriote, St.P., Green, H.J., & Ranney, D.A. (1983). Muscle performance, morphology and metabolic capacity during strength training and detraining: A one leg model. *European Journal of Applied Physiology and Occupational Physiology*, **51**, 25-35. [20]

Howald, H. (1982). Training-induced morphological and functional changes in skeletal muscle. *International Journal of Sports Medicine*, **3**, 1-12. [20]

Howald, H., Hoppeler, H., Claassen, H., Mathieu, O., & Straub, R. (1985). Influences of endurance training on the ultrastructural composition of the different muscle fibre types in humans. *Pflügers Archiv*, **403**, 369-376. [20]

Howard, J.E., McGill, K.C., & Dorfman, L.J. (1988). Age effects on properties of motor unit action potentials: ADEMG analysis. *Annals of Neurology*, **24**, 207-213. [22]

Huard, J., Fortier, L.-P., Dansereau, G., Labrecque, C., & Tremblay, J.P. (1992). A light and electron microscopic study of dystrophin at the mouse neuromuscular junction. *Synapse*, **10**, 83-93. [3]

Hubbard, J.I., & Schmidt, R.F. (1963). An electrophysiological investigation of mammalian motor nerve terminals. *Journal of Physiology*, **166**, 145-167. [10]

Hubbard, J.I., & Wilson, D.F. (1973). Neuromuscular transmission in a mammalian preparation in the absence of blocking drugs and the effect of D-tubocurarine. *Journal of Physiology*, **228**, 307-325. [10]

Hudlicka, O. (1990). The response of muscle to enhanced and reduced activity. *Endocrinology and Metabolism*, **4**, 417-439. [20]

Humphreys, P.W., & Lind, A.R. (1963). Blood flow through active and inactive muscles of the forearm during sustained hand-grip contractions. *Journal of Physiology*, **166**, 120-135. [13]

Hunt, C.C., Wilkinson, R.S., & Fukami, Y. (1978). Ionic basis of the receptor potential of mammalian muscle spindles. *Journal of General Physiology*, **71**, 683-698. [4]

Hursh, J.B. (1939). Conduction velocity and diameter of nerve fibers. *American Journal of Physiology*, **127**, 131-139. [9, 12]

Hutter, O.F., & Noble, D. (1960). The chloride conductance of frog skeletal muscle. *Journal of Physiology*, **151**, 89-102. [7, 9]

Huxley, A.F. (1959). Local activation in muscle. *Annals of the New York Academy of Sciences*, **81**, 446-452. [11]

Huxley, A.F. (1974). Muscular contraction (Review lecture). *Journal of Physiology*, **243**, 1-43. [11]

Huxley, A.F., & Niedergerke, R. (1954). Structural changes in muscle during contraction. Interference microscopy of living muscle fibres. *Nature*, **173**, 971-973. [11]

Huxley, A.F., & Simmons, R. (1971). Proposed mechanism of force generation in striated muscle. *Nature*, **233**, 533-538. [11]

Huxley, A.F., & Taylor, R.E. (1958). Local activation of striated muscle fibres. *Journal of Physiology*, **144**, 426-441. [11]

Huxley, H.E. (1958). The contraction of muscle. *Scientific American*, **199**, 67-82. [11]

Huxley, H.E. (1964). Evidence for continuity between the central elements of the triads and extracellular space in frog sartorius muscle. *Nature*, **202**, 1067-1071. [11]

Huxley, H.E. (1972). Molecular basis of contraction in cross-striated muscles. In G.H. Bourne (Ed.), *The structure and function of muscle* (pp. 302-387). New York: Academic Press. [1, 11]

Huxley, H.E. (1990). Sliding filaments and molecular motile systems. *Journal of Biological Chemistry*, **265**, 8347-8350. [11]

Huxley, H.E., & Hanson, J. (1954). Changes in the cross-striations of muscle during contraction and stretch and their structural interpretation. *Nature*, **173**, 973-976. [11]

Ianniruberto, A., & Tajani, E. (1981). Ultrasonographic study of fetal movements. *Seminars in Perinatology*, **5**, 175-181. [5]

Ibraghimov-Beskrovnaya, O., Ervasti, J.M., Leveille, C.J., Slaughter, C.A., Sermett, S.W., & Campbell, K.P. (1992). Primary structure of dystrophin-associated glycoprotein linking dystrophin to the extracellular matrix. *Nature*, **355**, 696-702. [1, 3]

Ikai, M., & Fukunaga, T. (1970). A study of training effect on strength per unit of cross-sectional area of muscle by means of ultrasonic measurement. *International Zeitschrift für Angewandte Physiologie Einschliesslich Arbeitsphysiologie*, **28**, 173-180. [20]

Inesi, G., & Kirtley, M.E. (1990). Coupling of catalytic and channel function in the Ca^{2+} transport ATPase. *Journal of Membrane Biology*, **116**, 1-8. [7]

Inestrosa, N.C. (1982). Differentiation of skeletal muscle cells in culture. *Cell Structure and Function*, **7**, 91-109. [5]

Ingemann-Hansen, T., & Halkjaer-Kristensen, J. (1983). Progressive resistance exercise training of the hypotrophic quadriceps in man. *Scandinavian Journal of Rehabilitation Medicine*, Suppl. **13**, 38-44. [19]

Ingjer, F. (1979). Effects of endurance training on muscle fibre ATP-ase activity, capillary supply and mitochondrial content in man. *Journal of Physiology*, **294**, 419-432. [20]

Inokuchi, S., Ishikawa, H., Iwamoto, S., & Kimura, T. (1975). Age-related changes in the histological composition of the rectus abdominis muscle of the adult human. *Human Biology*, **47**, 231-249. [22]

Isaacs, E.R., Bradley, W.G., & Henderson, G. (1973). Longitudinal fibre splitting in muscular dystrophy: A serial cinematographic study. *Journal of Neurology, Neurosurgery and Psychiatry*, **36**, 813-819. [20]

Isihara, A., & Araki, H. (1988). Effects of age on the number and histochemical properties of muscle fibers and motoneurons in the rat extensor digitorum longus muscle. *Mechanisms of Aging and Development*, **45**, 213-221. [22]

Isihara, A., Naitoh, H., & Katsuta, S. (1987). Effects of aging on the total number of muscle fibers and motoneurons of the tibialis anterior and soleus muscles in the rat. *Brain Research*, **435**, 355-358. [22]

Ito, Y., Miledi, R., Molenaar, P.C., Vincent, A., Polak, R.L., Van Gelder, M., & Davis, J.N. (1976). Acetylcholine in human muscle. *Proceedings of the Royal Society of London, Series B*, **192**, 475-480. [10]

Jablecki, C., & Brimijoin, S. (1974). Reduced axoplasmic transport of choline acetyltransferase activity in dystrophic mice. *Nature*, **250**, 151-154. [10]

Jackson, M.J., Jones, D.A., & Edwards, R.H.T. (1984). Experimental skeletal muscle damage: The nature of the calcium-activated degenerative process. *European Journal of Clinical Investigation*, **14**, 369-374. [21]

Jacobsen, M., & Rutishauser, U. (1986). Induction of neural cell adhesion molecule (N-CAM) in *Xenopus* embryos. *Developmental Biology*, **116**, 524-531. [6]

Jacobsen, S., & Guth, L. (1965). An electrophysiological study of the early stages of peripheral nerve regeneration. *Experimental Neurology*, **11**, 48-60. [17]

Jakobsson, F., Borg, K., Edström, L., & Grimby, L. (1988). Use of motor units in relation to muscle fiber type and size in man. *Muscle & Nerve*, **11**, 1211-1218. [22]

Jan, L.Y., & Jan, Y.N. (1989). Voltage-sensitive ion channels. *Cell*, **56**, 13-25. [7]

Jan, Y.N., Jan, L.Y., & Dennis, M.J. (1977). Two mutations of synaptic transmission in *Drosophila*. *Proceedings of the Royal Society, Series B*, **198**, 87-108. [7]

Jansen, J.K.S., Lømo, T., Nicolaysen, K., & Westgaard, R.H. (1973). Hyperinnervation of skeletal muscle fibers: Dependence on muscle activity. *Science*, **181**, 559-561. [19]

Jansson, E., Esbjörnsson, M., Holm, I., & Jacobs, I. (1990). Increase in the proportion of fast-twitch muscle fibres by sprint training in males. *Acta Physiologica Scandinavica*, **140**, 359-363. [20]

Jansson, E., & Kaijser, L. (1977). Muscle adaptation to extreme endurance training in man. *Acta Physiologica Scandinavica*, **100**, 315-324. [20]

Jasmin, B.J., Lavoie, P.-A., & Gardiner, P.F. (1988). Fast axonal transport of labeled proteins in motoneurons of exercise-trained rats. *American Journal of Physiology*, **255**, C731-C736. [20]

Jennekens, F.G.I., Tomlinson, B.E., & Walton, J.N. (1971). Data on the distribution of fibre types in five human limb muscles. An autopsy study. *Journal of the Neurological Sciences*, **14**, 245-257. [22]

Jentsch, T.J., Steinmeyer, K., & Schwarz, G. (1990). Primary structure of *Torpedo marmorata* chloride channel isolated by expression cloning in *Xenopus* oocytes. *Nature*, **348**, 510-514. [7]

Johns, T.R., & Thesleff, S. (1961). Effects of motor inactivation on the chemical sensitivity of skeletal muscle. *Acta Physiologica Scandinavica*, **51**, 136-141. [19]

Johnson, H.A., & Erner, S. (1972). Neuron survival in the aging mouse. *Experimental Gerontology*, **7**, 111-117. [22]

Johnson, M.A., Polgar, J., Weightman, D., & Appleton, D. (1973). Data on the distribution of fibre types in thirty-six human muscles. *Journal of the Neurological Sciences*, **18**, 111-129. [12]

Jokl, P., & Konstadt, S. (1983). The effect of limb immobilization on muscle function and protein composition. *Clinical Orthopedics and Related Research*, **174**, 222-228. [19]

Jones, D.A., Jackson, M.J., McPhail, G., & Edwards, R.H.T. (1984). Experimental mouse muscle damage: The importance of external calcium. *Clinical Science*, **66**, 317-322. [21]

Jones, D.A., Newham, D.J., Round, J.M., & Tolfree, S.E.J. (1986). Experimental muscle damage: Morphological changes in relation to other indices of damage. *Journal of Physiology*, **375**, 435-448. [21]

Jones, D.A., & Round, J.M. (1990). *Skeletal muscle in health and disease. A textbook of muscle physiology.* Manchester: Manchester University Press. [15, 21]

Jones, D.A., & Rutherford, O.M. (1987). Human muscle strength training: The effects of three different regimes and the nature of the resultant changes. *Journal of Physiology*, **391**, 1-11. [20]

Jones, D.A., Rutherford, O.M., & Parker, D.F. (1989). Physiological changes in skeletal muscle as a result of strength training. *Quarterly Journal of Experimental Physiology*, **74**, 233-256. [20]

Jones, S.F., & Kwanbunbumpen, S. (1970). The effects of nerve stimulation and hemicholinium on synaptic vesicles at the mammalian neuromuscular junction. *Journal of Physiology*, **207**, 31-50. [3]

Juel, C. (1986). Potassium and sodium shifts during *in vitro* isometric muscle contraction, and the time course of the ion-gradient recovery. *Pflügers Archiv*, **406**, 458-463. [9]

Juntunen, J., & Teravainen, H. (1972). Structure development of myoneural junctions in the human embryo. *Histochemie*, **32**, 107-112. [6]

Kaeser, H.E., & Lambert, E.H. (1962). Nerve function studies in experimental polyneuritis. *Electroencephalography and Clinical Neurophysiology*, Suppl. **22**, 29-35. [16]

Kalderon, N., & Gilula, N.B. (1979). Membrane events involved in myoblast fusion. *Journal of Cell Biology*, **81**, 411-425. [5]

Kallman, D.A., Plato, C.C., & Tobin, J.D. (1990). The role of muscle loss in the age-related decline of grip strength: Cross-sectional and longitudinal perspectives. *Journal of Gerontology: Medical Sciences*, **45**, M82-M88. [22]

Kalow, W.A., Britt, B.A., Terreau, M.E., & Haist, C. (1970). Metabolic error of muscle metabolism after recovery from malignant hyperthermia. *Lancet*, **ii**, 895-898. [11]

Kamb, A., Iverson, L.E., & Tanouye, M.A. (1987). Molecular characterization of *Shaker*, a Drosophila gene that encodes a potassium channel. *Cell*, **50**, 405-413. [7]

Kandel, E.R., & Siegelbaum, S.A. (1991). Directly gated transmission at the nerve-muscle synapse. In E.R. Kandel, J.H. Schwartz, & T.M. Jessel (Eds.), *Principles of neural science* (pp. 135-152). New York: Elsevier. [10]

Kao, C.Y. (1966). Tetrodotoxin, saxitoxin and their significance in the study of excitation phenomena. *Pharmacological Reviews*, **18**, 997-1049. [7]

Karpati, G., Carpenter, S., & Eisen, A.A. (1972). Experimental core-like lesions and nemaline rods: A correlative morphological and physiological study. *Archives of Neurology*, **27**, 237-251. [19]

Karpati, G., & Engel, W.K. (1968). Correlative histochemical study of skeletal muscle after suprasegmental denervation, peripheral nerve section and skeletal fixation. *Neurology*, **18**, 681-692. [19]

Kasprzak, H., & Salpeter, M.M. (1985). Recovery of acetylcholinesterase at intact neuromuscular junctions after *in vivo* inactivation with diisopropylflurophosphate. *Journal of Neurosciences*, **5**, 951-955. [3]

Katz, B. (1966). *Nerve, muscle and synapse.* New York: McGraw-Hill. [9]

Katz, B., & Miledi, R. (1965). Propagation of electric activity in motor nerve terminals. *Proceedings of the Royal Society of London, Series B*, **161**, 453-482. [10]

Katz, B., & Miledi, R. (1968). The role of calcium in neuromuscular facilitation. *Journal of Physiology*, **195**, 481-492. [10]

Katz, B., & Miledi, R. (1972). The statistical nature of the acetylcholine potential and its molecular components. *Journal of Physiology*, **224**, 665-699. [3]

Kaufman, T.L. (1985). Strength training effect in young and aged women. *Archives of Physical Medicine and Rehabilitation*, **65**, 223-226. [22]

Kawamura, Y., Okazaki, H., O'Brien, P.C., & Dyck, P.J. (1977). Lumbar motoneurons of man. I: Numbers and diameter histograms of alpha and gamma axons and ventral roots. *Journal of Neuropathology and Experimental Neurology*, **36**, 853-860. [22]

Keh-Evans, L., Rice, C.L., Noble, E.G., Paterson, D.H., Cunningham, D.A., & Taylor, A.W. (1992). Comparison of histochemical, biochemical and contractile properties of trained aged subjects. *Canadian Journal of Aging*, **11**, 412-425. [22]

Kellerth, J.O. (1973). Intracellular staining of cat spinal motoneurones with procion yellow for ultrastructural studies. *Brain Research*, **50**, 415-418. [2]

Kelly, A.M., & Schotland, D.L. (1972). The evolution of the "checkerboard" in a rat muscle. In B.Q. Banker, R.J. Przybylski, J.P. Van der Meulen, & M. Victor (Eds.), *Research in muscle development and the muscle spindle* (pp. 32-48). Amsterdam: Excerpta Medica. [12]

Kelly, A.M., & Zacks, S.I. (1969). The fine structure of motor end plate morphogenesis. *Journal of Cell Biology*, **42**, 154-169. [6]

Kennedy, J.M., Kamel, S., Tambone, W.W., Vrbová, G., & Zak, R. (1986). The expression of myosin heavy chain isoforms in normal and hypertrophied chicken slow muscle. *Journal of Cell Biology*, **103**, 977-983. [20]

Kennedy, T.E., Serafini, T., de la Torre, J.R., & Tessier-Lavigne, M. (1994). Netrins are diffusible chemotropic factors for commissural axons in the embryonic spinal cord. *Cell*, **78**, 425-435. [6]

Kennedy, W.R. (1970). Innervation of normal human muscle spindles. *Neurology*, **20**, 463-475. [4]

Kenny-Mobbs, T. (1985). Myogenic differentiation in early chick wing mesenchyme in the absence of the brachial somites. *Journal of Embryology and Experimental Morphology*, **90**, 415-436. [5]

Kereshi, S., Manzano, G., & McComas, A.J. (1983). Impulse conduction velocities in human biceps brachii muscles. *Experimental Neurology*, **80**, 652-662. [9]

Kernell, D. (1966). Input resistance, electrical excitability and size of ventral horn cells in cat spinal cord. *Science*, **152**, 1637-1640. [12]

Keynes, R.D., Ritchie, J.M., & Rojas, E. (1971). The binding of tetrodotoxin to nerve membranes. *Journal of Physiology*, **213**, 235-254. [9]

Kiens, B., Saltin, B., Wallye, L., & Wesche, J. (1989). Temporal relationship between blood flow changes and release of ions and metabolites from muscle upon single weak contractions. *Acta Physiologica Scandinavica*, **136**, 551-559. [13]

Kjeldsen, K., Everts, M.E., & Clausen, T. (1986). The effects of thyroid hormones on ³H-ouabain binding site concentration, Na,K-contents and ⁸⁶Rb-efflux in rat skeletal muscle. *Pflügers Archiv*, **406**, 529-535. [7]

Klaus, W., Lüllmann, H., & Muscholl, E. (1960). The binding of tetrodotoxin to nerve membranes. *Journal of Physiology*, **213**, 235-254. [16]

Klinkerfuss, G.H., & Haugh, M.J. (1970). Disuse atrophy of muscle. Histochemistry and electron microscopy. *Archives of Neurology*, **22**, 309-320. [19]

Klitgaard, H., Mantoni, M., Schiaffino, S., Ausoni, S., Gorza, L., Laurent-Winter, C., Schnohr, P., & Saltin, B. (1990). Function, morphology and protein expression of ageing skeletal muscle: A cross-sectional study of elderly men with different training backgrounds. *Acta Physiologica Scandinavica*, **140**, 41-54. [22]

Klitgaard, H., Zhou, M., & Richter, E.A. (1990). Myosin heavy chain composition of single fibres from m. biceps brachii of male body builders. *Acta Physiologica Scandinavica*, **140**, 175-180. [20]

Klitgaard, H., Zhou, M., Schiaffino, S., Betto, R., Salviati, G., & Saltin, B. (1990). Ageing alters the myosin heavy chain composition of single fibres from human skeletal muscle. *Acta Physiologica Scandinavica*, **140**, 55-62. [22]

Klitgaard, H.A. (1988). A model for quantitative strength training of hindlimb muscles of the rat. *Journal of Applied Physiology*, **64**, 1740-1745. [20]

Knochel, J.P., Blachley, J.D., Johnson, J.H., & Carter, N.W. (1985). Muscle cell electrical hyperpolarization and reduced exercise hyperkalemia in physically conditioned dogs. *Journal of Clinical Investigation*, **75**, 740-745. [7]

Knowles, M., Currie, S., Saunders, M., Walton, J.N., & Field, E.J. (1969). Lymphocyte transformation in the Guillain-Barré syndrome. *Lancet*, **ii**, 1168-1170. [16]

Koch, H.P. (1990). Comments on "Proposed mechanism of action in thalidomide embryopathy." *Teratology*, **41**, 243-244. [5]

Koch, M.C., Steinmeyer, K., Lorenz, C., Ricker, K., Wolf, F., Otto, M., Zoll, B., Lehmann-Horn, F., Grzechik, K.-H., & Jentsch, T.J. (1992). The skeletal muscle chloride channel in dominant and recessive human myotonia. *Science*, **257**, 797-800. [7]

Koenig, M., Monaco, A.P., & Kunkel, L.M. (1988). The complete sequence of dystrophin predicts a rod-shaped cytoskeletal protein. *Cell*, **53**, 219-228. [1]

Körner, G. (1960). Untersuchungen über Zahl, Anordnung und Länge der Muskelspindeln in einegen Schulterei, den Oberarmmunskeln und im Muskulus sternalis des Menschen. *Anatomischer Anzeiger*, **108**, 99-103. [4]

Kowalchuk, N., & McComas, A.J. (1987). Effects of impulse blockade on the contractile properties of rat skeletal muscle. *Journal of Physiology*, **382**, 255-266. [19]

Kraus, W.E., & Williams, R.S. (1990). Intracellular signals mediating contraction-induced changes in the oxidative capacity of skeletal muscle. In D. Pette (Ed.), *The dynamic state of muscle fibres* (pp. 601-615). Berlin: Walter de Gruyter. [20]

Krishnan, S., Lowrie, M.B., & Vrbová, G. (1985). The effect of reducing the peripheral field on motoneurone development in the rat. *Developmental Brain Research*, **19**, 11-20. [6]

Kristensson, K., & Olsson, Y. (1971). Retrograde axonal transport of protein. *Brain Research*, **29**, 363-365. [8, 18]

Krnjević, K., & Miledi, R. (1958). Failure of neuromuscular propagation in rats. *Journal of Physiology*, **140**, 440-461. [15]

Kucera, J. (1985). Characteristics of motor innervation of muscle spindles in the monkey. *American Journal of Physiology*, **173**, 113-125. [4]

Kuffler, S.W., & Yoshikami, D. (1975). The number of transmitter molecules in a quantum: An estimate from iontophoretic applications of acetylcholine at the neuromuscular synapse. *Journal of Physiology*, **251**, 465-482. [3]

Kugelberg, E. (1949). Electromyography in muscle dystrophies. Differentiation between dystrophies and chronic motor neurone lesions. *Journal of Neurology, Neurosurgery and Psychiatry*, **12**, 129-136. [13]

Kugelberg, E., Edström, L., & Abbruzzese, M. (1970). Mapping of motor units in experimentally reinnervated rat muscle. Interpretation of histochemical and atrophic fibre pattern in neurogenic lesions. *Journal of Neurology, Neurosurgery and Psychiatry*, **33**, 319-329. [17]

Kuiack, S., & McComas, A.J. (1992). Transient hyperpolarization of non-contracting muscle fibres in anaesthetized rats. *Journal of Physiology*, **454**, 609-618. [7, 15]

Kukulka, C.G., & Clamann, H.P. (1981). Comparison of the recruitment and discharge properties of motor units in human brachial biceps and adductor pollicis during isometric contractions. *Brain Research*, **219**, 45-55. [13]

Kuno, M., Miyata, Y., & Muñoz-Martinez, E.J. (1974a). Differential reaction of fast and slow α-motoneurones to axotomy. *Journal of Physiology*, **240**, 725-739. [18]

Kuno, M., Miyata, Y., & Muñoz-Martinez, E.J. (1974b). Properties of fast and slow alpha motoneurones following motor reinnervation. *Journal of Physiology*, **242**, 273-288. [18]

Kutay, U., Hartmann, E., & Rapoport, T.A. (1993). A class of membrane proteins with a C-terminal anchor. *Trends in Cell Biology*, **3**, 72-75. [10]

Laing, N.G., & Prestige, M.C. (1978). Prevention of spontaneous motoneuron death in chick embryos. *Journal of Physiology*, **282**, 33-34P. [6]

Lambert, E.H., & Elmqvist, D. (1971). Quantal components of end-plate potentials in the myasthenic syndrome. *Annals of the New York Academy of Sciences*, **183**, 183-199. [10]

Lampert, P.W. (1969). Mechanism of demyelination in experimental allergic neuritis. *Laboratory Investigation*, **20**, 127-138. [16]

Lance-Jones, C., & Landmesser, L.T. (1980a). Motoneuron projection patterns in embryonic chick limbs following partial deletions of the spinal cord. *Journal of Physiology*, **302**, 559-580. [6]

Lance-Jones, C., & Landmesser, L.T. (1980b). Motoneurone projection patterns in the chick hind limb following early partial reversals of the spinal cord. *Journal of Physiology*, **302**, 581-602. [6]

Landau, W.M. (1953). The duration of neuromuscular function after nerve section in man. *Journal of Neurosurgery*, **10**, 64-68. [16]

Landmesser, L. (1971). Contractile and electrical responses of vagus-innervated frog sartorius. *Journal of Physiology*, **213**, 707-752. [17]

Landmesser, L. (1972). Pharmacological properties, cholinesterase activity and anatomy of nerve-muscle junctions in vagus-innervated frog sartorius. *Journal of Physiology*, **220**, 243-256. [17]

Landon, D.N. (1982). The excitable apparatus of skeletal muscle. In W.J. Culp & J. Achoa (Eds.), *Abnormal nerves and muscles as impulse generators* (pp. 607-631). New York: Oxford University Press. [15]

Landon, D.N., & Langley, O.K. (1971). The local chemical environment of nodes of Ranvier: A study of cation binding. *Journal of Anatomy*, **108**, 419-432. [2]

Lang, B., Newsom-Davis, J., Prior, C., & Wray, D. (1983). Antibodies to motor nerve terminals: An electrophysiological study of a human myasthenic syndrome transferred to mouse. *Journal of Physiology*, **344**, 335-345. [10]

Langeland, J.A., & Carroll, S.B. (1993). Conservation of regulatory elements controlling hair pair-rule stripe formation. *Development*, **117**, 585-596. [5]

Lännergren, J., Larsson, L., & Westerblad, H. (1989). A novel type of delayed tension reduction observed in rat motor units after intense activity. *Journal of Physiology*, **412**, 267-276. [15]

Lännergren, J., & Westerblad, H. (1991). Force decline due to fatigue and intracellular acidification in isolated fibres from mouse skeletal muscle. *Journal of Physiology*, **434**, 307-322. [15]

Larsson, L. (1983). Histochemical characteristics of human skeletal muscle during ageing. *Acta Physiologica Scandinavica*, **117**, 469-471. [22]

Larsson, L., Ansved, T., Edström, L., Gorza, L., & Schiaffino, S. (1991). Effects of age on physiological, immunochemical and biochemical properties of fast-twitch single motor units in the rat. *Journal of Physiology*, **443**, 257-275. [22]

Larsson, L., Grimby, G., & Karlsson, J. (1979). Muscle strength and speed of movement in relation to age and muscle morphology. *Journal of Applied Physiology*, **46**, 451-456. [22]

Larsson, L., & Salviati, G. (1989). Effects of age on calcium transport activity of sarcoplasmic reticulum in fast- and slow-twitch rat muscle fibres. *Journal of Physiology*, **419**, 253-264. [22]

Lasek, R.J., Garner, J.A., & Brady, S.T. (1984). Axonal transport of the cytoplasmic matrix. *Journal of Cell Biology*, **99**, 212s-221s. [8]

Lazarides, E., & Capetanaki, Y.G. (1986). The striated muscle cytoskeleton: Expression and assembly in development. In C. Emerson, D. Fischman, B. Nadal-Ginard, & M.A.Q. Siddiqui (Eds.), *Molecular biology of muscle development* (pp. 749-772). New York: Alan R. Liss. [1]

Lee, J.A., Westerblad, H., & Allen, D.G. (1991). Changes in tetanic and resting $[Ca^{2+}]_i$ during fatigue and recovery of single muscle fibres from *Xenopus laevis*. *Journal of Physiology*, **433**, 307-326. [15]

Lee, R.G., Ashby, P., White, D.G., & Aguayo, A.J. (1975). Analysis of motor conduction velocity in the human median nerve by computer simulation of compound muscle action potentials. *Electroencephalography and Clinical Neurophysiology*, **39**, 225-237. [9, 12]

Lehman, C.F. (1850). *Lehrbuch der Physiologischen Chemie*. [15]

Lehmann-Horn, F., Küther, G., Ricker, K., Grafe, P., Ballanyi, K., & Rüdel, R. (1987). Adynamia episodica hereditaria with myotonia: A non-inactivating sodium current and the effect of extracellular pH. *Muscle & Nerve*, **10**, 363-374. [7]

Leibrock, J., Lottspeich, F., Hohn, A., Hofer, M., Hengerer, B., Masiakowski, P., Thoenen, H., & Barde, Y.A. (1989). Molecular cloning and expression of brain-derived neurotrophic factor. *Nature*, **341**, 149-152. [18]

Leksell, L. (1945). The action potential and excitatory effects of the small ventral root fibres to skeletal muscle. *Acta Physiologica Scandinavica*, **10**(Suppl. 31), 1-84. [12]

Lemire, R.J. (1969). Variations in development of the caudal neural tube in human embryos (Horizons XIV-XXI). *Teratology*, **2**, 361-370. [6]

Lenman, R.A.J., Tulley, F.M., Vrbová, G., Dimitrijevic, M.R., & Towle, J.A. (1989). Muscle fatigue in some neurological disorders. *Muscle & Nerve*, **12**, 938-942. [19]

Lennon, V.A., & Carnegie, P.R. (1971). Immunopharmacological disease: A break in tolerance to receptor sites. *Lancet*, **i**, 630-633. [10]

Lennon, V.A., Lindstrom, J.M., & Seybold, M.E. (1976). Experimental auto-immune myasthenia gravis: Cellular and humoral immune responses. *Annals of the New York Academy of Sciences*, **274**, 283-299. [10]

Leon, J., & McComas, A.J. (1984). Effects of vincristine sulfate on touch dome function in the rat. *Experimental Neurology*, **84**, 283-291. [8]

Levi-Montalcini, R. (1987). The nerve growth factor 35 years later. *Science*, **237**, 1154-1162. [18]

Lewis, D.M. (1972). The effect of denervation on the mechanical and electrical responses of fast and slow mammalian twitch muscle. *Journal of Physiology*, **222**, 51-95. [16]

Lewis, D.M., & Parry, D.J. (1979). Properties of motor units in mouse soleus. *Journal of Physiology*, **295**, 90P. [12]

Lexell, J. (1993). Ageing and human muscle: Observations from Sweden. *Canadian Journal of Applied Physiology*, **18**, 2-18. [22]

Lexell, J., Henriksson-Larsén, K., Winblad, B., & Sjöström, M. (1983). Distribution of different fiber types in human skeletal muscles: Effects of aging studied in whole muscle cross sections. *Muscle & Nerve*, **6**, 588-595. [22]

Lexell, J., Taylor, C.C., & Sjöström, M. (1988). What is the cause of the ageing atrophy? Total number, size and proportion of different fibre types studied in the whole vastus lateralis muscle from 15- to 83-year-old men. *Journal of Neurological Sciences*, **84**, 275-294. [22]

Libelius, R., & Tågerud, S. (1984). Uptake of horseradish peroxidase in denervated skeletal muscle occurs primarily at the endplate region. *Journal of the Neurological Sciences*, **66**, 273-281. [18]

Lichtman, J.W., & Balice-Gordon, R.J. (1990). Understanding synaptic competition in theory and in practice. *Journal of Neurobiology*, **21**, 99-106. [6]

Lieber, R.L. (1988). Time course and cellular control of muscle fiber transformation following chronic stimulation. *ISI Atlas of Science: Plants and Animals*, **1**, 189-194. [20]

Lindboe, C.F., & Platou, C.S. (1984). Effects of immobilization of short duration on the muscle fiber size. *Clinical Physiology*, **4**, 183-188. [19]

Lindsley, D.B. (1935). Electrical activity of human motor units during voluntary contraction. *American Journal of Physiology*, **114**, 90-99. [13]

Lipicky, R.J., & Bryant, S.H. (1973). A biophysical study of the human myotonias. In J.E. Desmedt (Ed.), *New developments in electromyography and clinical neurophysiology* (Vol. 1, pp. 451-463). Basel: Karger. [7]

Lipicky, R.J., Bryant, S.H., & Salmon, J.H. (1971). Cable parameters, sodium, potassium, chloride, and water content, and potassium efflux in isolated external intercostal muscle of normal volunteers and patients with myotonia congenita. *Journal of Clinical Investigation*, **50**, 2091-2103. [9]

Lissák, K., Dempsey, E.W., & Rosenblueth, A. (1939). The failure of transmission of motor nerve impulses in the course of Wallerian degeneration. *American Journal of Physiology*, **128**, 45-56. [16]

Loeb, G.E., Pratt, C.A., Chanaud, C.M., & Richmond, F.J.R. (1987). Distribution and innervation of short, interdigitated muscle fibers in parallel-fibered muscles of the cat hindlimb. *Journal of Morphology*, **191**, 1-15. [1, 12]

Lohof, A.M., Ip, N.Y., & Poo, M. (1993). Potentiation of developing neuromuscular synapses by the neurotrophins NT-3 and BDNF. *Nature*, **363**, 350-353. [18]

Lømo, T., & Gundersen, K. (1988). Trophic control of skeletal muscle membrane properties. In H.L. Fernandez & J.A. Donoso (Eds.), *Nerve-muscle cell trophic communication* (pp. 61-79). Boca Raton, FL: CRC Press, Inc. [18]

Lømo, T., & Rosenthal, J. (1972). Control of ACh sensitivity by muscle activity in the rat. *Journal of Physiology*, **221**, 493-513. [6, 18, 19]

Loughna, P.T., Izumo, S., Goldspink, G., & Nadal-Ginard, B. (1990). Disuse and passive stretch causes rapid alterations in expression of developmental and adult contractile protein genes in skeletal muscle. *Development*, **109**, 217-223. [19]

Lowey, S., Waller, G.S., & Trybus, K.M. (1993). Skeletal muscle myosin light chains are essential for physiological speeds of shortening. *Nature*, **365**, 454-456. [11]

Lubinska, L., & Niemierko, S. (1970). Velocity and intensity of bidirectional migration of acetylcholinesterase in transected nerves. *Brain Research*, **27**, 329-342. [8]

Luco, J.V., & Eyzaguirre, C. (1955). Fibrillation and hypersensitivity to ACh in denervated muscle: Effect of length of degenerating nerve fibers. *Journal of Neurophysiology*, **18**, 65-73. [18]

Ludin, H.P. (1970). Microelectrode study of dystrophic human skeletal muscle. *European Neurology*, **3**, 116-121. [9]

Luff, A.R., Hatcher, D.D., & Torkko, K. (1988). Enlarged motor units resulting from partial denervation of cat hindlimb muscles. *Journal of Neurophysiology*, **59**, 1377-1394. [17]

Lunn, E.R., Perry, V.H., Brown, M.C., Rosen, H., & Gordon, S. (1989). Absence of Wallerian degeneration does not hinder regeneration in peripheral nerve. *European Journal of Neuroscience*, **1**, 27-33. [16]

Luther, P., & Squire, J. (1978). Three-dimensional structure of the vertebrate muscle M-region. *Journal of Molecular Biology*, **125**, 313-324. [1]

Lüttgau, H.C. (1965). The effect of metabolic inhibitors on the fatigue of the action potential in single muscle fibres. *Journal of Physiology*, **178**, 45-67. [15]

Lux, H.D., Schubert, P., Kreutzberg, G.W., & Globus, A. (1970). Excitation and axonal flow: Autoradiographic study on motoneurons intracellularly injected with a ³H-amino acid. *Experimental Brain Research*, **10**, 197-204. [8]

Lymn, R.W., & Taylor, E.W. (1971). Mechanism of adenosine triphosphate hydrolysis by actomyosin. *Biochemistry*, **10**, 4617-4624. [11]

MacCallum, J.B. (1898). On the histogenesis of striated muscle fiber, and the growth of the human sartorius muscle. *Bulletin of Johns Hopkins Hospital*, **9**, 208-215. [1, 5]

MacDougall, J.D. (1986). Morphological changes in human skeletal muscle following strength training and immobilization. In N.L. Jones, N. McCartney, & A.J. McComas (Eds.), *Human muscle power* (pp. 269-288). Champaign, IL: Human Kinetics. [5]

MacDougall, J.D., Elder, G.C.B., Sale, D.G., Moroz, J.R., & Sutton, J.R. (1980). Effects of strength training and immobilization on human muscle fibres. *European Journal of Applied Physiology and Occupational Physiology*, **43**, 25-34. [19, 20]

MacDougall, J.D., Sale, D.G., Alway, S.E., & Sutton, J.R. (1984). Muscle fiber number in biceps brachii in body builders and control subjects. *Journal of Applied Physiology*, **57**, 1399-1403. [20]

MacDougall, J.D., Ward, G.R., Sale, D.G., & Sutton, J.R. (1977). Biochemical adaptation of human skeletal muscle to heavy resistance training and immobilization. *Journal of Applied Physiology*, **43**, 700-703. [20]

Macefield, G., Gandevia, S.C., & Burke, D. (1990). Perceptual responses to microstimulation of single afferents innervating joints, muscles and skin of the human hand. *Journal of Physiology*, **429**, 113-129. [4]

MacLennan, D.H., & Phillips, M.S. (1992). Malignant hyperthermia. *Nature*, **256**, 789-794. [11]

MacLennan, D.H., & Wong, P.T.S. (1971). Isolation of a calcium-sequestering protein from sarcoplasmic reticulum. *Proceedings of the National Academy of Sciences (U.S.)*, **68**, 1231-1235. [7]

Magill-Solc, C., & McMahan, U.J. (1990a). Agrin-like molecules in motor neurons. *Journale de Physiologie*, **84**, 78-81. [6]

Magill-Solc, C., & McMahan, U.J. (1990b). Synthesis and transport of agrin-like molecules in motor neurons. *Journal of Experimental Biology*, **153**, 1-10. [3]

Magleby, K.L., & Zengel, J.E. (1982). A quantitative description of stimulation-induced changes in transmitter release at the frog neuromuscular junction. *Journal of General Physiology*, **80**, 613-638. [10]

Manzano, G., & McComas, A.J. (1988). Longitudinal structure and innervation of two mammalian limb muscles. *Muscle & Nerve*, **11**, 1115-1122. [12]

Markelonis, G.J., & Oh, T.H. (1979). A sciatic nerve protein has a trophic effect on development and maintenance of skeletal muscle cells in culture. *Proceedings of the National Academy of Sciences (U.S.)*, **76**, 2470-2474. [18]

Marotte, L.R., & Mark, R.F. (1970). The mechanism of selective reinnervation of fish eye muscle. I. Evidence from muscle function during recovery. *Brain Research*, **19**, 41-51. [17]

Marsden, C.D., & Meadows, J.C. (1970). The effect of adrenaline on the contraction of human muscle. *Journal of Physiology*, **207**, 429-448. [11]

Marsden, C.D., Meadows, J.C., & Merton, P.A. (1969). Muscle wisdom. *Journal of Physiology*, **200**, 15P. [15]

Marsden, C.D., Meadows, J.C., & Merton, P.A. (1971). Isolated single motor units in human muscle and their rate of discharge during maximal voluntary effort. *Journal of Physiology*, **217**, 12-13P. [13]

Marsh, E., Sale, D., McComas, A.J., & Quinlan, J. (1981). Influence of joint position on ankle dorsiflexion in humans. *Journal of Applied Physiology*, **51**, 160-167. [12]

Martin, T.P., Edgerton, V.R., & Grindeland, R.E. (1988). Influence of spaceflight on rat skeletal muscle. *Journal of Applied Physiology*, **65**, 2318-2325. [19]

Mathiowetz, V., Kashman, N., Volland, G., Weber, K., Dowe, M., & Rogers, S. (1985). Grip and pinch strength: Normative data for adults. *Archives of Physical Medicine and Rehabilitation*, **66**, 69-74. [22]

Matsumura, K., & Campbell, K.P. (1994). Dystrophin-glycoprotein complex: Its role in the molecular pathogenesis of muscular dystrophies. *Muscle & Nerve*, **17**, 2-15. [1]

Matthews, P.B.C. (1962). The differentiation of two types of fusiform fibre by their effects on the dynamic response of múscle spindle primary endings. *Quarterly Journal of Experimental Physiology*, **47**, 324-333. [4]

Matthews, P.B.C. (1964). Muscle spindles and their motor control. *Physiological Reviews*, **44**, 219-288. [4]

Matthews, P.B.C. (1972). *Mammalian muscle receptors and their central actions*. London: Edward Arnold. [4]

Matzuk, M.M., Kumar, T.R., Vassalli, A., Bickenbach, J.R., Roop, D.R., Jaenisch, R., & Bradley, A. (1995). Functional analysis of activins during mammalian development. *Nature*, **374**, 354-356. [5]

Maxwell, M.H., & Kleeman, C.R. (1962). *Clinical disorders of fluid and electrolyte metabolism*. New York: McGraw-Hill. [9]

Mayer, R.F., Burke, R.E., Toop, J., Hudgson, J.A., Kanda, K., & Walnisley, B. (1981). The effect of long-term immobilization on the motor unit population of the cat medial gastrocnemius muscle. *Neuroscience*, **6**, 725-739. [19]

Mayer, R.F., & Doyle, A.M. (1970). Studies of the motor unit in the cat. Histochemistry and topology of anterior tibial and extensor digitorum longus muscles. In J.N. Walton, N. Canal, & G. Scarlato (Eds.), *Muscle diseases* (pp. 159-163). Amsterdam: Excerpta Medica. [12]

McArdle, B. (1951). Myopathy due to a defect in muscle glycogen breakdown. *Clinical Science*, **10**, 13-35. [14]

McArdle, J.J., & Albuquerque, E.X. (1973). A study of the reinnervation of fast and slow mammalian muscle. *Journal of General Physiology*, **61**, 1-23. [17]

McBride, W.G. (1961). Thalidomide and congenital abnormalities. *Lancet*, **2**, 1358. [5]

McBride, W.G. (1978). Role of neural crest and peripheral nerves in limb development. *Lancet*, **2**, 792-793. [5]

McCartney, N., Moroz, D., Garner, S.H., & McComas, A.J. (1988). The effects of strength training in patients with selected neuromuscular diseases. *Medicine and Science in Sports and Exercise*, **20**, 362-368. [20]

McComas, A.J. (1977). *Neuromuscular function and disorders*. London: Butterworths. [3, 5, 7, 8, 9, 10, 11, 12, 16, 17, 18, 19, 21, 22]

McComas, A.J. (1994). Human neuromuscular adaptations that accompany changes in activity. *Medicine and Science in Sports and Exercise*, **26**, 1499-1509. [19, 20]

McComas, A.J., De Bruin, H., & Quartly, C. (1991). Non-transmitting synapses in human neuromuscular disorders? In A. Wernig (Ed.), *Restorative neurology* (Vol. 5, pp. 223-229). Amsterdam: Elsevier Science. [16]

McComas, A.J., Fawcett, P.R.W., Campbell, M.J., & Sica, R.E.P. (1971). Electrophysiological estimation of the number of motor units within a human muscle. *Journal of Neurology, Neurosurgery and Psychiatry*, **34**, 121-131. [12]

McComas, A.J., Galea, V., Einhorn, R.W., Hicks, A.L., & Kuiack, S. (1993). The role of the Na⁺, K⁺-pump in delaying muscle fatigue. In A.J. Sargeant & D. Kernell (Eds.), *Neuromuscular fatigue* (pp. 35-43). Amsterdam: North-Holland. [15]

McComas, A.J., Jorgensen, P.B., & Upton, A.R.M. (1974). The neurapraxic lesion: A clinical contribution to the study of trophic mechanisms. *Canadian Journal of Neurological Sciences*, **1**, 170-179. [16]

McComas, A.J., & Mroźek, K. (1967). Denervated muscle fibres in hereditary mouse dystrophy. *Journal of Neurology, Neurosurgery and Psychiatry*, **30**, 526-530. [21]

McComas, A.J., Mroźek, K., Gardner-Medwin, D., & Stanton, W.H. (1968). Electrical properties of muscle fibre membranes in man. *Journal of Neurology, Neurosurgery and Psychiatry*, **31**, 434-440. [9]

McComas, A.J., Sica, R.E.P., & Banerjee, S. (1978). Central nervous system effects of limb amputation in man. *Nature*, **271**, 73-74. [16]

McComas, A.J., Sica, R.E.P., Campbell, M.J., & Upton, A.R.M. (1971). Functional compensation in partially denervated muscles. *Journal of Neurology, Neurosurgery and Psychiatry*, **34**, 453-460. [6, 17]

McComas, A.J., Sica, R.E.P., McNabb, A.R., Goldberg, W.M., & Upton, A.R.M. (1974). Evidence for reversible motoneurone dysfunction in thyrotoxicosis. *Journal of Neurology, Neurosurgery and Psychiatry*, **37**, 548-558. [16]

McComas, A.J., Sica, R.E.P., & Petito, F. (1973). Muscle strength in boys of different ages. *Journal of Neurology, Neurosurgery and Psychiatry*, **36**, 171-173. [5]

McComas, A.J., Sica, R.E.P., Upton, A.R.M., & Aguilera, N. (1973). Functional changes in motoneurones of hemiparetic muscles. *Journal of Neurology, Neurosurgery and Psychiatry*, **36**, 183-193. [18]

McComas, A.J., & Thomas, H.C. (1968). Fast and slow twitch muscles in man. *Journal of the Neurological Sciences*, **7**, 301-307. [12]

McCredie, J. (1975). The pathogenesis of congenital malformations. *Australian Radiology*, **19**, 348-355. [5]

McCully, K.K., & Faulkner, J.A. (1985). Injury to skeletal muscle fibers of mice following lengthening contraction. *Journal of Applied Physiology*, **59**, 119-126. [21]

McDonagh, J.C., Binder, M.D., Reinking, R.M., & Stuart, D.G. (1980). Tetrapartite classification of motor units of cat tibialis posterior. *Journal of Neurophysiology*, **44**, 696-712. [12]

McDonagh, M.J.N., White, M.J., & Davies, C.T.M. (1984). Different effects of ageing on the mechanical properties of arm and leg muscles. *Gerontology*, **30**, 49-54. [22]

McKenzie, D.K., Bigland-Ritchie, B., Gorman, R.B., & Gandevia, S.C. (1992). Central and peripheral fatigue of human diaphragm and limb muscles assessed by twitch interpolation. *Journal of Physiology*, **454**, 643-656. [15]

McLeod, K.G., & Wray, S.H. (1967). Conduction velocity and fibre diameter of the median and ulnar nerves of the baboon. *Journal of Neurology, Neurosurgery and Psychiatry*, **30**, 240-247. [9]

McMahan, U.J., & Slater, C.R. (1984). The influence of basal lamina on the accumulation of acetylcholine receptors at synaptic sites in regenerating muscle. *Journal of Cell Biology*, **98**, 1453-1473. [17]

McManaman, J.L., Oppenheim, R.W., Prevette, D., & Marchetti, D. (1990). Rescue of motoneurons from cell death by a purified skeletal muscle polypeptide: Effects of the ChAT development factor, CDF. *Neuron*, **4**, 891-898. [18]

McNamara, J.O., & Fridovich, I. (1993). Did radicals strike Lou Gehrig? *Nature*, **362**, 20-21. [22]

McPhedran, A.M., Wuerker, R.B., & Henneman, E. (1965). Properties of motor units in a homogeneous red muscle (soleus) of the cat. *Journal of Neurophysiology*, **28**, 71-84. [12]

Meakin, S.O., & Shooter, E.M. (1992). The nerve growth factor family of receptors. *Trends in Neurosciences*, **15**, 323-331. [18]

Medbø, J.I., & Sejersted, O.M. (1990). Plasma potassium changes with high intensity exercises. *Journal of Physiology*, **421**, 105-122. [15]

Melki, J., Abdelhak, S., Sheth, P., Bachelot, M.F., Burlet, P., Marcadet, A., Alcardi, J., Barois, A., Carriere, J.P., Fardeau, M., Fontan, D., Ponsot, G., Billsette, T., Angelini, C., Barbosa, C., Ferriere, G., Lanzi, G., Ottolini, A., Babron, M.C., Cohen, D., Hanauer, A., Colerget-Darpox, F., Lathrop, M., Munnich, A., & Frezal, J. (1990). Gene for chronic proximal spinal muscular atrophies maps to chromosome 5q. *Nature*, **344**, 767-768. [6]

Melnick, S.C. (1963). 38 cases of the Guillain-Barré syndrome: An immunological study. *British Medical Journal*, **1**, 368-373. [16]

Mense, S. (1986). Slowly conducting afferent fibers from deep tissues—Neurobiological properties and central nervous actions. In D. Ottoson (Ed.), *Progress in sensory physiology* (Vol. 6, pp. 139-219). Berlin: Springer-Verlag. [15]

Mense, S., & Meyer, H. (1985). Different types of slowly conducting afferent units in the cat skeletal muscle and tendon. *Journal of Physiology*, **363**, 403-417. [4]

Mense, S., & Stahnke, M. (1983). Responses in muscle afferent fibres of slow conduction velocity to contractions and ischaemia in the cat. *Journal of Physiology*, **342**, 383-397. [4]

Merton, P.A. (1954). Voluntary strength and fatigue. *Journal of Physiology*, **123**, 553-564. [13, 15]

Meryon, E. (1852). On granular and fatty degeneration of the voluntary muscles. *Medico-Chirurgical Transactions*, **35**, 73-84. [1]

Mesulam, M.-M. (1982). Principles of horseradish peroxidase neurohistochemistry and their application for tracing neural pathways-axonal transport, enzyme histochemistry and light microscopic analysis. In M.-M. Mesulam (Ed.), *Tracing neural connections with horseradish peroxidase* (pp. 1-151). New York: John Wiley & Sons. [12]

Miledi, R. (1960a). The acetylcholine sensitivity of frog muscle fibres after complete or partial denervation. *Journal of Physiology*, **151**, 1-23. [18]

Miledi, R. (1960b). Properties of regenerating neuromuscular synapses in the frog. *Journal of Physiology*, **154**, 190-205. [17]

Miledi, R., Molenaar, P.C., & Polak, R.L. (1983). Electrophysiological and chemical determination of acetylcholine release at the frog neuromuscular junction. *Journal of Physiology*, **334**, 245-254. [3]

Miledi, R., & Parker, I. (1981). Calcium transients recorded with arsenazo III in the presynaptic terminal of the squid giant synapse. *Proceedings of the Royal Society of London, Series B*, **212**, 197-211. [10]

Miledi, R., & Slater, C.R. (1969). Electron-microscopic structures of denervated skeletal muscle. *Proceedings of the Royal Society of London, Series B*, **174**, 253-269. [16]

Miledi, R., & Slater, C.R. (1970). On the degeneration of rat neuromuscular junctions after nerve section. *Journal of Physiology*, **207**, 507-528. [8, 16]

Miller, R.G., Boska, M.D., Moussavi, R.S., Carson, P.J., & Weiner, M.W. (1988). ^{31}P nuclear magnetic resonance studies of high energy phosphates and pH in human muscle fatigue. *Journal of Clinical Investigation*, **81**, 1190-1196. [15]

Miller, R.G., Giannini, D., Milner-Brown, H.S., Layzer, R.B., Koretsky, A.P., Hooper, D., & Weiner, M.W. (1987). Effects of fatiguing exercise on high-energy phosphates, force, and EMG: Evidence for three phases of recovery. *Muscle & Nerve*, **10**, 810-821. [15]

Miller, T.M., & Heuser, J.E. (1984). Endocytosis of synaptic vesicle membrane at the frog neuromuscular junction. *Journal of Cell Biology*, **98**, 685-698. [10]

Milner, T.E., & Stein, R.B. (1981). The effects of axotomy on the conduction of action potentials in peripheral sensory and motor nerve fibres. *Journal of Neurology, Neurosurgery and Psychiatry*, **44**, 485-496. [16]

Milner-Brown, H.S., Stein, R.B., & Lee, R.G. (1974a). Contractile and electrical properties of human motor units in neuropathies and motor neurone disease. *Journal of Neurology, Neurosurgery and Psychiatry*, **37**, 670-676. [17]

Milner-Brown, H.S., Stein, R.B., & Lee, R.G. (1974b). Pattern of recruiting human motor units in neuropathies and motor neurone disease. *Journal of Neurology, Neurosurgery and Psychiatry*, **37**, 665-669. [17]

Milner-Brown, H.S., Stein, R.B., & Yemm, R. (1973a). Changes in firing rate of human motor units during linearly changing voluntary contractions. *Journal of Physiology*, **230**, 371-390. [13]

Milner-Brown, H.S., Stein, R.B., & Yemm, R. (1973b). The orderly recruitment of human motor units during voluntary isometric contractions. *Journal of Physiology*, **230**, 359-370. [12, 13]

Mines, G.R. (1913). On functional analysis of the action of electrolytes. *Journal of Physiology*, **46**, 188-235. [11]

Mitchell, P. (1961). Coupling of phosphorylation to electron and hydrogen transfer by a chemi-osmotic type of mechanism. *Nature*, **191**, 144-148. [14]

Mitchell, S.W., Morehouse, G.R., & Keen, W.W. (1864). *Gunshot wounds and other injuries of nerves*. Philadelphia, PA: Lippincott. [19]

Mittal, K.R., & Logmani, F.H. (1987). Age-related reduction in 8th cervical ventral nerve root myelinated fiber diameters and numbers in man. *Journal of Gerontology*, **42**, 8-10. [22]

Mokri, B., & Engel, A.G. (1975). Duchenne dystrophy: Electron microscopic findings pointing to a basic or early abnormality in the plasma membrane of the muscle fiber. *Neurology*, **25**, 1111-1120. [1]

Molenaar, P.C., Newsom-Davis, J., Polak, R.L., & Vincent, A. (1982). Eaton-Lambert syndrome: Acetylcholine and choline acetyltransferase in skeletal muscle. *Neurology*, **32**, 1062-1065. [10]

Monaco, A.P., Bertelson, C.J., Liechti-Gallati, S., Moser, H., & Kunkel, L.M. (1988). An explanation for the phenotypic differences between patients bearing partial deletions of the DMD locus. *Genomics*, **2**, 90-95. [1]

Monster, A.W., & Chan, H. (1977). Isometric force production by motor units of extensor digitorum communis muscle in man. *Journal of Neurophysiology*, **40**, 1432-1443. [12, 13]

Moore, D.H., II. (1975). A study of age group track and field records to relate age and running speed. *Nature*, **253**, 264-265. [22]

Moore, J.W., Narahashi, T., & Shaw, T.I. (1967). An upper limit to the number of sodium channels in nerve membrane? *Journal of Physiology*, **188**, 99-105. [9]

Moore, R.L. & Stull, J.T. (1984). Myosin light chain phosphorylation in fast and slow skeletal muscles in situ. *American Journal of Physiology*, **247**, C462-C471. [11]

Mudge, A.W. (1993). Motor neurones find their factors. *Nature*, **363**, 213-214. [18]

Mullins, L.J., & Noda, K. (1964). The influence of sodium-free solutions on the membrane potential of frog muscle fibers. *Journal of General Physiology*, **47**, 117-132. [9]

Murray, M.P., Duthie, E.H., Gambert, S.T., Sepic, S.B., & Mollinger, L.A. (1985). Age related differences in knee muscle strength in normal women. *Journal of Gerontology*, **40**, 275-280. [22]

Murray, M.P., Gardner, G.M., Mollinger, L.A., & Sepic, S.B. (1980). Strength of isometric and isokinetic contractions: Knee muscles of men aged 20 to 86. *Physical Therapy*, **60**, 412-419. [22]

Musacchia, X.J., Deavers, D.R., Meininger, G.A., & Davis, T.P. (1980). A model for hypokinesia: Effects on muscle atrophy in the rat. *Journal of Applied Physiology*, **48**, 479-486. [19]

Musick, J., & Hubbard, J.I. (1972). Release of protein from mouse motor nerve terminals. *Nature*, **237**, 279-281. [8, 18]

Nabeshima, Y., Hanaoka, K., Hayasaka, M., Esumi, E., Li, S., Nonaka, I., & Nabeshima, Y. (1993). *Myogenin gene* disruption results in perinatal lethality because of severe muscle defect. *Nature*, **364**, 532-535. [5]

Natori, R. (1975). The electrical potential change of internal membrane during propagation of contraction in skinned fiber of toad skeletal muscle. *Japanese Journal of Physiology*, **25**, 51-63. [11]

Neher, E., & Sakmann, B. (1976a). Noise analysis of drug-induced voltage clamp currents in denervated frog muscle fibres. *Journal of Physiology*, **258**, 705-729. [10]

Neher, E., & Sakmann, B. (1976b). Single-channel currents recorded from membrane of denervated frog muscle fibres. *Nature*, **260**, 799-802. [7, 9]

New, H.V., & Mudge, A.W. (1986). Calcitonin gene-related protein regulates muscle acetylcholine receptor synthesis. *Nature*, **323**, 809-811. [3]

Newham, D.J., McPhail, G., Mills, K.R., & Edwards, R.H.T. (1983). Ultra-structural changes after concentric contractions of human muscles. *Journal of the Neurological Sciences*, **61**, 109-122. [21]

Newton, J.P., & Yemm, R. (1986). Changes in the contractile properties of the human first dorsal interosseous muscle with age. *Gerontology*, **32**, 98-104. [22]

Newton, J.P., Yemm, R., & McDonagh, M.J.N. (1988). Study of age changes in the motor units of the first dorsal interosseous muscle in man. *Gerontology*, **34**, 115-119. [22]

Nieuwkoop, P.D. (1973). The "organization center" of the amphibian embryo: Its origin, spatial organization, and morphogenetic action. *Advances in Morphogenesis*, **10**, 1-39. [5]

Nishizono, H., Saito, Y., & Miyashita, M. (1979). The estimation of conduction velocity in human skeletal muscle in situ with surface electrodes. *Electroencephalography and Clinical Neurophysiology*, **46**, 659-664. [9]

Nissl, F. (1892). Uber die veränderungen der Ganglienzellen am Facialiskern des kaninchens nach Aureissung der nerven. *Allgemeine Zeitschrift fur Psychiatrie*, **48**, 197-198. [18]

Niswander, L., Jeffrey, S., Martin, G.R., & Tickle, C. (1994). A positive feedback loop coordinates growth and patterning in the vertebrate limb. *Nature*, **371**, 609-612. [5]

Nixon, R.A., & Sihag, R.K. (1991). Neurofilament phosphorylation: A new look at regulation and function. *Trends in Neurosciences*, **14**, 501-506. [2, 8]

Noble, E.G., Dabrowski, B.L., & Ianuzzo, C.D. (1983). Myosin transformation in hypertrophied rat muscle. *Pflügers Archiv*, **396**, 260-262. [20]

Noda, M., Shimizu, S., Tanabe, T., Takai, T., Kayano, T., Ikeda, T., Takahashi, H., Nakayama, H., Kanaoka, Y., Minamino, N., Kangawa, K., Matsu, H., Raftery, M.A., Hirose, T., Inayama, S., Hayashida, H., Miyata, T., & Numa, S. (1984). Primary structure of *Electrophorus electricus* sodium channel deduced from cDNA sequence. *Nature*, **312**, 121-127. [3, 7, 9]

Nordstrom, M.A., & Miles, T.S. (1990). Fatigue of single motor units in human masseter. *Journal of Applied Physiology*, **68**, 26-34. [12]

Nornes, H.O., & Das, G.D. (1974). Temporal pattern of neurogenesis in spinal cord of rat. I. An autoradiographic study—Times and sites of origin and migration and settling patterns of neuroblasts. *Brain Research*, **73**, 121-138. [6]

Norris, A.H., Shock, N.W., & Wagman, I.H. (1953). Age changes in the maximum conduction velocity of motor fibers of the human ulnar nerve. *Journal of Applied Physiology*, **5**, 589-593. [22]

Norris, F.H., & Gasteiger, E.L. (1955). Action potentials of single motor units in normal muscle. *Electroencephalography and Clinical Neurophysiology*, **7**, 115-126. [13]

Nosek, T.M., Fender, K.Y., & Godt, R.E. (1987). It is diprotonated inorganic phosphate that depresses force in skinned skeletal muscle fibers. *Science*, **236**, 191-193. [15]

Nosek, T.M., Guo, N., Ginsburg, J.M., & Kolbeck, R.C. (1990). Inositol (1,4,5)triphosphate (IP_3) within diaphragm muscle increases upon depolarization. *Biophysical Journal*, **57**, 401a. [11]

Nüsslein-Volhard, C. (1991). Determination of the embryonic axes of Drosophila. *Development*, **1**(Suppl. 1), 1-10. [5]

O'Brien, R.A.D., Ostberg, A.J.C., & Vrbová, G. (1978). Observations on the elimination of polyneuronal innervation in developing mammalian skeletal muscle. *Journal of Physiology*, **282**, 571-582. [6]

Ochoa, J., Danta, G., Fowler, T.J., & Gilliatt, R.W. (1971). Nature of the nerve lesion caused by a pneumatic tourniquet. *Nature*, **233**, 265-266. [19]

Ochs, S. (1972). Fast transport of materials in mammalian nerve fibers. *Science*, **176**, 252-260. [8]

Ochs, S., & Ranish, R. (1969). Characteristics of the fast transport system in mammalian nerve fibers. *Journal of Neurobiology*, **1**, 247-261. [8]

Olson, E.N. (1990). MyoD family: A paradigm for development? *Genes & Development*, **4**, 1454-1461. [5]

Olsson, Y., & Sjöstrand, J. (1969). Origin of macrophages in Wallerian degeneration of peripheral nerves demonstrated autoradiographically. *Experimental Neurology*, **23**, 102-112. [16]

Ontko, J.A. (1986). Lipid metabolism in muscle. In A.G. Engel & B.Q. Banker (Eds.), *Myology. Basic and clinical* (pp. 697-720). New York: McGraw-Hill. [14]

Oppenheim, R.W. (1989). The neurotrophic theory and naturally occurring motoneuron death. *Trends in Neurosciences*, **12**, 252-255. [6]

Oppenheim, R.W., & Nunez, R. (1982). Electrical stimulation of hindlimb increases neuronal cell death in chick embryo. *Nature*, **295**, 57-59. [6]

Oppenheim, R.W., Prevette, D., Qin-Wei, Y., Collins, F., & MacDonald, J.C. (1991). Control of embryonic motoneuron survival in vitro by ciliary neurotrophic factor. *Science*, **251**, 1616-1618. [18]

Orlander, J., Kiessling, K.H., Larsson, L., Karlsson, J., & Aniansson, A. (1978). Skeletal muscle metabolism and ultrastructure in relation to age in sedentary men. *Acta Physiologica Scandinavica*, **104**, 249-261. [22]

Overend, T.J., Cunningham, D.A., Paterson, D.H., & Lefcoe, M.S. (1992). Thigh composition in young and elderly men determined by computed tomography. *Clinical Physiology*, **12**, 629-640. [22]

Pachter, B.R., & Eberstein, A. (1984). Neuromuscular plasticity following limb immobilization. *Journal of Neurocytology*, **13**, 1013-1025. [19]

Pachter, B.R., & Eberstein, A. (1992). Long-term effects of partial denervation on sprouting and muscle fiber area in rat plantaris. *Experimental Neurology*, **116**, 246-255. [17]

Padykula, H.A., & Gauthier, G.F. (1967). Ultrastructural features of three fiber types in the rat diaphragm. *Anatomical Record*, **157**, 296-297. [12]

Papazian, D.M., Schwarz, T.L., Tempel, B.L., Jan, Y.N., & Jan, L.Y. (1987). Cloning of genomic and complementary DNA from *Shaker*, a putative potassium channel gene from *Drosophila*. *Science*, **237**, 749-753. [7]

Paschal, B.M., Shpetner, H.S., & Vallee, R.B. (1987). MAP 1C is a microtubule-activated ATPase that translocates microtubules *in vitro* and has dynein-like properties. *Journal of Cell Biology*, **105**, 1273-1282. [8]

Patrick, J., & Lindstrom, J. (1973). Autoimmune response to acetylcholine receptor. *Science*, **180**, 871-872. [10]

Peachey, L.D. (1965). The sarcoplasmic reticulum and transverse tubules of the frog's sartorius. *Journal of Cell Biology*, **25**, 209-231. [1, 11]

Pearson, C.M., Rimer, D.G., & Mommaerts, W.F.H. (1961). A metabolic myopathy due to absence of muscle phosphorylase. *American Journal of Medicine*, **30**, 502-515. [14]

Pellegrino, C., & Franzini, C. (1963). An electron microscope study of denervation atrophy in red and white skeletal muscle fibers. *Journal of Cell Biology*, **17**, 327-349. [16]

Perkins, L.C., & Kaiser, H.L. (1961). Results of short-term isotonic and isometric exercise programs in persons over sixty. *Physical Therapy Review*, **41**, 633-635. [22]

Pernuš, F., & Eržen, I. (1991). Arrangement of fiber types within fascicles of human vastus lateralis muscle. *Muscle & Nerve*, **14**, 304-309. [12]

Person, R.S. (1974). Rhythmic activity of a group of human motoneurones during voluntary contraction of a muscle. *Electroencephalography and Clinical Neurophysiology*, **36**, 585-595. [13]

Person, R.S., & Kudina, L.P. (1972). Discharge frequency and discharge pattern of human motor units during voluntary contraction of muscle. *Electroencephalography and Clinical Neurophysiology*, **32**, 471-483. [13]

Persson, M.G., Hedqvist, P., & Gustafsson, L.E. (1991). Nerve-induced tachykinin-mediated vasodilatation in skeletal muscle is dependent on nitric oxide formation. *European Journal of Pharmacology*, **205**, 295-301. [13]

Pestronk, A., Drachman, D.B., & Griffin, J.W. (1976). Effect of botulinum toxin on trophic regulation of acetylcholine receptors. *Nature*, **264**, 787-788. [19]

Pestronk, A., Drachman, D.B., & Griffin, J.W. (1980). Effects of aging on nerve sprouting and regeneration. *Experimental Neurology*, **70**, 65-82. [22]

Petajan, J.H., & Philip, B.A. (1969). Frequency control of motor unit action potentials. *Electroencephalography and Clinical Neurophysiology*, **27**, 66-72. [13]

Peter, J.B., Barnard, R.J., Edgerton, V.R., Gillespie, C.A., & Stempel, K.E. (1972). Metabolic profiles of three fiber types of skeletal muscle in guinea pigs and rabbits. *Biochemistry*, **11**, 2627-2633. [12, 13]

Pette, D., Ramirez, B.U., Müller, W., Simon, R., Exner, G.U., & Hildebrand, R. (1975). Influence of intermittent long-term stimulation on contractile histochemical and metabolic properties of fibre populations in fast and slow rabbit muscles. *Pflügers Archiv*, **361**, 1-7. [18]

Pettigrew, F.P., & Gardiner, P.F. (1987). Changes in rat plantaris motor unit profiles with advanced age. *Mechanisms of Ageing and Development*, **40**, 243-259. [22]

Pinçon-Raymond, M., Rieger, F., Fosset, M., & Lazdunski, M. (1985). Abnormal transverse tubule system and abnormal amount of receptors for Ca^{2+} channel inhibitors of the dihydropyridine family in skeletal muscle from mice with embryonic muscular dysgenesis. *Developmental Biology*, **112**, 458-466. [11]

Pockett, S., & Slack, J.R. (1982). Source of the stimulus for nerve terminal sprouting in partially denervated muscle. *Neuroscience*, **7**, 3173-3176. [17]

Podolsky, R.J. (1964). The maximum sarcomere length for contraction of isolated myofibrils. *Journal of Physiology*, **170**, 110-123. [11]

Pollard, J.D., King, R.H.M., & Thomas, P.K. (1975). Recurrent experimental allergic neuritis. An electron microscope study. *Journal of Neurological Sciences*, **24**, 365-383. [16]

Popham, P., Band, D., & Linton, R. (1990). Potassium infusions cause release of adrenaline in anaesthetized cats. *Journal of Physiology*, **427**, 43P. [15]

Porayko, O., & Smith, R.S. (1968). Morphology of muscle spindles in the rat. *Experientia*, **24**, 588-589. [4]

Post, R.L., Kume, S., Tobin, T., Orcutt, B., & Sen, A.K. (1969). Flexibility of an active center in sodium-plus-potassium adenosine triphosphate. *Journal of General Physiology*, **54**, 306s-326s. [7]

Potter, L.T. (1970). Synthesis, storage and release of [^{14}C] acetylcholine in isolated rat diaphragm muscles. *Journal of Physiology*, **206**, 145-166. [3]

Powell, J.A., & Fambrough, D.M. (1973). Electrical properties of normal and dysgenic mouse skeletal muscle in culture. *Journal of Cell Physiology*, **82**, 21-38. [11]

Prahlad, K.V., Skala, G., Jones, D.G., & Briles, W.E. (1979). Limbless: A new genetic mutant in the chick. *Journal of Experimental Zoology*, **209**, 427-434. [5]

Provins, K.A. (1958). The effect of peripheral nerve block on the appreciation and execution of finger movements. *Journal of Physiology*, **143**, 55-67. [4]

Pumplin, D.W., Reese, T.S., & Llinas, R. (1981). Are the presynaptic membrane particles the calcium channels? *Proceedings of the National Academy of Sciences* (U.S.), **78**, 7210-7213. [10]

Purves, D., & Lichtman, J.W. (1980). Elimination of synapses in the developing nervous system. *Science*, **210**, 153-157. [6]

Purves, D., & Lichtman, J.W. (1985). *Principles of neural development* (pp. 13; 271-272). Sunderland, MA: Sinauer Associates. [6]

Purves, D., & Sakmann, B. (1974). Membrane properties underlying spontaneous activity of denervated muscle fibres. *Journal of Physiology*, **239**, 125-153. [16]

Raftery, M.A., Hunkapiller, M.W., Strader, C.D., & Hood, L.E. (1980). Acetylcholine receptor: Complex of homologous subunits. *Science*, **208**, 1454-1457. [3]

Ramsey, R.W., & Street, S.F. (1942). Absence of fatigue of the contractile mechanism in single muscle fibres. *Federation Proceedings*, **1**, 70. [15]

Ranvier, L. (1873). Propriétés et structures différents des muscles rouges et des muscles blancs, chez les lapins et chez les raies. *Comptes Rendus Hebdomadaires des Seances de l'Academie des Sciences: D. Sciences Naturelles* (Paris), **77**, 1030-1034. [12]

Rash, J.E., & Fambrough, D.M. (1973). Ultrastructural and electrophysiological correlates of cell coupling and cytoplasmic fusion during myogenesis *in vitro*. *Developmental Biology*, **30**, 166-186. [5]

Rasminsky, M., & Sears, T.A. (1972). Internodal conduction in undissected demyelinated nerve fibres. *Journal of Physiology*, **227**, 323-350. [9]

Ray, P.M., Belfall, B., Duff, C., Logan, C., Kean, V., Thompson, M.W., Sylvester, J.E., Gorski, J.L., Schmickel, R.D., & Worton, R.G. (1985). Cloning of the breakpoint of a X; 21 translocation associated with Duchenne muscular dystrophy. *Nature*, **318**, 672-675. [1]

Rayment, I., Holden, H.M., Whittaker, M., Yohn, C.B., Lorenz, M., Holmes, K.C., & Milligan, R.A. (1993). Structure of the actin-myosin complex and its implications for muscle contraction. *Science*, **261**, 58-65. [11]

Redfern, P.A. (1970). Neuromuscular transmission in newborn rats. *Journal of Physiology*, **209**, 701-709. [6]

Reichmann, H., Hoppeler, H., Mathieu-Costello, O., Von Bergen, F., & Pette, D. (1985). Biochemical and ultrastructural changes of skeletal muscle mitochondria after chronic electrical stimulation in rabbits. *Pflügers Archiv*, **404**, 1-9. [20]

Reicke, H., & Nelson, K.R. (1990). Duchenne de Boulogne: Electrodiagnosis of poliomyelitis. *Muscle & Nerve*, **13**, 56-62. [1]

Reiser, P.J., Moss, R.L., Giulian, G.G., & Geaser, M.L. (1985). Shortening velocity of single fibers from adult rabbit soleus muscles is correlated with myosin chain composition. *Journal of Biological Chemistry*, **260**, 9077-9080. [12]

Rich, M., & Lichtman, J.W. (1989). Motor nerve terminal loss from degenerating muscle fibers. *Neuron*, **3**, 677-688. [6]

Rieger, F., Bournaud, R., Shimahara, T., Garcia, L., Pinçon-Raymond, M., Romey, G., & Lazdunski, M. (1987). Restoration of dysgenic muscle contraction and calcium channel function by co-culture with normal spinal cord neurons. *Nature*, **330**, 563-566. [11]

Rifenberick, D.H., Gamble, J.G., & Max, S.R. (1973). Response of mitochondrial enzymes to increased muscular activity. *American Journal of Physiology*, **225**, 1295-1299. [19]

Ringer, S. (1883). A further contribution regarding the influence of different constituents of the blood on the contraction of the heart. *Journal of Physiology*, **4**, 29-42. [11]

Ríos, E., Ma, J.J., & González, A. (1991). The mechanical hypothesis of excitation-contraction (EC) coupling in skeletal muscle. *Journal of Muscle Research and Cell Motility*, **12**, 127-135. [11]

Ritchie, J.M., & Rogart, R.B. (1977). The binding of saxitoxin and tetrodotoxin to excitable tissue. *Reviews of Physiology, Biochemistry and Pharmacology*, **79**, 1-50. [7]

Robbins, N., & Yonezawa, T. (1971). Physiological studies during formation and development of rat neuromuscular junctions in tissue culture. *Journal of General Physiology*, **58**, 467-481. [6]

Robert, E.D., & Oester, Y.T. (1970). Absence of supersensitivity to acetylcholine in innervated muscle subjected to a prolonged pharmacologic block. *Journal of Pharmacology and Experimental Therapeutics*, **174**, 133-140. [19]

Robinson, G.A., Enoka, R.M., & Stuart, D.G. (1991). Immobilization-induced change in motor unit force and fatigability in the cat. *Muscle & Nerve*, **14**, 563-573. [16, 19]

Romanes, G.J. (1941). The development and significance of the cell columns in the ventral horn of the cervical and upper thoracic spinal cord of the rabbit. *Journal of Anatomy*, **76**, 112-130. [6]

Romanes, G.J. (1951). The motor-cell columns of the lumbosacral cord of the cat. *Journal of Comparative Neurology*, **94**, 313-364. [2, 6]

Romanul, F.C.A., & Hogan, E.L. (1965). Enzymatic changes in denervated muscle. I. Histological studies. *Archives of Neurology*, **13**, 263-273. [16]

Romanul, F.C.A., & Van Der Meulen, J.P. (1966). Reversal of the enzyme profiles of muscle fibres in fast and slow muscles by cross-innervation. *Nature*, **212**, 1369-1370. [17]

Romanul, F.C.A., & Van Der Meulen, J.P. (1967). Slow and fast muscles after cross innervation. Enzymatic and physiological changes. *Archives of Neurology*, **17**, 387-402. [18]

Rose, P.K., & Tourond, J. (1993). Structural remodelling of the dendritic trees of cat spinal motoneurons following permanent axotomy. *Physiology Canada*, **24**, 135. [18]

Rosenbleuth, A., & Dempsey, E.W. (1939). A study of Wallerian degeneration. *American Journal of Physiology*, **128**, 19-30. [16]

Rosenbleuth, J. (1974). Structure of amphibian motor endplate. Evidence for a granular component projecting from the outer surface of the receptive membrane. *Journal of Cell Biology*, **62**, 755-766. [3]

Rosenheimer, J.L. (1990). Factors affecting denervation-like changes at the neuromuscular junction during aging. *International Journal of Developmental Neuroscience*, **8**, 643-654. [22]

Rostami, A.M. (1993). Pathogenesis of immune-mediated neuropathics. *Pediatric Research*, **33**(Suppl.), S90-S94. [16]

Rotto, D.M., & Kaufman, M.P. (1988). Effect of metabolic products of muscular contraction on discharge of group III and IV afferents. *Journal of Applied Physiology*, **64**, 2306-2313. [15]

Round, J.M., Jones, D.A., Chapman, S.J., Edwards, R.H.T., Ward, P.S., & Fodden, D.L. (1984). The anatomy and fibre type composition of the human adductor pollicis in relation to its contractile properties. *Journal of the Neurological Sciences*, **66**, 263-292. [12]

Rowe, R.W.D. (1981). Morphology of perimysial and endomysial connective tissue in skeletal muscle. *Tissue & Cell*, **13**, 681-690. [1]

Rowe, R.W.D. (1969). The effect of senility on skeletal muscles in the mouse. *Experimental Gerontology*, **4**, 119-126. [22]

Roy, N., Mahadevan, M.S., McLean, J.M., Shutter, G., Yaraghi, Z., Farahani, R., Baird, S., Besner-Johnston, A., Lefebvre, C., Kang, X., Salih, M., Aubry, H., Tamai, K., Guan, X., Ioannou, P., Crawford, T.O., de Jong, P.J., Surh, L., Ikeda, J.-E., Korneluk, R.G., & MacKenzie, A. (1995). The gene for neuronal apoptosis inhibitory protein is partially deleted in individuals with spinal muscular atrophy. *Cell*, **80**, 167-178. [6]

Rubinstein, L.J. (1960). Ageing changes in muscles. In G.H. Bourne (Ed.), *The structure and function of muscle* (Vol. 3, pp. 209-226). New York: Academic Press. [22]

Rüdel, R., Ricker, K., & Lehmann-Horn, F. (1993). Genotype-phenotype correlations in human skeletal muscle sodium channel diseases. *Archives of Neurology*, **50**, 1241-1248. [7]

Rüdel, R., & Taylor, S.R. (1973). Aqueorin luminescence during contraction of amphibian skeletal muscle. *Journal of Physiology*, **233**, 5-6P. [11]

Rudnicki, M.A., Braun, T., Hinuma, S., & Jaenisch, R. (1992). Inactivation of Myo D in mice leads to up-regulation of the myogenic HLH gene myf-5 and results in apparently normal muscle development. *Cell*, **71**, 383-390. [5]

Rudy, B. (1988). Diversity and ubiquity of K channels. *Neuroscience*, **25**, 729-749. [7, 15]

Ruiz i Altaba, A., & Melton, D.A. (1990). Axial patterning and the establishment of polarity in the frog embryo. *Trends in Genetics*, **6**, 57-64. [5]

Rutherford, O.M., & Jones, D.A. (1986). The role of learning and coordination in strength training. *European Journal of Applied Physiology*, **55**, 100-105. [20]

Rutherford, O.M., & Jones, D.A. (1988). Contractile properties and fatigability of the human adductor pollicis and first dorsal interosseus: A comparison of the effects of two chronic stimulation patterns. *Journal of the Neurological Sciences*, **85**, 319-331. [20]

Sadoshima, J.-I., & Izumo, S. (1993). Mechanical stretch rapidly activates multiple signals transduction pathways in cardiac myocytes: Potential involvement of an autocrine/paracrine mechanism. *EMBO Journal*, **12**, 1681-1692. [20]

St. Johnston, D., & Nüsslein-Volhard, C. (1992). The origin of pattern and polarity in the *Drosophila* embryo. *Cell*, **68**, 201-219. [5]

St.-Pierre, D., & Gardiner, P.F. (1985). Effect of "disuse" on mammalian fast-twitch muscle: Joint fixation compared with neurally applied tetrodotoxin. *Experimental Neurology*, **90**, 635-651. [19]

St.-Pierre, D., & Gardiner, P.F. (1987). The effect of immobilization and exercise on muscle function: A review. *Physiotherapy Canada*, **39**, 24-36. [19]

Sale, D.G. (1988). Neural adaptation to resistance training. *Medicine and Science in Sports and Exercise*, **20**, S135-S145. [20]

Sale, D.G., McComas, A.J., MacDougall, J.D., & Upton, A.R.M. (1982). Neuromuscular adaptations in human thenar muscles following strength training and immobilization. *Journal of Applied Physiology*, **53**, 419-424. [19]

Sale, D.G., Quinlan, J., Marsh, E., McComas, A.J., & Bélanger, A.Y. (1982). Influence of joint position on ankle plantarflexion in humans. *Journal of Applied Physiology*, **52**, 1636-1642. [11, 12]

Salkoff, L. (1983). Genetic and voltage-clamp analysis of a *Drosophila* potassium channel. *Cold Spring Harbor Symposia on Quantitative Biology*, **48**, 221-231. [7]

Salmons, S., & Vrbová, G. (1969). The influence of activity on some contractile characteristics of mammalian fast and slow muscles. *Journal of Physiology*, **201**, 535-549. [18, 20]

Saltin, B., Henriksson, J., Nygaard, E., & Anderson, P. (1977). Fiber types and metabolic potentials of skeletal muscles in sedentary man and endurance runners. *Annals of the New York Academy of Science*, **301**, 3-29. [20]

Sanders, F.K., & Young, J.Z. (1946). The influence of peripheral connexion on the diameter of regenerating nerve fibres. *Experimental Biology*, **22**, 203-212. [18]

Sandow, A. (1965). Excitation-contraction coupling in skeletal muscle. *Pharmacological Reviews*, **17**, 265-320. [11]

Sanes, J.R., Marshall, L.M., & McMahan, U.J. (1978). Reinnervation of muscle fiber basal lamina after removal of myofibers. Differentiation of regenerating axons at original synaptic sites. *Journal of Cell Biology*, **78**, 176-198. [3, 17]

Sargeant, A.J., Davies, C.T.M., Edwards, R.H.T., Maunder, C., & Young, A. (1977). Functional and structural changes after disuse of human muscle. *Clinical Science and Molecular Medicine*, **52**, 337-342. [19]

Sargeant, A.J., Dolan, P., & Young, A. (1984). Optimal velocity for maximal short-term anaerobic power output in cycling. *International Journal of Sports Medicine*, **5**, 124-125. [11]

Schachat, F.H., Bronson, D.D., & McDonald, O.B. (1985). Heterogeneity of contractile proteins. A continuum of troponin-tropomyosin expression in mammalian skeletal muscle. *Journal of Biological Chemistry*, **260**, 1108-1113. [12]

Schantz, P., Randall-Fox, E., & Hutchison, W. (1983). Muscle fibre type distribution, muscle cross-sectional area and maximal voluntary strength in humans. *Acta Physiologica Scandinavica*, **117**, 219-226. [12]

Schantz, P.G., & Dhoot, G.K. (1987). Coexistence of slow and fast isoforms of contractile and regulatory proteins in human skeletal muscle fibres induced by endurance training. *Acta Physiologica Scandinavica*, **131**, 147-154. [20]

Schantz, P.G., & Henriksson, J. (1983). Increases in myofibrillar ATPase intermediate human skeletal muscle fibers in response to endurance training. *Muscle & Nerve*, **6**, 553-556. [20]

Schatzmann, H.J. (1989). The calcium pump of the surface membrane and of the sarcoplasmic reticulum. *Annual Reviews of Physiology*, **51**, 473-485. [7]

Schiaffino, S., Gorza, L., Ausoni, S., Bottinelli, R., Reggiani, C., Larson, L., Edström, L., Gundersen, K., & Lømo, T. (1990). Muscle fiber types expressing different myosin heavy chain isoforms. Their functional properties and adaptive capacity. In D. Pette (Ed.), *The dynamic state of muscle fibres* (pp. 329-341). Berlin: Walter de Gruyter. [12]

Schiaffino, S., Hanzlíková, V., & Pierobon, S. (1970). Relations between structure and function in rat skeletal muscle fibers. *Journal of Cell Biology*, **47**, 107-119. [12]

Schiavo, G., Benefinati, F., Poulain, B., Rossetto, O., Polverino de Laureto, P., Das Gupts, B.R., & Montecucco, C. (1992). Tetanus and botulinum-B neurotoxins block neurotransmitter release by proteolytic cleavage of synaptobrevin. *Nature*, **359**, 832-835. [10]

Schmalbruch, H., Al-Amood, W.S., & Lewis, D.M. (1991). Morphology of long-term denervated rat soleus muscle and the effect of chronic electrical stimulation. *Journal of Physiology*, **441**, 233-241. [16]

Schmid, S.L., Braell, W.A., Schossman, D.M., & Rothman, J.E. (1984). A role for clathrin light chains in the recognition of clathrin cages for "uncoating ATPase." *Nature*, **311**, 228-231. [3]

Schmidt, E.M., & Thomas, J.S. (1981). Motor unit recruitment order: Modification under volitional control. In J.E. Desmedt (Ed.), *Progress in neurophysiology* (Vol. 9, pp. 145-148). Basel: Karger. [13]

Schonberger, L.B., Bregman, D.J., Sullivan-Bolyai, J.Z., Keenlyside, R.A., Ziegler, D.W., Retailliau, H.F., Eddins, D.L., & Bryan, J.A. (1979). Guillain-Barré syndrome following vaccination in the national influenza immunization program, United States, 1976-1977. *American Journal of Epidemiology*, **110**, 105-123. [16]

Schröder, J.M. (1972). Altered ratio between axon diameter and myelin sheath thickness in regenerated nerve fibres. *Brain Research*, **45**, 49-65. [17]

Schulze, M.L. (1955). Die absolute und relative Zahl der Muskelspindeln in den Kurzen Daumenmuskeln des Menschen. *Anatomischer Anzeiger*, **102**, 290-291. [4]

Scott, S.A. (1977). Maintained function of foreign and appropriate junctions on reinnervated goldfish extraocular muscles. *Journal of Physiology*, **268**, 87-109. [17]

Seedorf, U., Leberer, E., Kirschbaum, B.J., & Pette, D. (1986). Neural control of gene expression in skeletal muscle. Effects of chronic stimulation on lactate dehydrogenase isoenzymes and citrate synthase. *Biochemical Journal*, **239**, 115-120. [20]

Sejersted, O.M., Vøllestad, N.K., & Medbø, J.I. (1986). Muscle fluid and electrolyte balance during and following exercise. *Acta Physiologica Scandinavica*, **128**(Suppl. 556), 119-127. [15]

Sendtner, M., Kreutzberg, G.W., & Thoenen, H. (1990). Ciliary neurotrophic factor prevents the degeneration of motor neurons after axotomy. *Nature*, **345**, 440-441. [18]

Sendtner, M., Schmalbruch, H., Stöckli, K.A., Carroll, P., Kreutzberg, G.W., & Thoenen, H. (1992). Ciliary neurotrophic factor prevents degeneration of motor neurons in mouse mutant progressive motor neuronopathy. *Nature*, **358**, 502-504. [18]

Serafini, T., Kennedy, T.E., Galko, M.J., Mirzayan, C., Jessell, T.M., & Tessier-Lavigne, M. (1994). The netrins define a family of axon outgrowth-promoting proteins homologous to C. elegans UNC-6. *Cell*, **78**, 409-424. [6]

Serratrice, G., Roux, H., & Aquaron, R. (1968). Proximal muscle weakness in elderly subjects. Report of 12 cases. *Journal of the Neurological Sciences*, **7**, 275-299. [22]

Shafiq, S.A., Lewis, S.G., Dimino, L.C., & Schutta, H.S. (1978). Electron microscopic study of skeletal muscle in elderly subjects. In G. Kaldor & W.J. Battista (Eds.), *Aging in muscle* (pp. 68-85). New York: Raven. [22]

Shainberg, A., Yagil, G., & Yaffe, D. (1969). Control of myogenesis *in vitro* by Ca^{2+} concentration in nutritional medium. *Experimental Cell Research*, **58**, 163-167. [5]

Sheetz, M.P., & Spudich, J. (1983). Movement of myosin-coated fluorescent beads on actin cables *in vitro*. *Nature*, **303**, 31-45. [11]

Sheetz, M.P., Steuer, E.R., & Schroer, T.A. (1989). The mechanism and regulation of fast axonal transport. *Trends in Neurosciences*, **12**, 474-479. [8]

Shelanski, M.L., & Wisniewski, H. (1969). Neurofibrillary degeneration: Induced by vincristine therapy. *Archives of Neurology*, **20**, 199-206. [8]

Shephard, R.J. (1969). The working capacity of the older employee. *Archives of Environmental Health*, **18**, 982-986. [22]

Sherrington, C.S. (1900). The muscular sense. In E.A. Schrafer, *Textbook of physiology* (Vol. 2, pp. 1002-1025). Edinburgh: Pentland. [4]

Sherrington, C.S. (1929). Some functional problems attaching to convergence. *Proceedings of the Royal Society of London, Series B*, **105**, 332-362. [2, 8, 12]

Shorey, M.L. (1909). The effect of the destruction of peripheral areas on the differentiation of the neuroblasts. *Journal of Experimental Zoology*, **7**, 25-63. [6]

Sica, R.E.P., & McComas, A.J. (1971). Fast and slow twitch units in a human muscle. *Journal of Neurology, Neurosurgery and Psychiatry*, **34**, 113-120. [12]

Sica, R.E.P., McComas, A.J., Upton, A.R.M., & Longmire, D. (1974). Estimations of motor units in small muscles of the hand. *Journal of Neurology, Neurosurgery and Psychiatry*, **37**, 55-67. [22]

Sica, R.E.P., Sanz, O.P., & Colombi, A. (1976). The effects of ageing upon the human soleus muscle. *Medicina (Buenos Aires)*, **36**, 443-446. [22]

Siddique, T., Ficlewicz, D.A., Pericak-Vance, M.A., Haines, J.L., Rouleau, G., Jeffers, A.J., Sapp, P., Hung, W.-Y., Bebout, J., McKenna-Yasek, D., Deng, G., Horvitz, H.R., Gusella, J.F., Brown, R.H., Roses, A.D., and others. (1991). Linkage of a gene causing familial amyotrophic lateral sclerosis to chromosome 21 and evidence of genetic-locus heterogeneity. *New England Journal of Medicine*, **324**, 1381-1384. [22]

Sigworth, F.J., & Neher E. (1980). Single Na^+ channel currents observed in cultured rat muscle cells. *Nature*, **287**, 447-449. [7]

Silinsky, E.M., & Hubbard, J.I. (1973). Release of ATP from rat motor nerve terminals. *Nature*, **243**, 404-405. [10]

Simoneau, J.A., Lortie, G., Bonlay, M.R., Marcotte, C.M., Thibault, M.C., & Bouchard, C. (1985). Human skeletal muscle fibre type alteration with high-intensity intermittent training. *European Journal of Applied Physiology*, **54**, 250-253. [20]

Sjøgaard, G. (1986). Water and electrolyte fluxes during exercise and their relation to muscle fatigue. *Acta Physiologica Scandinavica*, **128**(Suppl. 556), 129-136. [15]

Sjøgaard, G., Adams, R.P., & Saltin, B. (1985). Water and ion shifts in skeletal muscle of humans with intense dynamic knee extension. *American Journal of Physiology*, **248**, R190-196. [15]

Skene, J.H.P. (1989). Axonal growth-associated proteins. *Annual Review of Neuroscience*, **12**, 127-156. [18]

Slack, J.R., & Pockett, S. (1982). Motor neurotrophic factor in denervated muscle. *Brain Research*, **247**, 138-140. [17]

Slater, C.R., Lyons, P.R., Walls, T.J., Fawcett, P.R.W., & Young, C. (1992). Structure and function of neuromuscular junctions in the *vastus lateralis* of man. A motor point biopsy study of two groups of patients. *Brain*, **115**, 451-478. [10]

Slomic, A., Rosenfalck, A., & Buchthal, F. (1968). Electrical and mechanical responses of normal and myasthenic muscle with particular reference to the staircase phenomenon. *Brain Research*, **10** (special issue), 1-78. [11, 12]

Small, D.H. (1990). Non-cholinergic actions of acetylcholinesterases: Proteases regulating cell growth and development? *Trends in Biochemical Sciences*, **15**, 213-216. [3]

Smith, D.O. (1984). Acetylcholine storage, release and leakage at the neuromuscular junction of mature adult and aged rats. *Journal of Physiology*, **347**, 161-176. [22]

Smith, J.C., Price, B.M.J., Van Nimmen, K., & Huylebroeck, D. (1990). Identification of a potent *Xenopus* mesoderm-inducing factor as a homologue of activin A. *Nature*, **345**, 729-731. [5]

Smith, J.L., Betts, B., Edgerton, V.R., & Zernicke, R.F. (1980). Rapid ankle extension during paw shakes: Selective recruitment of fast ankle extensors. *Journal of Neurophysiology*, **43**, 612-620. [13]

Smith, R.S. (1966). Properties of intrafusal muscle fibres. In R. Granit (Ed.), *Muscular afferents and motor control* (pp. 69-80). Stockholm: Almqvist & Wiksell. [4]

Soderberg, G.L., Minor, S.D., & Nelson, R.M. (1991). A comparison of motor unit behaviour in young and aged subjects. *Age/Ageing*, **20**, 8-15. [22]

Sola, O.M., Christensen, D.L., & Martin, A.W. (1973). Hypertrophy and hyperplasia in adult chicken anterior latissimus dorsi muscle following stretch with and without denervation. *Experimental Neurology*, **41**, 76-100. [20]

Solandt, D.Y. (1936). The effect of potassium on the excitability and resting metabolism of frog's muscle. *Journal of Physiology*, **86**, 162-170. [15]

Söllner, T., Whiteheart, S.W., Brummer, M., Erdjument-Bromage, H., Geromanos, S., Tempst, P., & Rothman, J.E. (1993). SNAP receptors implicated in vesicle targeting and fusion. *Nature*, **362**, 318-324. [10]

Spemann, H., & Mangold, H. (1924). Uber induction von embryonalagen durch implantation artfremder organisatoren. *Roux's Archives of Developmental Biology*, **100** 599-638. [6]

Spencer, P.S. (1972). Reappraisal of the model for "bulk axoplasmic flow." *Nature (New Biology)*, **240**, 283-285. [8]

Spudich, J.A. (1994). How molecular motors work. *Nature*, **372**, 515-518. [11]

Spudich, J.A., Kron, S.J., & Sheetz, M.P. (1985). Movement of myosin-coated beads on oriented filaments reconstituted from purified actin. *Nature*, **315**, 584-586. [11]

Sreter, F.A., Gergely, J., Salmons, S., & Romanul, F. (1973). Synthesis by fast muscle of myosin light chains characteristic of slow muscle in response to long-term stimulation. *Nature (New Biology)*, **241**, 17-18. [20]

Stacey, M.J. (1969). Free nerve endings in skeletal muscle of the cat. *Journal of Anatomy*, **105**, 231-254. [4]

Stålberg, E. (1966). Propagation velocity in human muscle fibres *in situ*. *Acta Physiologica Scandinavica*, **70**(Suppl. 287), 1-112. [9]

Stålberg, E., & Antoni, L. (1980). Electrophysiological cross section of the motor unit. *Journal of Neurology, Neurosurgery and Psychiatry*, **43**, 469-474. [12]

Stålberg, E., Borges, O., Ericsson, M., Essén-Gustavsson, B., Fawcett, P.R.W., Nordesjö, L.O., Nordgren, B., & Uhlin, R. (1989). The quadriceps femoris muscle in 20–70-year-old subjects: Relationship between knee extension torque, electrophysiological parameters, and muscle fiber characteristics. *Muscle & Nerve*, **12**, 382-389. [22]

Stålberg, E., & Fawcett, P.R.W. (1982). Macro EMG in healthy subjects of different ages. *Journal of Neurology, Neurosurgery and Psychiatry*, **45**, 870-878. [22]

Stålberg, E., & Trontelj, J.V. (1979). *Single fibre electromyography*. Surrey: Mirvalle Press. [22]

Stanley, E.F., & Drachman, D.B. (1978). Effect of myasthenic immunoglobulin on acetylcholine receptors of intact mammalian neuromuscular junctions. *Science*, **200**, 1285-1287. [10]

Staron, R.S., Malicky, E.S., Leonard, M.J., Falkel, J.E., Hagerman, F.C., & Dudley, G.A. (1990). Muscle hypertrophy and fast fiber type conversions in heavy resistance–trained women. *European Journal of Physiology and Occupational Physiology*, **60**, 71-79. [20]

Staron, R.S., & Pette, D. (1990). The multiplicity of myosin light and heavy chain combinations in muscle fibers. In D. Pette (Ed.), *The dynamic state of muscle fibres* (pp. 315-328). Berlin: Walter de Gruyter. [12]

Steinbach, J.H., Schubert, D., & Eldridge, L. (1980). Changes in cat muscle contractile proteins after prolonged muscle inactivity. *Experimental Neurology*, **67**, 655-669. [19]

Steinmeyer, K., Lorenz, C., Pusch, M., Koch, M.C., & Jentsch, T.J. (1994). Multimeric structure of Cl C-1 chloride channel revealed by mutations in dominant myotonia congenita (Thomsen). *Embo Journal*, **13**, 737-743. [7]

Steinmeyer, K., Ortland, C., & Jentsch, T.J. (1991). Primary structure and functional expression of a developmentally regulated skeletal muscle chloride channel. *Nature*, **354**, 301-303. [7]

Stephens, J.A., & Stuart, D.G. (1975). The motor units of cat medial gastrocnemius: Speed-size relations and their significance for the recruitment order of motor units. *Brain Research*, **91**, 177-195. [13]

Stephens, J.A., & Taylor, A. (1972). Fatigue of maintained voluntary contraction in man. *Journal of Physiology*, **220**, 1-18. [15]

Stephens, J.A., & Usherwood, T.P. (1977). The mechanical properties of human motor units with special reference to their fatigability and recruitment threshold. *Brain Research*, **125**, 91-97. [12]

Stevens, C.F. (1991). Making a submicroscopic hole in one. *Nature*, **349**, 657-658. [7]

Stickland, N.C. (1981). Muscle development in the human faetus as exemplified by m. sartorius: A quantitative study. *Journal of Anatomy*, **132**, 557-579. [5]

Stonnington, H.H., & Engel, A.G. (1973). Normal and denervated muscle. A morphometric study of fine structure. *Neurology*, **23**, 714-724. [16]

Strecker, T.R., & Stephens, T.D. (1983). Peripheral nerves do not play a trophic role in limb skeletal morphogenesis. *Teratology*, **27**, 159-167. [5]

Street, S.F. (1983). Lateral transmission of tension in frog myofibers: A myofibrillar network and transverse cytoskeletal connections are possible transmitters. *Journal of Cellular Physiology*, **114**, 346-364. [1]

Street, S.F., & Ramsey, R.W. (1965). Sarcolemma: Transmitter of active tension in frog skeletal muscle. *Science*, **149**, 1379-1380. [1]

Strehler, E.E., Carlsson, E., Eppenberger, H.M., & Thornell, L.-E. (1983). Ultrastructural localization of M-band proteins in chicken breast muscle as revealed by combined immunocytochemistry and ultramicrotomy. *Journal of Molecular Biology*, **166**, 141-158. [1]

Strickholm, A. (1974). Intracellular generated potentials during excitation coupling in muscle. *Journal of Neurobiology*, **5**, 161-187. [11]

Stuart, A., McComas, A.J., Goldspink, G., & Elder, G. (1981). Electrophysiological features of muscle regeneration. *Experimental Neurology*, **74**, 148-159. [21]

Studitsky, A.N. (1952). [The restoration of muscle by means of transplantation of minced muscle tissue] (Russian). *Dolk. Akad. Nauk SSSR*, **84**, 389-392. [21]

Suarez-Isla, B.A., Orozco, C., Heller, P.F., & Froehlich, J.P. (1986). Single calcium channels in native sarcoplasm reticulum membranes from skeletal muscle. *Proceedings of the National Academy of Sciences* (U.S.), **83**, 7741-7745. [11]

Sulik, K.K., & Dehart, D.B. (1988). Retinoic-acid-induced limb malformations resulting from apical ectodermal ridge cell death. *Teratology*, **37**, 527-537. [5]

Summerbell, D., & Maden, M. (1990). Retinoic acid, a developmental signalling molecule. *Trends in Neurosciences*, **13**, 142-147. [5]

Sunderland, S. (1947). Rate of regeneration in human, peripheral nerves. Analysis of the interval between injury and onset of recovery. *Archives of Neurology and Psychiatry*, **58**, 251-295. [17]

Sunderland, S., & Ray, L.J. (1950). Denervation changes in mammalian striated muscle. *Journal of Neurology, Neurosurgery and Psychiatry*, **13**, 159-177. [16]

Suter, U., Welcher, A.A., & Snipes, G.J. (1993). Progress in the molecular understanding of hereditary peripheral neuropathies reveals new insights into the biology of the peripheral nervous system. *Trends in Neuroscience*, **16**, 50-56. [2]

Suzuki, H., Suzimoto, H.T., Ishiko, T., Kasuga, N., Taguchi, S., & Ishihari, A. (1991). Effect of endurance training on the oxidative enzyme activity of soleus motoneurons in rats. *Acta Physiologica Scandinavica*, **143**, 127-128. [20]

Svoboda, K., Schmidt, C.F., Schapp, B.J., & Black, S.M. (1993). Direct observation of kinesin stepping by optical trapping interferometry. *Nature*, **365**, 721-727. [8]

Swett, J.E., & Schoultz, T.W. (1975). Mechanical transduction in the Golgi tendon organ. A hypothesis. *Archives Italiennes de Biologie*, **113**, 374-382. [4]

Tabary, J.C., Tabary, C., Tardieu, C., Tardieu, G., & Goldspink, G. (1972). Physiological and structural changes in the cat's soleus muscle due to immobilization at different lengths by plaster casts. *Journal of Physiology*, **224**, 231-244. [5, 11, 19]

Tanabe, T., Beam, K.G., Powell, J.A., & Numa, S. (1988). Restoration of excitation-contraction coupling and slow calcium current in dysgenic muscle by dihydropyridine receptor complementary DNA. *Nature*, **336**, 134-139. [11]

Tanji, J., & Kato, M. (1972). Discharge of single motor units at voluntary contraction of abductor digiti minimi muscle in man. *Brain Research*, **45**, 590-593. [13]

Tanji, J., & Kato, M. (1973). Firing rate of individual motor units in voluntary contraction of abductor digiti minimi muscle in man. *Experimental Neurology*, **40**, 771-783. [13]

Tanouye, M.A., & Ferris, A. (1985). Action potentials in normal and *Shaker* mutant *Drosophila*. *Journal of Neurogenetics*, **2**, 253-271. [7]

Taormino, J.P., & Fambrough, D.M. (1990). Pre-translational regulation of the $(Na^+ + K^+)$-ATPase in response to demand for ion transport in cultured chicken skeletal muscle. *Journal of Biological Chemistry*, **265**, 4116-4123. [7]

Thesleff, S. (1960). Supersensitivity of skeletal muscle produced by botulinum toxin. *Journal of Physiology*, **151**, 598-607. [19]

Thesleff, S. (1963). Spontaneous electrical activity in denervated rat skeletal muscle. In E. Gutmann & P. Hník (Eds.), *The effect of use and disuse on neuromuscular function* (pp. 41-51). Prague: Czechoslovak Academy of Sciences. [16]

Thesleff, S., Molgó, J., & Tågerud, S. (1990). Trophic interrelations at the neuromuscular junction as revealed by the use of botulinum neurotoxins. *Journal de Physiologie*, **84**, 167-173. [18, 19]

Thesleff, S., & Ward, M.R. (1975). Studies on the mechanism of fibrillation potentials in denervated muscle. *Journal of Physiology*, **244**, 313-323. [16]

Thimm, F., & Baum, K. (1987). Response of chemosensitive nerve fibers of group III and IV to metabolic changes in rat muscles. *Pflügers Archiv*, **410**, 143-152. [4]

Thoenen, H. (1991). The changing scene of neurotrophic factors. *Trends in Neurosciences*, **14**, 165-170. [18]

Thomas, C.K., Bigland-Ritchie, B., Westling, G., & Johansson, R.S. (1990). A comparison of human thenar motor unit properties studied by intraneural motor-axon stimulation and spike-triggered averaging. *Journal of Neurophysiology*, **64**, 1347-1351. [12]

Thomas, C.K., Johansson, R.S., Westling, G., & Bigland-Ritchie, B. (1990). Twitch properties of human thenar motor units measured in response to intraneural motor-axon stimulation. *Journal of Neurophysiology*, **64**, 1339-1346. [12]

Thomas, C.K., Ross, B.H., & Calancie, B. (1987). Human motor-unit recruitment during isometric contractions and repeated dynamic movements. *Journal of Neurophysiology*, **57**, 311-324. [12]

Thomas, C.K., Ross, B.H., & Stein, R.B. (1986). Motor-unit recruitment in human first dorsal interosseous muscle for static contractions in three different directions. *Journal of Neurophysiology*, **55**, 1017-1029. [12]

Thomas, C.K., Stein, R.B., Gordon, T., Lee, R.G., & Elleker, M.G. (1987). Patterns of reinnervation and motor unit recruitment in human hand muscles after complete ulnar and median nerve section and resuture. *Journal of Neurology, Neurosurgery and Psychiatry*, **250**, 259-268. [17]

Thomas, C.K., Woods, J.J., & Bigland-Ritchie, B. (1989). Impulse propagation and muscle activation in long voluntary contractions. *Journal of Applied Physiology*, **67**, 1835-1842. [15]

Thomas, P.K. (1970). The cellular response to nerve injury. 3. The effect of repeated crush injuries. *Journal of Anatomy*, **106**, 463-470. [17]

Thomas, R.C. (1972). Intracellular sodium activity and the sodium pump in snail neurons. *Journal of Physiology*, **220**, 55-71. [15]

Thomsen, G., Woolf, T., Whitman, M., Sokol, S., Vaughan, J., Vale, W., & Melton, D.A. (1990). Activins are expressed early in *Xenopus* embryogenesis and can induce axial mesoderm and anterior structures. *Cell*, **63**, 485-493. [5]

Thomsen, P., & Luco, J.V. (1944). Changes in weight and neuromuscular transmission in muscles of immobilized joints. *Journal of Neurophysiology*, **7**, 246-251. [20]

Tibes, U., Hemmer, B., Schweigart, U., Bóning, D., & Fotescu, D. (1974). Exercise acidosis as cause of electrolyte changes in femoral venous blood of trained and untrained man. *Pflügers Archiv*, **347**, 145-158. [7]

Tidball, J.G. (1983). The geometry of actin filament-membrane interactions can modify adhesive strength of the myotendinous junction. *Cell Motility*, **3**, 439-447. [1]

Timpe, L.C., Schwarz, T.L., Tempel, B.L., Papazian, D.M., Jan, Y.N., & Jan, L.Y. (1988). Expression of functional potassium channels from *Shaker* cDNA in *Xenopus* oocytes. *Nature*, **331**, 143-145. [7]

Timson, B.F. (1990). Evaluation of animal models for the study of exercise-induced muscle enlargement. *Journal of Applied Physiology*, **69**, 1935-1945. [20]

Timson, B.F., Bowlin, B.K., Dudenhoeffer, G.A., & George, J.B. (1985). Fiber number, area, and composition of mouse soleus muscle following enlargement. *Journal of Applied Physiology*, **58**, 619-624. [20]

Titmus, M.J., & Faber, D.S. (1990). Axotomy-induced alterations in the electrophysiological characteristics of neurons. *Progress in Neurobiology*, **35**, 1-51. [18]

Todd, I.C., Wosornu, D., Stewart, I., & Wild, T. (1992). Cardiac rehabilitation following myocardial infarction. A practical approach. *Sports Medicine*, **14**, 243-259. [20, 22]

Tomanek, R.J., & Lund, D.D. (1974). Degeneration of different types of muscle fibers. II. Immobilization. *Journal of Anatomy*, **118**, 531-541. [19]

Tomlinson, B.E., & Irving, D. (1977). The number of limb motor neurons in the human lumbosacral cord throughout life. *Journal of the Neurological Sciences*, **34**, 213-219. [22]

Tomlinson, B.E., Walton, J.N., & Rebeiz, J.J. (1969). The effects of ageing and of cachexia upon skeletal muscle. *Journal of the Neurological Sciences*, **9**, 321-346. [22]

Tomonaga, M. (1977). Histochemical and ultrastructural changes in senile human skeletal muscle. *Journal of the American Geriatrics Society*, **25**, 125-131. [22]

Tonge, D.A. (1974). Physiological characteristics of re-innervation of skeletal muscle in the mouse. *Journal of Physiology*, **241**, 141-153. [17]

Toop, J. (1975). The histochemical development of human skeletal muscle and its motor innervation. In W.G. Bradley, D. Gardner-Medwin, & J.N. Walton (Eds.), *Recent advances in myology* (pp. 322-329). Amsterdam: Excerpta Medica. [5, 6]

Tower, S. (1939). The reaction of muscle to denervation. *Physiological Reviews*, **19**, 1-48. [16]

Tower, S.S. (1937). Trophic control of non-nervous tissues by the nervous system: A study of muscle and bone innervated from an isolated and quiescent region of spinal cord. *Journal of Comparative Neurology*, **67**, 241-267. [19]

Toyoshima, C., & Unwin, N. (1988). Ion channel of acetylcholine receptor reconstructed from images of post synaptic membranes. *Nature*, **336**, 247-250. [3]

Trimmer, J.S., Cooperman, S.S., Tomiko, S.A., Zhou, J., Crean, S.M., Boyle, M.B., Kallen, R.G., Sheng, Z., Barchi, R.L., Sigworth, F.J., Goodman, R.H., Agnew, W.S., & Mandel, G. (1989). Primary structure and functional expression of a mammalian skeletal muscle sodium channel. *Neuron*, **3**, 33-49. [7]

Trinick, J. (1991). Elastic filaments and giant proteins in muscle. *Current Opinion in Cell Biology*, **3**, 112-119. [1]

Tuffery, A.R. (1971). Growth and degeneration of motor end-plates in normal cat hind limb muscles. *Journal of Anatomy*, **110**, 221-247. [22]

Turner, P.R., Westwood, T., Regen, C.M., & Steinhardt, R.A. (1988). Increased protein degradation results from elevated free calcium levels found in muscle from *mdx* mice. *Nature*, **335**, 735-738. [15]

Tweedle, C.D., & Kabara, J. (1977). Lipophilic nerve sprouting factor(s) isolated from denervated muscle. *Neuroscience Letters*, **6**, 41-46. [17]

Tzartos, S.J., Langeberg, L., Hochschwender, S., & Lindstrom, J. (1983). Demonstration of a main immunogenic region on acetylcholine receptors from human muscle using monoclonal antibodies to human receptor. *FEBS Letters*, **158**, 116-118. [10]

Uchida, S., Yamamoto, H., Iio, S., Matsumoto, N., Wang, X.B., Yonehara, N., Imal, Y., Inokl, R., & Yoshida, H. (1990). Release of calcitonin gene-related peptide-like immunoreactive substance from neuromuscular junction by nerve excitation and its action on striated muscle. *Journal of Neurochemistry*, **54**, 1000-1003. [6]

Ulfhake, B., & Kellerth, J.-O. (1982). Does alpha-motoneurone size correlate with motor unit type in cat triceps surae? *Brain Research*, **251**, 201-209. [12]

Ungewickell, E. (1984). First clue to biological role of clathrin light chains. *Nature*, **311**, 213. [3]

Usdin, T.B., & Fischbach, G.D. (1986). Purification and characterization of a polypeptide from chick brain that promotes the accumulation of acetylcholine receptors in chick myotubes. *Journal of Cell Biology*, **103**, 493-507. [3]

Vale, R.D., Reese, T.S., & Sheetz, M.P. (1985a). Different axoplasmic proteins generate movement in opposite direction along microtubules *in vitro*. *Cell*, **43**, 623-632. [8]

Vale, R.D., Reese, T.S., & Sheetz, M.P. (1985b). Identification of a novel force-generating protein, kinesin, involved in microtubule-based motility. *Cell*, **42**, 39-50. [8]

Vallee, R.B., & Bloom, G.S. (1991). Mechanisms of fast and slow axonal transport. *Annual Review of Neurosciences*, **14**, 59-92. [8]

Vallee, R.B., Shpetner, H.S., & Paschal, B.M. (1989). The role of dynein in retrograde axonal transport. *Trends in Neurosciences*, **12**, 66-70. [8]

Vandenburgh, H.H., Hatfalundy, S., Karlisch, P., & Shansky, J. (1991). Mechanically induced alterations in cultured skeletal muscle growth. *Journal of Biomechanics*, **24**(Suppl. 1), 91-99. [20]

Vandervoort, A.A., Hayes, K.C., & Bélanger, A.Y. (1986). Strength and endurance of skeletal muscle in the elderly. *Physiotherapy Canada*, **38**, 167-173. [20, 22]

Vandervoort, A.A., & McComas, A.J. (1986). Contractile changes in opposing muscles of the human ankle joint with aging. *Journal of Applied Physiology*, **61**, 361-367. [22]

Vandervoort, A.A., Quinlan, J., & McComas, A.J. (1983). Twitch potentiation after voluntary contraction. *Experimental Neurology*, **81**, 141-152. [11]

Van Essen, D., & Jansen, J.K.S. (1974). Re-innervation of the rat diaphragm during perfusion with α-bungarotoxin. *Acta Physiologica Scandinavica*, **91**, 571-573. [17]

Van Harreveld, A. (1947). On the mechanism of the spontaneous reinnervation in paretic muscles. *American Journal of Physiology*, **150**, 670-676. [17]

Vassilev, P., Scheuer, T., & Catterall, W. (1988). Identification of an intracellular peptide segment involved in sodium channel inactivation. *Science*, **241**, 1658-1661. [7]

Veratti, E. (1902). Richerche sulla fine struttura della fibra muscolare striata. *Memorie Reale Istituto Lombardi*, **19**, 87-133. [1]

Vibert, P., & Cohen, C. (1988). Domains, motions and regulation in the myosin head. *Journal of Muscle Research and Cell Motility*, **9**, 296-305. [11]

Vincent, A., Lang, B., & Newsom-Davis, J. (1989). Autoimmunity to the voltage-gated calcium channel underlies the Lambert-Eaton myasthenic syndrome, a paraneoplastic disorder. *Trends in Neurosciences*, **12**, 496-502. [10]

Vizoso, A.D. (1950). The relationship between internodal length and growth in human nerves. *Journal of Anatomy*, **84**, 342-353. [2]

Vøllestad, N.K., Sejersted, O.M., Bahr, R., Woods, J.J., & Bigland-Ritchie, B. (1988). Motor drive and metabolic responses during repeated submaximal contractions in humans. *Journal of Applied Physiology*, **64**, 1421-1427. [15]

Voss, H. (1937). Untersuchungen über Zahl, Anordnung und Länge der Muskelspindeln in den Lumbricalmuskeln des Menschen und einiger. *Zeitschrift fuer Mikroskopisch-Anatomische Forschung*, **42**, 509-524. [4]

Voss, H. (1956). Zahl und Anordnung der Muskelspindeln in den oberen Zungennbeinmuskeln, im M. trapezius und M. Latissimus dorsi. *Anatomischer Anzeiger*, **103**, 443-446. [4]

Voss, H. (1958). Zahl und Anordnung der Muskelspindeln in den unteren Zungenbeinmuskeln, dem M. sternocleido-mastoideus und den Brauch- und tiefen Nackenmuskeln. *Anatomischer Anzeiger*, **104**, 345-355. [4]

Vrbová, G. (1963). The effect of motoneurone activity on the speed of contraction of striated muscle. *Journal of Physiology*, **169**, 513-526. [19]

Vrbová, G., & Lowrie, M. (1989). Role of activity in developing synapses, search for molecular mechanisms. *News in Physiological Sciences*, **4**, 75-78. [6]

Vyskočil, F., Hník, P., Rehfeldt, H., Vejapada, R., & Ujec, E. (1983). The measurement of K⁺ concentration changes in human muscles during volitional contractions. *Pflügers Archiv*, **399**, 235-237. [15]

Wagenknecht, T., Grassucci, R., Frank, J., Saito, A., Inui, M., & Fleischer, S. (1989). Three-dimensional architecture of the calcium channel/foot structure of sarcoplasmic reticulum. *Nature*, **338**, 167-170. [11]

Wainman, P., & Shipounoff, G.C. (1941). The effects of castration and testosterone propionate on the striated peroneal musculature in the rat. *Endocrinology*, **29**, 975-978. [5]

Wakayama, Y. (1976). Electron microscopic study on the satellite cell in the muscle of Duchenne muscular dystrophy. *Journal of Neuropathology and Experimental Neurology*, **35**, 532-540. [21]

Wakelam, M.J.O. (1985). The fusion of myoblasts. *Biochemical Journal*, **228**, 1-12. [5]

Waksman, B.H., & Adams, R.D. (1955). Allergic neuritis: An experimental disease of rabbits induced by the injection of peripheral nervous tissue and adjuvants. *Journal of Experimental Medicine*, **102**, 213-236. [16]

Wallace, B.G., Qu, Z., & Huganir, R.L. (1991). Agrin induces phosphorylation of the nicotinic acetylcholine receptor. *Neuron*, **6**, 869-878. [6]

Waller, A. (1850). Experiments on the section of the glossopharyngeal and hypoglossal nerves of the frog, and observation on the alteration produced thereby in the structure of their primitive fibres. *Philosophical Transactions of the Royal Society, London*, **140**, 423. [16]

Wang, J., & Best, P.M. (1992). Inactivation of the sarcoplasmic reticulum calcium channel by protein kinase. *Nature*, **359**, 739-741. [7]

Warren, G. (1993). Bridging the gap. *Nature*, **362**, 297-298. [10]

Wasserschaff, M. (1990). Coordination of reinnervated muscle and reorganization of spinal cord motoneurons after nerve transection in mice. *Brain Research*, **515**, 241-246. [12]

Watson, W.E. (1968a). Centripetal passage of labelled molecules along mammalian motor axons. *Journal of Physiology*, **196**, 122P-123P. [18]

Watson, W.E. (1968b). Observations on the nucleolar and total cell body nucleic acid of injured nerve cells. *Journal of Physiology*, **196**, 655-676. [18]

Watson, W.E. (1969). The response of motor neurones to intramuscular injection of botulinum toxin. *Journal of Physiology*, **202**, 611-630. [18]

Watson, W.E. (1970). Some metabolic responses of axotomized neurones to contact between their axons and denervated muscle. *Journal of Physiology*, **210**, 321-343. [18]

Watson, W.E. (1972). Some quantitative observations upon the responses of neuroglial cells which follow axotomy of adjacent neurones. *Journal of Physiology*, **225**, 415-435. [18]

Weintraub, H., Tapscott, S.J., Davis, R.L., Thayer, M.J., Adam, M.A., Lassar, A.B., & Miller, A.D. (1989). Activation of muscle-specific genes in pigment, nerve, fat, liver, and fibroblast cell lines by forced expression of MyoD. *Proceedings of the National Academy of Sciences* (U.S.), **86**, 5434-5438. [5]

Weiss, P. (1944). Damming of axoplasm in constricted nerve: A sign of perpetual growth in nerve fibers. *Anatomical Record*, **88**, 464. [8]

Weiss, P. (1969). Neuronal dynamics. In F.O. Schmitt et al. (Eds.), *Neurosciences Research Symposium summaries* (Vol. 3, pp. 255-299). Cambridge, MA: M.I.T. Press. [2, 8]

Weiss, P., & Cavanaugh, M.W. (1959). Further evidence of perpetual growth of nerve fibers: Recovery of fiber diameter after the release of prolonged constrictions. *Journal of Experimental Zoology*, **142**, 461-473. [8]

Weiss, P., & Davis, H. (1943). Pressure block in nerves provided with arterial sleeves. *Journal of Neurophysiology*, **6**, 269-286. [8]

Weiss, P., & Edds, M.V., Jr. (1945). Spontaneous recovery of muscle following partial denervation. *American Journal of Physiology*, **145**, 587-607. [17]

Weiss, P., & Hiscoe, H. (1948). Experiments on the mechanism of nerve growth. *Journal of Experimental Zoology*, **107**, 315-395. [8]

Wernig, A., Carmody, J.J., Anzil, A.P., Hansert, E., Marchiniak, M., & Zucker, H. (1984). Persistence of nerve sprouting with features of synapse remodelling in soleus muscles of adult mice. *Neuroscience*, **11**, 241-253. [22]

Westall, F., Robinson, A.B., Caccam, J., Jackson, J., & Eylar, E.H. (1971). Essential chemical requirements for induction of allergic encephalomyelitis. *Nature*, **229**, 22-24. [16]

Westerblad, H., & Allen, D.G. (1991). Changes of myoplasmic calcium concentration during fatigue in single mouse muscle fibers. *Journal of General Physiology*, **98**, 615-635. [11]

Westerfield, M., & Eisen, J.S. (1988). Neuromuscular specificity: Pathfinding by identified motor growth cones in a vertebrate embryo. *Trends in Neurosciences*, **11**, 18-22. [6]

Westgaard, R.H., & Lømo, T. (1988). Control of contractile properties within adaptive ranges by patterns of impulse activity in the rat. *Journal of Neuroscience*, **8**, 4415-4426. [18, 20]

Westling, G., Johansson, R.S., Thomas, C.K., & Bigland-Ritchie, B. (1990). Measurement of contractile and electrical properties of single human thenar motor units in response to intraneural motor-axon stimulation. *Journal of Neurophysiology*, **64**, 1331-1337. [12]

White, M.J., & Davies, C.T.M. (1984). The effects of immobilization, after lower leg fracture, on the contractile properties of human triceps surae. *Clinical Science*, **66**, 277-282. [19]

Whittaker, V.P., Michaelson, I.A., & Kirkland, R.J.A. (1964). The separation of synaptic vesicles from nerve-ending particles (''synaptosomes''). *Biochemical Journal*, **90**, 293-303. [3]

Wickiewicz, T.L., Roy, R.R., Powell, P.L., & Edgerton, V.R. (1983). Muscle architecture of the human lower limb. *Clinical Orthopaedics and Related Research*, **179**, 275-283. [1]

Widdas, W.F., & Baker, G.F. (1992). The theory for the electrical potential control and the chemical control of bistable polyguanidinium cationic gates. *Biomedical Letters*, **47**, 259-267. [7]

Williams, P.E., & Goldspink, G. (1971). Longitudinal growth of striated muscle fibres. *Journal of Cell Science*, **9**, 751-767. [5]

Williams, P.E., & Goldspink, G. (1978). Changes in sarcomere length and physiological properties in immobilized muscle. *Journal of Anatomy*, **127**, 459-468. [19]

Williams, P.L., & Hall, S.M. (1971a). Chronic Wallerian degeneration—An *in vivo* and ultrastructural study. *Journal of Anatomy*, **109**, 487-503. [16]

Williams, P.L., & Hall, S.M. (1971b). Prolonged *in vivo* observations of normal peripheral nerve fibres and their acute reactions to crush and deliberate trauma. *Journal of Anatomy*, **108**, 397-408. [16]

Williams, P.L., & Landon, D.N. (1967). In *Gray's anatomy* (34th ed., p. 62). London: Longmans Green. [2]

Willis, T. (1672). *De anima brutorum*. Amsterdam: Blaeus. [10]

Willison, H.J., & Kennedy, P.G.E. (1993). Gangliosides and bacterial toxins in Guillain-Barré syndrome. *Journal of Neuroimmunology*, **46**, 105-112. [16]

Willison, R.G. (1978). Preservation of bulk and strength in muscles affected by neurogenic lesions. *Muscle & Nerve*, **1**, 404-406. [6, 12]

Wines, M.M., & Hall-Craggs, E.C.B. (1986). Neuromuscular relationships in a muscle having segregated motor end-plate zones. I. Anatomical and physiological considerations. *Journal of Comparative Neurology*, **249**, 147-151. [12]

Wiśniewski, H.M., Brostoff, S.W., Carter, H., & Eylar, E.H. (1974). Recurrent experimental allergic polyganglioradiculoneuritis. *Archives of Neurology*, **30**, 347-358. [16]

Witzmann, F.A., Kim, D.H., & Fitts, R.H. (1983). Effect of hindlimb immobilization on the fatigability of skeletal muscle. *Journal of Applied Physiology*, **54**, 1242-1248. [19]

Wolpert, L. (1969). Positional information and the spatial pattern of cellular differentiation. *Journal of Theoretical Biology*, **25**, 1-47. [5]

Wright, E.A., & Spink, J.M. (1959). A loss of nerve cells in the central nervous system in relation to age. *Gerontologica*, **3**, 277-287. [22]

Wrogemann, K., & Pena, S.D.J. (1976). Mitochondrial calcium overload. A general mechanism for cell necrosis in muscle diseases. *Lancet*, **i**, 672. [15]

Wuerker, R.B., McPhedran, A.M., & Henneman, E. (1965). Properties of motor units in a heterogeneous pale muscle (m. gastrocnemius) of the cat. *Journal of Neurophysiology*, **28**, 85-99. [12]

Xie, Z.-P., & Poo, M.-M. (1986). Initial events in the formation of neuromuscular synapse: Rapid induction of acetylcholine release from embryonic neuron. *Proceedings of the National Academy of Sciences* (U.S.), **83**, 7069-7073. [6]

Yan, Q., Elliot, J., & Snider, W.D. (1992). Brain-derived neurotrophic factor rescues spinal motor neurons from axotomy-induced cell death. *Nature*, **360**, 753-755. [18]

Yan, Q., Matheson, C., & Lopez, O.T. (1995). *In vivo* neurotrophic effects of GDNF on neonatal and adult facial motor neurons. *Nature*, **373**, 341-344. [18]

Yemm, R. (1977). The orderly recruitment of motor units of the masseter and temporal muscles during voluntary isometric contraction in man. *Journal of Physiology*, **265**, 163-174. [12, 13]

Young, A., Hughes, I., Round, J.M., & Edwards, R.H.T. (1982). The effect of knee injury on the number of muscle fibers in the human quadriceps femoris. *Clinical Sciences*, **62**, 227-234. [19]

Young, A., Stokes, M., & Crowe, M. (1984). Size and strength of the quadriceps muscles of old and young women. *European Journal of Clinical Investigation*, **14**, 282-287. [22]

Young, A., Stokes, M., & Crowe, M. (1985). The size and strength of the quadriceps muscles of old and young men. *Clinical Physiology*, **5**, 145-154. [22]

Young, J.L., & Mayer, R.F. (1981). Physiological properties and classification of single motor units activated by intramuscular microstimulation in the first dorsal interosseous muscle in man. In J. Desmedt (Ed.), *Progress in clinical neurophysiology: Vol. 9. Motor unit types, recruitment and plasticity in health and disease* (pp. 17-25). Basel: Karger. [12]

Young, J.Z. (1936). The giant nerve fibres and epistellar body of cephalopods. *Quarterly Journal of Microscopic Science*, **78**, 367-386. [9]

Zacks, S.I., & Saito, A. (1969). Uptake of exogenous horseradish peroxidase by coated vesicles in mouse neuromuscular junctions. *Journal of Histochemistry and Cytochemistry*, **17**, 161-170. [18]

Zajac, F.E., & Faden, J.S. (1985). Relationship among recruitment order, axonal conduction velocity, and muscle-unit properties of type-identified motor units in cat plantaris muscle. *Journal of Neurophysiology*, **53**, 1303-1322. [13]

Zajac, F.E., & Young, J. (1975). *Motor unit discharge patterns during treadmill walking and trotting in the cat* (Abstract No. 255). Fifth annual meeting of the Society for Neuroscience, New York. [13]

Zebra, E., Komorowski, T.E., & Faulkner, J.A. (1990). Free radical injury to skeletal muscles of young, adult and old mice. *American Journal of Physiology*, **258**, C429-C435. [21]

Glossary

(Abbreviations, symbols, and key words)

A-band — Dark (highly refractive), regularly repeating region of muscle fiber, associated with the presence of myosin filaments.

ACh — Acetylcholine; chemical transmitter released by motor nerve terminals.

AChE — Acetylcholinesterase; enzyme that splits acetylcholine.

AChR — Acetylcholine receptor; protein in muscle fiber membrane that combines with acetylcholine to produce excitation.

actin — Globular protein that forms strengthening filaments in nerve and muscle fibers. In muscle fiber, contraction is produced by movement of actin filaments over myosin.

action potential — Impulse. Brief electrical signal that travels along nerve and muscle fibers.

anion — Negatively charged atom or molecule in solution.

ADP — Adenosine diphosphate.

AMP — Adenosine monophosphate

ATP — Adenosine triphosphate, an energy-providing molecule.

ARIA — Acetylcholine-receptor-inducing activity. Trophic factor released by motor nerve terminal.

axis cylinder — Central part of an axon.

axon — Nerve fiber.

axoplasmic transport — Process for moving proteins and organelles along the axon.

basement membrane — Complex sheath investing nerve and muscle fibers.

BDNF — Brain-derived neurotrophic factor.

brackets [ ] — May be used to indicate concentration of an ion in solution.

BTX — Bungarotoxin; powerful snake toxin that combines with the acetylcholine receptor.

C — Symbol for capacitance, the ability of a "device" (e.g., cell membrane) to store electric charge.

Ca^{2+} — Calcium ion.

cAMP — Cyclic AMP (adenosine monophosphate); messenger molecule in cell interior.

cation — Positively charged atom or molecule in solution.

CGRP — Calcitonin gene-related peptide. Trophic factor released by motor nerve terminal.

Cl^- — Chloride ion.

CNS — Central nervous system.

CNTF — Ciliary neurotrophic factor.

contraction time — Time elapsing between the start of a contraction and the moment of peak tension (force).

cross-bridge — Projection from the myosin filament that interacts with actin to produce force or movement by muscle fibers.

dendrite — Branching structure extending from nerve cell body.

depolarization — Reduction in the potential across a cell membrane.

DHP channel — Type of Ca^{2+} channel.

DNA — Deoxyribonucleic acid, the genetic material in the cell nucleus.

E-C coupling — Excitation-contraction coupling. The process linking the action potential to the contractile machinery in the muscle fiber.

E_K, E_{Na}, E_{Cl} — Equilibrium potentials for potassium, sodium, and chloride ions respectively.

E_m — Resting membrane potential.

end-plate — Region of muscle fiber underlying the motor nerve terminals.

end-plate potential — Depolarization of muscle fiber by acetylcholine released at nerve terminal.

F — Faraday constant (in Nernst equation).

fast — Associated with a fast (brief) twitch of the muscle fiber.

FF — Fast-twitch fatigable type of motor unit.

FG — Fast-twitch glycolytic type of motor unit.

FGF — Fibroblast growth factor. A peptide that stimulates proliferation of connective tissue fibroblasts, and that is also involved in muscle development and neurotrophic mechanisms.

FInt — Fast-twitch intermediate type of motor unit.

FOG — Fast-twitch oxidative glycolytic type of motor unit.

FR — Fast-twitch fatigue-resistant type of motor unit.

g — Symbol for conductance; the ease with which electric current flows across a structure (e.g., ions across a cell membrane).

gene — Unit of inheritance; each gene codes for a single protein.

H⁺ — Hydrogen ion.

H-zone — Pale area in the center of the muscle fiber A-band.

i — Symbol for electric current.

I-band — Light (poorly refractive) region of muscle fiber, which alternates with A-band and is associated with the presence of actin.

impulse — See action potential.

ion — Electrically charged atom or molecule, responsible for current flow in biological fluids.

K⁺ — Potassium ion.

MEPPs — Miniature end-plate potentials. Small depolarizations of muscle fibers produced by single packets (vesicles) of acetylcholine.

MHC — Myosin heavy chain. Part of the myosin molecule.

microtubule — Narrow, linear structure in axon.

M-line — Narrow, dark line in center of H-zone.

motoneuron — Motor nerve cell that supplies muscle fibers.

motor unit — A single motoneuron and the colony of muscle fibers to which it is connected.

MW — Molecular weight.

M-wave — Muscle compound action potential.

myelin — Lipid that forms sheath around axis cylinder in a nerve fiber.

myofibril — Narrow contractile cylinder in muscle fiber, formed of actin and myosin.

myonucleus — Nucleus in a muscle fiber.

myosin — Contractile protein that interacts with actin through its cross-bridges.

Na⁺ — Sodium ion.

N-CAM — Neural-cell adhesion molecule. Protein involved in the development of the nervous system.

neurofilament — Very fine linear structure inside axon.

neurotrophin — Molecule released by one cell that helps to sustain another.

neuromuscular junction — Connection between motor nerve terminal and muscle fiber.

NGF — Nerve growth factor.

P — Symbol for permeability. The ease with which ions cross a membrane.

P — Phosphorus, also used as symbol for phosphate (e.g., P_i = inorganic phosphate).

PCr — Phosphocreatine. An energy-providing molecule.

plasmalemma — Membrane lining nerve and muscle fibers, and some organelles.

Power — Force × Velocity.

R — Symbol for resistance to electrical current.

R — Universal gas constant (in Nernst equation).

resting potential — Difference in potential across the membrane of a quiescent nerve or muscle fiber.

RNA — Ribonucleic acid. Genetic material formed from DNA and used to manufacture proteins in cytoplasm.

ryanodine receptor — Calcium-releasing channel in the sarcoplasmic reticulum of the muscle fiber.

RYR — Ryanodine receptor.

S — Slow-twitch type of motor unit.

sarcolemma — Plasmalemma and basement membrane of a muscle fiber.

sarcomere — Region of muscle fiber between adjacent Z-lines.

satellite cell — Small cell at periphery of muscle fiber, responsible for regeneration of the fiber.

slow — Associated with a slow (prolonged) twitch of the muscle fiber.

SR — Sarcoplasmic reticulum. The calcium-storing and -releasing structure within the muscle fiber.

STX — Saxitonin. Paralytic poison manufactured by certain algae.

tetanus — A train of electrical stimuli.

torque — Force (tension) × Distance from joint.

trophic — Sustaining action of a molecule or cellular activity.

T-tubule — Narrow channel that conducts impulses into the interior of a muscle fiber.

TTX — Tetrodotoxin; a paralytic poison secreted by the Japanese puffer fish.

twitch — Muscle contraction in response to a single excitation.

type I — Type of muscle fiber, recognized by histochemical staining and having slow-twitch properties.

type II — Type of muscle fiber, recognized by histochemical staining and having fast-twitch properties.

Z-disc or Z-line — Dark structure in the center of the muscle fiber I-band, to which actin filaments are attached.

(Units)

kD — Kilodalton; unit of molecular weight.

μm — Micrometer. A millionth part of a meter and a thousandth part of a millimeter.

mA — Milliampere; unit of electric current.

mm — Millimeter. Thousandth part of a meter.

mN — Millinewton. Unit of force (tension), one thousandth part of a newton.

ms — Millisecond. One thousandth part of a second.

m/s — Meters per second. Unit of velocity, used for conduction of the impulse in a nerve or muscle fiber.

mV — Millivolt. One thousandth part of a volt.

N — Newton. Unit of force (tension). 9.8 N = 1 kilogram.

nm — Nanometer. 10^{-9} meter, or a thousandth part of a micrometer.

pS — Picosiemen. Unit of electric conductance adopted for single ion channel studies.

Index